SI Units (Système international d'unités)

Selected SI units			Commonly used SI prefixes		
Quantity	Name	SI symbol	Factor	Prefix	SI Symbol
Energy	joule	J (1 J = 1 N·m)	10^9	giga	G
Force	newton	N (1 N = 1 kg·m/s^2)	10^6	mega	M
Length	meter*	m	10^3	kilo	k
Mass	kilogram*	kg	10^{-3}	milli	m
Moment (torque)	newton meter	N·m	10^{-6}	micro	μ
Rotational frequency	revolution per second	r/s	10^{-9}	nano	n
	hertz	Hz (1 Hz = 1 r/s)			
Stress (pressure)	pascal	Pa (1 Pa = 1 N/m^2)			
Time	second*	s			
Power	watt	W (1 W = 1 J/s)			

* SI base unit

Selected Rules and Suggestions for SI Usage

1. Be careful in the use of capital and lowercase for symbols, units, and prefixes (e.g., m for meter or milli, M for mega).
2. For numbers having five or more digits, the digits should be placed in groups of three separated by a small space, counting both to the left and to the right of the decimal point (e.g., 61 354.982 03). The space is not required for four-digit numbers. Spaces are used instead of commas to avoid confusion—many countries use the comma as the decimal marker.
3. In compound units formed by multiplication, use the product dot (e.g., N·m).
4. Division may be indicated by a slash (m/s), or a negative exponent with a product dot (m·s^{-1}).
5. Avoid the use of prefixes in the denominator (e.g., km/s is preferred over m/ms). The exception to this rule is the prefix k in the base unit kg (kilogram).

Equivalence of U.S. Customary and SI Units (Asterisks indicate exact values; others are approximations.)

	U.S. Customary to SI	SI to U.S. Customary
1. Length	1 in. = 25.4* mm = 0.0254* m 1 ft = 304.8* mm = 0.3048* m	1 mm = 0.039 370 in. 1 m = 39.370 in. = 3.281 ft
2. Area	1 in.^{2} = 645.16* mm^2 1 ft^2 = 0.092 903 04* m^2	1 mm^2 = 0.001 550 in.^{2} 1 m^2 = 1550.0 in.^{2} = 10.764 ft^2
3. Volume	1 in.^{3} = 16 387.064* mm^3 1 ft^3 = 0.028 317 m^3	1 mm^3 = 0.000 061 024 in.^{3} 1 m^3 = 61 023.7 in.^{3} = 35.315 ft^3
4. Force	1 lb = 4.448 N 1 lb/ft = 14.594 N/m	1 N = 0.2248 lb 1 N/m = 0.068 522 lb/ft
5. Mass	1 lbm = 0.453 59 kg 1 slug = 14.593 kg	1 kg = 2.205 lbm 1 kg = 0.068 53 slugs
6. Moment of a force	1 lb·in. = 0.112 985 N·m 1 lb·ft = 1.355 82 N·m	1 N·m = 8.850 75 lb·in. 1 N·m = 0.737 56 lb·ft
7. Power	1 hp (550 lb·ft/s) = 0.7457 kW	1 kW = 1.3410 hp

Area Moments of Inertia

Rectangle

$\bar{I}_x = \dfrac{bh^3}{12} \quad \bar{I}_y = \dfrac{b^3 h}{12} \quad \bar{I}_{xy} = 0$

$I_x = \dfrac{bh^3}{3} \quad I_y = \dfrac{b^3 h}{3} \quad I_{xy} = \dfrac{b^2 h^2}{4}$

Circle

$I_x = I_y = \dfrac{\pi R^4}{4} \quad I_{xy} = 0$

Half parabolic complement

$\bar{I}_x = \dfrac{37 b h^3}{2100} \quad I_x = \dfrac{bh^3}{21}$

$\bar{I}_y = \dfrac{b^3 h}{80} \quad I_y = \dfrac{b^3 h}{5}$

$\bar{I}_{xy} = \dfrac{b^2 h^2}{120} \quad I_{xy} = \dfrac{b^2 h^2}{12}$

Right triangle

$\bar{I}_x = \dfrac{bh^3}{36} \quad \bar{I}_y = \dfrac{b^3 h}{36} \quad \bar{I}_{xy} = -\dfrac{b^2 h^2}{72}$

$I_x = \dfrac{bh^3}{12} \quad I_y = \dfrac{b^3 h}{12} \quad I_{xy} = \dfrac{b^2 h^2}{24}$

Semicircle

$\bar{I}_x = 0.1098 R^4 \quad \bar{I}_{xy} = 0$

$I_x = I_y = \dfrac{\pi R^4}{8} \quad I_{xy} = 0$

Half parabola

$\bar{I}_x = \dfrac{8 b h^3}{175} \quad I_x = \dfrac{2 b h^3}{7}$

$\bar{I}_y = \dfrac{19 b^3 h}{480} \quad I_y = \dfrac{2 b^3 h}{15}$

$\bar{I}_{xy} = \dfrac{b^2 h^2}{60} \quad I_{xy} = \dfrac{b^2 h^2}{6}$

Isosceles triangle

$\bar{I}_x = \dfrac{bh^3}{36} \quad \bar{I}_y = \dfrac{b^3 h}{48} \quad \bar{I}_{xy} = 0$

$I_x = \dfrac{bh^3}{12} \quad \quad I_{xy} = 0$

Quarter circle

$\bar{I}_x = \bar{I}_y = 0.05488 R^4 \quad I_x = I_y = \dfrac{\pi R^4}{16}$

$\bar{I}_{xy} = -0.01647 R^4 \quad I_{xy} = \dfrac{\pi R^4}{8}$

Circular sector

$I_x = \dfrac{R^4}{8}(2\alpha - \sin 2\alpha)$

$I_y = \dfrac{R^4}{8}(2\alpha + \sin 2\alpha)$

$I_{xy} = 0$

Triangle

$\bar{I}_x = \dfrac{bh^3}{36} \quad\quad I_x = \dfrac{bh^3}{12}$

$\bar{I}_y = \dfrac{bh}{36}(a^2 - ab + b^2) \quad I_y = \dfrac{bh}{12}(a^2 + ab + b^2)$

$\bar{I}_{xy} = \dfrac{bh^2}{72}(2a - b) \quad I_{xy} = \dfrac{bh^2}{24}(2a + b)$

Quarter ellipse

$\bar{I}_x = 0.05488 ab^3 \quad I_x = \dfrac{\pi a b^3}{16}$

$\bar{I}_y = 0.05488 a^3 b \quad I_y = \dfrac{\pi a^3 b}{16}$

$\bar{I}_{xy} = -0.01647 a^2 b^2 \quad I_{xy} = \dfrac{a^2 b^2}{8}$

高等学校原版经典系列教材

材料力学(含光盘)

Mechanics of Materials

Andrew Pytel
The Pennsylvania State University

Jaan Kiusalaas
The Pennsylvania State University

中国建筑工业出版社
China Architecture & Building Press

著作权合同登记图字：01-2004-1451 号

图书在版编目(CIP)数据

材料力学 /(英)皮特尔(Pytel),(英)基乌萨拉斯(Kiusalaas)编著.—影印本.
北京：中国建筑工业出版社，2004
ISBN 978-7-112-06414-4

Ⅰ.材… Ⅱ.①皮…②基… Ⅲ.材料力学－高等学校－教学参考资料－英文
Ⅳ.TB301

中国版本图书馆 CIP 数据核字（2004）第 027469 号

COPYRIGHT © 2003 Brooks/Cole, a division of Thomson Learning, Inc.
Thomson Learning ™ is a trademark used herein under license.
© 2004 China Architecture & Building Press

ALL RIGHTS RESERVED. No part of this work covered by the copyright hereon may be reproduced or used in any form or by any means—graphic, electronic, or mechanical, including but not limited to photocopying, recording, taping, Web distribution, information networks, or information storage and retrieval systems — without the written permission of the publisher.

Mechanics of Materials/Andrew Pytel, Jaan Kiusalaas

本书经 Thomson Learning 出版公司正式授权我社出版、发行本书原文版图书

责任编辑：董苏华
责任设计：郑秋菊
责任校对：赵明霞

高等学校原版经典系列教材
材料力学(含光盘)
Pytel · Kiusalaas
*
中国建筑工业出版社出版、发行(北京西郊百万庄)
各地新华书店、建筑书店经销
北京嘉泰利德制版公司
北京建筑工业印刷厂印刷
*
开本：880×1230毫米 1/16 印张：35 字数：1200千字
2004年8月第一版 2008年1月第二次印刷
定价：68.00元(含光盘)
ISBN 978-7-112-06414-4
(12428)

版权所有 翻印必究
如有印装质量问题，可寄本社退换
(邮政编码 100037)

To Jean, Leslie, Lori, John, Nicholas

and

To Judy, Nicholas, Jennifer, Timothy

Contents

CHAPTER 1
Stress 1

- 1.1 Introduction 1
- 1.2 Analysis of Internal Forces; Stress 2
- 1.3 Axially Loaded Bars 4
  - a. Centroidal (axial) loading 4
  - b. Saint Venant's principle 5
  - c. Stresses on inclined planes 6
  - d. Procedure for stress analysis 7
- 1.4 Shear Stress 17
- 1.5 Bearing Stress 18

CHAPTER 2
Strain 29

- 2.1 Introduction 29
- 2.2 Axial Deformation; Stress-Strain Diagram 30
  - a. Normal (axial) strain 30
  - b. Tension test 31
  - c. Working stress and factor of safety 34
- 2.3 Axially Loaded Bars 34
- 2.4 Generalized Hooke's Law 44
  - a. Uniaxial loading; Poisson's ratio 44
  - b. Multiaxial loading 44
  - c. Shear loading 45
- 2.5 Statically Indeterminate Problems 51
- 2.6 Thermal Stresses 60

CHAPTER 3
Torsion 73

- 3.1 Introduction 73
- 3.2 Torsion of Circular Shafts 74
  - a. Simplifying assumptions 74
  - b. Compatibility 75
  - c. Equilibrium 75
  - d. Torsion formulas 76
  - e. Power transmission 77
  - f. Statically indeterminate problems 78
- 3.3 Torsion of Thin-Walled Tubes 88

CHAPTER 4
Shear and Moment in Beams 99

- 4.1 Introduction 99
- 4.2 Supports and Loads 100
- 4.3 Shear-Moment Equations and Shear-Moment Diagrams 101
  - a. Sign conventions 101
  - b. Procedure for determining shear force and bending moment diagrams 102
- 4.4 Area Method for Drawing Shear-Moment Diagrams 111
  - a. Distributed loading 112
  - b. Concentrated forces and couples 114
  - c. Summary 116
- 4.5 Moving Loads 125

CHAPTER 5
Stresses in Beams 135

- 5.1 Introduction 135
- 5.2 Bending Stress 136
  - a. Simplifying assumptions 136
  - b. Compatibility 137
  - c. Equilibrium 138
  - d. Flexure formula; section modulus 139
  - e. Procedures for determining bending stresses 140
- 5.3 Economic Sections 154
  - a. Standard structural shapes 155
  - b. Procedure for selecting standard shapes 156
- 5.4 Shear Stress in Beams 160
  - a. Analysis of Flexure Action 160
  - b. Horizontal shear stress 161
  - c. Vertical shear stress 163
  - d. Discussion and limitations of the shear stress formula 163
  - e. Rectangular and wide-flange sections 164
  - f. Procedure for analysis of shear stress 165

5.5 Design for Flexure and Shear 173
5.6 Design of Fasteners in Built-up Beams 180

CHAPTER 6
Deflection of Beams 191

6.1 Introduction 191
6.2 Double-Integration Method 192
    a. Differential equation of the elastic curve 192
    b. Double integration of the differential equation 194
    c. Procedure for double integration 194
6.3 Double Integration Using Bracket Functions 205
*6.4 Moment-Area Method 215
    a. Moment-area theorems 216
    b. Bending moment diagrams by parts 218
    c. Application of the moment-area method 221
6.5 Method of Superposition 231

CHAPTER 7
Statically Indeterminate Beams 243

7.1 Introduction 243
7.2 Double-Integration Method 244
7.3 Double Integration Using Bracket Functions 250
*7.4 Moment-Area Method 254
7.5 Method of Superposition 260

CHAPTER 8
Stresses Due to Combined Loads 271

8.1 Introduction 271
8.2 Thin-Walled Pressure Vessels 272
    a. Cylindrical vessels 272
    b. Spherical vessels 274
8.3 Combined Axial and Lateral Loads 278
8.4 State of Stress at a Point (Plane Stress) 289
    a. Reference planes 289
    b. State of stress at a point 290
    c. Sign convention and subscript notation 290

8.5 Transformation of Plane Stress 291
    a. Transformation equations 291
    b. Principal stresses and principal planes 293
    c. Maximum in-plane shear stress 294
    d. Summary of stress transformation procedures 295
8.6 Mohr's Circle for Plane Stress 301
    a. Construction of Mohr's circle 302
    b. Properties of Mohr's circle 303
    c. Verification of Mohr's circle 304
8.7 Absolute Maximum Shear Stress 310
    a. Plane state of stress 311
    b. General state of stress 312
8.8 Applications of Stress Transformation to Combined Loads 315
8.9 Transformation of Strain; Mohr's Circle for Strain 325
    a. Review of strain 326
    b. Transformation equations for plane strain 327
    c. Mohr's circle for strain 328
8.10 The Strain Rosette 333
    a. Strain gages 333
    b. Strain rosette 334
    c. The 45° strain rosette 335
    d. The 60° strain rosette 335
8.11 Relationship Between Shear Modulus and Modulus of Elasticity 337

CHAPTER 9
Composite Beams 343

9.1 Introduction 343
9.2 Flexure Formula for Composite Beams 344
9.3 Shear Stress and Deflection in Composite Beams 349
    a. Shear stress 349
    b. Deflection 350
9.4 Reinforced Concrete Beams 353
    a. Analysis 354
    b. Balanced-stress reinforcement 355

CHAPTER 10
Columns 363

10.1 Introduction 363
10.2 Critical Load 364
    a. Definition of critical load 364
    b. Euler's formula 365

---
*Indicates optional articles.

10.3	Discussion of Critical Loads 367		13.3	Limit Moment 458
10.4	Design Formulas for Intermediate Columns 372		13.4	Residual Stresses 463

10.3 Discussion of Critical Loads 367
10.4 Design Formulas for Intermediate Columns 372
    a. Tangent modulus theory 372
    b. AISC column specifications 373
10.5 Eccentric Loading: Secant Formula 378
    a. Derivation of the secant formula 379
    b. Application of the secant formula 380

CHAPTER 11
Additional Beam Topics   389

11.1 Introduction 389
11.2 Shear Flow in Thin-Walled Beams 390
11.3 Shear Center 392
11.4 Unsymmetrical Bending 399
    a. Review of symmetrical bending 399
    b. Symmetrical sections 400
    c. Inclination of the neutral axis 401
    d. Unsymmetrical sections 402
11.5 Curved Beams 407
    a. Background 407
    b. Compatibility 408
    c. Equilibrium 409
    d. Curved beam formula 410

CHAPTER 12
Special Topics   417

12.1 Introduction 417
12.2 Energy Methods 418
    a. Work and strain energy 418
    b. Strain energy of bars and beams 418
    c. Deflections by Castigliano's theorem 420
12.3 Dynamic Loading 429
    a. Assumptions 429
    b. Mass-spring model 430
    c. Elastic bodies 431
    d. Modulus of resilience; modulus of toughness 431
12.4 Theories of Failure 436
    a. Brittle materials 437
    b. Ductile materials 438
12.5 Stress Concentration 444
12.6 Fatigue under Repeated Loading 450

CHAPTER 13
Inelastic Action   455

13.1 Introduction 455
13.2 Limit Torque 456
13.3 Limit Moment 458
13.4 Residual Stresses 463
    a. Loading-unloading cycle 463
    b. Torsion 463
    c. Bending 464
    d. Elastic spring-back 465
13.5 Limit Analysis 469
    a. Axial loading 469
    b. Torsion 470
    c. Bending 471

APPENDIX A
Review of Properties of Plane Areas   479

A.1 First Moments of Area; Centroid 479
A.2 Second Moments of Area 480
    a. Moments and product of inertia 480
    b. Parallel-axis theorems 481
    c. Radii of gyration 483
    d. Method of composite areas 483
A.3 Transformation of Second Moments of Area 492
    a. Transformation equations for moments and products of inertia 492
    b. Comparison with stress transformation equations 493
    c. Principal moments of inertia and principal axes 493
    d. Mohr's circle for second moments of area 494

APPENDIX B
Tables   501

B.1 Average Physical Properties of Common Metals 502
B.2 Properties of Wide-Flange Sections (W-Shapes): SI Units 504
B.3 Properties of I-Beam Sections (S-Shapes): SI Units 510
B.4 Properties of Channel Sections: SI Units 511
B.5 Properties of Equal Angle Sections: SI Units 512
B.6 Properties of Unequal Angle Sections: SI Units 514
B.7 Properties of Wide-Flange Sections (W-Shapes): U.S. Customary Units 516
B.8 Properties of I-Beam Sections (S-Shapes): U.S. Customary Units 524

B.9 Properties of Channel Sections: U.S. Customary Units 526
B.10 Properties of Equal and Unequal Angle Sections: U.S. Customary Units 527

Answers to Even-Numbered Problems   **531**

Photo Credits   **538**

Index   **539**

Preface

This textbook is intended for use in a first course in mechanics of materials. Programs of instruction relating to the mechanical sciences, such as mechanical, civil, and aerospace engineering, often require that students take this course in the second or third year of studies. Because of the fundamental nature of the subject matter, mechanics of materials is often a required course, or an acceptable technical elective in many other curricula. Students must have completed courses in statics of rigid bodies and mathematics through integral calculus as prerequisites to the study of mechanics of materials.

To place this book in context for engineering education, the user should know that it is an extensive revision of the fourth edition of *Strength of Materials* by Pytel and Singer. The contents have been thoroughly modernized to reflect the realities and trends in contemporary engineering education. In addition to eliminating a few of the specialized topics normally taught in more advanced civil engineering courses, the coverage of fundamental topics has been expanded. All of the illustrations have been redrawn and improved, with the addition of a second color for clarity and as an aid to understanding complex structures. Many new diagrams aid the visualization of concepts and improve the comprehension of derivations. Fully 60% of the homework problems are new or modified versions of previous problems. A new feature is the computer problems found at the end of each chapter.

Every effort has been made to maintain the conciseness and organization that were the hallmarks of the earlier editions of Pytel and Singer. In the first eight chapters, emphasis is placed exclusively on elastic analysis through the presentation of stress, strain, torsion, bending, and combined loading. An instructor can easily teach these topics within the time constraints of a two- or three-credit course. The remaining five chapters of the text cover material that can be omitted from an introductory course. Because these more advanced topics are not interwoven in the early chapters on the basic theory, the core material can efficiently be taught without skipping over topics within chapters. Once the instructor has covered the material on elastic analysis, he or she can freely choose topics from the more advanced later chapters, as time permits. Organizing the material in this manner has created a significant savings in the number of pages without sacrificing topics that are usually found in an introductory text.

Features The most notable features of the organization of this text include the following:

- Chapter 1 introduces the concept of stress (including stresses acting on inclined planes). However, the general stress transformation equations and Mohr's circle are deferred until Chapter 8. Engineering instructors often hold off teaching the concept of state of stress at a point due to combined loading until students have gained sufficient experience analyzing axial, torsional, and bending loads. However, if instructors wish to teach the general transformation equations and Mohr's circle at the beginning of the course, they may go to the freestanding discussion in Chapter 8 and use it whenever they see fit.
- Advanced beam topics, such as composite and curved beams, unsymmetrical bending, and shear center appear in chapters that are distinct from the basic beam theory. This makes it convenient for instructors to choose only those topics that they wish to present in their course.
- Chapter 12, entitled "Special Topics," consolidates topics that are important but not essential to an introductory course, including energy methods, theories of failure, stress concentrations, and fatigue. Some, but not all, of this material is commonly covered in a three-credit course at the discretion of the instructor.
- Chapter 13, the final chapter of the text, discusses the fundamentals of inelastic analysis. Positioning this topic at the end of the book enables the instructor to present an efficient and coordinated treatment of elastoplastic deformation, residual stress, and limit analysis after students have learned the basics of elastic analysis.

The text contains an equal number of problems using SI and U.S. Customary units. Homework problems strive to present a balance between directly relevant engineering-type problems and "teaching" problems that illustrate the principles in a straightforward manner. An outline of the applicable problem-solving procedure is included in the text to help students make the sometimes difficult transition from theory to problem analysis. Throughout the text and the sample problems, free-body diagrams are used to identify the unknown quantities and to recognize the number of independent equations. The three basic concepts of mechanics—equilibrium, compatibility, and constitutive equations—are continually reinforced in statically indeterminate problems. The problems are arranged in the following manner:

- Virtually every article in the text is followed by sample problems and homework problems that illustrate the principles and the problem-solving procedure introduced in the article.
- Every chapter contains review problems, with the exception of optional topics. In this way, the review problems test the students' comprehension of the material presented in the entire chapter, since it is not always obvious which of the principles presented in the chapter apply to the problem at hand.
- Most chapters conclude with computer problems, the majority of which are design oriented. Students should solve these problems using a high-level language, such as MathCad® or MATLAB®, which minimizes the programming effort and permits them to concentrate on the organization and presentation of the solution.

Acknowledgments We would like to acknowledge the following reviewers for their suggestions and comments: Daniel O. Adams, University of Utah; Patricia D. Brackin, Rose-Hulman Institute of Technology; Harvey L. Hoy, Milwaukee School of Engineering; Joe Iannelli, The University of Tennessee; M. Sathyamoorthy, Clarkson University; and Bruce Whelchel, Bradley University.

We are indebted to Dr. Christine Masters for checking solutions to the homework problems.

Andrew Pytel
Jaan Kiusalaas

List of Symbols

A	area
A'	partial area of beam cross section
b	width; distance from origin to center of Mohr's circle
c	distance from neutral axis to extreme fiber
C	centroid of area; couple
C_c	critical slenderness ratio of column
D, d	diameter
d	distance
E	modulus of elasticity
e	eccentricity of load; spacing of connectors
f	frequency
F	force
G	shear modulus
g	gravitational acceleration
H	horizontal force
h	height; depth of beam
I	moment of inertia of area
$\bar{I}$	centroidal moment of inertia of area
I_1, I_2	principal moments of inertia of area
J	polar moment of inertia of area
$\bar{J}$	centroidal polar moment of inertia of area
k	stress concentration factor; radius of gyration of area; spring stiffness
L	length
L_e	effective length of column
M	bending moment
M_L	limit moment
M_{yp}	yield moment
m	mass
N	factor of safety; normal force; number of load cycles
n	impact factor; ratio of moduli of elasticity
P	force; axial force in bar
P_{cr}	critical (buckling) load of column
$\mathscr{P}$	power
p	pressure
Q	first moment of area; dummy load
q	shear flow
R	radius; reactive force; resultant force
r	radius; least radius of gyration of cross-sectional area of column
S	section modulus; length of median line
s	distance
T	kinetic energy; temperature; tensile force; torque

T_L	limit torque
T_{yp}	yield torque
t	thickness; tangential deviation; torque per unit length
$\mathbf{t}$	stress vector
U	strain energy; work
u, v	rectangular coordinates
v	deflection of beam; velocity
V	vertical shear force
W	weight or load
w	load intensity
x, y, z	rectangular coordinates
$\bar{x}, \bar{y}, \bar{z}$	coordinates of centroid of area or center of gravity
α	coefficient of thermal expansion
α, β	angles
γ	shear strain; weight density
δ	elongation or contraction of bar; displacement
δ_s	static displacement
Δ	prescribed displacement
ϵ	normal strain
$\epsilon_1, \epsilon_2, \epsilon_3$	principal strains
θ	angle; slope angle of elastic curve
θ_1, θ_2	angles between x-axis and principal directions
ν	Poisson's ratio
ρ	radius of curvature; variable radius; mass density
σ	normal stress
$\sigma_1, \sigma_2, \sigma_3$	principal stresses
σ_a	stress amplitude in cyclic loading
σ_b	bearing stress
σ_c	circumferential stress
σ_ℓ	longitudinal stress
σ_{pl}	normal stress at proportional limit
σ_{ult}	ultimate stress
σ_w	working (allowable) normal stress
σ_{yp}	normal stress at yield point
τ	shear stress
τ_w	working (allowable) shear stress
τ_{yp}	shear stress at yield point
ω	angular velocity

1
Stress

Truss of a highway bridge. The members of a truss carry loading by direct tension or compression; there is very little bending. A truss is an efficient structure in the sense that it has a high load/structural weight ratio.

1.1 Introduction

The three fundamental areas of engineering mechanics are statics, dynamics, and mechanics of materials. Statics and dynamics are devoted primarily to the study of the *external effects upon rigid bodies*—that is, bodies for which the change in shape (deformation) can be neglected. In contrast, *mechanics of materials* deals with the *internal effects and deformations* that are caused by the applied loads. Both considerations are of paramount importance in design. A machine part or structure must be strong enough to carry the applied load without breaking and, at the same time, the deformations must not be excessive.

FIG. 1.1 Equilibrium analysis will determine the force P, but not the strength or the rigidity of the bar.

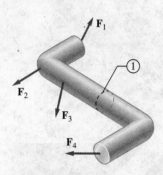

FIG. 1.2 External forces acting on a body.

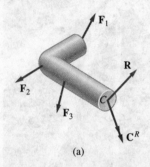

FIG. 1.3(a) Free-body diagram for determining the internal force system acting on section ①.

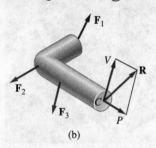

FIG. 1.3(b) Resolving the internal force $\mathbf{R}$ into the axial force P and the shear force V.

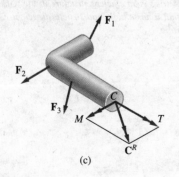

FIG. 1.3(c) Resolving the internal couple $\mathbf{C}^R$ into the torque T and the bending moment M.

The differences between rigid-body mechanics and mechanics of materials can be appreciated if we consider the bar shown in Fig. 1.1. The force P required to support the load W in the position shown can be found easily from equilibrium analysis. After we draw the free-body diagram of the bar, summing moments about the pin at O determines the value of P. In this solution, we assume that the bar is both rigid (the deformation of the bar is neglected) and strong enough to support the load W. In mechanics of materials, the statics solution is extended to include an analysis of the forces acting *inside* the bar to be certain that the bar will neither break nor deform excessively.

1.2 Analysis of Internal Forces; Stress

The equilibrium analysis of a rigid body is concerned primarily with the calculation of external reactions (forces that act external to a body) and internal reactions (forces that act at internal connections). In mechanics of materials, we must extend this analysis to determine *internal forces*—that is, forces that act on cross sections that are *internal* to the body itself. In addition, we must investigate the manner in which these internal forces are distributed within the body. Only after these computations have been made can the design engineer select the proper dimensions for a member and select the material from which the member should be fabricated.

If the external forces that hold a body in equilibrium are known, we can compute the internal forces by straightforward equilibrium analysis. For example, consider the bar in Fig. 1.2 that is loaded by the external forces $\mathbf{F}_1$, $\mathbf{F}_2$, $\mathbf{F}_3$, and $\mathbf{F}_4$. To determine the internal force system acting on the cross section labeled ①, we must first isolate the segments of the bar lying on either side of section ①. The free-body diagram of the segment to the left of section ① is shown in Fig. 1.3(a). In addition to the external forces $\mathbf{F}_1$, $\mathbf{F}_2$, and $\mathbf{F}_3$, this free-body diagram shows the resultant force-couple system of the internal forces that are distributed over the cross section: the resultant force $\mathbf{R}$, acting at the centroid C of the cross section, and $\mathbf{C}^R$, the resultant couple[1] (we use double-headed arrows to represent couple-vectors). If the external forces are known, the equilibrium equations $\Sigma \mathbf{F} = \mathbf{0}$ and $\Sigma \mathbf{M}_C = \mathbf{0}$ can be used to compute $\mathbf{R}$ and $\mathbf{C}^R$.

It is conventional to represent both $\mathbf{R}$ and $\mathbf{C}^R$ in terms of two components: one perpendicular to the cross section and the other lying in the cross section, as shown in Figs. 1.3(b) and (c). These components are given the

---

[1] The resultant force $\mathbf{R}$ can be located at any point, provided that we introduce the correct resultant couple. The reason for locating $\mathbf{R}$ at the centroid of the cross section will be explained shortly.

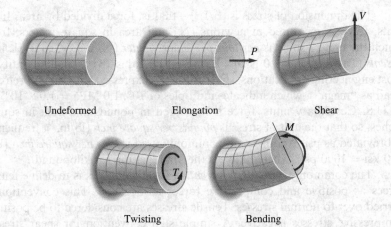

FIG. 1.4 Deformations produced by the components of internal forces and couples.

following physically meaningful names:

- P: The force component that is perpendicular to the cross section, tending to elongate or shorten the bar, is called the *normal force*.
- V: The force component lying in the plane of the cross section, tending to shear (slide) one segment of the bar relative to the other segment, is called the *shear force*.
- T: The component of the resultant couple that tends to twist (rotate) the bar is called the *twisting moment* or *torque*.
- M: The component of the resultant couple that tends to bend the bar is called the *bending moment*.

The deformations produced by these internal forces and internal couples are shown in Fig. 1.4.

Up to this point, we have been concerned only with the resultant of the internal force system. However, in design, the manner in which the internal forces are distributed is equally important. This consideration leads us to introduce the force intensity at a point, called *stress*, which plays a central role in the design of load-bearing members.

Figure 1.5(a) shows a small area element ΔA of the cross section located at the arbitrary point O. We assume that $\Delta \mathbf{R}$ is that part of the resultant force that is transmitted across ΔA, with its normal and shear components being ΔP and ΔV, respectively. The *stress vector* acting on the cross section at point O is defined as

$$\mathbf{t} = \lim_{\Delta A \to 0} \frac{\Delta \mathbf{R}}{\Delta A} \qquad (1.1)$$

Its normal component σ (lowercase Greek *sigma*) and shear component τ (lowercase Greek *tau*), shown in Fig. 1.5(b), are

$$\sigma = \lim_{\Delta A \to 0} \frac{\Delta P}{\Delta A} = \frac{dP}{dA} \qquad \tau = \lim_{\Delta A \to 0} \frac{\Delta V}{\Delta A} = \frac{dV}{dA} \qquad (1.2)$$

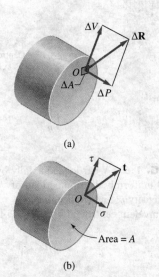

FIG. 1.5 Normal and shear stresses acting on the section at point O are defined in Eq. (2).

The dimension of stress is $[F/L^2]$—that is, force divided by area. In SI units, force is measured in newtons (N) and area in square meters, from which the unit of stress is *newtons per square meter* (N/m^2) or, equivalently, *pascals* (Pa): $1.0 \text{ Pa} = 1.0 \text{ N/m}^2$. Because 1 pascal is a very small quantity in most engineering applications, stress is usually expressed with the SI prefix M (read as "mega"), which indicates multiples of 10^6: $1.0 \text{ MPa} = 1.0 \times 10^6 \text{ Pa}$. In U.S. Customary units, force is measured in pounds and area in square inches, so that the unit of stress is *pounds per square inch* ($lb/in.^2$), frequently abbreviated as psi. Another unit commonly used is *kips per square inch* (ksi) (1.0 ksi = 1000 psi), where "kip" is the abbreviation for kilopound.

The commonly used *sign convention* for axial forces is to define tensile forces as positive and compressive forces as negative. This convention is carried over to normal stresses: Tensile stresses are considered to be positive, compressive stresses negative. A simple sign convention for shear stresses does not exist; a convention that depends on a coordinate system will be introduced later in the text. If the stresses are *uniformly distributed*, they can be computed from

$$\sigma = \frac{P}{A} \qquad \tau = \frac{V}{A} \tag{1.3}$$

where A is the area of the cross section. If the stress distribution is not uniform, then Eqs. (1.3) should be viewed as the *average stress* acting on the cross section.

1.3 Axially Loaded Bars

a. Centroidal (axial) loading

Figure 1.6(a) shows a bar of constant cross-sectional area A. The ends of the bar carry uniformly distributed normal loads of intensity p (units: Pa or psi). We know from statics that when the loading is uniform, its resultant passes through the centroid of the loaded area. Therefore, the resultant $P = pA$ of each end load acts along the centroidal axis (the line connecting the centroids of cross sections) of the bar, as shown in Fig. 1.6(b). The loads shown in Fig. 1.6 are called *axial* or *centroidal loads*.

Although the loads in Figs. 1.6(a) and (b) are statically equivalent, they do not result in the same stress distribution in the bar. In the case of the uniform loading in Fig. 1.6(a), the internal forces acting on all cross sections are also uniformly distributed. Therefore, the normal stress acting at any point on a cross section is

$$\boxed{\sigma = \frac{P}{A}} \tag{1.4}$$

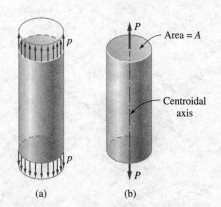

FIG. 1.6 A bar loaded axially by (a) uniformly distributed load of intensity p; and (b) a statically equivalent centroidal force $P = pA$.

The stress distribution caused by the concentrated loading in Fig. 1.6(b) is more complicated. Advanced methods of analysis show that on cross sections close to the ends, the maximum stress is considerably higher than the average stress P/A. As we move away from the ends, the stress becomes more uniform, reaching the constant value P/A in a relatively short

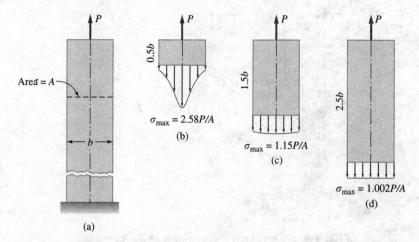

FIG. 1.7 Normal stress distribution in a strip caused by a concentrated load.

distance from the ends. In other words, the stress distribution is approximately uniform in the bar, except in the regions close to the ends.

As an example of concentrated loading, consider the thin strip of width b shown in Fig. 1.7(a). The strip is loaded by the centroidal force P. Figures 1.7(b)–(d) show the stress distribution on three different cross sections. Note that at a distance $2.5b$ from the loaded end, the maximum stress differs by only 0.2% from the average stress P/A.

b. Saint Venant's principle

About 150 years ago the French mathematician Saint Venant studied the effects of statically equivalent loads on the twisting of bars. His results led to the following observation, called *Saint Venant's principle*:

> *The difference between the effects of two different but statically equivalent loads becomes very small at sufficiently large distances from the load.*

The example in Fig. 1.7 is an illustration of Saint Venant's principle. The principle also applies to the effects caused by abrupt changes in the cross section. Consider, as an example, the grooved cylindrical bar of radius R shown in Fig. 1.8(a). The loading consists of the force P that is uniformly distributed over the end of the bar. If the groove were not present, the normal stress acting at all points on a cross section would be P/A. Introduction of the groove disturbs the uniformity of the stress, but this effect is confined to the vicinity of the groove, as seen in Figs. 1.8(b) and (c).

Most analysis in mechanics of materials is based on simplifications that can be justified with Saint Venant's principle. We often replace loads (including support reactions) by their resultants and ignore the effects of holes, grooves, and fillets on stresses and deformations. Many of the simplifications are not only justified but necessary. Without simplifying assumptions, analysis would be exceedingly difficult. However, we must always keep in mind the approximations that were made, and make allowances for them in the final design.

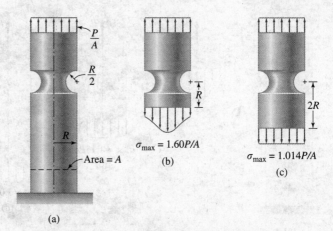

FIG. 1.8 Normal stress distribution in a grooved bar.

c. Stresses on inclined planes

When a bar of cross-sectional area A is subjected to an axial load P, the normal stress P/A acts on the cross section of the bar. Let us now consider the stresses that act on plane a-a that is inclined at the angle θ to the cross section, as shown in Fig. 1.9(a). Note that the area of the inclined plane is $A/\cos\theta$. To investigate the forces that act on this plane, we consider the free-body diagram of the segment of the bar shown in Fig. 1.9(b). Because the segment is a two-force body, the resultant internal force acting on the inclined plane must be the axial force P, which can be resolved into the normal component $P\cos\theta$ and the shear component $P\sin\theta$. Therefore, the corresponding stresses, shown in Fig. 1.9(c), are

$$\sigma = \frac{P\cos\theta}{A/\cos\theta} = \frac{P}{A}\cos^2\theta \qquad (1.5\text{a})$$

$$\tau = \frac{P\sin\theta}{A/\cos\theta} = \frac{P}{A}\sin\theta\cos\theta = \frac{P}{2A}\sin 2\theta \qquad (1.5\text{b})$$

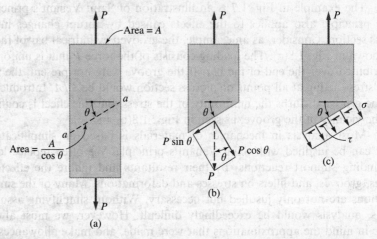

FIG. 1.9 Determining the stresses acting on an inclined section of a bar.

From these equations we see that the maximum normal stress is P/A, and it acts on the cross section of the bar (that is, on the plane $\theta = 0$). The shear stress is zero when $\theta = 0$, as would be expected. The maximum shear stress is $P/2A$, which acts on the planes inclined at $\theta = 45°$ to the cross section.

In summary, an axial load causes not only normal stress but also shear stress. The magnitudes of both stresses depend on the orientation of the plane on which they act.

By replacing θ with $\theta + 90°$ in Eqs. (1.5), we obtain the stresses acting on plane a'-a', which is perpendicular to a-a, as illustrated in Fig. 10(a):

$$\sigma' = \frac{P}{A}\sin^2\theta \qquad \tau' = -\frac{P}{2A}\sin 2\theta \qquad (1.6)$$

where we used the identities $\cos(\theta + 90°) = -\sin\theta$ and $\sin 2(\theta + 90°) = -\sin 2\theta$. Because the stresses in Eqs. (1.5) and (1.6) act on mutually perpendicular, or "complementary" planes, they are called *complementary stresses*. The traditional way to visualize complementary stresses is to draw them on a small (infinitesimal) element of the material, the sides of which are parallel to the complementary planes, as in Fig. 10(b). When labeling the stresses, we made use of the following important result that follows from Eqs. (1.5) and (1.6):

$$\tau' = -\tau \qquad (1.7)$$

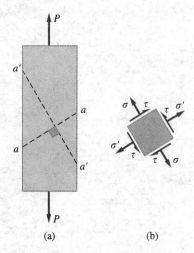

FIG. 1.10 Stresses acting on two mutually perpendicular inclined sections of a bar.

In other words,

> The shear stresses that act on complementary planes have the same magnitude but opposite sense.

Although Eq. (1.7) was derived for axial loading, we will show later that it also applies to more complex loadings.

The design of axially loaded bars is usually based on the maximum normal stress in the bar. This stress is commonly called simply the *normal stress* and denoted by σ, a practice that we follow in this text. The design criterion thus is that $\sigma = P/A$ must not exceed the *working stress* of the material from which the bar is to be fabricated. The working stress, also called the *allowable stress*, is the largest value of stress that can be safely carried by the material. Working stress, denoted by σ_w, will be discussed more fully in Art. 2.2.

d. Procedure for stress analysis

In general, finding the normal stress in an axially loaded member of a structure involves the following steps.

Equilibrium Analysis

- If necessary, find the external reactions using a free-body diagram (FBD) of the entire structure.
- Compute the axial force P in the member using the method of sections. This method introduces an imaginary cutting plane that isolates a segment of the structure. The cutting plane must include the cross section of the member of interest. The axial force acting in the member can

then be found from the FBD of the isolated segment because it now appears as an external force on the FBD.

Computation of Stresses

- After the axial force has been found by equilibrium analysis, the *average* normal stress in the member can be obtained from $\sigma = P/A$, where A is the cross-sectional area of the member at the cutting plane.
- In slender bars, $\sigma = P/A$ is the *actual* normal stress if the section is sufficiently far from applied loads and abrupt changes in the cross section (Saint Venant's principle).

Note on the Analysis of Trusses The usual assumptions made in the analysis of trusses are: (1) weights of the member are negligible in comparison to the applied loads; (2) joints behave as smooth pins; and (3) all loads are applied at the joints. Under these assumptions, each member of the truss is an axially loaded bar. The internal forces in the bars can be obtained by the method of sections or the method of joints (utilizing the free-body diagrams of the joints).

Sample Problem 1.1

The bar *ABCD* in Fig. (a) consists of three cylindrical steel segments, each with a different cross-sectional area. Axial loads are applied as shown. Calculate the normal stress in each segment.

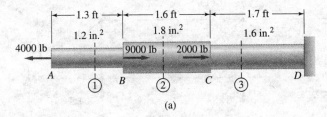

(a)

Solution

To calculate the normal stresses, we must first use equilibrium analysis to compute the axial load in each segment of the bar (recall that equilibrium analysis is the first step in stress analysis). The required free-body diagrams (FBDs), shown in Fig. (b), were drawn by isolating the portions of the bar lying to the left of sections ①, ②, and ③. From the FBDs, we see that the internal forces in the three segments of the bar are $P_{AB} = 4000$ lb (T), $P_{BC} = 5000$ lb (C), and $P_{CD} = 7000$ lb (C), where (T) denotes tension and (C) indicates compression. Note that the force in the bar varies from segment to segment, but the force within each segment is constant.

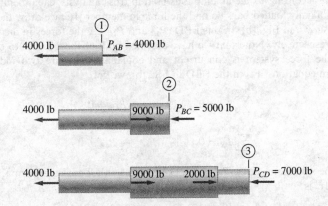

(b) Free-body diagrams (FBDs)

The normal stresses in the three segments are

$$\sigma_{AB} = \frac{P_{AB}}{A_{AB}} = \frac{4000 \text{ lb}}{1.2 \text{ in.}^2} = 3330 \text{ psi (T)} \qquad \textbf{Answer}$$

$$\sigma_{BC} = \frac{P_{BC}}{A_{BC}} = \frac{5000 \text{ lb}}{1.8 \text{ in.}^2} = 2780 \text{ psi (C)} \qquad \textbf{Answer}$$

$$\sigma_{CD} = \frac{P_{CD}}{A_{CD}} = \frac{7000 \text{ lb}}{1.6 \text{ in.}^2} = 4380 \text{ psi (C)} \qquad \textbf{Answer}$$

Observe that the lengths of the segments do not affect the calculation of stresses. In addition, the fact that the bar is made of steel is irrelevant; the stresses would be the same if the bar, or any of its segments, were fabricated from another material.

Sample Problem 1.2

For the truss shown in Fig. (a), calculate the normal stresses in (1) member AC; and (2) member BD. The cross-sectional area of each member is 900 mm^2.

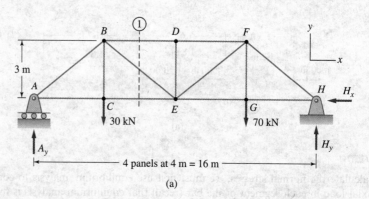

(a)

Solution

Equilibrium analysis using the FBD of the entire truss in Fig. (a) gives the following values for the external reactions: $A_y = 40$ kN, $H_y = 60$ kN, and $H_x = 0$.

Part 1

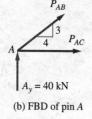

(b) FBD of pin A

Recall that according to the assumptions used in truss analysis, each member of the truss is an axially loaded bar. To find the force in member AC, we draw the FBD of pin A, as shown in Fig. (b). In this FBD, P_{AB} and P_{AC} are the forces in members AB and AC, respectively. Note that we have assumed both of these forces to be tensile. Because the force system is concurrent and coplanar, there are two independent equilibrium equations. From the FBD in Fig. (b), we get

$$\sum F_y = 0 \quad +\uparrow \quad 40 + \frac{3}{5} P_{AB} = 0$$

$$\sum F_x = 0 \quad \xrightarrow{+} \quad P_{AC} + \frac{4}{5} P_{AB} = 0$$

Solving the equations gives $P_{AC} = 53.33$ kN (tension). Thus, the normal stress in member AC is

$$\sigma_{AC} = \frac{P_{AC}}{A_{AC}} = \frac{53.33 \text{ kN}}{900 \text{ mm}^2} = \frac{53.33 \times 10^3 \text{ N}}{900 \times 10^{-6} \text{ m}^2}$$

$$= 59.3 \times 10^6 \text{ N/m}^2 = 59.3 \text{ MPa (T)} \qquad \text{Answer}$$

Part 2

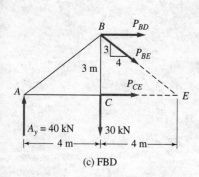

(c) FBD

To determine the force in member BD, we see that section ① in Fig. (a) cuts through members BD, BE, and CE. Because three equilibrium equations are available for a portion of the truss separated by this section, we can find the forces in all three members, if needed.

The FBD of the portion of the truss lying to the left of section ① is shown in Fig. (c) (the portion lying to the right could also be used). We have again assumed that the forces in the members are tensile. To calculate the force in member BD, we use the equilibrium equation

$$\sum M_E = 0 \quad +\circlearrowleft \quad -40(8) + 30(4) - P_{BD}(3) = 0$$

10

which yields
$$P_{BD} = -66.67 \text{ kN} = 66.67 \text{ kN (C)}$$

Therefore, the normal stress in member BD is

$$\sigma_{BD} = \frac{P_{BD}}{A_{BD}} = \frac{-66.67 \text{ kN}}{900 \text{ mm}^2} = \frac{-66.67 \times 10^3 \text{ N}}{900 \times 10^{-6} \text{ m}^2}$$
$$= -74.1 \times 10^6 \text{ N/m}^2 = 74.1 \text{ MPa (C)} \qquad \text{Answer}$$

Sample Problem 1.3

Figure (a) shows a two-member truss supporting a block of weight W. The cross-sectional areas of the members are 800 mm² for AB and 400 mm² for AC. Determine the maximum safe value of W if the working stresses are 110 MPa for AB and 120 MPa for AC.

Solution

Being members of a truss, AB and AC can be considered to be axially loaded bars. The forces in the bars can be obtained by analyzing the free-body diagram (FBD) of pin A in Fig. (b). The equilibrium equations are

$$\sum F_x = 0 \quad \xrightarrow{+} \quad P_{AC} \cos 60° - P_{AB} \cos 40° = 0$$

$$\sum F_y = 0 \quad +\uparrow \quad P_{AC} \sin 60° + P_{AB} \sin 40° - W = 0$$

Solving simultaneously, we get

$$P_{AB} = 0.5077 W \qquad P_{AC} = 0.7779 W$$

The value of W that will cause the normal stress in bar AB to equal its working stress is given by

$$P_{AB} = \sigma_{AB} A_{AB}$$
$$0.5077 W = (110 \times 10^6 \text{ N/m}^2)(800 \times 10^{-6} \text{ m}^2)$$
$$W = 173.3 \times 10^3 \text{ N} = 173.3 \text{ kN}$$

The value of W that will cause the normal stress in bar AC to equal its working stress is found from

$$P_{AC} = \sigma_{AC} A_{AC}$$
$$0.7779 W = (120 \times 10^6 \text{ N/m}^2)(400 \times 10^{-6} \text{ m}^2)$$
$$W = 61.7 \times 10^3 \text{ N} = 61.7 \text{ kN}$$

The maximum safe value of W is the smaller of the preceding two values—namely,

$$W = 61.7 \text{ kN} \qquad \text{Answer}$$

We see that the stress in bar AC determines the safe value of W. The other "solution," $W = 173.3$ kN, must be discarded because it would cause the stress in AC to exceed its working stress of 120 MPa.

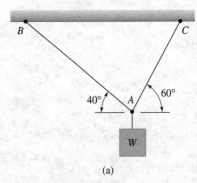

(a)

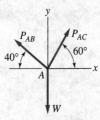

(b) FBD of pin A

Problems

1.1 A hollow steel tube with an inside diameter of 100 mm must carry an axial tensile load of 400 kN. Determine the smallest allowable outside diameter of the tube if the working stress is 120 MN/m².

1.2 The cross-sectional area of bar $ABCD$ is 600 mm². Determine the maximum normal stress in the bar.

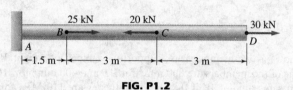

FIG. P1.2

1.3 Determine the largest weight W that can be supported by the two wires AB and AC. The working stresses are 100 MPa for AB and 150 MPa for AC. The cross-sectional areas of AB and AC are 400 mm² and 200 mm², respectively.

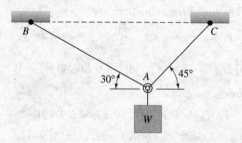

FIG. P1.3

1.4 Axial loads are applied to the compound rod that is composed of an aluminum segment rigidly connected between steel and bronze segments. What is the stress in each material given that $P = 3000$ lb?

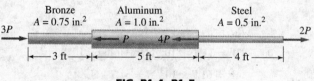

FIG. P1.4, P1.5

1.5 Axial loads are applied to the compound rod that is composed of an aluminum segment rigidly connected between steel and bronze segments. Find the largest safe value of P if the working stresses are 18 000 psi for steel, 10 000 psi for aluminum, and 16 000 psi for bronze.

1.6 The wood pole is supported by two cables of 1/4-in. diameter. The turnbuckles in the cables are tightened until the stress in the cables reaches 60 000 psi. If the working compressive stress for wood is 200 psi, determine the smallest permissible diameter of the pole.

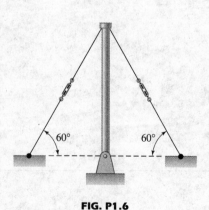

FIG. P1.6

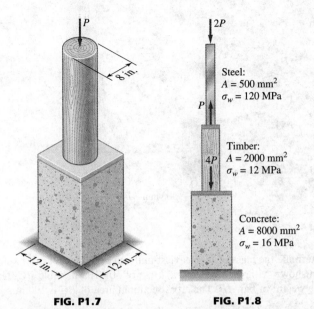
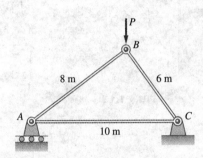

FIG. P1.7 **FIG. P1.8** **FIG. P1.9**

1.7 The column consists of a wooden post and a concrete footing, separated by a steel bearing plate. Find the maximum safe value of the axial load P if the working stresses are 1000 psi for wood and 450 psi for concrete.

1.8 Find the maximum allowable value of P for the column. The cross-sectional areas and working stresses (σ_w) are shown in the figure.

1.9 Each bar of the truss has a rectangular cross section, 30 mm by 60 mm. Determine the maximum vertical load P that can be applied at B if the working stresses are 100 MPa in tension and 80 MPa in compression. (A reduced stress in compression is specified to reduce the danger of buckling.)

1.10 The homogeneous 800-kg bar AB is supported at either end by a steel cable. Calculate the smallest safe area of each cable if the working stress is 120 MPa for steel.

1.11 The homogeneous 6000-lb bar ABC is supported by a pin at C and a cable that runs from A to B around the frictionless pulley at D. Find the stress in the cable if its diameter is 0.6 in.

1.12 Determine the largest weight W that can be supported safely by the structure shown in the figure. The working stresses are 16 000 psi for the steel cable AB and 720 psi for the wood strut BC. Neglect the weight of the structure.

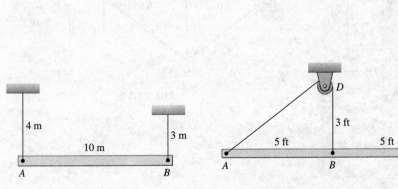
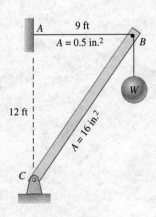

FIG. P1.10 **FIG. P1.11** **FIG. P1.12**

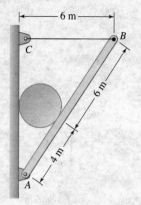

FIG. P1.13

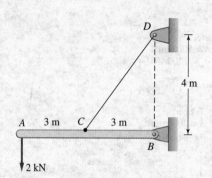

FIG. P1.14

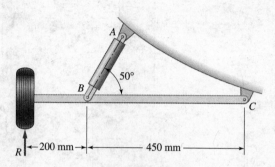

FIG. P1.15

1.13 Determine the mass of the heaviest uniform cylinder that can be supported in the position shown without exceeding a stress of 50 MPa in cable BC. Neglect friction and the weight of bar AB. The cross-sectional area of BC is 100 mm^2.

1.14 The uniform 150-kg bar AB carries a 2-kN vertical force at A. The bar is supported by a pin at B and the 10-mm-diameter cable CD. Find the stress in the cable.

1.15 The figure shows the landing gear of a light airplane. Determine the compressive stress in strut AB caused by the landing reaction $R = 20$ kN. Neglect the weights of the members. The strut is a hollow tube, with 40-mm outer diameter and 30-mm inner diameter.

1.16 The 1000-kg uniform bar AB is suspended from two cables AC and BD, each with cross-sectional area 400 mm^2. Find the magnitude P and location x of the largest additional vertical force that can be applied to the bar. The stresses in AC and BD are limited to 100 MPa and 50 MPa, respectively.

1.17 The cross-sectional area of each member of the truss is 1.8 in.^{2}. Calculate the stresses in members CE, DE, and DF. Indicate tension or compression.

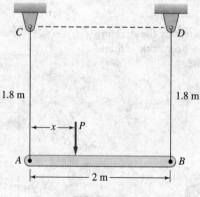

FIG. P1.16

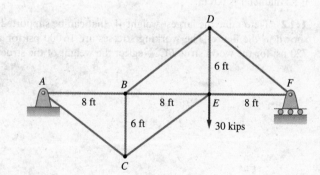

FIG. P1.17

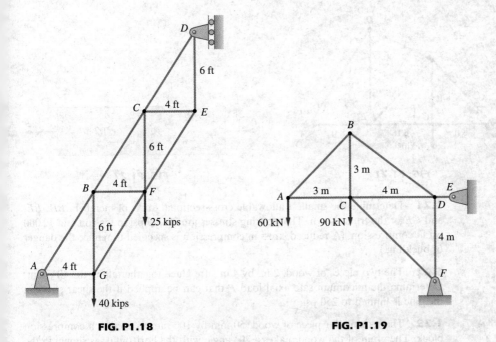

FIG. P1.18

FIG. P1.19

1.18 Determine the smallest safe cross-sectional areas of members AG and BC for the truss shown in the figure. The working stresses are 20 ksi in tension and 14 ksi in compression. (The working stress in compression is smaller to reduce the danger of buckling.)

1.19 Find the stresses in members BC, BD, and CF for the truss shown. Indicate tension or compression. The cross-sectional area of each member is 1600 mm^2.

1.20 For the truss shown, determine the smallest allowable cross-sectional areas of members BE, BF, and CF if the working stresses are 100 MPa in tension and 80 MPa in compression. (A reduced stress in compression is specified to reduce the danger of buckling.)

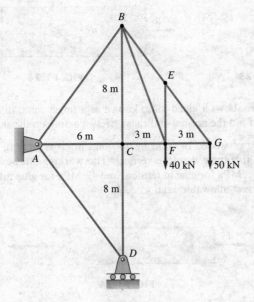

FIG. P1.20

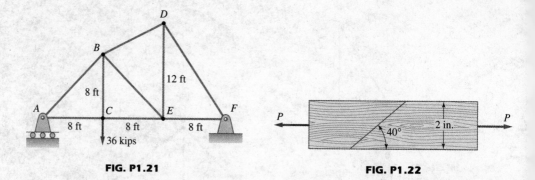

FIG. P1.21   FIG. P1.22

1.21 Determine the smallest allowable cross-sectional areas of members BD, BE, and CE of the truss shown. The working stresses are 20 000 psi in tension and 12 000 psi in compression. (A reduced stress in compression is specified to reduce the danger of buckling.)

1.22 The two pieces of wood, 2 in. by 4 in., are glued together along the 40° joint. Determine the maximum safe axial load P that can be applied if the shear stress in the glue is limited to 250 psi.

1.23 The rectangular piece of wood, 50 mm by 100 mm, is used as a compression block. The grain of the wood makes a 20° angle with the horizontal, as shown in the figure. Determine the largest axial force P that can be applied safely if the allowable stresses on the plane of the grain are 20 MPa for compression and 5 MPa for shear.

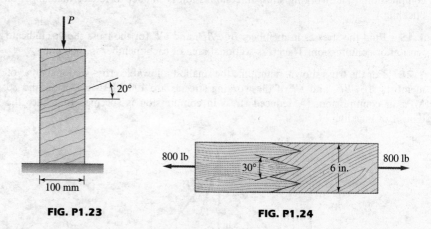

FIG. P1.23   FIG. P1.24

1.24 The figure shows a glued joint, known as a finger joint, in a 6-in. by 3/4-in. piece of lumber. Find the normal and shear stresses acting on the surface of the joint.

1.25 The piece of wood, 100 mm by 100 mm in cross section, contains a glued joint inclined at the angle θ to the vertical. The working stresses are 20 MPa for wood in tension, 8 MPa for glue in tension, and 12 MPa for glue in shear. If $\theta = 50°$, determine the largest allowable axial force P.

FIG. P1.25

1.4 Shear Stress

By definition, normal stress acting on an interior surface is directed perpendicular to that surface. Shear stress, on the other hand, is tangent to the surface on which it acts. Shear stress arises whenever the applied loads cause one section of a body to slide past its adjacent section. In Art. 1.3, we examined how shear stress occurs in an axially loaded bar. Three other examples of shear stress are illustrated in Fig. 1.11. Figure 1.11(a) shows two plates that are joined by a rivet. As seen in the FBD, the rivet must carry the shear force $V = P$. Because only one cross section of the rivet resists the shear, the rivet is said to be in *single shear*. The bolt of the clevis in Fig. 1.11(b) carries the load P across two cross-sectional areas, the shear force being $V = P/2$ on each cross section. Therefore, the bolt is said to be in a state of *double shear*. In Fig. 1.11(c) a circular slug is being punched out of a metal sheet. Here the shear force is P and the shear area is similar to the milled edge of a coin. The loads shown in Fig. 1.11 are sometimes referred to as *direct shear*, to distinguish them from the *induced shear* illustrated in Fig. 1.9.

The distribution of direct shear stress is usually complex and not easily determined. It is common practice to assume that the shear force V is uniformly distributed over the shear area A, so that the shear stress can be computed from

$$\tau = \frac{V}{A} \tag{1.8}$$

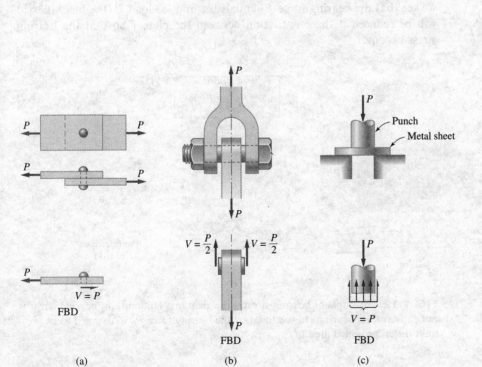

FIG. 1.11 Examples of direct shear: (a) single shear in a rivet; (b) double shear in a bolt; and (c) shear in a metal sheet produced by a punch.

Strictly speaking, Eq. (1.8) must be interpreted as the *average* shear stress. It is often used in design to evaluate the strength of connectors, such as rivets, bolts, and welds.

1.5 Bearing Stress

If two bodies are pressed against each other, compressive forces are developed on the area of contact. The pressure caused by these surface loads is called *bearing stress*. Examples of bearing stress are the soil pressure beneath a pier and the contact pressure between a rivet and the side of its hole. If the bearing stress is large enough, it can locally crush the material, which in turn can lead to more serious problems. In order to reduce bearing stresses, engineers sometimes employ bearing plates, the purpose of which is to distribute the contact forces over a larger area.

As an illustration of bearing stress, consider the lap joint formed by the two plates that are riveted together as shown in Fig. 1.12(a). The bearing stress caused by the rivet is not constant; it actually varies from zero at the sides of the hole to a maximum behind the rivet as illustrated in Fig. 1.12(b). The difficulty inherent in such a complicated stress distribution is avoided by the common practice of assuming that the bearing stress σ_b is uniformly distributed over a reduced area. The reduced area A_b is taken to be the *projected area* of the rivet:

$$A_b = td$$

where t is the thickness of the plate and d represents the diameter of the rivet, as shown in the FBD of the upper plate in Fig. 1.12(c). From this FBD we see that the bearing force P_b equals the applied load P (the bearing load will be reduced if there is friction between the plates), so that the bearing stress becomes

$$\sigma_b = \frac{P_b}{A_b} = \frac{P}{td} \tag{1.9}$$

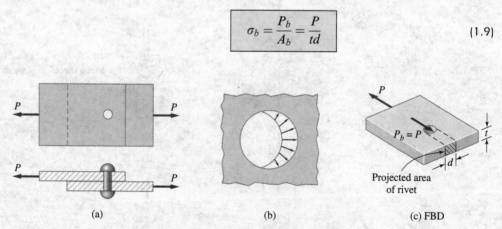

FIG. 1.12 Example of bearing stress: (a) a rivet in a lap joint; (b) bearing stress is not constant; (c) bearing stress caused by the bearing force P_b is assumed to be uniform on projected area td.

Sample Problem 1.4

The lap joint shown in Fig. (a) is fastened by four rivets of 3/4-in. diameter. Find the maximum load P that can be applied if the working stresses are 14 ksi for shear in the rivet and 18 ksi for bearing in the plate. Assume that the applied load is distributed evenly among the four rivets, and neglect friction between the plates.

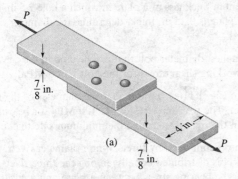

(a)

Solution

We will calculate P using each of the two design criteria. The largest safe load will be the smaller of the two values. Figure (b) shows the free-body diagram (FBD) of the lower plate. In this FBD, the lower halves of the rivets are in the plate, having been isolated from their top halves by a cutting plane. This cut exposes the shear forces V that act on the cross sections of the rivets. We see that the equilibrium condition is $V = P/4$.

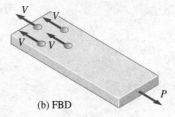

(b) FBD

Design for Shear Stress in Rivets

The value of P that would cause the shear stress in the rivets to reach its working value is found as follows:

$$V = \tau A$$

$$\frac{P}{4} = (14 \times 10^3)\left[\frac{\pi(3/4)^2}{4}\right]$$

$$P = 24\,700 \text{ lb}$$

Design for Bearing Stress in Plate

The shear force $V = P/4$ that acts on the cross section of one rivet is equal to the bearing force P_b due to the contact between the rivet and the plate. The value of P that would cause the bearing stress to equal its working value is computed from Eq. (1.9):

$$P_b = \sigma_b t d$$

$$\frac{P}{4} = (18 \times 10^3)(7/8)(3/4)$$

$$P = 47\,300 \text{ lb}$$

Choose the Correct Answer

Comparing the above solutions, we conclude that the maximum safe load P that can be applied to the lap joint is

$$P = 24\,700 \text{ lb} \qquad \text{Answer}$$

with the shear stress in the rivets being the governing design criterion.

Problems

1.26 What force is required to punch a 20-mm-diameter hole in a plate that is 25 mm thick? The shear strength of the plate is 350 MN/m².

1.27 A circular hole is to be punched in a plate that has a shear strength of 40 ksi—see Fig. 1.11(c). The working compressive stress for the punch is 50 ksi. (a) Compute the maximum thickness of a plate in which a hole 2.5 in. in diameter can be punched. (b) If the plate is 0.25 in. thick, determine the diameter of the smallest hole that can be punched.

1.28 Find the smallest diameter bolt that can be used in the clevis in Fig. 1.11(b) if $P = 400$ kN. The working shear stress for the bolt is 300 MPa.

1.29 Referring to Fig. 1.11(a), assume that the diameter of the rivet that joins the plates is $d = 20$ mm. The working stresses are 120 MPa for bearing in the plate and 60 MPa for shear in the rivet. Determine the minimum safe thickness of each plate.

1.30 The lap joint is connected by three 20-mm-diameter rivets. Assuming that the axial load $P = 50$ kN is distributed equally among the three rivets, find (a) the shear stress in a rivet; (b) the bearing stress between a plate and a rivet; and (c) the maximum average tensile stress in each plate.

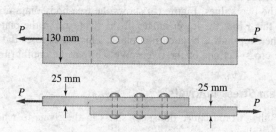

FIG. P1.30, P1.31

1.31 Assume that the axial load P applied to the lap joint is distributed equally among the three 20-mm-diameter rivets. What is the maximum load P that can be applied if the allowable stresses are 60 MPa for shear in rivets, 110 MPa for bearing between a plate and a rivet, and 140 MPa for tension in the plates.

1.32 A key prevents relative rotation between the shaft and the pulley. If the torque $T = 2200$ N·m is applied to the shaft, determine the smallest safe dimension b if the working shear stress for the key is 60 MPa.

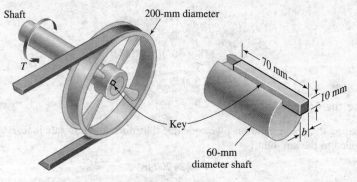

FIG. P1.32

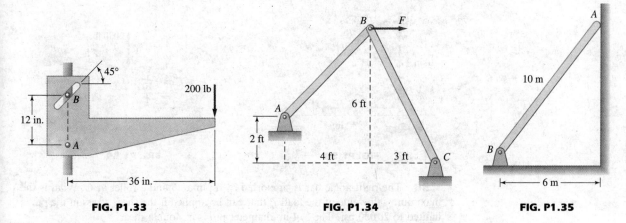

FIG. P1.33 **FIG. P1.34** **FIG. P1.35**

1.33 The bracket is supported by 1/2-in.-diameter pins at A and B (the pin at B fits in the 45° slot in the bracket). Neglecting friction, determine the shear stresses in the pins, assuming single shear.

1.34 The 7/8-in.-diameter pins at A and C that support the structure are in single shear. Find the largest force F that can be applied to the structure if the working shear stress for these pins is 5000 psi. Neglect the weights of the members.

1.35 The uniform 2-Mg bar is supported by a smooth wall at A and by a pin at B that is in double shear. Determine the diameter of the smallest pin that can be used if its working shear stress is 60 MPa.

1.36 The bell crank, which is in equilibrium under the forces shown in the figure, is supported by a 20-mm-diameter pin at D that is in double shear. Determine (a) the required diameter of the connecting rod AB, given that its tensile working stress is 100 MPa; and (b) the shear stress in the pin.

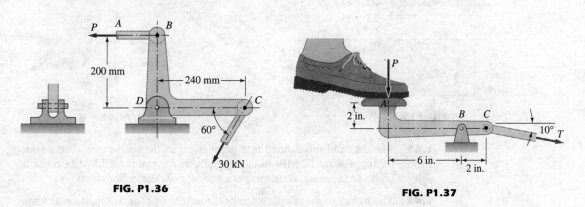

FIG. P1.36 **FIG. P1.37**

1.37 Compute the maximum force P that can be applied to the foot pedal. The 1/4-in.-diameter pin at B is in single shear, and its working shear stress is 4000 psi. The cable attached at C has a diameter of 1/8 in. and a working normal stress of 20 000 psi.

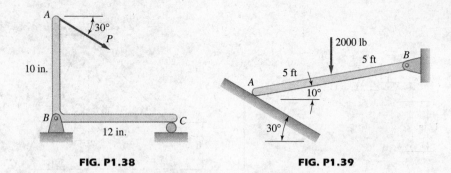

FIG. P1.38 **FIG. P1.39**

1.38 The right-angle bar is supported by a pin at B and a roller at C. What is the maximum safe value of the load P that can be applied if the shear stress in the pin is limited to 20 000 psi? The 3/4-in.-diameter pin is in double shear.

1.39 The bar AB is supported by a frictionless inclined surface at A and a 7/8-in.-diameter pin at B that is in double shear. Determine the shear stress in the pin when the vertical 2000-lb force is applied. Neglect the weight of the bar.

1.40 A joint is made by gluing two plywood gussets of thickness t to wood boards. The tensile working stresses are 1200 psi for the plywood and 700 psi for the boards. The working shear stress for the glue is 50 psi. Determine the dimensions b and t so that the joint is as strong as the boards.

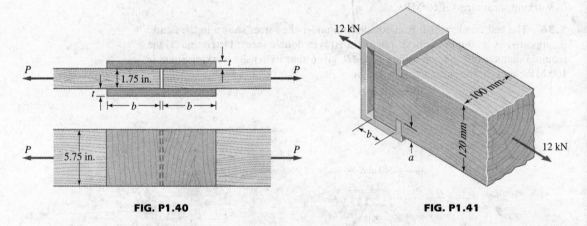

FIG. P1.40 **FIG. P1.41**

1.41 The steel end-cap is fitted into grooves cut in the timber post. The working stresses for the post are 1.8 MPa in shear parallel to the grain and 5.5 MPa in bearing perpendicular to the grain. Determine the smallest safe dimensions a and b.

1.42 The halves of the coupling are held together by four 5/8-in.-diameter bolts. The working stresses are 12 ksi for shear in the bolts and 15 ksi for bearing in the coupling. Find the largest torque T that can be safely transmitted by the coupling. Assume that the forces in the bolts have equal magnitudes.

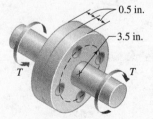

FIG. P1.42

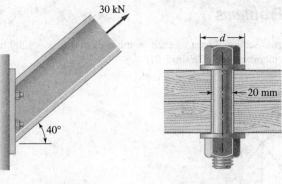

FIG. P1.43   **FIG. P1.44**

1.43 The plate welded to the end of the I-beam is fastened to the support with four 10-mm-diameter bolts (two on each side). Assuming that the load is equally divided among the bolts, determine the normal and shear stresses in a bolt.

1.44 The 20-mm-diameter bolt fastens two wooden planks together. The nut is tightened until the tensile stress in the bolt is 150 MPa. Find the smallest safe diameter d of the washers if the working bearing stress for wood is 13 MPa.

1.45 The figure shows a roof truss and the detail of the connection at joint B. Members BC and BE are angle sections with the thicknesses shown in the figure. The working stresses are 70 MPa for shear in the rivets and 140 MPa for bearing stress due to the rivets. How many 19-mm-diameter rivets are required to fasten the following members to the gusset plate: (a) BC; and (b) BE?

1.46 Repeat Prob. 1.45 if the rivet diameter is 22 mm, with all other data remaining unchanged.

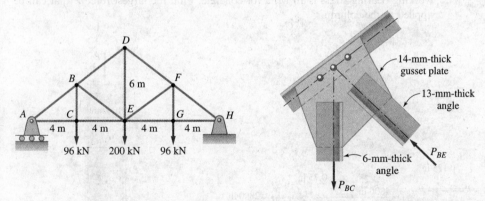

FIG. P1.45, P1.46

Review Problems

1.47 The cross-sectional area of each member of the truss is 1200 mm². Calculate the stresses in members *DF*, *CE*, and *BD*.

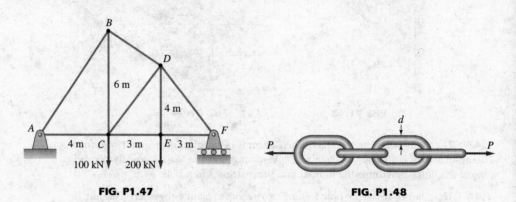

FIG. P1.47

FIG. P1.48

1.48 The links of the chain are made of steel that has a working stress of 300 MPa in tension. If the chain is to support the force $P = 45$ kN, determine the smallest safe diameter *d* of the links.

1.49 Segment *AB* of the bar is a tube with an outer diameter of 1.5 in. and a wall thickness of 0.125 in. Segment *BC* is a solid rod of diameter 0.75 in. Determine the normal stress in each segment.

1.50 The cylindrical steel column has an outer diameter of 100 mm and inner diameter of 85 mm. The column is separated from the concrete foundation by a square bearing plate. The working compressive stress is 180 MPa for the column, and the working bearing stress is 10 MPa for concrete. Find the largest force *P* that can be applied to the column.

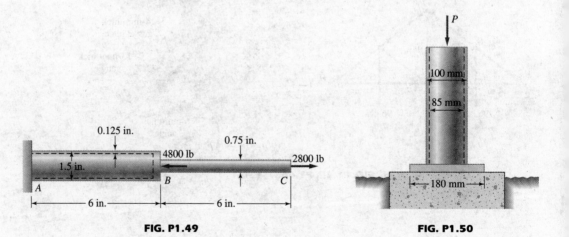

FIG. P1.49

FIG. P1.50

1.51 The tubular tension member is fabricated by welding a steel strip into a 12° helix. The cross-sectional area of the resulting tube is 2.75 in.² If the normal stress acting on the plane of the weld is 12 ksi, determine (a) the axial force P; and (b) the shear stress acting on the plane of the weld.

1.52 An aluminum cable of 6-mm-diameter is suspended from a high-altitude balloon. The density of aluminum is 2700 kg/m³, and its breaking stress is 390 MPa. Determine the largest length of cable that can be suspended without breaking.

1.53 The 20-mm-diameter steel bolt is placed in the aluminum sleeve. The nut is tightened until the normal stress in the bolt is 80 MPa. Determine the normal stress in the sleeve.

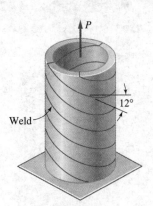

FIG. P1.51

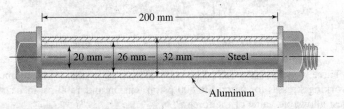

FIG. P1.53

1.54 For the joint shown in the figure, calculate (a) the largest bearing stress between the pin and the members; (b) the average shear stress in the pin; and (c) the largest average normal stress in the members.

1.55 The lap joint is fastened with four 3/4-in.-diameter rivets. The working stresses are 14 ksi for the rivets in shear and 18 ksi for the plates in bearing. Find the maximum safe axial load P that can be applied to the joint. Assume that the load is equally distributed among the rivets.

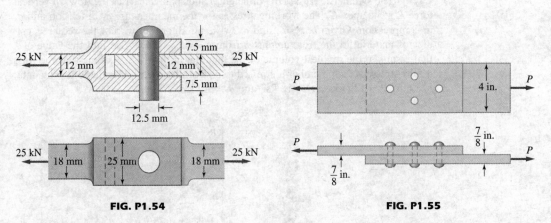

FIG. P1.54   **FIG. P1.55**

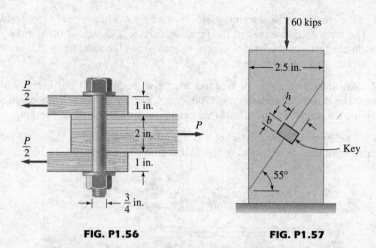

FIG. P1.56   FIG. P1.57

1.56 Three wood boards, each 4 in. wide, are joined by the 3/4-in.-diameter bolt. If the working stresses for wood are 800 psi in tension and 1500 psi in bearing, find the largest allowable value of the force P.

1.57 The cast iron block with cross-sectional dimensions of 2.5 in. by 2.5 in. consists of two pieces. The pieces are prevented from sliding along the 55° inclined joint by the steel key, which is 2.5 in. long. Determine the smallest safe dimensions b and h of the key if the working stresses are 40 ksi for cast iron in bearing and 50 ksi for the key in shear.

Computer Problems

C1.1 The symmetric truss ABC of height h and span $2b$ carries the upward vertical force P at its apex C. The working stresses for the members are σ_t in tension and σ_c in compression. Given b, P, σ_t, and σ_c, write an algorithm to plot the required volume of material in the truss against h from $h = 0.5b$ to $4b$. Also find the value of h that results in the smallest volume of the material in the truss. Assume that the truss is fully stressed (each member is stressed to its working stress). Use the following data: $b = 6$ ft, $P = 120$ kips, $\sigma_t = 18$ ksi, and $\sigma_c = 12$ ksi.

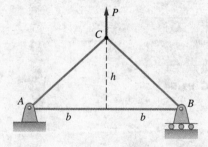

FIG. C1.1, C1.2

C1.2 Solve Prob. C1.1 assuming that P acts vertically downward.

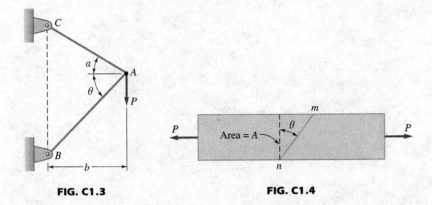

FIG. C1.3 **FIG. C1.4**

C1.3 The truss ABC has an overhang b, and its two members are inclined at angles α and θ to the horizontal, both angles being positive. A downward vertical force P acts at A. The working stresses for the members are σ_t in tension and σ_c in compression. Given b, P, α, σ_t, and σ_c, construct an algorithm to plot the required volume of material in the truss against θ from $\theta = 0°$ to $75°$. Assume that each member of the truss is stressed to its working stress. What is the value of θ that results in the smallest material volume? Use the following data: $b = 1.8$ m, $P = 530$ kN, $\alpha = 30°$, $\sigma_t = 125$ MPa, and $\sigma_c = 85$ MPa.

C1.4 A high-strength adhesive is used to join two halves of a metal bar of cross-sectional area A along the plane m-n, which is inclined at the angle θ to the cross section. The working stresses for the adhesive are σ_w in tension and τ_w in shear. Given A, σ_w, and τ_w, write an algorithm that plots the maximum allowable axial force P that can be applied to the bar as a function of θ in the range $0° \leq \theta \leq 60°$. Assume that the metal is much stronger than the adhesive, so that P is determined by the stresses in the adhesive. Use the following data: $A = 4$ in.^{2}, $\sigma_w = 3500$ psi, and $\tau_w = 1800$ psi.

C1.5 The concrete cooling tower with a constant wall thickness of 1.5 ft is loaded by its own weight. The outer diameter of the tower varies as

$$d = 20 \text{ ft} - 0.1x + (0.35 \times 10^{-3} \text{ ft}^{-1})x^2$$

where x and d are in feet. Write an algorithm to plot the axial stress in the tower as a function of x. What is the maximum stress and where does it occur? Use 150 lb/ft^3 for the weight density of concrete.

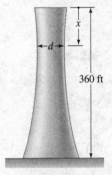

FIG. C1.5

2 Strain

A metal specimen that was broken in a testing machine. Tensile testing is a standard procedure for determining the mechanical properties of structural materials.

2.1 Introduction

So far, we have dealt mainly with the strength, or load-carrying capacity, of structural members. Here we begin our study of an equally important topic of mechanics of materials—deformations, or strains. In general terms, *strain* is a geometric quantity that measures the deformation of a body. There are two types of strain: *normal strain*, which characterizes dimensional changes, and *shear strain*, which describes distortion (changes in angles). Stress and strain are two fundamental concepts of mechanics of materials. Their relationship to each other defines the mechanical properties of a material, the knowledge of which is of the utmost importance in design.

Although our emphasis in this chapter will be on axially loaded bars, the principles and methods developed here apply equally well to more complex cases of loading discussed later. Among other topics, we will learn how to use force-deformation relationships in conjunction with equilibrium analysis to solve statically indeterminate problems.

2.2 Axial Deformation; Stress-Strain Diagram

The strength of a material is not the only criterion that must be considered when designing machine parts or structures. The stiffness of a material is often equally important, as are mechanical properties such as hardness, toughness, and ductility. These properties are determined by laboratory tests. Many materials, particularly metals, have established standards that describe the test procedures in detail. We will confine our attention to only one of the tests—the tensile test of steel—and use its results to illustrate several important concepts of material behavior.

a. Normal (axial) strain

Before describing the tensile test, we must formalize the definition of normal (axial) strain. We begin by considering the elongation of the prismatic bar of length L in Fig. 2.1. The elongation δ may be caused by an applied axial force, or an expansion due to an increase in temperature, or even a force and a temperature increase acting simultaneously. Strain describes the geometry of deformation, independent of what actually causes the deformation. The normal strain ϵ (lowercase Greek *epsilon*) is defined as the *elongation per unit length*. Therefore, the *normal strain* in the bar in the axial direction, also known as the *axial strain*, is

$$\epsilon = \frac{\delta}{L} \qquad (2.1)$$

If the bar deforms uniformly, then Eq. (2.1) represents the axial strain everywhere in the bar. Otherwise, this expression should be viewed as the *average axial strain*. Note that normal strain, being elongation per unit length, is

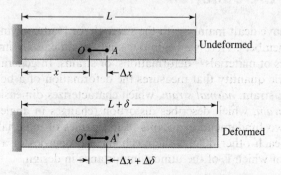

FIG. 2.1 Deformation of a prismatic bar.

a dimensionless quantity. However, "units" such as in./in. or mm/mm are frequently used for normal strain.

If the deformation is not uniform, we must define strain *at a point*. In Fig. 2.1, we let O be a point in the bar located at the distance x from the fixed end. To determine the axial strain at point O, we consider the deformation of an imaginary line element (fiber) OA of length Δx that is embedded in the bar at O. Denoting the elongation of OA by $\Delta \delta$, we define the *axial strain* at point O as

$$\epsilon = \lim_{\Delta x \to 0} \frac{\Delta \delta}{\Delta x} = \frac{d\delta}{dx} \quad (2.2)$$

Observe that normal strain, like normal stress, is defined *at a point in a given direction*.

We note that if the distribution of the axial strain is known, the elongation of the bar can be computed from

$$\delta = \int_0^L d\delta = \int_0^L \epsilon \, dx \quad (2.3)$$

For uniform strain distribution (the axial strain is the same at all points), Eq. (2.3) yields $\delta = \epsilon L$, which agrees with Eq. (2.1).

Although the preceding discussion assumed elongation, the results are also applicable to compression. By convention, compression (shortening) carries a negative sign. For example $\epsilon = -0.001$ means a compressive strain of magnitude 0.001.

b. Tension test

In the standard tension test, the specimen shown in Fig. 2.2 is placed in the grips of a testing machine. The grips are designed so that the load P applied by the machine is axial. Two gage marks are scribed on the specimen to define the *gage length L*. These marks are located away from the ends to avoid the local effects caused by the grips and to ensure that the stress and strain are uniform in the material between the marks.

The testing machine elongates the specimen at a slow, constant rate until the specimen ruptures. During the test, continuous readings are taken of the applied load and the elongation of the gage length. These data are then converted to stress and strain. The stress is obtained from $\sigma = P/A$, where P is the load and A represents the original cross-sectional area of the specimen. The strain is computed from $\epsilon = \delta/L$, where δ is the elongation

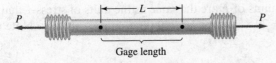

FIG. 2.2 Specimen used in the standard tension test.

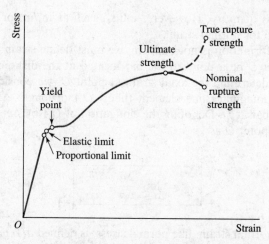

FIG. 2.3 Stress-strain diagram obtained from the standard tension test on a structural steel specimen.

between the gage marks and L is the original gage length. These results, which are based on the *original area* and the *original gage length*, are referred to as *nominal stress* and *nominal strain*.

As the bar is being stretched, its cross-sectional area is reduced and the length between the gage marks increases. Dividing the load by the actual (current) area of the specimen, we get the *true stress*. Similarly, the *true strain* is obtained by dividing the elongation δ by the current gage length. The nominal and true measures are essentially the same in the working range of metals. They differ only for very large strains, such as occur in rubberlike materials or in ductile metals just before rupture. With only a few exceptions, engineering applications use nominal stress and strain.

Plotting axial stress versus axial strain results in a *stress-strain diagram*. If the test is carried out properly, the stress-strain diagram for a given material is independent of the dimensions of the test specimen. That is, the characteristics of the diagram are determined solely by the mechanical properties of the material. A stress-strain diagram for structural steel is shown in Fig. 2.3. The following mechanical properties can be determined from the diagram.

Proportional Limit and Hooke's Law As seen in Fig. 2.3, the stress-strain diagram is a straight line from the origin O to a point called the *proportional limit*. This plot is a manifestation of *Hooke's law*:[1] Stress is proportional to strain; that is,

$$\sigma = E\epsilon \qquad (2.4)$$

where E is a material property known as the *modulus of elasticity* or *Young's modulus*. The units of E are the same as the units of stress—that is, Pa or psi. For steel, $E = 29 \times 10^6$ psi, or 200 GPa, approximately. Note that Hooke's

---

[1] This law was first postulated by Robert Hooke in 1678.

law does not apply to the entire diagram; its validity ends at the proportional limit. Beyond this point, stress is no longer proportional to strain.[2]

Elastic Limit A material is said to be *elastic* if, after being loaded, the material returns to its original shape when the load is removed. The *elastic limit* is, as its name implies, the stress beyond which the material is no longer elastic. The permanent deformation that remains after the removal of the load is called the *permanent set*. The elastic limit is slightly larger than the proportional limit. However, because of the difficulty in determining the elastic limit accurately, it is usually assumed to coincide with the proportional limit.

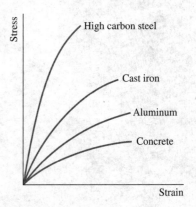

FIG. 2.4 Stress-strain diagrams for various materials that fail without significant yielding.

Yield Point The point where the stress-strain diagram becomes almost horizontal is called the *yield point*, and the corresponding stress is known as the *yield stress* or *yield strength*. Beyond the yield point there is an appreciable elongation, or yielding, of the material without a corresponding increase in load. Indeed, the load may actually decrease while the yielding occurs. However, the phenomenon of yielding is unique to structural steel. Other grades of steel, steel alloys, and other materials do not yield, as indicated by the stress-strain curves of the materials shown in Fig. 2.4. Incidentally, these curves are typical for a first loading of materials that contain appreciable residual stresses produced by manufacturing or aging processes. After repeated loading, these residual stresses are removed and the stress-strain curves become practically straight lines.

For materials that do not have a well-defined yield point, yield stress is determined by the *offset method*. This method consists of drawing a line parallel to the initial tangent of the stress-strain curve; this line starts at a prescribed offset strain, usually 0.2% ($\epsilon = 0.002$). The intersection of this line with the stress-strain curve, shown in Fig. 2.5, is called the *yield point at 0.2% offset*.

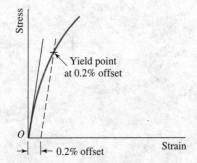

FIG. 2.5 Determining the yield point by the 2% offset method.

Ultimate Stress The *ultimate stress* or *ultimate strength*, as it is often called, is the highest stress on the stress-strain curve.

Rupture Stress The *rupture stress* or *rupture strength* is the stress at which failure occurs. For structural steel, the nominal rupture strength is considerably lower than the ultimate strength because the nominal rupture strength is computed by dividing the load at rupture by the original cross-sectional area. The true rupture strength is calculated using the reduced area of the cross section where the fracture occurred. The difference in the two values results from a phenomenon known as *necking*. As failure approaches, the material stretches very rapidly, causing the cross section to narrow, as shown in Fig. 2.6. Because the area where rupture occurs is smaller than the original area, the true rupture strength is larger than the ultimate strength. However, the ultimate strength is commonly used as the maximum stress that the material can carry.

FIG. 2.6 Failed tensile test specimen showing necking, or narrowing, of the cross section.

---

[2] The stress-strain diagram of many materials is actually a curve on which there is no definite proportional limit. In such cases, the stress-strain proportionality is assumed to exist up to a stress at which the strain increases at a rate 50% greater than shown by the initial tangent to the stress-strain diagram.

c. Working stress and factor of safety

The *working stress* σ_w, also called the *allowable stress*, is the maximum safe axial stress used in design. In most designs, the working stress should be limited to values not exceeding the proportional limit so that the stresses remain in the elastic range (the straight-line portion of the stress-strain diagram). However, because the proportional limit is difficult to determine accurately, it is customary to base the working stress on either the yield stress σ_{yp} or the ultimate stress σ_{ult}, divided by a suitable number N, called the *factor of safety*. Thus,

$$\sigma_w = \frac{\sigma_{yp}}{N} \quad \text{or} \quad \sigma_w = \frac{\sigma_{ult}}{N} \tag{2.5}$$

The yield point is selected as the basis for determining σ_w in structural steel because it is the stress at which a prohibitively large permanent set may occur. For other materials, the working stress is usually based on the ultimate strength.

Many factors must be considered when selecting the working stress. This selection should not be made by the novice; usually the working stress is set by a group of experienced engineers and is embodied in building codes and specifications. A discussion of the factors governing the selection of a working stress starts with the observation that in many materials the proportional limit is about one-half the ultimate strength. To avoid accidental overloading, a working stress of one-half the proportional limit is usually specified for dead loads that are gradually applied. (The term *dead load* refers to the weight of the structure and other loads that, once applied, are not removed.) A working stress set in this way corresponds to a factor of safety of 4 with respect to σ_{ult} and is recommended for materials that are known to be uniform and homogeneous. For other materials, such as wood, in which unpredictable nonuniformities (such as knotholes) may occur, larger factors of safety are used. The dynamic effect of suddenly applied loads also requires higher factors of safety.

2.3 Axially Loaded Bars

Figure 2.7 shows a bar of length L and constant cross-sectional area A that is loaded by an axial force P. We assume that the stress caused by P is below the proportional limit, so that Hooke's law $\sigma = E\epsilon$ is applicable. Because the

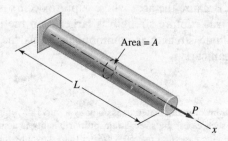

FIG. 2.7 Axially loaded bar.

bar deforms uniformly, the axial strain is $\epsilon = \delta/L$, which upon substitution into Hooke's law yields $\sigma = E(\delta/L)$. Therefore, the elongation of the bar is

$$\delta = \frac{\sigma L}{E} = \frac{PL}{EA} \qquad (2.6)$$

where in the last step we substituted $\sigma = P/A$. If the strain (or stress) in the bar is not uniform, then Eq. (2.6) is invalid. In the case where the axial strain varies with the x-coordinate, the elongation of the bar can be obtained by integration, as stated in Eq. (2.3): $\delta = \int_0^L \epsilon\, dx$. Using $\epsilon = \sigma/E = P/(EA)$, where P is the *internal* axial force, we get

$$\delta = \int_0^L \frac{\sigma}{E}\, dx = \int_0^L \frac{P}{EA}\, dx \qquad (2.7)$$

We see that Eq. (2.7) reduces to Eq. (2.6) only if P, E, and A are constants.

Notes on the Computation of Deformation

- The magnitude of the internal force P in Eqs. (2.6) and (2.7) must be found from equilibrium analysis. Note that a positive (tensile) P results in positive δ (elongation); conversely, a negative P (compression) gives rise to negative δ (shortening).
- Care must be taken to use consistent units in Eqs. (2.6) and (2.7). It is common practice to let the units of E determine the units to be used for P, L, and A. In the U.S. Customary system, E is expressed in psi (lb/in.^{2}), so that the units of the other variables should be P [lb], L [in.], and A [in.^{2}]. In the SI system, where E is in Pa (N/m^2), the consistent units are P [N], L [m], and A [m^2].
- As long as the axial stress is in the elastic range, the elongation (or shortening) of a bar is very small compared to its length. This property can be utilized to simplify the computation of displacements in structures containing axially loaded bars, such as trusses.

Sample Problem 2.1

The steel propeller shaft $ABCD$ carries the axial loads shown in Fig. (a). Determine the change in the length of the shaft caused by these loads. Use $E = 29 \times 10^6$ psi for steel.

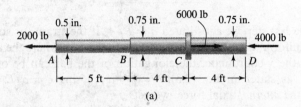

(a)

Solution

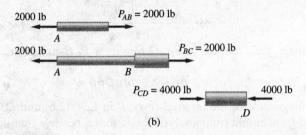

(b)

From the free-body diagrams in Fig. (b) we see that the internal forces in the three segments of the shaft are

$$P_{AB} = P_{BC} = 2000 \text{ lb (T)} \qquad P_{CD} = 4000 \text{ lb (C)}$$

Because the axial force and the cross-sectional area are constant within each segment, the changes in the lengths of the segments can be computed from Eq. (2.6): $\delta = PL/(EA)$. The change in the length of the shaft is obtained by adding the contributions of the segments. Noting that tension causes elongation and compression results in shortening, we obtain for the elongation of the shaft

$$\delta = \sum \frac{PL}{EA} = \frac{1}{E}\left[\left(\frac{PL}{A}\right)_{AB} + \left(\frac{PL}{A}\right)_{BC} - \left(\frac{PL}{A}\right)_{CD}\right]$$

$$= \frac{1}{29 \times 10^6}\left[\frac{2000(5 \times 12)}{\pi(0.5)^2/4} + \frac{2000(4 \times 12)}{\pi(0.75)^2/4} - \frac{4000(4 \times 12)}{\pi(0.75)^2/4}\right]$$

$$= \frac{(611.2 + 217.3 - 434.6) \times 10^3}{29 \times 10^6}$$

$$= 0.013\ 58 \text{ in.} \quad \text{(elongation)} \qquad \textit{Answer}$$

Sample Problem 2.2

The cross section of the 10-m-long flat steel bar AB has a constant thickness of 20 mm, but its width varies as shown in the figure. Calculate the elongation of the bar due to the 100-kN axial load. Use $E = 200$ GPa.

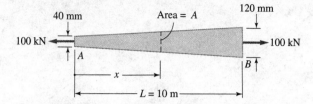

Solution

Equilibrium requires that the internal axial force $P = 100$ kN is constant along the entire length of the bar. However, the cross-sectional area A of the bar varies with the x-coordinate, so that the elongation of the bar must be computed from Eq. (2.7).

We start by determining A as a function of x. The cross-sectional areas at A and B are $A_A = 20 \times 40 = 800$ mm² and $A_B = 20 \times 120 = 2400$ mm². Between A and B the cross-sectional area is a linear function of x:

$$A = A_A + (A_B - A_A)\frac{x}{L} = 800 \text{ mm}^2 + (1600 \text{ mm}^2)\frac{x}{L}$$

Converting the areas from mm² to m² and substituting $L = 10$ m, we get

$$A = (800 + 160x) \times 10^{-6} \text{ m}^2 \qquad (a)$$

Substituting Eq. (a) together with $P = 100 \times 10^3$ N and $E = 200 \times 10^9$ Pa into Eq. (2.7), we obtain for the elongation of the rod

$$\delta = \int_0^L \frac{P}{EA}\,dx = \int_0^{10 \text{ m}} \frac{100 \times 10^3}{(200 \times 10^9)[(800 + 160x) \times 10^{-6}]}\,dx$$

$$= 0.5 \int_0^{10 \text{ m}} \frac{dx}{800 + 160x} = \frac{0.5}{160}[\ln(800 + 160x)]_0^{10}$$

$$= \frac{0.5}{160} \ln \frac{2400}{800} = 3.43 \times 10^{-3} \text{ m} = 3.43 \text{ mm} \qquad \textbf{Answer}$$

Sample Problem 2.3

The rigid bar BC in Fig. (a) is supported by the steel rod AC of cross-sectional area 0.25 in.². Find the vertical displacement of point C caused by the 2000-lb load. Use $E = 29 \times 10^6$ psi for steel.

Solution

We begin by computing the axial force in rod AC. Noting that bar BC is a two-force body, the FBD of joint C in Fig. (b) yields

$$\Sigma F_y = 0 \quad +\uparrow \quad P_{AC} \sin 40° - 2000 = 0 \qquad P_{AC} = 3111 \text{ lb}$$

The elongation of AC can now be obtained from Eq. (2.6). Noting that the length of the rod is

$$L_{AC} = \frac{L_{BC}}{\cos 40°} = \frac{8 \times 12}{\cos 40°} = 125.32 \text{ in.}$$

we get

$$\delta_{AC} = \left(\frac{PL}{EA}\right)_{AC} = \frac{3111(125.32)}{(29 \times 10^6)(0.25)} = 0.053\,78 \text{ in.} \quad \text{(elongation)}$$

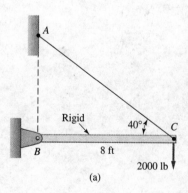

(a)

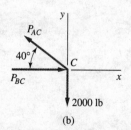

(b)

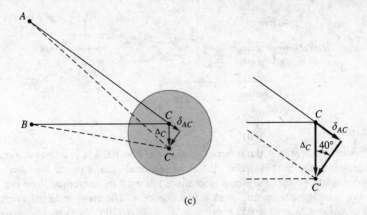

(c)

The geometric relationship between δ_{AC} and the displacement Δ_C of C is illustrated in the displacement diagram in Fig. (c). Because bar BC is rigid, the movement of point C is confined to a circular arc centered at B. Observing that the displacements are very small relative to the lengths of the bars, we note that this arc is practically the straight line CC', perpendicular to BC. Having established the direction of Δ_C, we now resolve Δ_C into components that are parallel and perpendicular to AC. The perpendicular component is due to the rotation of bar AC about A, whereas the parallel component is the elongation of AC. From geometry, the enlarged portion of the displacement diagram in Fig. (c) yields

$$\Delta_C = \frac{\delta_{AC}}{\sin 40°} = \frac{0.053\,78}{\sin 40°} = 0.0837 \text{ in.} \downarrow \qquad \textit{Answer}$$

Problems

2.1 The following data were recorded during a tensile test of a 14.0-mm-diameter mild steel rod. The gage length was 50.0 mm.

Load (N)	Elongation (mm)	Load (N)	Elongation (mm)
0	0	46 200	1.25
6 310	0.010	52 400	2.50
12 600	0.020	58 500	4.50
18 800	0.030	65 400	7.50
25 100	0.040	69 000	12.50
31 300	0.050	67 800	15.50
37 900	0.060	65 000	20.00
40 100	0.163	61 500	Fracture
41 600	0.433		

Plot the stress-strain diagram and determine the following mechanical properties: (a) proportional limit; (b) modulus of elasticity; (c) yield stress; (d) ultimate stress; and (e) nominal rupture stress.

2.2 The following data were obtained during a tension test of an aluminum alloy. The initial diameter of the test specimen was 0.505 in., and the gage length was 2.0 in.

Load (lb)	Elongation (in.)	Load (lb)	Elongation (in.)
0	0	14 000	0.020
2 310	0.0022	14 400	0.025
4 640	0.0044	14 500	0.060
6 950	0.0066	14 600	0.080
9 290	0.0088	14 800	0.100
11 600	0.0110	14 600	0.120
13 000	0.0150	13 600	Fracture

Plot the stress-strain diagram and determine the following mechanical properties: (a) proportional limit; (b) modulus of elasticity; (c) yield stress at 0.2% offset; (d) ultimate stress; and (e) nominal rupture stress.

2.3 A uniform bar of length L, cross-sectional area A, and mass density ρ is suspended vertically from one end. (a) Show that the elongation of the bar is $\delta = \rho g L^2 / (2E)$, where g is the gravitational acceleration and E is the modulus of elasticity. (b) If the mass of the bar is M, show that $\delta = MgL/(2EA)$.

2.4 A steel rod having a cross-sectional area of 300 mm^2 and a length of 150 m is suspended vertically from one end. The rod supports a tensile load of 20 kN at its free end. Given that the mass density of steel is 7850 kg/m^3 and $E = 200$ GPa, find the total elongation of the rod. (*Hint:* Use the results of Prob. 2.3.)

2.5 Determine the elongation of the tapered cylindrical aluminum bar caused by the 30-kN axial load. Use $E = 72$ GPa.

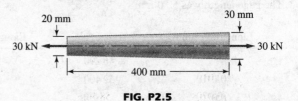

FIG. P2.5

2.6 A 0.2-in.-diameter steel wire, 10 ft long, carries an axial tensile load P. Find the maximum safe value of P if the allowable normal stress is 40 ksi and the elongation of the wire is limited to 0.15 in. Use $E = 29 \times 10^6$ psi.

2.7 The steel rod is placed inside the copper tube, the length of each being exactly 15 in. If the assembly is compressed by 0.0075 in., determine the stress in each component and the applied force P. The moduli of elasticity are 29×10^6 psi for steel and 17×10^6 psi for copper.

2.8 A steel hoop, 10 mm thick and 80 mm wide, with inside diameter 1500.0 mm, is heated and shrunk onto a steel cylinder 1500.5 mm in diameter. What is the normal force in the hoop after it has cooled? Neglect the deformation of the cylinder, and use $E = 200$ GPa for steel.

2.9 The timber member has a cross-sectional area of 1750 mm^2 and its modulus of elasticity is 12 GPa. Compute the change in the total length of the member after the loads shown are applied.

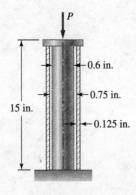

FIG. P2.7

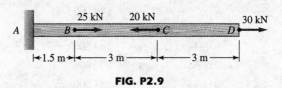

FIG. P2.9

2.10 The steel member is made by attaching the tube AB to the solid circular bar BC. Determine the change in the length of the member caused by the loads shown in the figure. Use $E = 29 \times 10^6$ psi.

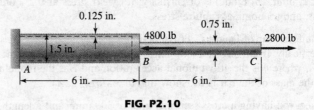

FIG. P2.10

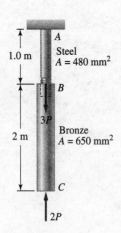

FIG. P2.11

2.11 The member consists of the steel rod AB that is screwed into the end of the bronze rod BC. Find the largest value of P that meets the following design criteria: (i) the overall length of the member is not to change by more than 3 mm; and (ii) the stresses are not to exceed 140 MPa in steel and 120 MPa in bronze. The moduli of elasticity are 200 GPa for steel and 70 GPa for aluminum.

2.12 The compound bar carries the axial forces P and $2P$. Find the maximum allowable value of P if the working stresses are 40 ksi for steel and 20 ksi for aluminum, and the total elongation of the bar is not to exceed 0.2 in.

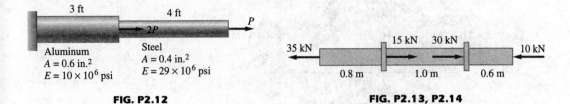

FIG. P2.12   **FIG. P2.13, P2.14**

2.13 The aluminum bar with a cross-sectional area of 160 mm² carries the axial loads at the positions shown in the figure. Given that $E = 70$ GPa, compute the total change in length of the bar.

2.14 Solve Prob. 2.13 if the magnitudes of the end loads are interchanged—that is, if the load at the left end is 10 kN and the load at the right end is 35 kN.

2.15 The compound bar containing steel, bronze, and aluminum segments carries the axial loads shown in the figure. The properties of the segments and the working stresses are listed in the table.

	A (in.²)	E (psi)	σ_w (psi)
Steel	0.75	30×10^6	20 000
Bronze	1.00	12×10^6	18 000
Aluminum	0.50	10×10^6	12 000

Determine the maximum allowable value of P if the change in length of the entire bar is limited to 0.08 in. and the working stresses are not to be exceeded.

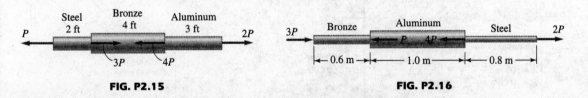

FIG. P2.15   **FIG. P2.16**

2.16 A compound bar consisting of bronze, aluminum, and steel segments is loaded axially as shown in the figure. Determine the maximum allowable value of P if the change in length of the bar is limited to 2 mm and the working stresses prescribed in the table are not to be exceeded.

	A (mm²)	E (GPa)	σ_w (MPa)
Bronze	450	83	120
Aluminum	600	70	80
Steel	300	200	140

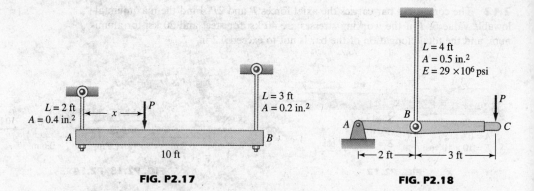

FIG. P2.17   FIG. P2.18

2.17 The rigid bar AB is supported by two rods made of the same material. If the bar is horizontal before the load P is applied, find the distance x that locates the position where P must act if the bar is to remain horizontal. Neglect the weight of bar AB.

2.18 The rigid bar ABC is supported by a pin at A and a steel rod at B. Determine the largest vertical load P that can be applied at C if the stress in the steel rod is limited to 30 ksi and the vertical movement of end C must not exceed 0.10 in. Neglect the weights of the members.

2.19 The rigid bar AB, attached to aluminum and steel rods, is horizontal before the load P is applied. Find the vertical displacement of point C caused by the load $P = 50$ kN. Neglect all weights.

2.20 The rigid bars ABC and CD are supported by pins at A and D and by a steel rod at B. There is a roller connection between the bars at C. Compute the vertical displacement of point C caused by the 50-kN load.

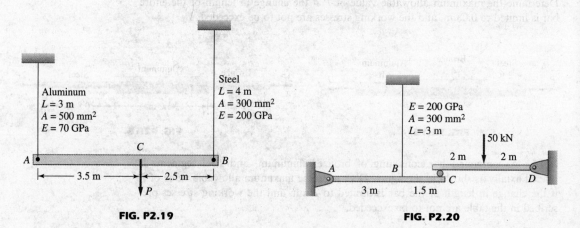

FIG. P2.19   FIG. P2.20

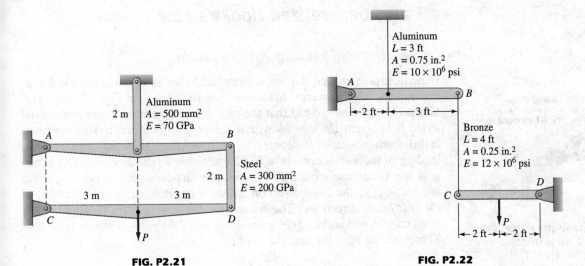

FIG. P2.21   **FIG. P2.22**

2.21 The structure in the figure is composed of two rigid bars (*AB* and *CD*) and two vertical rods made of aluminum and steel. All connections are pin joints. Determine the maximum force *P* that can be applied to the structure if the vertical displacement of its point of application is limited to 5 mm. Neglect the weights of the members.

2.22 The rigid bars *AB* and *CD* are supported by pins at *A* and *D*. The vertical rods are made of aluminum and bronze. Determine the vertical displacement of the point where the force $P = 10$ kips is applied. Neglect the weights of the members.

2.23 The uniform 2200-lb bar *BC* is supported by a pin at *C* and the aluminum wire *AB*. The cross-sectional area of the wire is 0.165 in.^{2}. Assuming bar *BC* to be rigid, find the vertical displacement of *B* due to the weight of the bar. Use $E = 10.6 \times 10^6$ psi for aluminum.

2.24 The steel bars *AC* and *BC*, each of cross-sectional area 120 mm^2, are joined at *C* with a pin. Determine the displacement of point *C* caused by the 15-kN load. Use $E = 200$ GPa for steel.

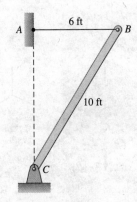

FIG. P2.23

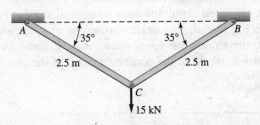

FIG. P2.24, P2.25

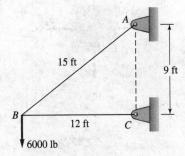

FIG. P2.26

2.25 Solve Prob. 2.24 if the 15-kN load acts horizontally to the right.

2.26 The steel truss supports a 6000-lb load. The cross-sectional areas of the members are 0.5 in.^{2} for *AB* and 0.75 in.^{2} for *BC*. Compute the horizontal displacement of *B* using $E = 29 \times 10^6$ psi.

2.27 The uniform rod of length *L*, cross-sectional area *A*, mass density ρ, and modulus of elasticity *E* is rotating in a horizontal plane about a vertical axis through one end. When the rod is rotating at a constant angular velocity of ω rad/s, show that the elongation of the rod is $\rho \omega^2 L^3 / (3E)$.

FIG. P2.27

2.4 Generalized Hooke's Law

a. Uniaxial loading; Poisson's ratio

Experiments show that if a bar is stretched by an axial force, there is a contraction in the transverse dimensions, as illustrated in Fig. 2.8. In 1811, Siméon D. Poisson showed that the ratio of the transverse strain to the axial strain is constant for stresses within the proportional limit. This constant, called *Poisson's ratio*, is denoted by v (lowercase Greek *nu*). For uniaxial loading in the x-direction, as in Fig 2.8, Poisson's ratio is $v = -\epsilon_t/\epsilon_x$, where ϵ_t is the transverse strain. The minus sign indicates that a positive strain (elongation) in the axial direction causes a negative strain (contraction) in the transverse directions. The transverse strain is uniform throughout the cross section and is the *same in any direction in the plane of the cross section*. Therefore, we have for uniaxial loading

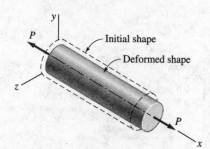

FIG. 2.8 Transverse dimensions contract as the bar is stretched by an axial force P.

$$\epsilon_y = \epsilon_z = -v\epsilon_x \qquad (2.8)$$

Poisson's ratio is a dimensionless quantity that ranges between 0.25 and 0.33 for metals.

Using $\sigma_x = E\epsilon_x$ in Eq. (2.8) yields the generalized Hooke's law for uniaxial loading ($\sigma_y = \sigma_z = 0$):

$$\epsilon_x = \frac{\sigma_x}{E} \qquad \epsilon_y = \epsilon_z = -v\frac{\sigma_x}{E} \qquad (2.9)$$

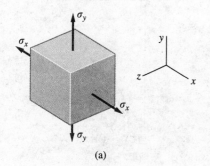

b. Multiaxial Loading

Biaxial Loading Poisson's ratio permits us to extend Hooke's law for uniaxial loading to biaxial and triaxial loadings. Consider an element of the material that is subjected simultaneously to normal stresses in the x- and y-directions, as in Fig. 2.9(a). The strains caused by σ_x alone are given in Eqs. (2.9). Similarly, the strains due to σ_y are $\epsilon_y = \sigma_y/E$ and $\epsilon_x = \epsilon_z = -v\sigma_y/E$. Using superposition, we write the combined effect of the two normal stresses as

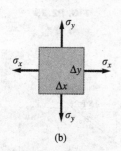

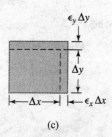

$$\epsilon_x = \frac{1}{E}(\sigma_x - v\sigma_y) \qquad \epsilon_y = \frac{1}{E}(\sigma_y - v\sigma_x) \qquad \epsilon_z = -\frac{v}{E}(\sigma_x + \sigma_y) \qquad (2.10)$$

which is Hooke's law for biaxial loading in the xy-plane ($\sigma_z = 0$). The first two of Eqs. (2.10) can be inverted to express the stresses in terms of the strains:

FIG. 2.9 (a) Stresses acting on a material element in biaxial loading; (b) two-dimensional view of stresses; (c) deformation of the element.

$$\sigma_x = \frac{(\epsilon_x + v\epsilon_y)E}{1 - v^2} \qquad \sigma_y = \frac{(\epsilon_y + v\epsilon_x)E}{1 - v^2} \qquad (2.11)$$

Two-dimensional views of the stresses and the resulting deformation in the *xy*-plane are shown in Figs. 2.9(b) and (c). Note that Eqs. (2.10) show that for biaxial loading ϵ_z is not zero; that is, the strain is triaxial rather than biaxial.

Triaxial Loading Hooke's law for the triaxial loading in Fig. 2.10 is obtained by adding the contribution of σ_z, $\epsilon_z = \sigma_z/E$ and $\epsilon_x = \epsilon_y = -\nu\sigma_z/E$, to the strains in Eqs. (2.10), which yields

$$\epsilon_x = \frac{1}{E}[\sigma_x - \nu(\sigma_y + \sigma_z)]$$

$$\epsilon_y = \frac{1}{E}[\sigma_y - \nu(\sigma_z + \sigma_x)] \quad (2.12)$$

$$\epsilon_z = \frac{1}{E}[\sigma_z - \nu(\sigma_x + \sigma_y)]$$

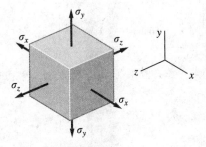

FIG. 2.10 Stresses acting on a material element in triaxial loading.

Equations (2.8)–(2.12) are valid for both tensile and compressive effects. It is only necessary to assign positive signs to elongations and tensile stresses and, conversely, negative signs to contractions and compressive stresses.

c. Shear loading

Shear stress causes the deformation shown in Fig. 2.11. The lengths of the sides of the element do not change, but the element undergoes a distortion from a rectangle to a parallelogram. The *shear strain*, which measures the amount of distortion, is the angle γ (lowercase Greek *gamma*), always expressed in radians. It can be shown that the relationship between shear stress τ and shear strain γ is linear within the elastic range; that is,

$$\tau = G\gamma \quad (2.13)$$

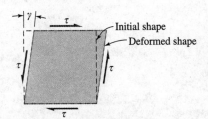

FIG. 2.11 Deformation of a material element caused by shear stress.

which is Hooke's law for shear. The material constant G is called the *shear modulus of elasticity* (or simply *shear modulus*), or the *modulus of rigidity*. The shear modulus has the same units as the modulus of elasticity (Pa or psi). We will prove later that G is related to the modulus of elasticity E and Poisson's ratio ν by

$$G = \frac{E}{2(1 + \nu)} \quad (2.14)$$

Sample Problem 2.4

The 50-mm-diameter rubber rod is placed in a hole with rigid, lubricated walls. There is no clearance between the rod and the sides of the hole. Determine the change in the length of the rod when the 8-kN load is applied. Use $E = 40$ MPa and $\nu = 0.45$ for rubber.

Solution

Lubrication allows the rod to contract freely in the axial direction, so that the axial stress throughout the bar is

$$\sigma_x = -\frac{P}{A} = -\frac{8000}{\frac{\pi}{4}(0.05)^2} = -4.074 \times 10^6 \text{ Pa}$$

(the negative sign implies compression). Because the walls of the hole prevent transverse strain in the rod, we have $\epsilon_y = \epsilon_z = 0$. The tendency of the rubber to expand laterally (Poisson's effect) is resisted by the uniform contact pressure p between the walls and the rod, so that $\sigma_y = \sigma_z = -p$. If we use the second of Eqs. (2.12) (the third equation would yield the same result), the condition $\epsilon_y = 0$ becomes

$$\frac{\sigma_y - \nu(\sigma_z + \sigma_x)}{E} = \frac{-p - \nu(-p + \sigma_x)}{E} = 0$$

which yields

$$p = -\frac{\nu \sigma_x}{1 - \nu} = -\frac{0.45(-4.074 \times 10^6)}{1 - 0.45} = 3.333 \times 10^6 \text{ Pa}$$

The axial strain is given by the first of Eqs. (2.12):

$$\epsilon_x = \frac{\sigma_x - \nu(\sigma_y + \sigma_z)}{E} = \frac{\sigma_x - \nu(-2p)}{E}$$

$$= \frac{[-4.074 - 0.45(-2 \times 3.333)] \times 10^6}{40 \times 10^6} = -0.02686$$

The corresponding change in the length of the rod is

$$\delta = \epsilon_x L = -0.02686(300)$$

$$= -8.06 \text{ mm} = 8.06 \text{ mm} \quad \text{(contraction)} \qquad \textbf{Answer}$$

For comparison, note that if the constraining effect of the hole were neglected, the deformation would be

$$\delta = -\frac{PL}{EA} = -\frac{8000(0.3)}{(40 \times 10^6)\left[\frac{\pi}{4}(0.05)^2\right]} = -0.0306 \text{ m} = -30.6 \text{ mm}$$

Sample Problem 2.5

Two 1.75-in.-thick rubber pads are bonded to three steel plates to form the shear mount shown in Fig. (a). Find the displacement of the middle plate when the 1200-lb load is applied. Consider the deformation of rubber only. Use $E = 500$ psi and $\nu = 0.48$.

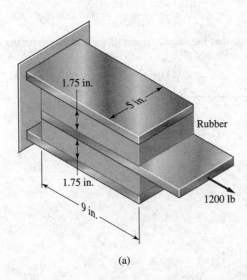

(a)

Solution

To visualize the deformation of the rubber pads, we introduce a grid drawn on the edge of the upper pad—see Fig. (b). When the load is applied, the grid deforms as shown in the figure. Observe that the deformation represents uniform shear, except for small regions at the edges of the pad (Saint Venant's principle).

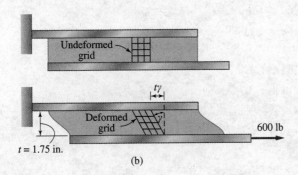

(b)

Each rubber pad has a shear area of $A = 5 \times 9 = 45$ in.^{2} that carries half the 1200-lb load. Hence, the average shear stress in the rubber is

$$\tau = \frac{V}{A} = \frac{600}{45} = 13.333 \text{ psi}$$

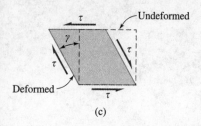

(c)

This stress is shown acting on the sides of a grid element in Fig. (c). The corresponding shear strain is $\gamma = \tau/G$, where from Eq. (2.14),

$$G = \frac{E}{2(1+\nu)} = \frac{500}{2(1+0.48)} = 168.92 \text{ psi}$$

Therefore,

$$\gamma = \frac{\tau}{G} = \frac{13.333}{168.92} = 0.07893$$

From Fig. (b) we see that the displacement of the middle plate (the lower plate in the figure) is

$$t\gamma = 1.75(0.07893) = 0.1381 \text{ in.} \qquad \textit{Answer}$$

Problems

2.28 A solid cylinder of diameter d carries an axial load P. Show that the change in diameter is $4Pv/(\pi E d)$.

2.29 The 3/4-in.-thick steel plate is loaded biaxially as shown in the figure. Determine the change in the thickness of the plate. For steel, use $E = 29 \times 10^6$ psi and $v = 0.28$.

2.30 A sheet of copper is stretched biaxially in the xy-plane. If the strains in the sheet are $\epsilon_x = 0.50 \times 10^{-3}$ and $\epsilon_y = 0.25 \times 10^{-3}$, determine σ_x and σ_y. Use $E = 110$ GPa and $v = 0.35$.

2.31 The normal stresses at a point in a steel member are $\sigma_x = 6$ ksi, $\sigma_y = -5$ ksi, and $\sigma_z = 9$ ksi. Using $E = 29 \times 10^3$ ksi and $v = 0.3$, determine the normal strains at this point.

2.32 The rectangular block of material of length L and cross-sectional area A fits snugly between two rigid, lubricated walls. Derive the expression for the change in length of the block due to the axial load P.

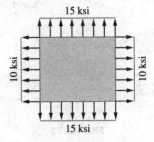

FIG. P2.29

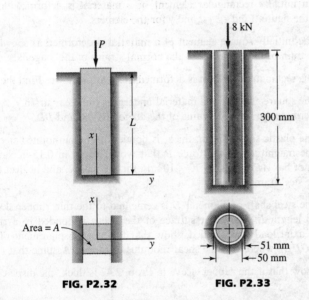

FIG. P2.32   FIG. P2.33

2.33 The plastic 50-mm-diameter rod is placed in a 51-mm-diameter hole with rigid walls. Determine the change in the length of the rod after the 8-kN load is applied. Use $E = 40$ MPa and $v = 0.45$ for the rod.

2.34 A material specimen is subjected to a uniform, triaxial compressive stress (hydrostatic pressure) of magnitude p. Show that the volumetric strain of the material is $\Delta V/V = -3p(1-2v)/E$, where ΔV is the volume change and V is the initial volume.

2.35 A rubber sheet of thickness t and area A is compressed as shown in the figure. All contact surfaces are sufficiently rough to prevent slipping. Show that the change in the thickness of the rubber sheet caused by the load P is

$$\delta = \frac{(1+v)(1-2v)}{(1-v)} \frac{Pt}{EA}$$

(*Hint*: The roughness of the surfaces prevents transverse expansion of the sheet.)

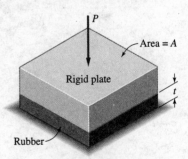

FIG. P2.35

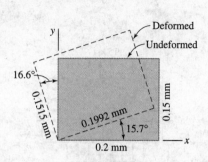

FIG. P2.37

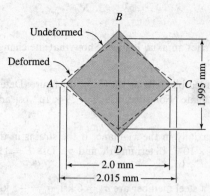

FIG. P2.38

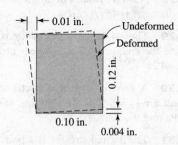

FIG. P2.39

2.36 A torsion test shows that the shear modulus of an aluminum specimen is 4.60×10^6 psi. When the same specimen is used in a tensile test, the modulus of elasticity is found to be 12.2×10^6 psi. Find Poisson's ratio for the specimen.

2.37 An initially rectangular element of a material is deformed into the shape shown in the figure. Find ϵ_x, ϵ_y, and γ for the element.

2.38 The initially square element of a material is deformed as shown. Determine the shear strain of the element and the normal strains of the diagonals AC and BD.

2.39 The rectangular element is deformed in shear as shown. Find the shear strain.

2.40 The square element of a material undergoes the shear strain γ. Assuming that $\gamma \ll 1$, determine the normal strains of the diagonals AC and BD.

2.41 The plastic sheet, 1/2 in. thick, is bonded to the pin-jointed steel frame. Determine the magnitude of the force P that would result in 0.15-in. horizontal displacement of bar AB. Use $G = 70 \times 10^3$ psi for the plastic, and neglect the deformation of the steel frame.

2.42 The steel shaft of diameter D is cemented to the thin rubber sleeve of thickness t and length L. The outer surface of the sleeve is bonded to a rigid support. When the axial load P is applied, show that the axial displacement of the shaft is $\delta = Pt/(\pi GDL)$, where G is the shear modulus of rubber. Assume that $t \ll D$.

2.43 Show that if the rubber sleeve in Prob. 2.42 is thick, the displacement of the shaft is

$$\delta = \frac{P}{2\pi GL} \ln \frac{D+2t}{D}$$

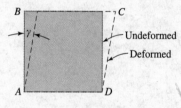

FIG. P2.40

FIG. P2.41

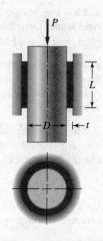

FIG. P2.42, P2.43

2.5 Statically Indeterminate Problems

If the equilibrium equations are sufficient to calculate all the forces (including support reactions) that act on a body, these forces are said to be *statically determinate*. In statically determinate problems, the number of unknown forces is always equal to the number of independent equilibrium equations. If the number of unknown forces exceeds the number of independent equilibrium equations, the problem is said to be *statically indeterminate*.

Static indeterminacy does not imply that the problem cannot be solved; it simply means that the solution cannot be obtained from the equilibrium equations alone. A statically indeterminate problem always has geometric restrictions imposed on its deformation. The mathematical expressions of these restrictions, known as the *compatibility equations*, provide us with the additional equations needed to solve the problem (the term *compatibility* refers to the geometric compatibility between deformation and the imposed constraints). Because the source of the compatibility equations is deformation, these equations contain as unknowns either strains or elongations. We can, however, use Hooke's law to express the deformation measures in terms of stresses or forces. The equations of equilibrium and compatibility can then be solved for the unknown forces.

Procedure for Solving Statically Indeterminate Problems In summary, the solution of a statically indeterminate problem involves the following steps:

- Draw the required free-body diagrams and derive the equations of **equilibrium**.
- Derive the **compatibility** equations. To visualize the restrictions on deformation, it is often helpful to draw a sketch that exaggerates the magnitudes of the deformations.
- Use **Hooke's law** to express the deformations (strains) in the compatibility equations in terms of forces (or stresses).
- Solve the equilibrium and compatibility equations for the unknown forces.

Sample Problem 2.6

The concrete post in Fig. (a) is reinforced axially with four symmetrically placed steel bars, each of cross-sectional area 900 mm². Compute the stress in each material when the 1000-kN axial load is applied. The moduli of elasticity are 200 GPa for steel and 14 GPa for concrete.

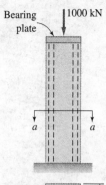

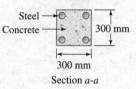

(a)

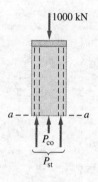

(b) FBD

Solution

Equilibrium The FBD in Fig. (b) was drawn by isolating the portion of the post above section a-a, where P_{co} is the force in concrete and P_{st} denotes the total force carried by the steel rods. For equilibrium, we must have

$$\Sigma F = 0 \quad +\uparrow \quad P_{st} + P_{co} - 1.0 \times 10^6 = 0$$

which, written in terms of stresses, becomes

$$\sigma_{st} A_{st} + \sigma_{co} A_{co} = 1.0 \times 10^6 \text{ N} \qquad \text{(a)}$$

Equation (a) is the only independent equation of equilibrium that is available in this problem. Because there are two unknown stresses, we conclude that the problem is statically indeterminate.

Compatibility For the deformations to be compatible, the changes in lengths of the steel rods and the concrete must be equal; that is, $\delta_{st} = \delta_{co}$. Because the lengths of steel and concrete are identical, the compatibility equation, written in terms of strains, is

$$\epsilon_{st} = \epsilon_{co} \qquad \text{(b)}$$

Hooke's Law From Hooke's law, Eq. (b) becomes

$$\frac{\sigma_{st}}{E_{st}} = \frac{\sigma_{co}}{E_{co}} \qquad \text{(c)}$$

Equations (a) and (c) can now be solved for the stresses. From Eq. (c) we obtain

$$\sigma_{st} = \frac{E_{st}}{E_{co}} \sigma_{co} = \frac{200}{14} \sigma_{co} = 14.286 \sigma_{co} \qquad \text{(d)}$$

Substituting the cross-sectional areas

$$A_{st} = 4(900 \times 10^{-6}) = 3.6 \times 10^{-3} \text{ m}^2$$

$$A_{co} = 0.3^2 - 3.6 \times 10^{-3} = 86.4 \times 10^{-3} \text{ m}^2$$

and Eq. (d) into Eq. (a) yields

$$(14.286\sigma_{co})(3.6 \times 10^{-3}) + \sigma_{co}(86.4 \times 10^{-3}) = 1.0 \times 10^6$$

Solving for the stress in concrete, we get

$$\sigma_{co} = 7.255 \times 10^6 \text{ Pa} = 7.255 \text{ MPa} \qquad \textbf{Answer}$$

From Eq. (d), the stress in steel is

$$\sigma_{st} = 14.286(7.255) = 103.6 \text{ MPa} \qquad \textbf{Answer}$$

Sample Problem 2.7

Let the allowable stresses in the post described in Sample Problem 2.6 be $\sigma_{st} = 120$ MPa and $\sigma_{co} = 6$ MPa. Compute the maximum safe axial load P that may be applied.

Solution

The unwary student may attempt to obtain the forces by substituting the allowable stresses into the equilibrium equation—see Eq. (a) in Sample Problem 2.6. This approach is incorrect because it ignores the compatibility condition—that is, the equal strains of the two materials. From Eq. (d) in Sample Problem 2.6, we see that equal strains require the following relationship between the stresses:

$$\sigma_{st} = 14.286\sigma_{co}$$

Therefore, if the concrete were stressed to its limit of 6 MPa, the corresponding stress in the steel would be

$$\sigma_{st} = 14.286(6) = 85.72 \text{ MPa}$$

which is below the allowable stress of 120 MPa. The maximum safe axial load is thus found by substituting $\sigma_{co} = 6$ MPa and $\sigma_{st} = 85.72$ MPa into the equilibrium equation:

$$P = \sigma_{st}A_{st} + \sigma_{co}A_{co}$$
$$= (85.72 \times 10^6)(3.6 \times 10^{-3}) + (6 \times 10^6)(86.4 \times 10^{-3})$$
$$= 827 \times 10^3 \text{ N} = 827 \text{ kN} \qquad \textit{Answer}$$

Sample Problem 2.8

Figure (a) shows a copper rod that is placed in an aluminum sleeve. The rod is 0.005 in. longer than the sleeve. Find the maximum safe load P that can be applied to the bearing plate, using the following data:

	Copper	Aluminum
Area (in.^{2})	2	3
E (psi)	17×10^6	10×10^6
Allowable stress (ksi)	20	10

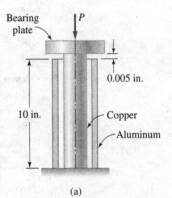

(a)

Solution

Equilibrium We assume that the rod deforms enough so that the bearing plate makes contact with the sleeve, as indicated in the FBD in Fig. (b). From this FBD we get

$$\Sigma F = 0 \quad +\uparrow \quad P_{cu} + P_{al} - P = 0 \qquad (a)$$

Because no other equations of equilibrium are available, the forces P_{cu} and P_{al} are statically indeterminate.

Compatibility Figure (c) shows the changes in the lengths of the two materials (the deformations have been greatly exaggerated). We see that the compatibility equation is

$$\delta_{cu} = \delta_{al} + 0.005 \text{ in.} \qquad (b)$$

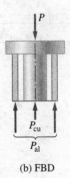

(b) FBD

Hooke's Law Substituting $\delta = \sigma L/E$ into Eq. (b), we get

$$\left(\frac{\sigma L}{E}\right)_{cu} = \left(\frac{\sigma L}{E}\right)_{al} + 0.005 \text{ in.}$$

or

$$\frac{\sigma_{cu}(10.005)}{17 \times 10^6} = \frac{\sigma_{al}(10)}{10 \times 10^6} + 0.005$$

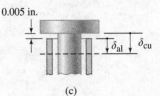

(c)

53

which reduces to

$$\sigma_{cu} = 1.6992\sigma_{al} + 8496 \qquad (c)$$

From Eq. (c) we find that if $\sigma_{al} = 10\,000$ psi, the copper will be overstressed to 25 500 psi. Therefore, the allowable stress in the copper (20 000 psi) is the limiting condition. The corresponding stress in the aluminum is found from Eq. (c):

$$20\,000 = 1.6992\sigma_{al} + 8496$$

which gives

$$\sigma_{al} = 6770 \text{ psi}$$

From Eq. (a), the safe load is

$$P = P_{cu} + P_{al} = \sigma_{cu}A_{cu} + \sigma_{al}A_{al}$$
$$= 20\,000(2) + 6770(3) = 60\,300 \text{ lb} = 60.3 \text{ kips} \qquad \textit{Answer}$$

Sample Problem 2.9

Figure (a) shows a rigid bar that is supported by a pin at A and two rods, one made of steel and the other of bronze. Neglecting the weight of the bar, compute the stress in each rod caused by the 50-kN load, using the following data:

	Steel	Bronze
Area (mm²)	600	300
E (GPa)	200	83

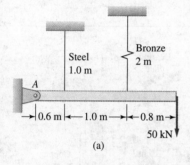

(a)

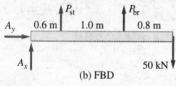

(b) FBD

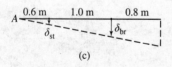

(c)

Solution

Equilibrium The free-body diagram of the bar, shown in Fig. (b), contains four unknown forces. Since there are only three independent equilibrium equations, these forces are statically indeterminate. The equilibrium equation that does not involve the pin reactions at A is

$$\Sigma M_A = 0 \quad +\circlearrowleft \quad 0.6P_{st} + 1.6P_{br} - 2.4(50 \times 10^3) = 0 \qquad (a)$$

Compatibility The displacement of the bar, consisting of a rigid-body rotation about A, is shown greatly exaggerated in Fig. (c). From similar triangles, we see that the elongations of the supporting rods must satisfy the compatibility condition

$$\frac{\delta_{st}}{0.6} = \frac{\delta_{br}}{1.6} \qquad (b)$$

Hooke's Law When we substitute $\delta = PL/(EA)$ into Eq. (b), the compatibility equation becomes

$$\frac{1}{0.6}\left(\frac{PL}{EA}\right)_{st} = \frac{1}{1.6}\left(\frac{PL}{EA}\right)_{br}$$

Using the given data, we obtain

$$\frac{1}{0.6}\frac{P_{st}(1.0)}{(200)(600)} = \frac{1}{1.6}\frac{P_{br}(2)}{(83)(300)}$$

which simplifies to

$$P_{st} = 3.614P_{br} \qquad (c)$$

Note that the we did not convert the areas from mm² to m², and we omitted the factor 10^9 from the moduli of elasticity. Since these conversion factors appear on both sides of the equation, they would cancel out.

Solving Eqs. (a) and (c), we obtain

$$P_{st} = 115.08 \times 10^3 \text{ N} \qquad P_{br} = 31.84 \times 10^3 \text{ N}$$

The stresses are

$$\sigma_{st} = \frac{P_{st}}{A_{st}} = \frac{115.08 \times 10^3}{600 \times 10^{-6}} = 191.8 \times 10^6 \text{ Pa} = 191.8 \text{ MPa} \qquad \textit{Answer}$$

$$\sigma_{br} = \frac{P_{br}}{A_{br}} = \frac{31.84 \times 10^3}{300 \times 10^{-6}} = 106.1 \times 10^6 \text{ Pa} = 106.1 \text{ MPa} \qquad \textit{Answer}$$

■

Problems

2.44 A steel bar 50 mm in diameter and 2 m long is encased in a shell of cast iron 5 mm thick. Compute the stress in each material if a compressive axial load of 200 kN is applied to the assembly. For steel, $E = 200$ GPa, and for cast iron, $E = 100$ GPa.

2.45 A reinforced concrete column 200 mm in diameter is designed to carry an axial compressive load of 300 kN. Determine the required cross-sectional area of the reinforcing steel if the allowable stresses are 6 MPa for concrete and 120 MPa for steel. Use $E_{co} = 14$ GPa and $E_{st} = 200$ GPa.

2.46 A timber column, 8 in. by 8 in. in cross section, is reinforced on all four sides by steel plates, each plate being 8 in. wide and t in. thick. Determine the smallest value of t for which the column can support an axial load of 300 kips if the working stresses are 1200 psi for timber and 20 ksi for steel. The moduli of elasticity are 1.5×10^6 psi for timber and 29×10^6 psi for steel.

2.47 The rigid block of mass M is supported by the three symmetrically placed rods. The ends of the rods were level before the block was attached. Determine the largest allowable value of M if the properties of the rods are as listed (σ_w is the working stress):

	E (GPa)	A (mm^2)	σ_w (MPa)
Copper	120	900	70
Steel	200	1200	140

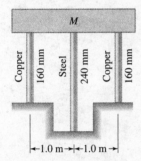

FIG. P2.47, P2.48

2.48 In Prob. 2.47, determine the lengths of the two copper rods so that the stresses in all three rods reach their allowable limits simultaneously.

2.49 The lower ends of the three vertical steel bars were at the same level before the homogeneous, rigid block of weight W was attached. The cross-sectional areas of the bars at A and C are 1.0 in.^{2}. Determine the cross-sectional area of the bar at B so that the axial force in each bar is $W/3$.

2.50 Before the 400-kN load is applied, the rigid platform rests on two steel bars, each of cross-sectional area 1200 mm^2, as shown in the figure. The cross-sectional area of the aluminum bar is 2400 mm^2. Compute the stress in the aluminum bar after the 400-kN load is applied. Use $E = 200$ GPa for steel and $E = 70$ GPa for aluminum. Neglect the weight of the platform.

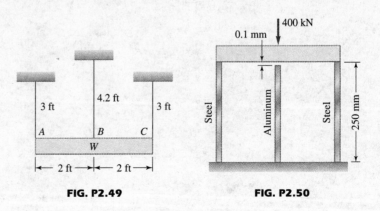

FIG. P2.49 FIG. P2.50

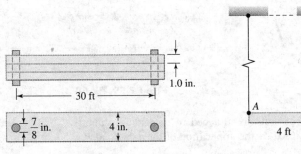

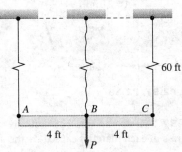
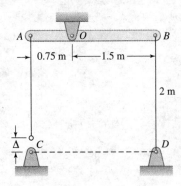

FIG. P2.51 **FIG. P2.52** **FIG. P2.53, P2.54**

2.51 The three steel ($E = 29 \times 10^6$ psi) eye-bars, each 4 in. by 1.0 in. in cross section, are assembled by driving 7/8-in.-diameter drift pins through holes drilled in the ends of the bars. The distance between the holes is 30 ft in the two outer bars, but 0.045 in. less in the middle bar. Find the shear stress developed in the drift pins. Neglect local deformation at the holes.

2.52 The rigid bar ABC of negligible weight is suspended from three steel wires, each of cross-sectional area 0.3 in.^{2}. Before the load P is applied, the middle wire is slack, being 0.2 in. longer than the other two wires. Determine the largest safe value of P if the working stress for the wires is 20 ksi. Use $E = 29 \times 10^6$ psi for steel.

2.53 The rigid bar AB of negligible weight is supported by a pin at O. When the two steel rods are attached to the ends of the bar, there is a gap $\Delta = 5$ mm between the lower end of the left rod and its pin support at C. Compute the stress in the left rod after its lower end is attached to the support. The cross-sectional areas are 250 mm^2 for rod AC and 300 mm^2 for rod BD. Use $E = 200$ GPa for steel.

2.54 The rigid bar AB of negligible weight is supported by a pin at O. When the two steel rods are attached to the ends of the bar, there is a gap Δ between the lower end of the left rod and its pin support at C. After attachment, the strain in the left rod is 1.5×10^{-3}. What is the length of the gap Δ? The cross-sectional areas are 250 mm^2 for rod AC and 300 mm^2 for rod BD. Use $E = 200$ GPa for steel.

2.55 The homogeneous rod of constant cross section is attached to unyielding supports. The rod carries an axial load P, applied as shown in the figure. Show that the reactions are given by $R_1 = Pb/L$ and $R_2 = Pa/L$.

2.56 The homogeneous bar with a cross-sectional area of 500 mm^2 is attached to rigid supports. The bar carries the axial loads $P_1 = 25$ kN and $P_2 = 50$ kN, as shown. Determine the stress in segment BC. (*Hint*: Use the results of Prob. 2.55 to compute the reactions caused by P_1 and P_2 acting separately. Then use superposition to compute the reactions when both loads are applied.)

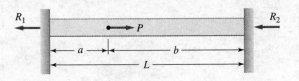
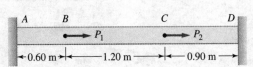

FIG. P2.55 **FIG. 2.56**

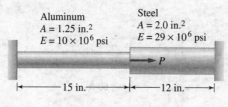
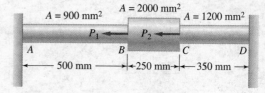

FIG. P2.57, P2.58 **FIG. P2.59, P2.60**

2.57 The composite bar is firmly attached to unyielding supports. Compute the stress in each material caused by the application of the axial load $P = 50$ kips.

2.58 The composite bar, firmly attached to unyielding supports, is initially stress-free. What maximum axial load P can be applied if the allowable stresses are 10 ksi for aluminum and 18 ksi for steel?

2.59 The steel rod is stress-free before the axial loads $P_1 = 150$ kN and $P_2 = 90$ kN are applied to the rod. Assuming that the walls are rigid, calculate the axial force in each segment after the loads are applied. Use $E = 200$ GPa.

2.60 Solve Prob. 2.59 if the wall at D yields (moves to the left) 0.8 mm.

2.61 The steel column of circular cross section is attached to rigid supports at A and C. Find the maximum stress in the column caused by the 25-kN load.

2.62 The assembly consists of a bronze tube and a threaded steel bolt. The pitch of the thread is 1/32 in. (one turn of the nut advances it 1/32 in.). The cross-sectional areas are 1.5 in.^{2} for the tube and 0.75 in.^{2} for the bolt. The nut is turned until there is a compressive stress of 4000 psi in the tube. Find the stresses in the bolt and the tube if the nut is given one additional turn. Use $E = 12 \times 10^6$ psi for bronze and $E = 29 \times 10^6$ psi for steel.

2.63 The two vertical rods attached to the rigid bar are identical except for length. Before the 6600-lb weight was attached, the bar was horizontal. Determine the axial force in each bar caused by the application of the weight. Neglect the weight of the bar.

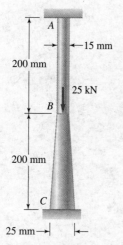

FIG. P2.61

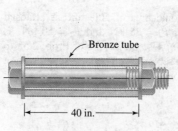

FIG. P2.62

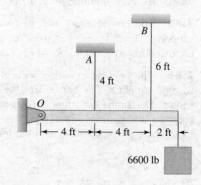

FIG. P2.63

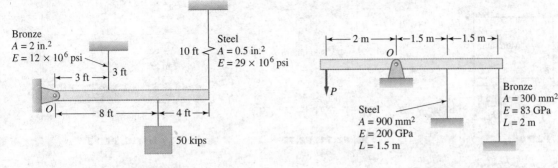

FIG. P2.64   FIG. P2.65

2.64 The rigid beam of negligible weight is supported by a pin at O and two vertical rods. Find the vertical displacement of the 50-kip weight.

2.65 The rigid bar of negligible weight is pinned at O and attached to two vertical rods. Assuming that the rods were initially stress-free, what is the largest load P that can be applied without exceeding stresses of 150 MPa in the steel rod and 70 MPa in the bronze rod?

2.66 The rigid, homogeneous slab weighing 600 kN is supported by three rods of identical material and cross section. Before the slab was attached, the lower ends of the rods were at the same level. Compute the axial force in each rod.

2.67 The rigid bar BCD of negligible weight is supported by two steel cables of identical cross section. Determine the force in each cable caused by the applied weight W.

2.68 The three steel rods, each of cross-sectional area 250 mm², jointly support the 7.5-kN load. Assuming that there was no slack or stress in the rods before the load was applied, find the force in each rod. Use $E = 200$ GPa for steel.

2.69 The bars AB, AC, and AD are pinned together as shown in the figure. Horizontal movement of the pin at A is prevented by the rigid horizontal strut AE. Calculate the axial force in the strut caused by the 10-kip load. For each steel bar, $A = 0.3$ in.² and $E = 29 \times 10^6$ psi. For the aluminum bar, $A = 0.6$ in.² and $E = 10 \times 10^6$ psi.

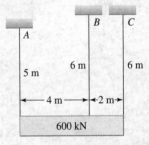

FIG. P2.66

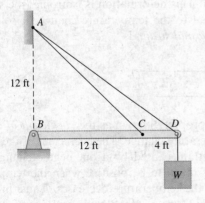

FIG. P2.67

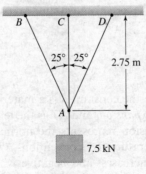

FIG. P2.68

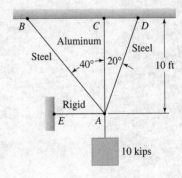

FIG. P2.69

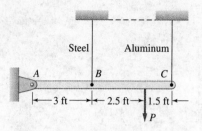

FIG. P2.70

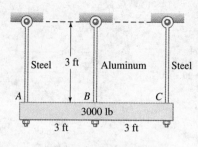

FIG. P2.71, P2.72

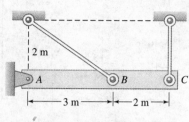

FIG. P2.73

2.70 The horizontal bar ABC is supported by a pin at A and two rods with identical cross-sectional areas. The rod at B is steel and the rod at C is aluminum. Neglecting the weight of the bar, determine the force in each rod when the force $P = 10$ kips is applied. Use $E_{st} = 29 \times 10^6$ psi and $E_{al} = 10 \times 10^6$ psi.

2.71 The lower ends of the three vertical rods were at the same level before the uniform, rigid bar ABC weighing 3000 lb was attached. Each rod has a cross-sectional area of 0.5 in.^{2}. The two outer rods are steel and the middle rod is aluminum. Find the force in the middle rod. Use $E_{st} = 29 \times 10^6$ psi and $E_{al} = 10 \times 10^6$ psi.

2.72 Solve Prob. 2.71 if the steel rod attached at C is replaced by an aluminum rod of the same size.

2.73 The uniform rigid bar ABC of weight W is supported by two rods that are identical except for their lengths. Assuming that the bar was held in the horizontal position when the rods were attached, determine the force in each rod after the attachment.

2.6 Thermal Stresses

It is well known that changes in temperature cause dimensional changes in a body: An increase in temperature results in expansion, whereas a temperature decrease produces contraction. This deformation is isotropic (the same in every direction) and proportional to the temperature change. It follows that the associated strain, called *thermal strain*, is

$$\epsilon_T = \alpha(\Delta T) \qquad (2.15)$$

where the constant α is a material property known as the *coefficient of thermal expansion*, and ΔT is the temperature change. The coefficient of thermal expansion represents the normal strain caused by a one-degree change in temperature. By convention, ΔT is taken to be positive when the temperature increases, and negative when the temperature decreases. Thus, in Eq. (2.15), positive ΔT produces positive strain (elongation) and negative ΔT produces negative strain (contraction). The units of α are $1/°C$ (per degree Celsius) in the SI system, and $1/°F$ (per degree Fahrenheit) in the U.S. Customary system. Typical values of α are $23 \times 10^{-6}/°C$ ($13 \times 10^{-6}/°F$) for aluminum and $12 \times 10^{-6}/°C$ ($6.5 \times 10^{-6}/°F$) for steel.

If the temperature change is uniform throughout the body, the thermal strain is also uniform. Consequently, the change in any dimension L of the body is given by

$$\delta_T = \epsilon_T L = \alpha(\Delta T)L \qquad (2.16)$$

If thermal deformation is permitted to occur freely (by using expansion joints or roller supports, for example), no internal forces will be induced in the body—there will be strain, but no stress. In cases where the deformation of a body is restricted, either totally or partially, internal forces will develop that oppose the thermal expansion or contraction. The stresses caused by these internal forces are known as *thermal stresses*.

The forces that result from temperature changes cannot be determined by equilibrium analysis alone; that is, these forces are statically indeterminate. Consequently, the analysis of thermal stresses follows the same principles that we used in Art. 2.5: equilibrium, compatibility, and Hooke's law. The only difference here is that we must now include thermal expansion in the analysis of deformation.

Procedure for Deriving Compatibility Equations
We recommend the following procedure for deriving the equations of compatibility:

- Remove all constraints from the body so that the thermal deformation can occur freely (this procedure is sometimes referred to as "relaxing the supports"). Show the thermal deformation on a sketch using an exaggerated scale.
- Apply the forces that are necessary to restore the specified conditions of constraint. Add the deformations caused by these forces to the sketch that was drawn in the previous step. (Draw the magnitudes of the deformations so that they are compatible with the geometric constraints.)
- By inspection of the sketch, write the relationships between the thermal deformations and the deformations due to the constraint forces.

Sample Problem 2.10

The horizontal steel rod, 2.5 m long and 1200 mm² in cross-sectional area, is secured between two walls as shown in Fig. (a). If the rod is stress-free at 20°C, compute the stress when the temperature has dropped to −20°C. Assume that (1) the walls do not move and (2) the walls move together a distance $\Delta = 0.5$ mm. Use $\alpha = 11.7 \times 10^{-6}/°C$ and $E = 200$ GPa.

Solution

Part 1

Compatibility We begin by assuming that the rod has been disconnected from the right wall, as shown in Fig. (b), so that the contraction δ_T caused by the temperature drop ΔT can occur freely. To reattach the rod to the wall, we must stretch it to the original length by applying the tensile force P. Compatibility of deformations requires that the resulting elongation δ_P, shown in Fig. (c), must be equal to δ_T; that is,

$$\delta_T = \delta_P$$

Hooke's Law If we substitute $\delta_T = \alpha(\Delta T)L$ and $\delta_P = PL/(EA) = \sigma L/E$, the compatibility equation becomes

$$\frac{\sigma L}{E} = \alpha(\Delta T)L$$

Therefore, the stress in the rod is

$$\sigma = \alpha(\Delta T)E = (11.7 \times 10^{-6})(40)(200 \times 10^9)$$
$$= 93.6 \times 10^6 \text{ Pa} = 93.6 \text{ MPa} \qquad \text{Answer}$$

Note that L canceled out in the preceding equation, which indicates that the stress is independent of the length of the rod.

Part 2

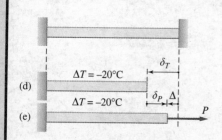

Compatibility When the walls move together a distance Δ, we see from Figs. (d) and (e) that the free thermal contraction δ_T is related to Δ and the elongation δ_P caused by the axial force P by

$$\delta_T = \delta_P + \Delta$$

Hooke's Law Substituting for δ_T and δ_P as in Part 1, we obtain

$$\alpha(\Delta T)L = \frac{\sigma L}{E} + \Delta$$

The solution for the stress σ is

$$\sigma = E\left[\alpha(\Delta T) - \frac{\Delta}{L}\right]$$

$$= (200 \times 10^9)\left[(11.7 \times 10^{-6})(40) - \frac{0.5 \times 10^{-3}}{2.5}\right]$$

$$= 53.6 \times 10^6 \text{ Pa} = 53.6 \text{ MPa} \qquad \text{Answer}$$

We see that the movement of the walls reduces the stress considerably. Also observe that the length of the rod does not cancel out as in Part 1.

Sample Problem 2.11

Figure (a) shows a homogeneous, rigid block weighing 12 kips that is supported by three symmetrically placed rods. The lower ends of the rods were at the same level before the block was attached. Determine the stress in each rod after the block is attached and the temperature of all bars increases by 100°F. Use the following data:

	A (in.^{2})	E (psi)	α (/°F)
Each steel rod	0.75	29×10^6	6.5×10^{-6}
Bronze rod	1.50	12×10^6	10.0×10^{-6}

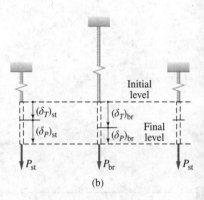

(a)

Solution

Compatibility Note that the block remains horizontal because of the symmetry of the structure. Let us assume that the block is detached from the rods, as shown in Fig. (b). With the rods unconstrained, a temperature rise will cause the elongations $(\delta_T)_{st}$ in the steel rods and $(\delta_T)_{br}$ in the bronze rod. To reattach the block to the rods, the rods must undergo the additional deformations $(\delta_P)_{st}$ and $(\delta_P)_{br}$, both assumed to be elongations. From the deformation diagram in Fig. (b), we obtain the following compatibility equation (recall that the block remains horizontal):

$$(\delta_T)_{st} + (\delta_P)_{st} = (\delta_T)_{br} + (\delta_P)_{br}$$

Hooke's Law Using Hooke's law, we can write the compatibility equation as

$$[\alpha(\Delta T)L]_{st} + \left[\frac{PL}{EA}\right]_{st} = [\alpha(\Delta T)L]_{br} + \left[\frac{PL}{EA}\right]_{br}$$

Substituting the given data, we have

$$(6.5 \times 10^{-6})(100)(2 \times 12) + \frac{P_{st}(2 \times 12)}{(29 \times 10^6)(0.75)}$$

$$= (10.0 \times 10^{-6})(100)(3 \times 12) + \frac{P_{br}(3 \times 12)}{(12 \times 10^6)(1.50)}$$

If we rearrange terms and simplify, the compatibility equation becomes

$$0.091\,95 P_{st} - 0.1667 P_{br} = 1700 \qquad (a)$$

Equilibrium From the free-body diagram in Fig. (c) we obtain

$$\Sigma F = 0 \quad +\uparrow \quad 2P_{st} + P_{br} - 12\,000 = 0 \qquad (b)$$

Solving Eqs. (a) and (b) simultaneously yields

$$P_{st} = 8700 \text{ lb} \quad \text{and} \quad P_{br} = -5400 \text{ lb}$$

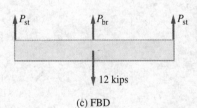

(c) FBD

The negative sign for P_{br} means that the force in the bronze rod is compressive (it acts in the direction opposite to that shown in the figures). The stresses in the rods are:

$$\sigma_{st} = \frac{P_{st}}{A_{st}} = \frac{8700}{0.75} = 11\,600 \text{ psi (T)} \qquad \textbf{Answer}$$

$$\sigma_{br} = \frac{P_{br}}{A_{br}} = \frac{-5400}{1.50} = -3600 \text{ psi} = 3600 \text{ psi (C)} \qquad \textbf{Answer}$$

Sample Problem 2.12

Using the data in Sample Problem 2.11, determine the temperature increase that would cause the entire weight of the block to be carried by the steel rods.

Solution

Equilibrium The problem statement implies that the bronze rod is stress-free. Thus, each steel rod carries half the weight of the rigid block, so that $P_{st} = 6000$ lb.

Compatibility The temperature increase causes the elongations $(\delta_T)_{st}$ and $(\delta_T)_{br}$ in the steel and bronze rods, respectively, as shown in the figure. Because the bronze rod is to carry no load, the ends of the steel rods must be at the same level as the end of the unstressed bronze rod before the rigid block can be reattached. Therefore, the steel rods must elongate by $(\delta_P)_{st}$ due to the tensile forces $P_{st} = 6000$ lb, which gives

$$(\delta_T)_{br} = (\delta_T)_{st} + (\delta_P)_{st}$$

Hooke's Law From Hooke's law, the compatibility equation becomes

$$[\alpha(\Delta T)L]_{br} = [\alpha(\Delta T)L]_{st} + \left[\frac{PL}{EA}\right]_{st}$$

$$(10 \times 10^{-6})(\Delta T)(3 \times 12) = (6.5 \times 10^{-6})(\Delta T)(2 \times 12) + \frac{6000(2 \times 12)}{(29 \times 10^6)(0.75)}$$

which yields

$$\Delta T = 32.5°F \qquad \text{Answer}$$

as the temperature increase at which the bronze rod would be unstressed.

Problems

2.74 A steel rod with a cross-sectional area of 0.25 in.^{2} is stretched between two fixed points. The tensile force in the rod at 70°F is 1200 lb. (a) What will be the stress at 0°F? (b) At what temperature will the stress be zero? Use $\alpha = 6.5 \times 10^{-6}/°F$ and $E = 29 \times 10^6$ psi.

2.75 A steel rod is stretched between two walls. At 20°C, the tensile force in the rod is 5000 N. If the stress is not to exceed 130 MPa at −20°C, find the minimum allowable diameter of the rod. Use $\alpha = 11.7 \times 10^{-6}/°C$ and $E = 200$ GPa.

2.76 Steel railroad rails 10 m long are laid with end-to-end clearance of 3 mm at a temperature of 15°C. (a) At what temperature will the rails just come in contact? (b) What stress would be induced in the rails at that temperature if there were no initial clearance? Use $\alpha = 11.7 \times 10^{-6}/°C$ and $E = 200$ GPa.

2.77 A steel rod 3 ft long with a cross-sectional area of 0.3 in.^{2} is stretched between two fixed points. The tensile force in the rod is 1200 lb at 40°F. Using $\alpha = 6.5 \times 10^{-6}/°F$ and $E = 29 \times 10^6$ psi, calculate the temperature at which the stress in the rod will be (a) 10 ksi; and (b) zero.

2.78 The bronze bar 3 m long with a cross-sectional area of 320 mm^2 is placed between two rigid walls. At a temperature of −20°C, there is a gap $\Delta = 2.5$ mm, as shown in the figure. Find the temperature at which the compressive stress in the bar will be 35 MPa. Use $\alpha = 18.0 \times 10^{-6}/°C$ and $E = 80$ GPa.

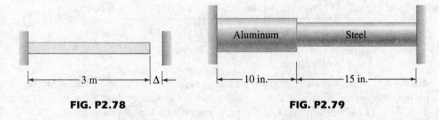

FIG. P2.78 **FIG. P2.79**

2.79 Calculate the increase in stress in each segment of the compound bar if the temperature is increased by 100°F. Assume that the supports are unyielding and use the following data:

	A (in.^{2})	E (psi)	α (/°F)
Aluminum	2.0	10×10^6	12.8×10^{-6}
Steel	1.5	29×10^6	6.5×10^{-6}

2.80 A prismatic bar of length L fits snugly between two rigid walls. If the bar is given a temperature increase that varies linearly from ΔT_A at one end to ΔT_B at the other end, show that the resulting stress in the bar is $\sigma = \alpha E (\Delta T_A + \Delta T_B)/2$.

2.81 The rigid bar ABC is supported by a pin at B and two vertical steel rods. Initially the bar is horizontal and the rods are stress-free. Determine the stress in the each rod if the temperature of the rod at A is decreased by 40°C. Neglect the weight of bar ABC. Use $\alpha = 11.7 \times 10^{-6}/°C$ and $E = 200$ GPa for steel.

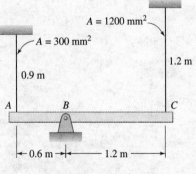

FIG. P2.81

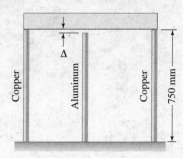

FIG. P2.82

2.82 The rigid, horizontal slab is attached to two identical copper rods. There is a gap $\Delta = 0.18$ mm between the middle bar, which is made of aluminum, and the slab. Neglecting the mass of the slab, calculate the stress in each rod when the temperature in the assembly is increased by 85°C. Use the following data:

	A (mm^2)	α (/°C)	E (GPa)
Each copper rod	500	16.8×10^{-6}	120
Aluminum rod	400	23.1×10^{-6}	70

2.83 A bronze sleeve is slipped over a steel bolt and held in place by a nut that is tightened to produce an initial stress of 2000 psi in the bronze. Find the stress in each material after the temperature of the assembly is increased by 100°F. The properties of the components are listed in the table.

	A (in.^{2})	α (/°F)	E (psi)
Bronze sleeve	1.50	10.5×10^{-6}	12×10^6
Steel bolt	0.75	6.5×10^{-6}	29×10^6

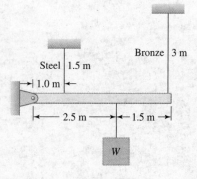

FIG. P2.84, P2.85

2.84 The rigid bar of negligible weight is supported as shown in the figure. If $W = 80$ kN, compute the temperature change of the assembly that will cause a tensile stress of 55 MPa in the steel rod. Use the following data:

	A (mm^2)	α (/°C)	E (GPa)
Steel rod	320	11.7×10^{-6}	200
Bronze rod	1300	18.9×10^{-6}	83

2.85 The rigid bar of negligible weight is supported as shown. The assembly is initially stress-free. Find the stress in each rod if the temperature rises 30°C after a load $W = 120$ kN is applied. Use the properties of the bars given in Prob. 2.84.

2.86 The composite bar is firmly attached to unyielding supports. The bar is stress-free at 60°F. Compute the stress in each material after the 50-kip force is applied and the temperature is increased to 120°F. Use $\alpha = 6.5 \times 10^{-6}$/°F for steel and $\alpha = 12.8 \times 10^{-6}$/°F for aluminum.

2.87 At what temperature will the aluminum and steel segments in Prob. 2.86 have stresses of equal magnitude after the 50-kip force is applied?

2.88 All members of the steel truss have the same cross-sectional area. If the truss is stress-free at 10°C, determine the stresses in the members at 90°C. For steel, $\alpha = 11.7 \times 10^{-6}$/°C and $E = 200$ GPa.

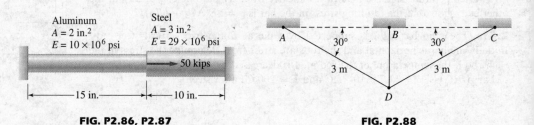

FIG. P2.86, P2.87   FIG. P2.88

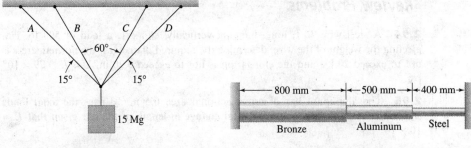

FIG. P2.89   **FIG. P2.90**

2.89 The truss is made of four steel bars, each of cross-sectional area 600 mm². Find the axial force in each bar caused by the 15-Mg load and a temperature rise of 50°C applied simultaneously. Use $\alpha = 11.7 \times 10^{-6}/°C$ and $E = 200$ GPa.

2.90 The compound bar, composed of the three segments shown, is initially stress-free. Compute the stress in each material if the temperature drops 30°C. Assume that the walls do not yield and use the following data:

	A (mm²)	α (/°C)	E (GPa)
Bronze segment	2400	19.0×10^{-6}	83
Aluminum segment	1200	23.0×10^{-6}	70
Steel segment	600	11.7×10^{-6}	200

2.91 The rigid bar AOB is pinned at O and connected to aluminum and steel rods. If the bar is horizontal at a given temperature, determine the ratio of the areas of the two rods so that the bar will be horizontal at any temperature. Neglect the mass of the bar.

2.92 The aluminum and bronze cylinders are centered and secured between two rigid end-plates by tightening the two steel bolts. There is no axial load in the assembly at a temperature of 50°F. Find the stress in the steel bolts when the temperature is increased to 200°F. Use the following data:

	A (in.²)	α (/°F)	E (psi)
Aluminum cylinder	2.00	12.8×10^{-6}	10×10^6
Bronze cylinder	3.00	10.5×10^{-6}	12×10^6
Each steel bolt	0.75	6.5×10^{-6}	29×10^6

2.93 At what temperature will the steel bolts in Prob. 2.92 be stressed to 18 000 psi?

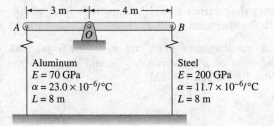   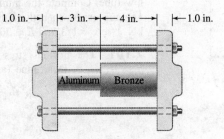

FIG. P2.91   **FIG. P2.92, P2.93**

Review Problems

2.94 A steel wire 30 ft long, hanging vertically, supports a load of 500 lb. Neglecting the weight of the wire, determine the required diameter if the normal stress is not to exceed 20 ksi and the elongation is not to exceed 0.20 in. Use $E = 29 \times 10^6$ psi.

2.95 The aluminum bar of cross-sectional area 0.5 in.^{2} carries the axial loads shown in the figure. Compute the total change in length of the bar given that $E = 10 \times 10^6$ psi.

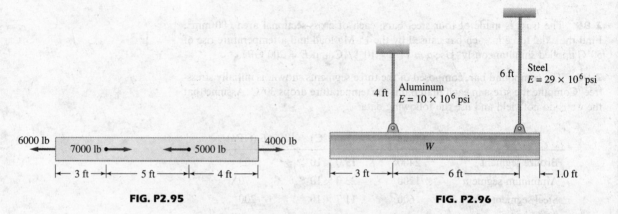

FIG. P2.95   FIG. P2.96

2.96 The uniform beam of weight W is to be supported by the two rods, the lower ends of which were initially at the same level. Determine the ratio of the areas of the rods so that the beam will be horizontal after it is attached to the rods. Neglect the deformation of the beam.

2.97 A round bar of length L, modulus of elasticity E, and weight density γ tapers uniformly from a diameter $2D$ at one end to a diameter D at the other end. If the bar is suspended vertically from the larger end, find the elongation of the bar caused by its own weight.

2.98 The timber member BC, inclined at angle $\theta = 60°$ to the vertical, is supported by a pin at B and the 0.75-in.-diameter steel bar AC. (a) Determine the cross-sectional area of BC for which the displacement of C will be vertical when the 5000-lb force is applied. (b) Compute the corresponding displacement of C. The moduli of elasticity are 1.8×10^6 psi for timber and 29×10^6 psi for steel. Neglect the weight of BC.

2.99 Solve Prob. 2.98 if $\theta = 30°$.

2.100 A solid aluminum shaft of diameter 80 mm fits concentrically inside a hollow tube. Compute the minimum internal diameter of the tube so that no contact pressure exists when the aluminum shaft carries an axial compressive force of 400 kN. Use $\nu = 1/3$ and $E = 70$ GPa for aluminum.

2.101 The normal stresses in an aluminum block are $\sigma_x = -6000$ psi and $\sigma_y = \sigma_z = -p$. Determine (a) the value of p for which $\epsilon_x = 0$; and (b) the corresponding value of ϵ_y. Use $E = 10 \times 10^6$ psi and $\nu = 0.33$.

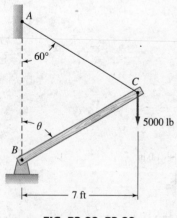

FIG. P2.98, P2.99

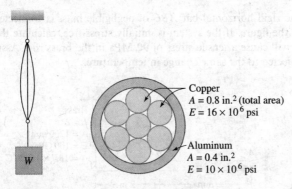

FIG. P2.102 **FIG. P2.103, P2.104**

2.102 The three steel wires, each of cross-sectional area 0.05 in.², support the weight W. Their unstressed lengths are 74.98 ft, 74.99 ft, and 75.00 ft. (a) Find the stress in the longest wire if $W = 1500$ lb. (b) Determine the stress in the shortest wire if $W = 500$ lb. Use $E = 29 \times 10^6$ psi.

2.103 The figure shows the cross section of a composite bar consisting of copper rods encased in an aluminum sleeve. The working stresses are 10 ksi for copper and 15 ksi for aluminum. Assuming that the ends of the bar are attached to rigid endplates, determine the largest safe axial load that can be applied to the bar.

2.104 The composite bar described in Prob. 2.103 is initially stress-free. Determine the largest temperature increase (no applied load) for which the stresses remain below the working stresses. Use $\alpha = 9.5 \times 10^{-6}/°F$ for copper and $13.0 \times 10^{-6}/°F$ for aluminum.

2.105 The rigid bar ACE is supported by a pin at A and two horizontal aluminum rods, each of cross-sectional area 50 mm². When the 200-kN load is applied at point E, determine (a) the axial force in rod DE and (b) the vertical displacement of point E. Use $E = 70$ GPa for aluminum.

2.106 The two vertical steel rods that support the rigid bar $ABCD$ are initially stress-free. Determine the stress in each rod after the 20-kip load is applied. Neglect the weight of the bar and use $E = 29 \times 10^6$ psi for steel.

2.107 The rigid bar $ABCD$ of negligible weight is initially horizontal, and the steel rods attached at A and C are stress-free. The 20-kip load is then applied and the temperature of the steel rods is changed by ΔT. Find ΔT for which the stresses in the two steel rods will be equal. Use $\alpha = 6.5 \times 10^{-6}/°F$ and $E = 29 \times 10^6$ psi for steel.

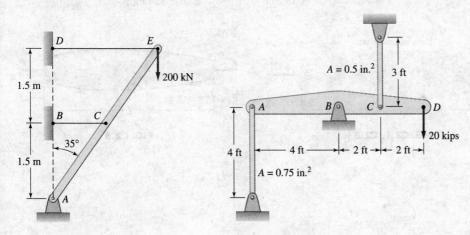

FIG. P2.105 **FIG. P2.106, P2.107**

2.108 The rigid horizontal bar ABC of negligible mass is connected to two rods as shown in the figure. If the system is initially stress-free, calculate the temperature change that will cause a tensile stress of 90 MPa in the brass rod. Assume that both rods are subjected to the same change in temperature.

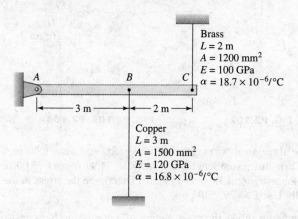

FIG. P2.108

Computer Problems

C2.1 The figure shows an aluminum bar of circular cross section with variable diameter. Use numerical integration to compute the elongation of the bar caused by the 6-kN axial force. Use $E = 70 \times 10^9$ Pa for aluminum.

C2.2 The flat aluminum bar shown in profile has a constant thickness of 10 mm. Determine the elongation of the bar caused by the 6-kN axial load using numerical integration. For aluminum $E = 70 \times 10^9$ Pa.

C2.3 The shaft of length L has diameter d that varies with the axial coordinate x. Given L, $d(x)$, and the modulus of elasticity E, write an algorithm to compute the axial stiffness $k = P/\delta$ of the bar. Use (a) $L = 500$ mm and

$$d = (25 \text{ mm})\left(1 + 3.8\frac{x}{L} - 3.6\frac{x^2}{L^2}\right)$$

and (b) $L = 200$ mm and

$$d = \begin{cases} (24 - 0.05x) \text{ mm} & \text{if } x \leq 120 \text{ mm} \\ 18 \text{ mm} & \text{if } x \geq 120 \text{ mm} \end{cases}$$

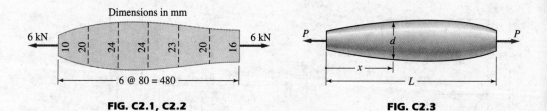

FIG. C2.1, C2.2   **FIG. C2.3**

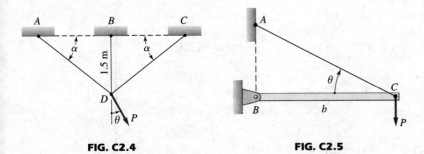

FIG. C2.4 **FIG. C2.5**

C2.4 The symmetric truss carries a force P inclined at the angle θ to the vertical. Given P and the angle α, write an algorithm to plot the axial force in each member as a function of θ from $\theta = -90°$ to $90°$. Assume the cross-sectional areas of members are the same. Use $P = 10$ kN and (a) $\alpha = 30°$; and (b) $\alpha = 60°$. (*Hint*: Compute the effects of the horizontal and vertical components of P separately, and then superimpose the effects.)

C2.5 The rigid bar BC of length b and negligible weight is supported by the wire AC of cross-sectional area A and modulus of elasticity E. The vertical displacement of point C can be expressed in the form

$$\Delta_C = \frac{Pb}{EA} f(\theta)$$

where θ is the angle between the wire and the rigid bar. (a) Derive the function $f(\theta)$ and plot it from $\theta = 20°$ to $85°$. (b) What value of θ yields the smallest vertical displacement of C?

C2.6 The steel bolt of cross-sectional area A_0 is placed inside the aluminum tube, also of cross-sectional area A_0. The assembly is completed by making the nut "fingertight." The dimensions of the reduced segment of the bolt (length b and cross-sectional area A) are designed so that the segment will yield when the temperature of the assembly is increased by $200°$F. Write an algorithm that determines the relationship between A/A_0 and b/L that satisfies this design requirement. Plot A/A_0 against b/L from $b/L = 0$ to 1.0. Use the properties of steel and aluminum shown in the figure.

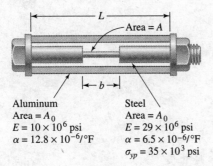

Aluminum
Area = A_0
$E = 10 \times 10^6$ psi
$\alpha = 12.8 \times 10^{-6}/°$F

Steel
Area = A_0
$E = 29 \times 10^6$ psi
$\alpha = 6.5 \times 10^{-6}/°$F
$\sigma_{yp} = 35 \times 10^3$ psi

FIG. C2.6

3 Torsion

A turbine at a hydroelectric power plant. The power of the turbine is transmitted to the generator by a shaft that must carry very large torques.

3.1 Introduction

In many engineering applications, members are required to carry torsional loads. In this chapter, we consider the torsion of circular shafts. Because a circular cross section is an efficient shape for resisting torsional loads, circular shafts are commonly used to transmit power in rotating machinery. We also discuss another important application—torsion of thin-walled tubes.

Torsion is our introduction to problems in which the stress is not uniform, or assumed to be uniform, over the cross section of the member. Another problem in this category, which we will treat later, is the bending of beams. Derivation of the equations used in the analysis of both torsion and bending follows these steps:

- Make simplifying assumptions about the deformation based on experimental evidence.
- Determine the strains that are geometrically compatible with the assumed deformations.
- Use Hooke's law to express the equations of compatibility in terms of stresses.
- Derive the equations of equilibrium. (These equations provide the relationships between the stresses and the applied loads.)

3.2 Torsion of Circular Shafts

a. Simplifying assumptions

Figure 3.1 shows the deformation of a circular shaft that is subjected to a twisting couple (torque) T. To visualize the deformation, we scribe the straight line AB on the surface of the shaft before the torque is applied. After loading, this line deforms into the helix AB' as the free end of the shaft rotates through the angle θ. During the deformation, the cross sections are not distorted in any manner—they remain plane, and the radius r does not change. In addition, the length L of the shaft remains constant. Based on these observations, we make the following assumptions:

- Circular cross sections remain plane (do not warp) and perpendicular to the axis of the shaft.
- Cross sections do not deform (there is no strain in the plane of the cross section).
- The distances between cross sections do not change (the axial normal strain is zero).

The deformation that results from the above assumptions is relatively simple: *Each cross section rotates as a rigid entity about the axis of the shaft.* Although this conclusion is based on the observed deformation of a cylindrical shaft carrying a constant internal torque, we assume that the result remains valid even if the diameter of the shaft or the internal torque varies along the length of the shaft.

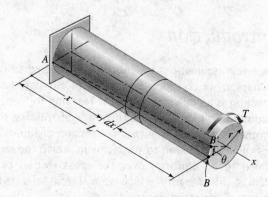

FIG. 3.1 Deformation of a circular shaft caused by the torque T. The initially straight line AB deforms into a helix.

b. Compatibility

To analyze the deformation in the interior of the shaft in Fig. 3.1, we consider the portion of the shaft shown in Fig. 3.2(a). We first isolate a segment of the shaft of infinitesimal length dx and then "peel" off its outer layer, leaving us with the cylindrical core of radius ρ. As the shaft deforms, the two cross sections of the segment rotate about the x-axis. Because the cross sections are separated by an infinitesimal distance, the difference in their rotations, denoted by the angle $d\theta$, is also infinitesimal. We now imagine that the straight line CD has been drawn on the cylindrical surface. As the cross sections undergo the relative rotation $d\theta$, CD deforms into the helix CD'. By observing the distortion of the shaded element, we recognize that the helix angle γ is the *shear strain* of the element.

From the geometry of Fig. 3.2(a), we obtain $\overline{DD'} = \rho\, d\theta = \gamma\, dx$, from which the shear strain is

$$\gamma = \frac{d\theta}{dx}\rho \qquad (3.1)$$

The quantity $d\theta/dx$ is the *angle of twist per unit length*, where θ is expressed in radians. The corresponding shear stress, illustrated in Fig. 3.2(b), is determined from Hooke's law:

$$\tau = G\gamma = G\frac{d\theta}{dx}\rho \qquad (3.2)$$

Note that $G(d\theta/dx)$ in Eq. (3.2) is independent of the radial distance ρ. Therefore, *the shear stress varies linearly with the radial distance ρ from the axis of the shaft*. The variation of the shear stress acting on the cross section is illustrated in Fig. 3.3. The maximum shear stress, denoted by τ_{max}, occurs at the surface of the shaft.

Note that the above derivations assume neither a constant internal torque nor a constant cross section along the length of the shaft.

c. Equilibrium

For the shaft to be in equilibrium, the resultant of the shear stress acting on a cross section must be equal to the internal torque T acting on that cross section. Figure 3.4 shows a cross section of the shaft containing a differential element of area dA located at the radial distance ρ from the axis of the shaft. The shear force acting on this area is $dP = \tau\, dA = G(d\theta/dx)\rho\, dA$, directed perpendicular to the radius. Hence, the moment (torque) of dP about the center O is $\rho\, dP = G(d\theta/dx)\rho^2\, dA$. Summing the contributions of all the differential elements across the cross-sectional area A and equating the result to the internal torque yields $\int_A \rho\, dP = T$, or

$$G\frac{d\theta}{dx}\int_A \rho^2\, dA = T$$

Recognizing that $\int_A \rho^2\, dA = J$ is (by definition) the *polar moment of inertia* of the cross-sectional area, we can write this equation as $G(d\theta/dx)J = T$, or

$$\frac{d\theta}{dx} = \frac{T}{GJ} \qquad (3.3)$$

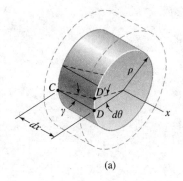
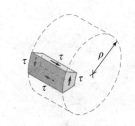

FIG. 3.2 (a) Shear strain of a material element caused by twisting of the shaft; (b) the corresponding shear stress.

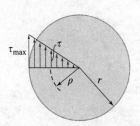

FIG. 3.3 Distribution of shear stress along the radius of a circular shaft.

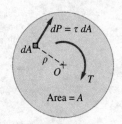

FIG. 3.4 Calculating the resultant of the shear stress acting on the cross section. Resultant is a couple equal to the internal torque T.

The rotation of the cross section at the free end of the shaft, called the *angle of twist*, is obtained by integration:

$$\theta = \int_0^L d\theta = \int_0^L \frac{T}{GJ}\,dx \qquad (3.4\text{a})$$

If the integrand is independent of x, as in the case of a prismatic bar carrying a constant torque, then Eq. (3.4a) reduces to the *torque-twist relationship*

$$\theta = \frac{TL}{GJ} \qquad (3.4\text{b})$$

Note the similarity between Eqs. (3.4) and the corresponding formulas for axial deformation: $\delta = \int_0^L (P/EA)\,dx$ and $\delta = PL/(EA)$.

Notes on the Computation of Angle of Twist

- It is common practice to let the units of G determine the units of the other terms in Eqs. (3.4). In the U.S. Customary system, the consistent units are G [psi], T [lb·in.], L [in.], and J [in.^{4}]; in the SI system, the consistent units are G [Pa], T [N·m], L [m], and J [m^4].
- The unit of θ in Eqs. (3.4) is radians, regardless of which system of units is used in the computation.
- In problems where it is convenient to use a sign convention for torques and angles of twist, we represent torques as vectors (we use double-headed arrows to represent couples and rotations) using the right-hand rule, as illustrated in Fig. 3.5. A torque vector is considered positive if it points away from the cross section, and negative if it points toward the cross section. The same sign convention applies to the angle of twist θ.

d. Torsion formulas

From Eq. (3.3) we see that $G(d\theta/dx) = T/J$, which, upon substitution into Eq. (3.2), gives the shear stress acting at the distance ρ from the center of the shaft:

$$\tau = \frac{T\rho}{J} \qquad (3.5\text{a})$$

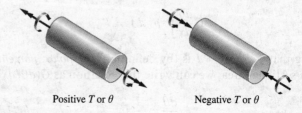

Positive T or θ \qquad Negative T or θ

FIG. 3.5 Sign conventions for torque T and angle of twist θ.

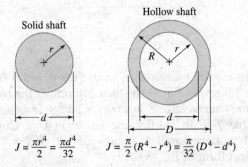

FIG. 3.6 Polar moments of inertia of circular areas.

The maximum shear stress is found by replacing ρ by the radius r of the shaft:

$$\tau_{max} = \frac{Tr}{J} \quad (3.5b)$$

Because Hooke's law was used in the derivation of Eqs. (3.2)–(3.5), these formulas are valid only if the shear stresses do not exceed the proportional limit of the material in shear.[1] Furthermore, these formulas are applicable only to circular shafts, either solid or hollow.

The expressions for the polar moments of circular areas are given in Fig. 3.6. Substituting these formulas into Eq. (3.5b), we obtain:

$$\text{Solid shaft:} \quad \tau_{max} = \frac{2T}{\pi r^3} = \frac{16T}{\pi d^3} \quad (3.5c)$$

$$\text{Hollow shaft:} \quad \tau_{max} = \frac{2TR}{\pi(R^4 - r^4)} = \frac{16TD}{\pi(D^4 - d^4)} \quad (3.5d)$$

Equations (3.5c) and (3.5d) are called the *torsion formulas*.

e. Power transmission

In many practical applications, shafts are used to transmit power. The power $\mathscr{P}$ transmitted by a torque T rotating at the angular speed ω is given by $\mathscr{P} = T\omega$, where ω is measured in radians per unit time. If the shaft is rotating with a frequency of f revolutions per unit time, then $\omega = 2\pi f$, which gives $\mathscr{P} = T(2\pi f)$. Therefore, the torque can be expressed as

---

[1] Equation (3.5b) is sometimes used to determine the "shear stress" corresponding to the torque at rupture, although the proportional limit is exceeded. The value so obtained is called the *torsional modulus of rupture*. It is used to compare the ultimate strengths of different materials and diameters.

$$T = \frac{\mathscr{P}}{2\pi f} \quad (3.6a)$$

In SI units, $\mathscr{P}$ in usually measured in watts (1.0 W = 1.0 N·m/s) and f in hertz (1.0 Hz = 1.0 rev/s); Eq. (3.6a) then determines the torque T in N·m. In U.S. Customary units with $\mathscr{P}$ in lb·in./s and f in hertz, Eq. (3.6a) calculates the torque T in lb·in. Because power in U.S. Customary units is often expressed in horsepower (1.0 hp = 550 lb·ft/s = 396×10^3 lb·in./min), a convenient form of Eq. (3.6a) is

$$T \text{ (lb·in.)} = \frac{\mathscr{P} \text{ (hp)}}{2\pi f \text{ (rev/min)}} \times \frac{396 \times 10^3 \text{ (lb·in./min)}}{1.0 \text{ (hp)}}$$

which simplifies to

$$T \text{ (lb·in.)} = 63.0 \times 10^3 \frac{\mathscr{P} \text{ (hp)}}{f \text{ (rev/min)}} \quad (3.6b)$$

f. Statically indeterminate problems

The procedure for solving statically indeterminate torsion problems is similar to the steps presented in Art. 2.5 for axially loaded bars:

- Draw the required free-body diagrams and write the equations of **equilibrium**.
- Derive the **compatibility** equations from the restrictions imposed on the angles of twist.
- Use the **torque-twist relationships** in Eqs. (3.4) to express the angles of twist in the compatibility equations in terms of the torques.
- Solve the equations of equilibrium and compatibility for the torques.

Sample Problem 3.1

A solid steel shaft in a rolling mill transmits 20 kW of power at 2 Hz. Determine the smallest safe diameter of the shaft if the shear stress is not to exceed 40 MPa and the angle of twist is limited to 6° in a length of 3 m. Use $G = 83$ GPa.

Solution

This problem illustrates a design that must possess sufficient strength as well as rigidity. We begin by applying Eq. (3.6a) to determine the torque:

$$T = \frac{\mathscr{P}}{2\pi f} = \frac{20 \times 10^3}{2\pi(2)} = 1591.5 \text{ N} \cdot \text{m}$$

To satisfy the strength condition, we apply the torsion formula, Eq. (3.5c):

$$\tau_{max} = \frac{16T}{\pi d^3} \qquad 40 \times 10^6 = \frac{16(1591.5)}{\pi d^3}$$

which yields $d = 58.7 \times 10^{-3}$ m $= 58.7$ mm.

We next apply the torque-twist relationship, Eq. (3.4b), to determine the diameter necessary to satisfy the requirement of rigidity (remembering to convert θ from degrees to radians):

$$\theta = \frac{TL}{GJ} \qquad 6\left(\frac{\pi}{180}\right) = \frac{1591.5(3)}{(83 \times 10^9)(\pi d^4/32)}$$

from which we obtain $d = 48.6 \times 10^{-3}$ m $= 48.6$ mm.

To satisfy both strength and rigidity requirements, we must choose the larger diameter—namely,

$$d = 58.7 \text{ mm} \qquad \qquad \text{Answer}$$

Sample Problem 3.2

The shaft in Fig. (a) consists of a 3-in.-diameter aluminum segment that is rigidly joined to a 2-in.-diameter steel segment. The ends of the shaft are attached to rigid supports. Calculate the maximum shear stress developed in each segment when the torque $T = 10$ kip·in. is applied. Use $G = 4 \times 10^6$ psi for aluminum and $G = 12 \times 10^6$ psi for steel.

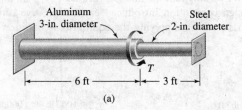

(a)

Solution

Equilibrium From the FBD of the entire shaft in Fig. (b), the equilibrium equation is

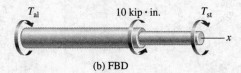

(b) FBD

$$\Sigma M_x = 0 \qquad (10 \times 10^3) - T_{st} - T_{al} = 0 \qquad \text{(a)}$$

This problem is statically indeterminate because there are two unknown torques (T_{st} and T_{al}) but only one independent equilibrium equation.

Compatibility A second relationship between the torques is obtained by noting that the right end of the aluminum segment must rotate through the same angle as the left end of the steel segment. Therefore, the two segments must have the same angle of twist; that is, $\theta_{st} = \theta_{al}$. From Eq. (3.4b), this condition becomes

$$\left(\frac{TL}{GJ}\right)_{st} = \left(\frac{TL}{GJ}\right)_{al}$$

$$\frac{T_{st}(3 \times 12)}{(12 \times 10^6)\frac{\pi}{32}(2)^4} = \frac{T_{al}(6 \times 12)}{(4 \times 10^6)\frac{\pi}{32}(3)^4}$$

from which

$$T_{st} = 1.1852\, T_{al} \qquad \text{(b)}$$

Solving Eqs. (a) and (b), we obtain

$$T_{al} = 4576 \text{ lb} \cdot \text{in.} \qquad T_{st} = 5424 \text{ lb} \cdot \text{in.}$$

From the torsion formula, Eq. (3.5c), the maximum shear stresses are

$$(\tau_{max})_{al} = \left(\frac{16T}{\pi d^3}\right)_{al} = \frac{16(4576)}{\pi(3)^3} = 863 \text{ psi} \qquad \textbf{Answer}$$

$$(\tau_{max})_{st} = \left(\frac{16T}{\pi d^3}\right)_{st} = \frac{16(5424)}{\pi(2)^3} = 3450 \text{ psi} \qquad \textbf{Answer}$$

Sample Problem **3.3**

The four rigid gears, loaded as shown in Fig. (a), are attached to a 2-in.-diameter steel shaft. Compute the angle of rotation of gear A relative to gear D. Use $G = 12 \times 10^6$ psi for the shaft.

Solution

It is convenient to represent the torques as vectors (using the right-hand rule) on the FBDs in Fig. (b). We assume that the internal torques T_{AB}, T_{BC}, and T_{CD} are positive according to the sign convention introduced earlier (positive torque vectors point away from the cross section). Applying the equilibrium condition $\Sigma M_x = 0$ to each FBD, we obtain

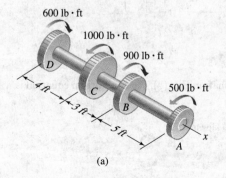

(a)

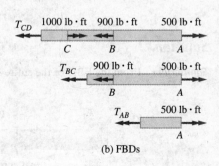

(b) FBDs

$$500 - 900 + 1000 - T_{CD} = 0$$
$$500 - 900 - T_{BC} = 0$$
$$500 - T_{AB} = 0$$

which yield

$$T_{AB} = 500 \text{ lb} \cdot \text{ft} \qquad T_{BC} = -400 \text{ lb} \cdot \text{ft} \qquad T_{CD} = 600 \text{ lb} \cdot \text{ft}$$

The minus sign indicates that the sense of T_{BC} is opposite to that shown on the FBD.

The rotation of gear A relative to gear D can be viewed as the rotation of gear A if gear D were fixed. This rotation is obtained by summing the angles of twist of the three segments:

$$\theta_{A/D} = \theta_{A/B} + \theta_{B/C} + \theta_{C/D}$$

Using Eq. (3.4b), we obtain (converting the lengths to inches and torques to pound-inches)

$$\theta_{A/D} = \frac{T_{AB}L_{AB} + T_{BC}L_{BC} + T_{CD}L_{CD}}{GJ}$$
$$= \frac{(500 \times 12)(5 \times 12) - (400 \times 12)(3 \times 12) + (600 \times 12)(4 \times 12)}{[\pi(2)^4/32](12 \times 10^6)}$$
$$= 0.02827 \text{ rad} = 1.620° \qquad \text{Answer}$$

The positive result indicates that the rotation vector of A relative to D is in the positive x-direction; that is, θ_{AD} is directed counterclockwise when viewed from A toward D.

∎

Sample Problem 3.4

Figure (a) shows a steel shaft of length $L = 1.5$ m and diameter $d = 25$ mm that carries a distributed torque of intensity (torque per unit length) $t = t_B(x/L)$, where $t_B = 200$ N · m/m. Determine (1) the maximum shear stress in the shaft; and (2) the angle of twist. Use $G = 80$ GPa for steel.

Solution

Part 1

Figure (b) shows the FBD of the shaft. The applied torque acting on a length dx of the shaft is $t\,dx$, so that the total torque applied to the shaft is $\int_0^L t\,dx$. The maximum torque in the shaft is T_A, which occurs at the fixed support. From the FBD we get

$$\Sigma M_x = 0 \qquad \int_0^L t\,dx - T_A = 0$$

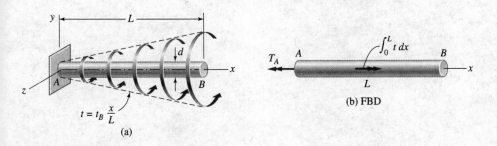

Therefore,

$$T_A = \int_0^L t\,dx = \int_0^L t_B \frac{x}{L}\,dx = \frac{t_B L}{2}$$

$$= \frac{1}{2}(200)(1.5) = 150 \text{ N}\cdot\text{m}$$

From Eq. (3.5c), the maximum stress in the shaft is

$$\tau_{max} = \frac{16 T_A}{\pi d^3} = \frac{16(150)}{\pi(0.025)^3} = 48.9 \times 10^6 \text{ Pa} = 48.9 \text{ MPa} \qquad \text{Answer}$$

Part 2

The torque T acting on a cross section located at the distance x from the fixed end can be found from the FBD in Fig. (c):

$$\Sigma M_x = 0 \qquad T + \int_0^x t\,dx - T_A = 0$$

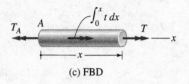

(c) FBD

which gives

$$T = T_A - \int_0^x t\,dx = \frac{t_B L}{2} - \int_0^x t_B \frac{x}{L}\,dx = \frac{t_B}{2L}(L^2 - x^2)$$

From Eq. (3.4a), the angle of twist of the shaft is

$$\theta = \int_0^L \frac{T}{GJ}\,dx = \frac{t_B}{2LGJ}\int_0^L (L^2 - x^2)\,dx = \frac{t_B L^2}{3GJ}$$

$$= \frac{200(1.5)^2}{3(80 \times 10^9)[(\pi/32)(0.025)^4]} = 0.0489 \text{ rad} = 2.80° \qquad \text{Answer}$$

Problems

3.1 A solid steel shaft transmits 30 hp while running at 100 rev/min. Find the smallest safe diameter of the shaft if the shear stress is limited to 6000 psi and the angle of twist of the shaft is not to exceed 6° in a length of 10 ft. Use $G = 12 \times 10^6$ psi for steel.

3.2 A hollow steel shaft, 6 ft long, has an outer diameter of 3 in. and an inner diameter of 1.5 in. The shaft is transmitting 200 hp at 120 rev/min. Determine (a) the maximum shear stress in the shaft; and (b) the angle of twist in degrees. Use $G = 12 \times 10^6$ psi for steel.

3.3 A hollow steel propeller shaft, 18 ft long with 14-in. outer diameter and 10-in. inner diameter, transmits 5000 hp at 189 rev/min. Use $G = 12 \times 10^6$ psi for steel. Calculate (a) the maximum shear stress; and (b) the angle of twist.

3.4 A solid steel propeller shaft is to transmit 4.5 MW at 3 Hz without exceeding a shear stress of 50 MPa or twisting through more than 1.0° in a length of 26 diameters. Compute the smallest acceptable diameter of the shaft. For steel, use $G = 83$ GPa.

3.5 The steel shaft with two different diameters rotates at 4 Hz. The power supplied to gear C is 55 kW, of which 35 kW is removed by gear A and 20 kW is removed by gear B. Find (a) the maximum shear stress in the shaft; and (b) the angle of rotation of gear A relative to gear C. Use $G = 83$ GPa for steel.

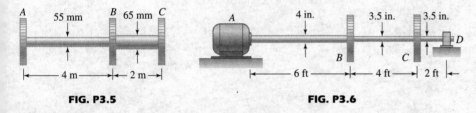

FIG. P3.5   FIG. P3.6

3.6 The motor A delivers 3000 hp to the shaft at 1500 rev/min, of which 1000 hp is removed by gear B and 2000 hp is removed by gear C. Determine (a) the maximum shear stress in the shaft; and (b) the angle of twist of end D relative to end A. Use $G = 12 \times 10^6$ psi for steel, and assume that friction at bearing D is negligible.

3.7 A steel shaft, 3 ft long and 4 in. in diameter, carries a torque of 15 kip·ft. Determine (a) the maximum shear stress; and (b) the angle of twist. Use $G = 12 \times 10^6$ psi for steel.

3.8 A solid steel shaft must not twist through more than 3° in a 6-m length when subjected to a torque of 12 kN·m. (a) Find the diameter of the smallest shaft that can be used. (b) What is the maximum shear stress in the shaft of this diameter? Use $G = 83$ GPa for steel.

3.9 A torque of 100×10^3 lb·ft produces a maximum shear stress of 8000 psi in a hollow steel shaft. The inner diameter of the shaft is two-thirds as large as the outer diameter. (a) Determine the outer diameter. (b) Find the angle of twist in a 16-ft length. Use $G = 12 \times 10^6$ psi for steel.

3.10 The inner diameter of a hollow shaft is equal to half its outer diameter D. Show that the torsional strength of the shaft is equal to 15/16 of the strength of a solid shaft of diameter D.

3.11 A 16-ft solid steel shaft is twisted through 4°. If the maximum shear stress is 8000 psi, determine the diameter of the shaft. Use $G = 12 \times 10^6$ psi for steel.

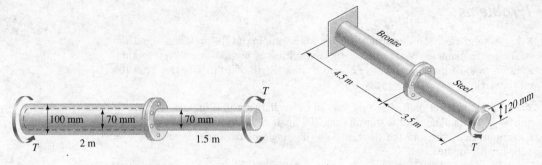

FIG. P3.12   FIG. P3.13

3.12 The steel shaft is formed by attaching a hollow shaft to a solid shaft. Determine the maximum torque T that can be applied to the ends of the shaft without exceeding a shear stress of 70 MPa or an angle of twist of 2.5° in the 3.5-m length. Use $G = 83$ GPa for steel.

3.13 The compound shaft consists of bronze and steel segments, both having 120-mm diameters. If the torque T causes a maximum shear stress of 80 MPa in the bronze segment, determine the angle of rotation of the free end. Use $G = 83$ GPa for steel and $G = 35$ GPa for bronze.

3.14 The stepped steel shaft carries the torque T. Determine the maximum allowable magnitude of T if the working shear stress is 14 MPa and the rotation of the free end is limited to 3.5°. Use $G = 83$ GPa for steel.

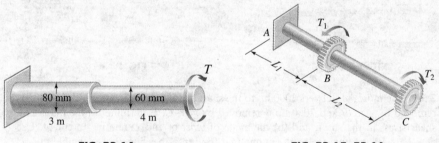

FIG. P3.14   FIG. P3.15, P3.16

3.15 The 2-in.-diameter aluminum shaft carries the torques $T_1 = 200$ lb·ft and $T_2 = 800$ lb·ft. Using $L_1 = 2$ ft, $L_2 = 3$ ft, and $G = 4 \times 10^6$ psi, determine (a) the maximum shear stress in each segment; and (b) the angle of rotation of the free end.

3.16 The solid steel shaft carries the torques $T_1 = 750$ N·m and $T_2 = 1200$ N·m. Using $L_1 = L_2 = 2.5$ m and $G = 83$ GPa, determine the smallest allowable diameter of the shaft if the shear stress is limited to 60 MPa and the angle of rotation of the free end is not to exceed 4°.

3.17 A 40-mm-diameter hole is drilled 3 m deep into the steel shaft. When the two torques are applied, determine (a) the maximum shear stress in the shaft; and (b) the angle of rotation of the free end of the shaft. Use $G = 83$ GPa for steel.

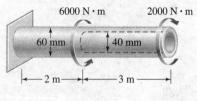

FIG. P3.17

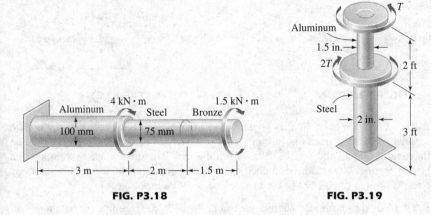

FIG. P3.18   **FIG. P3.19**

3.18 The solid compound shaft, made of three different materials, carries the two torques shown. (a) Calculate the maximum shear stress in each material. (b) Find the angle of rotation of the free end of the shaft. The shear moduli are 28 GPa for aluminum, 83 GPa for steel, and 35 GPa for bronze.

3.19 The shaft consisting of steel and aluminum segments carries the torques T and $2T$. Find the largest allowable value of T if the working shear stresses are 12 000 psi for steel and 8000 psi for aluminum, and the angle of rotation at the free end must not exceed $6°$. Use $G = 12 \times 10^6$ psi for steel and $G = 4 \times 10^6$ psi for aluminum.

3.20 Four pulleys are attached to the 50-mm-diameter aluminum shaft. If torques are applied to the pulleys as shown in the figure, determine the angle of rotation of pulley D relative to pulley A. Use $G = 28$ GPa for aluminum.

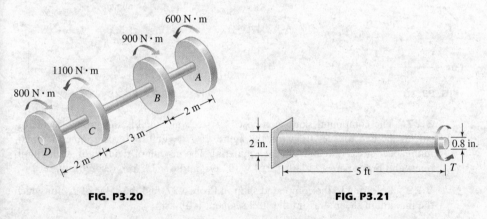

FIG. P3.20   **FIG. P3.21**

3.21 The tapered, wrought iron shaft carries the torque $T = 2000$ lb·in. at its free end. Determine the angle of twist of the shaft. Use $G = 10 \times 10^6$ psi for wrought iron.

3.22 The shaft carries a total torque T_0 that is uniformly distributed over its length L. Determine the angle of twist of the shaft in terms of T_0, L, G, and J.

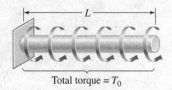

FIG. P3.22

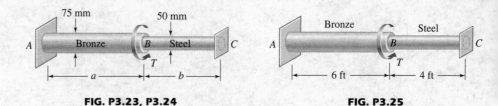

FIG. P3.23, P3.24   FIG. P3.25

3.23 The compound shaft is attached to a rigid wall at each end. For the bronze segment AB, the diameter is 75 mm and $G = 35$ GPa. For the steel segment BC, the diameter is 50 mm and $G = 83$ GPa. Given that $a = 2$ m and $b = 1.5$ m, compute the largest torque T that can be applied as shown in the figure if the maximum shear stress is limited to 60 MPa in the bronze and 80 MPa in the steel.

3.24 For the compound shaft described in Prob. 3.23, determine the torque T and the ratio b/a so that each material is stressed to its permissible limit.

3.25 The ends of the compound shaft are attached to rigid walls. The maximum shear stress is limited to 8000 psi for the bronze segment AB and 12 000 psi for the steel segment BC. Determine the diameter of each segment so that each material is simultaneously stressed to its permissible limit when the torque $T = 12$ kip·ft is applied as shown. The shear moduli are 6×10^6 psi for bronze and 12×10^6 psi for steel.

3.26 Both ends of the steel shaft are attached to rigid supports. Find the distance a where the torque T must be applied so that the reactive torques at A and B are equal.

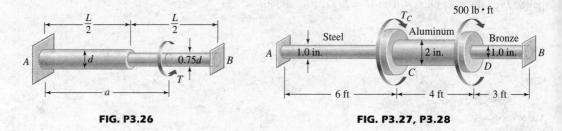

FIG. P3.26   FIG. P3.27, P3.28

3.27 The compound shaft, composed of steel, aluminum, and bronze segments, carries the two torques shown in the figure. If $T_C = 250$ lb·ft, determine the maximum shear stress developed in each material. The moduli of rigidity for steel, aluminum, and bronze are 12×10^6 psi, 4×10^6 psi, and 6×10^6 psi, respectively.

3.28 Referring to the compound shaft in Prob. 3.27, find the value of T_C for which the maximum shear stress in the steel segment is 10 ksi.

3.29 The steel rod fits loosely inside the aluminum sleeve. Both components are attached to a rigid wall at A and joined together by a pin at B. Because of a slight misalignment of the pre-drilled holes, the torque $T_0 = 750$ N·m was applied to the steel rod before the pin could be inserted into the holes. Determine the torque in each component after T_0 was removed. Use $G = 80$ GPa for steel and $G = 28$ GPa for aluminum.

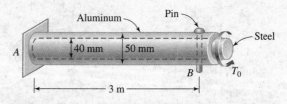

FIG. P3.29

3.30 A composite shaft is made by slipping a bronze tube of 3-in. outer diameter and 2-in. inner diameter over a solid steel shaft of the same length and 2-in.-diameter. The two components are then fastened rigidly together at their ends. What is the largest torque that can be carried by the composite shaft if the working shear stresses are 8 ksi for bronze and 12 ksi for the steel? For bronze, $G = 6 \times 10^6$ psi, and for steel, $G = 12 \times 10^6$ psi.

3.31 If the composite shaft described in Prob. 3.30 carries a 2500-lb·ft torque, determine the maximum shear stress in each material.

3.32 The aluminum shaft, composed of three segments, is fastened to rigid supports at A and D. Calculate the maximum shear stress in each segment when the two torques are applied.

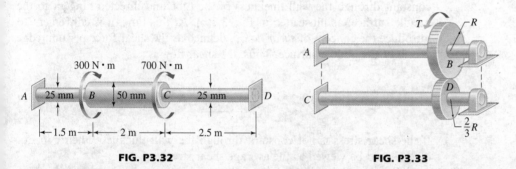

FIG. P3.32   FIG. P3.33

3.33 Each of the two identical shafts is attached to a rigid wall at one end and supported by a bearing at the other end. The gears attached to the shafts are in mesh. Determine the reactive torques at A and C when the torque T is applied to gear B.

3.34 The two steel shafts, each with one end built into a rigid support, have flanges attached to their free ends. The flanges are to be bolted together. However, initially there is a 6° mismatch in the location of the bolt holes as shown in the figure. Determine the maximum shear stress in each shaft after the flanges have been bolted together. The shear modulus of elasticity for steel is 12×10^6 psi. Neglect deformations of the bolts and the flanges.

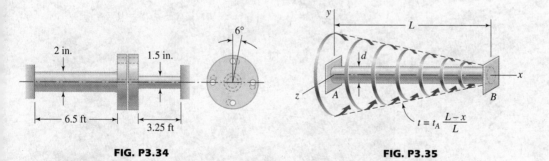

FIG. P3.34   FIG. P3.35

3.35 The steel shaft of length $L = 1.5$ m and diameter $d = 25$ mm is attached to rigid walls at both ends. A distributed torque of intensity $t = t_A(L - x)/L$ is acting on the shaft, where $t_A = 200$ N·m/m. Determine the maximum shear stress in the shaft.

3.3 Torsion of Thin-Walled Tubes

Although torsion of noncircular shafts requires advanced methods of analysis, fairly simple approximate formulas are available for thin-walled tubes. Such members are common in construction where light weight is of paramount importance, such as automobile and aircraft structures.

Consider the thin-walled tube subjected to the torque T shown in Fig. 3.7(a). We assume the tube to be prismatic (constant cross section), but the wall thickness t is allowed to vary within the cross section. The surface that lies midway between the inner and outer boundaries of the tube is called the *middle surface*. If t is small compared to the overall dimensions of the cross section, the shear stress τ induced by torsion can be shown to be almost constant through the wall thickness of the tube and directed tangent to the middle surface, as illustrated in Fig. 3.7(b). At this time, it is convenient to introduce the concept of *shear flow* q, defined as the shear force per unit edge length of the middle surface. Thus, the shear flow is

$$q = \tau t \qquad (3.7)$$

If the shear stress is not constant through the wall thickness, then τ in Eq. (3.7) should be viewed as the average shear stress.

We now show that the *shear flow is constant throughout the tube*. This result can be obtained by considering equilibrium of the element shown in Fig. 3.7(c). In labeling the shear flows, we assume that q varies in the longitudinal (x) as well as the circumferential (s) directions. Thus, the terms $(\partial q/\partial x)\,dx$ and $(\partial q/\partial s)\,ds$ represent the changes in the shear flow over the distances dx and ds, respectively. The force acting on each side of the element is equal to the shear flow multiplied by the edge length, resulting in the equilibrium equations

$$\Sigma F_x = 0 \qquad \left(q + \frac{\partial q}{\partial s}\,ds\right)dx - q\,dx = 0$$

$$\Sigma F_s = 0 \qquad \left(q + \frac{\partial q}{\partial x}\,dx\right)ds - q\,ds = 0$$

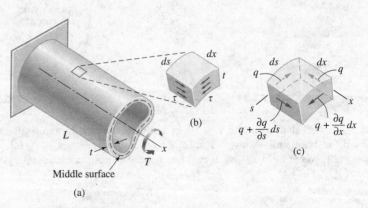

FIG. 3.7 (a) Thin-walled tube in torsion; (b) shear stress in the wall of the tube; (c) shear flows on wall element.

which yield $\partial q/\partial x = \partial q/\partial s = 0$, thereby proving that the shear flow is constant throughout the tube.

To relate the shear flow to the applied torque T, consider the cross section of the tube in Fig. 3.8. The shear force acting over the infinitesimal edge length ds of the middle surface is $dP = q\,ds$. The moment of this force about an arbitrary point O in the cross section is $r\,dP = (q\,ds)r$, where r is the perpendicular distance of O from the line of action of dP. Equilibrium requires that the sum of these moments must be equal to the applied torque T; that is,

$$T = \oint_S qr\,ds \qquad \text{(a)}$$

where the integral is taken over the closed curve formed by the intersection of the middle surface and the cross section, called the *median line*.

The integral in Eq. (a) need not be evaluated formally. Recalling that q is constant, we can take it outside the integral sign, so that Eq. (a) can be written as $T = q\oint_S r\,ds$. But from Fig. 3.8 we see that $r\,ds = 2\,dA_0$, where dA_0 is the area of the shaded triangle. Therefore, $\oint_S r\,ds = 2A_0$, where A_0 is the area of the cross section that is enclosed by the median line. Consequently, Eq. (a) becomes

$$T = 2A_0 q \qquad \text{(3.8a)}$$

from which the shear flow is

$$\boxed{q = \frac{T}{2A_0}} \qquad \text{(3.8b)}$$

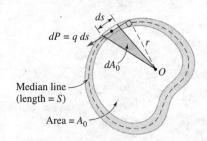

FIG. 3.8 Calculating the resultant of the shear flow acting on the cross section of the tube. Resultant is a couple equal to the internal torque T.

We can find the angle of twist of the tube by equating the work done by the shear stress in the tube to the work of the applied torque T. Let us start by determining the work done by the shear flow acting on the element in Fig. 3.7(c). The deformation of the element is shown in Fig. 3.9, where γ is the shear strain of the element. We see that work is done on the element by the shear force $dP = q\,ds$ as it moves through the distance $\gamma\,dx$. If we assume that γ is proportional to τ (Hooke's law), this work is

$$dU = \frac{1}{2}(\text{force} \times \text{distance}) = \frac{1}{2}(q\,ds)(\gamma\,dx)$$

Substituting $\gamma = \tau/G = q/Gt$ yields

$$dU = \frac{q^2}{2Gt}\,ds\,dx \qquad \text{(b)}$$

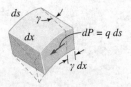

FIG. 3.9 Deformation of element caused by shear flow.

The work U of the shear flow for the entire tube is obtained by integrating Eq. (b) over the middle surface of the tube. Noting that q and G are constants and t is independent of x, we obtain

$$U = \frac{q^2}{2G}\int_0^L \left(\oint_S \frac{ds}{t}\right) dx = \frac{q^2 L}{2G} \oint_S \frac{ds}{t} \qquad \text{(c)}$$

Conservation of energy requires U to be equal to the work of the applied torque; that is, $U = T\theta/2$. After substituting the expression for q from Eq. (3.8b) into Eq. (c), we obtain

$$\left(\frac{T}{2A_0}\right)^2 \frac{L}{2G} \oint_S \frac{ds}{t} = \frac{1}{2} T\theta$$

from which the angle of twist of the tube is

$$\theta = \frac{TL}{4GA_0^2} \oint_S \frac{ds}{t} \qquad (3.9a)$$

If t is constant, we have $\oint_S (ds/t) = S/t$, where S is the length of the median line. Therefore, Eq. (3.9a) becomes

$$\theta = \frac{TLS}{4GA_0^2 t} \quad \text{(constant } t\text{)} \qquad (3.9b)$$

If the tube is not cylindrical, its cross sections do not remain plane but tend to warp. When the ends of the tube are attached to rigid plates or supports, the end sections cannot warp. As a result, the torsional stiffness of the tube is increased and the state of stress becomes more complicated—there are normal stresses in addition to the shear stress. However, if the tube is slender (length much greater than the cross-sectional dimensions), warping is confined to relatively small regions near the ends of the tube (Saint Venant's principle).

Tubes with very thin walls can fail by buckling (the walls "fold" like an accordion) while the stresses are still within their elastic ranges. For this reason, the use of very thin walls is not recommended. In general, the shear stress that results in buckling depends on the shape of the cross section and the material properties. For example, steel tubes of circular cross section require $r/t < 50$ to forestall buckling due to torsion.

Sharp re-entrant corners in the cross section of the tube should also be avoided because they cause stress concentration. It has been found that the shear stress at the inside boundary of a corner can be considerably higher than the average stress. The stress concentration effect diminishes as the radius a of the corner is increased, becoming negligible when $a/t > 2.5$, approximately.

Sample Problem 3.5

A steel tube with the cross section shown carries a torque T. The tube is 6 ft long and has a constant wall thickness of 3/8 in. (1) Compute the torsional stiffness $k = T/\theta$ of the tube. (2) If the tube is twisted through 0.5°, determine the shear stress in the wall of the tube. Use $G = 12 \times 10^6$ psi, and neglect stress concentrations at the corners.

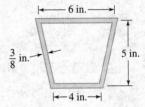

Solution

Part 1

Because the wall thickness is constant, the angle of twist is given by Eq. (3.9b):

$$\theta = \frac{TLS}{4GA_0^2 t}$$

Therefore, the torsional stiffness of the tube can be computed from

$$k = \frac{T}{\theta} = \frac{4GA_0^2 t}{LS}$$

The area enclosed by the median line is

$$A_0 = \text{average width} \times \text{height} = \left(\frac{6+4}{2}\right)(5) = 25 \text{ in.}^2$$

and the length of the median line is

$$S = 6 + 4 + 2\sqrt{1^2 + 5^2} = 20.20 \text{ in.}$$

Consequently, the torsional stiffness becomes

$$k = \frac{4(12 \times 10^6)(25)^2(3/8)}{(6 \times 12)(20.20)} = 7.735 \times 10^6 \text{ lb} \cdot \text{in./rad}$$

$$= 135.0 \times 10^3 \text{ lb} \cdot \text{in./deg} \qquad \text{Answer}$$

Part 2

The torque required to produce an angle of twist of 0.5° is

$$T = k\theta = (135.0 \times 10^3)(0.5) = 67.5 \times 10^3 \text{ lb} \cdot \text{in.}$$

which results in the shear flow

$$q = \frac{T}{2A_0} = \frac{67.5 \times 10^3}{2(25)} = 1350 \text{ lb/in.}$$

The corresponding shear stress is

$$\tau = \frac{q}{t} = \frac{1350}{3/8} = 3600 \text{ psi} \qquad \text{Answer}$$

Sample Problem 3.6

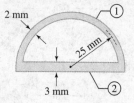

An aluminum tube, 1.2 m long, has the semicircular cross section shown in the figure. If stress concentrations at the corners are neglected, determine (1) the torque that causes a maximum shear stress of 40 MPa, and (2) the corresponding angle of twist of the tube. Use $G = 28$ GPa for aluminum.

Solution

Part 1

Because the shear flow is constant in a prismatic tube, the maximum shear stress occurs in the thinnest part of the wall, which is the semicircular portion with $t = 2$ mm. Therefore, the shear flow that causes a maximum shear stress of 40 MPa is

$$q = \tau t = (40 \times 10^6)(0.002) = 80 \times 10^3 \text{ N/m}$$

The cross-sectional area enclosed by the median line is

$$A_0 = \frac{\pi r^2}{2} = \frac{\pi (0.025)^2}{2} = 0.9817 \times 10^{-3} \text{ m}^2$$

which results in the torque—see Eq. (3.8a):

$$T = 2A_0 q = 2(0.9817 \times 10^{-3})(80 \times 10^3) = 157.07 \text{ N} \cdot \text{m} \qquad \textit{Answer}$$

Part 2

The cross section consists of two parts, labeled ① and ② in the figure, each having a constant thickness. Hence, we can write

$$\oint_S \frac{ds}{t} = \frac{1}{t_1} \int_{S_1} ds + \frac{1}{t_2} \int_{S_2} ds = \frac{S_1}{t_1} + \frac{S_2}{t_2}$$

where S_1 and S_2 are the lengths of the median lines of parts ① and ②, respectively. Therefore,

$$\oint_S \frac{ds}{t} = \frac{\pi r}{t_1} + \frac{2r}{t_2} = \frac{\pi(25)}{2} + \frac{2(25)}{3} = 55.94$$

and Eq. (3.9a) yields for the angle of twist

$$\theta = \frac{TL}{4GA_0^2} \oint_S \frac{ds}{t} = \frac{157.07(1.2)}{4(28 \times 10^9)(0.9817 \times 10^{-3})^2}(55.94)$$

$$= 0.0977 \text{ rad} = 5.60° \qquad \textit{Answer}$$

Problems

Neglect stress concentrations at the corners of the tubes in the following problems.

3.36 Consider a thin cylindrical tube of mean radius $\bar{r}$, constant thickness t, and length L. (a) Show that the polar moment of inertia of the cross-sectional area can be approximated by $J = 2\pi\bar{r}^3 t$. (b) Use this approximation to show that Eqs. (3.8b) and (3.9b) are equivalent to $\tau = T\bar{r}/J$ and $\theta = TL/(GJ)$, respectively.

3.37 A cylindrical metal tube of mean radius $r = 6$ in., length $L = 15$ ft, and shear modulus $G = 11 \times 10^6$ psi carries the torque $T = 340$ kip·in. Determine the smallest allowable constant wall thickness t if the shear stress is limited to 12 ksi and the angle of twist is not to exceed $2°$.

3.38 A cylindrical tube of constant wall thickness t and inside radius $r = 10t$ carries a torque T. Find the expression for the maximum shear stress in the tube using (a) the torsion formula for a hollow shaft in Eq. (3.5d); and (b) the thin-walled tube formula in Eq. (3.8b). What is the percentage error in the thin-walled tube approximation?

3.39 A torque of 600 N·m is applied to a tube with the rectangular cross section shown in the figure. Determine the smallest allowable constant wall thickness t if the shear stress is not to exceed 80 MPa.

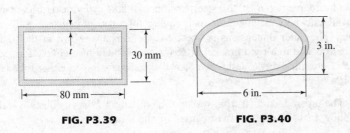

FIG. P3.39   FIG. P3.40

3.40 The constant wall thickness of a tube with the elliptical cross section shown is 0.10 in. What torque will cause a shear stress of 8000 psi?

3.41 The constant wall thickness of a steel tube with the cross section shown is 2 mm. If a 600-N·m torque is applied to the tube, find (a) the shear stress in the wall of the tube; and (b) the angle of twist per meter of length. Use $G = 80$ GPa for steel.

3.42 Two identical metal sheets are formed into tubes with the circular and square cross sections shown. If the same torque is applied to each tube, determine the ratios (a) $\tau_{\text{circle}}/\tau_{\text{square}}$ of the shear stresses; and (b) $\theta_{\text{circle}}/\theta_{\text{square}}$ of the angles of twist.

3.43 A steel tube with the cross section shown carries a 50-kip·in. torque. Determine (a) the maximum shear stress in the tube; and (b) the angle of twist per foot of length. Use $G = 11 \times 10^6$ psi for steel.

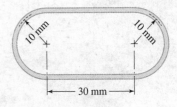

FIG. P3.41

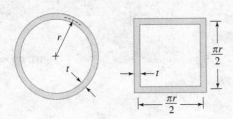

FIG. P3.42

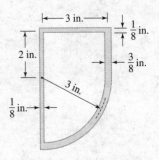

FIG. P3.43

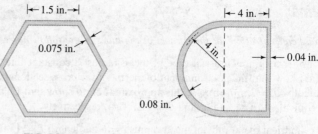

FIG. P3.44 **FIG. P3.45, P3.46**

3.44 An aluminum tube with the hexagonal cross section shown is 2 ft long and has a constant wall thickness of 0.075 in. Find (a) the largest torque that the tube can carry if the shear stress is limited to 8000 psi; and (b) the angle of twist caused by this torque. Use $G = 4 \times 10^6$ psi for aluminum.

3.45 A 4-ft-long tube with the cross section shown in the figure is made of aluminum. Find the torque that will cause a maximum shear stress of 10 000 psi. Use $G = 4 \times 10^6$ psi for aluminum.

3.46 A 3-ft-long tube with the cross section shown is made of aluminum. Determine the angle of twist if a 50-kip·in. torque is applied to the tube. For aluminum, use $G = 4 \times 10^6$ psi.

3.47 The segment AB of the steel torsion bar is a cylindrical tube of constant 2-mm wall thickness. Segment BC is a square tube with a constant wall thickness of 3 mm. The outer dimensions of the cross sections are shown in the figure. The tubes are attached to a rigid bracket at B, which is loaded by a couple formed by the forces P. Determine the largest value of P if the shear stress in either tube is limited to 60 MPa.

3.48 The tapered, circular, thin-walled tube of length L has a constant wall thickness t. Show that the angle of twist caused by the torque T is

$$\theta = \frac{20}{9\pi} \frac{TL}{Gtd_A^3}$$

(*Hint*: Apply Eq. (3.9b) to an infinitesimal length dx of the shaft.)

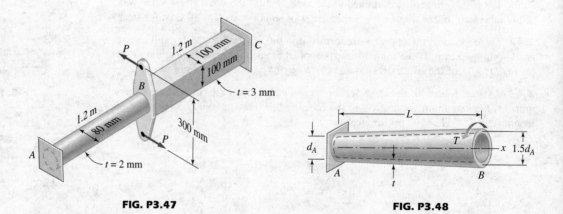

FIG. P3.47 **FIG. P3.48**

Review Problems

3.49 A solid steel shaft 5 m long is stressed to 80 MPa when twisted through 4°. (a) Given that $G = 83$ GPa, find the diameter of the shaft. (b) What power does this shaft transmit when running at 20 Hz?

3.50 Determine the maximum torque that can be applied to a hollow circular steel shaft of 100-mm outer diameter and 80-mm inner diameter. The shear stress is limited to 60 MPa, and the angle of twist must not exceed 0.5° in a length of 1 m. Use $G = 83$ GPa for steel.

3.51 A 2-in.-diameter steel shaft rotates at 240 rev/min. If the shear stress is limited to 12 ksi, determine the maximum horsepower that can be transmitted at that speed.

3.52 The compound shaft, consisting of steel and aluminum segments, carries the two torques shown in the figure. Determine the maximum permissible value of T subject to the following design conditions: $\tau_{st} \leq 83$ MPa, $\tau_{al} \leq 55$ MPa, and $\theta \leq 6°$ (θ is the angle of rotation of the free end). Use $G = 83$ GPa for steel and $G = 28$ GPa for aluminum.

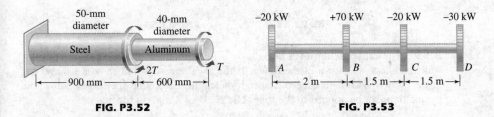

FIG. P3.52   **FIG. P3.53**

3.53 The four gears are attached to a steel shaft that is rotating at 2 Hz. Gear B supplies 70 kW of power to the shaft. Of that power, 20 kW are used by gear A, 20 kW by gear C, and 30 kW by gear D. (a) Find the uniform shaft diameter if the shear stress in the shaft is not to exceed 60 MPa. (b) If a uniform shaft diameter of 100 mm is specified, determine the angle by which one end of the shaft lags behind the other end. Use $G = 83$ GPa for steel.

3.54 The composite shaft consists of a copper rod that fits loosely inside an aluminum sleeve. The two components are attached to a rigid wall at one end and joined with an end-plate at the other end. Determine the maximum shear stress in each material when the 2-kN·m torque is applied to the end-plate. Use $G = 26$ GPa for aluminum and $G = 47$ GPa for copper.

3.55 The torque T is applied to the solid shaft with built-in ends. (a) Show that the reactive torques at the walls are $T_A = Tb/L$ and $T_C = Ta/L$. (b) How would the results of Part (a) change if the shaft were hollow?

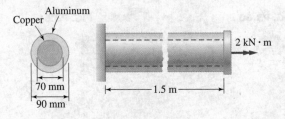

FIG. P3.54

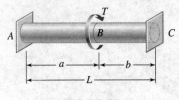

FIG. P3.55

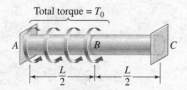

FIG. P3.57

3.56 A flexible shaft consists of a 0.20-in.-diameter steel rod encased in a stationary tube that fits closely enough to impose a torque of intensity 0.50 lb·in./in. on the rod. (a) Determine the maximum length of the shaft if the shear stress in the rod is not to exceed 20 ksi. (b) What will be the relative angular rotation between the ends of the rod? Use $G = 12 \times 10^6$ psi for steel.

3.57 The shaft ABC is attached to rigid walls at A and C. The torque T_0 is distributed uniformly over segment AB of the shaft. Determine the reactions at A and C.

3.58 A torque of 450 lb·ft is applied to the square tube with constant 0.10-in. wall thickness. Determine the smallest permissible dimension a if the shear stress is limited to 6000 psi.

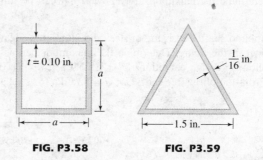

FIG. P3.58   **FIG. P3.59**

3.59 The cross section of a brass tube is an equilateral triangle with a constant wall thickness, as shown in the figure. If the shear stress is limited to 8 ksi and the angle of twist is not to exceed 2° per foot length, determine the largest allowable torque that can be applied to the tube. Use $G = 5.7 \times 10^6$ psi for brass.

3.60 A torsion member is made by placing a circular tube inside a square tube, as shown, and joining their ends by rigid end-plates. The tubes are made of the same material and have the same constant wall thickness $t = 5$ mm. If a torque T is applied to the member, what fraction of T is carried by each component?

3.61 A 3-m aluminum tube with the cross section shown carries a 200-N·m torque. Determine (a) the maximum shear stress in the tube; and (b) the relative angle of rotation of the ends of the tube. For aluminum, use $G = 28$ GPa.

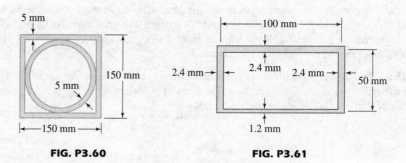

FIG. P3.60   **FIG. P3.61**

Computer Problems

C3.1 An aluminum bar of circular cross section and the profile specified in Prob. C2.1 is subjected to a 15-N·m torque. Use numerical integration to compute the angle of twist of the bar. For aluminum, use $G = 30$ GPa.

C3.2 A steel bar of circular cross section has the profile shown in the figure. Use numerical integration to compute the torsional stiffness $k = T/\theta$ of the bar. For steel, use $G = 12 \times 10^6$ psi.

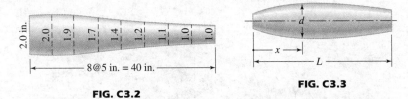

FIG. C3.2

FIG. C3.3

C3.3 The diameter d of the solid shaft of length L varies with the axial coordinate x. Given L and $d(x)$, write an algorithm to calculate the constant diameter D of a shaft that would have the same torsional stiffness (assume that the two shafts have the same length and are made of the same material). Use (a) $L = 500$ mm and

$$d = (25 \text{ mm})\left(1 + 3.8\frac{x}{L} - 3.6\frac{x^2}{L^2}\right)$$

and (b) $L = 650$ mm and

$$d = \begin{cases} 20 \text{ mm} & \text{if } x \leq 200 \text{ mm} \\ 20 \text{ mm} + \dfrac{x - 200 \text{ mm}}{10} & \text{if } 200 \text{ mm} \leq x \leq 350 \text{ mm} \\ 35 \text{ mm} & \text{if } x \geq 350 \text{ mm} \end{cases}$$

C3.4 The solid shaft ABC of length L and variable diameter d is attached to rigid supports at A and C. A torque T acts at the distance b from end A. Given L, b, and $d(x)$, write an algorithm to compute the fraction of T that is carried by segments AB and BC. Use (a) $L = 200$ mm, $b = 110$ mm, and

$$d = 30 \text{ mm} - (20 \text{ mm})\sin\frac{\pi x}{L}$$

and (b) $L = 400$ mm, $b = 275$ mm, and

$$d = \begin{cases} 25 \text{ mm} & \text{if } x \leq 200 \text{ mm} \\ 25 \text{ mm} + \dfrac{(x - 200 \text{ mm})^2}{250 \text{ mm}} & \text{if } 200 \text{ mm} \leq x \leq 250 \text{ mm} \\ 35 \text{ mm} & \text{if } x \geq 250 \text{ mm} \end{cases}$$

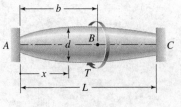

FIG. C3.4

C3.5 An extruded tube of length L has the cross section shown in the figure. The radius of the median line is $r = 75$ mm, and the wall thickness varies with the angle α as

$$t = t_1 + (t_2 - t_1) \sin \frac{\alpha}{2}$$

Given L, r, t_1, t_2, and G, write an algorithm to compute the angle of twist required to produce the maximum shear stress τ_{max}. Use $L = 1.8$ m, $r = 75$ mm, $t_1 = 2$ mm, $t_2 = 4$ mm, $G = 40$ GPa (brass), and $\tau_{max} = 110$ MPa.

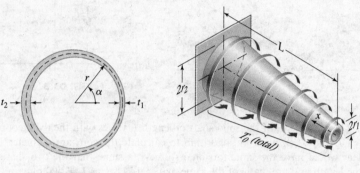

FIG. C3.5 **FIG. C3.6**

C3.6 The thin-walled tube in the shape of a truncated cone carries a torque T_0 that is uniformly distributed over its length L. The radius of the median line varies linearly from r_1 to r_2 over the length of the tube. The wall thickness t is constant. Given L, r_1, r_2, t, T_0, and G, construct an algorithm that (a) plots the shear stress in the tube as a function of the axial distance x; and (b) computes the angle of rotation at the free end of the tube. Use $L = 10$ ft, $r_1 = 3$ in., $r_2 = 12$ in., $t = 0.2$ in., $T_0 = 60$ kip·ft, and $G = 12 \times 10^6$ psi (steel).

4 Shear and Moment in Beams

Roof beams at a stadium under construction. Beams carry loads mainly by a combination of shearing and bending. Determination of the shear forces and bending moments is the first step in beam analysis.

4.1 Introduction

The term *beam* refers to a slender bar that carries transverse loading; that is, the applied forces are perpendicular to the bar. In a beam, the internal force system consists of a shear force and a bending moment acting on the cross section of the bar. As we have seen in previous chapters, axial and torsional loads often result in internal forces that are constant in the bar, or over portions of the bar. The study of beams, however, is complicated by the fact that the shear force and the bending moment usually vary continuously along the length of the beam.

The internal forces give rise to two kinds of stresses on a transverse section of a beam: (1) normal stress that is caused by the bending moment and (2) shear stress due to the shear force. This chapter is concerned only with the variation of the shear force and the bending moment under various combinations of loads and types of supports. Knowing the distribution of the shear force and the bending moment in a beam is essential for the computation of stresses and deformations, which will be investigated in subsequent chapters.

4.2 Supports and Loads

Beams are classified according to their supports. A *simply supported beam*, shown in Fig. 4.1(a), has a pin support at one end and a roller support at the other end. The pin support prevents displacement of the end of the beam, but not its rotation. The term *roller support* refers to a pin connection that is free to move parallel to the axis of the beam; hence, this type of support suppresses only the transverse displacement. A *cantilever beam* is built into a rigid support at one end, with the other end being free, as shown in Fig. 4.1(b). The built-in support prevents displacements as well as rotations of the end of the beam. An *overhanging beam*, illustrated in Fig. 4.1(c), is supported by a pin and a roller support, with one or both ends of the beam extending beyond the supports. The three types of beams are statically determinate because the support reactions can be found from the equilibrium equations.

A *concentrated load*, such as P in Fig. 4.1(a), is an approximation of a force that acts over a very small area. In contrast, a *distributed load* is applied over a finite area. If the distributed load acts on a very narrow area, the load may be approximated by a *line load*. The intensity w of this loading is expressed as force per unit length (lb/ft, N/m, etc.). The load distribution may be uniform, as shown in Fig. 4.1(b), or it may vary with distance along the beam, as in Fig. 4.1(c). The weight of the beam is an example of distributed loading, but its magnitude is usually small compared to the loads applied to the beam.

Figure 4.2 shows other types of beams. These beams are *oversupported* in the sense that each beam has at least one more reaction than is necessary for support. Such beams are statically indeterminate; the presence of these *redundant supports* requires the use of additional equations obtained by considering the deformation of the beam. The analysis of statically indeterminate beams will be discussed in Chapter 7.

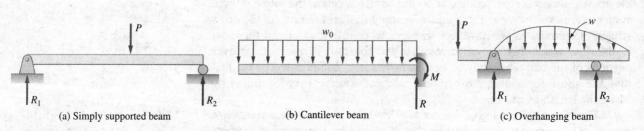

FIG. 4.1 Statically determinate beams.

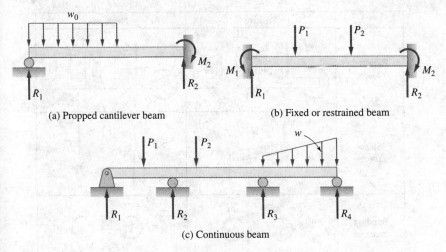

FIG. 4.2 Statically indeterminate beams.

4.3 Shear-Moment Equations and Shear-Moment Diagrams

The determination of the internal force system acting at a *given* section of a beam is straightforward: We draw a free-body diagram that exposes these forces and then compute the forces using equilibrium equations. However, the goal of beam analysis is more involved—we want to determine the shear force V and the bending moment M at *every* cross section of the beam. To accomplish this task, we must derive the expressions for V and M in terms of the distance x measured along the beam. By plotting these expressions to scale, we obtain the *shear force and bending moment diagrams* for the beam. The shear force and bending moment diagrams are convenient visual references to the internal forces in a beam; in particular, they identify the maximum values of V and M.

a. Sign conventions

For consistency, it is necessary to adopt sign conventions for applied loading, shear forces, and bending moments. We will use the conventions shown in Fig. 4.3, which assume the following to be *positive*:

- External forces that are directed downward; external couples that are directed clockwise.
- Shear forces that tend to rotate a beam element clockwise.
- Bending moments that tend to bend a beam element concave upward (the beam "smiles").

The main disadvantage of the above conventions is that they rely on such adjectives as "downward," "clockwise," and so on. To eliminate this obstacle, a convention based upon a Cartesian coordinate system is sometimes used.[1]

---

[1] See, for example, Andrew Pytel and Jaan Kiusalaas, *Statics*, 2d ed. (Pacific Grove, CA: Brooks/Cole, 1991), p. 257.

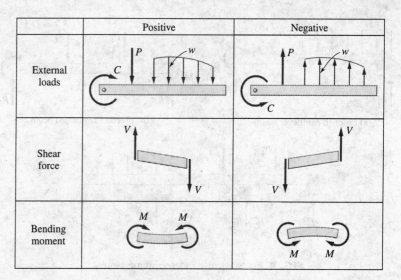

FIG. 4.3 Sign conventions for external loads, shear force, and bending moment.

b. Procedure for determining shear force and bending moment diagrams

The following is a general procedure for obtaining shear force and bending moment diagrams of a statically determinate beam:

- Compute the support reactions from the free-body diagram (FBD) of the entire beam.
- Divide the beam into segments so that the loading within each segment is continuous. Thus, the end-points of the segments are discontinuities of loading, including concentrated loads and couples.

Perform the following steps for each segment of the beam:

- Introduce an imaginary cutting plane within the segment, located at a distance x from the left end of the beam, that cuts the beam into two parts.
- Draw a FBD for the part of the beam lying either to the left or to the right of the cutting plane, whichever is more convenient. At the cut section, show V and M acting in their positive directions.
- Determine the expressions for V and M from the equilibrium equations obtainable from the FBD. These expressions, which are usually functions of x, are the shear force and bending moment equations for the segment.
- Plot the expressions for V and M for the segment. It is visually desirable to draw the V-diagram below the FBD of the entire beam, and then draw the M-diagram below the V-diagram.

The bending moment and shear force diagrams of the beam are composites of the V- and M-diagrams of the segments. These diagrams are usually discontinuous, or have discontinuous slopes, at the end-points of the segments due to discontinuities in loading.

Sample Problem 4.1

The simply supported beam in Fig. (a) carries two concentrated loads. (1) Derive the expressions for the shear force and the bending moment for each segment of the beam. (2) Sketch the shear force and bending moment diagrams. Neglect the weight of the beam. Note that the support reactions at A and D have been computed and are shown in Fig. (a).

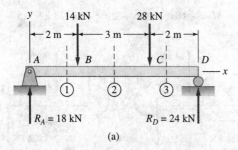

(a)

Solution

Part 1

The determination of the expressions for V and M for each of the three beam segments (AB, BC, and CD) is explained below.

Segment AB ($0 < x < 2$ m) Figure (b) shows the FBDs for the two parts of the beam that are separated by section ①, located within segment AB. Note that we show V and M acting in their positive directions according to the sign conventions in Fig. 4.3. Because V and M are equal in magnitude and oppositely directed on the two FBDs, they can be computed using either FBD. The analysis of the FBD of the part to the left of section ① yields

$$\Sigma F_y = 0 \quad +\uparrow \quad 18 - V = 0$$
$$V = +18 \text{ kN} \qquad \text{Answer}$$
$$\Sigma M_E = 0 \quad +\circlearrowleft \quad -18x + M = 0$$
$$M = +18x \text{ kN} \cdot \text{m} \qquad \text{Answer}$$

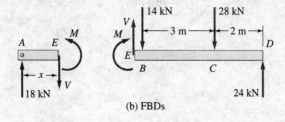

(b) FBDs

103

Segment BC (2 m < x < 5 m) Figure (c) shows the FBDs for the two parts of the beam that are separated by section ②, an arbitrary section within segment *BC*. Once again, *V* and *M* are assumed to be positive according to the sign conventions in Fig. 4.3. The analysis of the part to the left of section ② gives

$$\Sigma F_y = 0 \quad +\uparrow \quad 18 - 14 - V = 0$$

$$V = +18 - 14 = +4 \text{ kN} \qquad \text{Answer}$$

$$\Sigma M_F = 0 \quad +\circlearrowleft \quad -18x + 14(x-2) + M = 0$$

$$M = +18x - 14(x-2) = 4x + 28 \text{ kN} \cdot \text{m} \qquad \text{Answer}$$

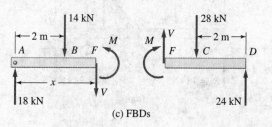

(c) FBDs

Segment CD (5 m < x < 7 m) Section ③ is used to find the shear force and bending moment in segment *CD*. The FBDs in Fig. (d) again show *V* and *M* acting in their positive directions. Analyzing the portion of the beam to the left of section ③, we obtain

$$\Sigma F_y = 0 \quad +\uparrow \quad 18 - 14 - 28 - V = 0$$

$$V = +18 - 14 - 28 = -24 \text{ kN} \qquad \text{Answer}$$

$$\Sigma M_G = 0 \quad +\circlearrowleft \quad -18x + 14(x-2) + 28(x-5) + M = 0$$

$$M = +18x - 14(x-2) - 28(x-5) = -24x + 168 \text{ kN} \cdot \text{m} \qquad \text{Answer}$$

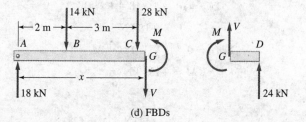

(d) FBDs

Part 2

The shear force and bending moment diagrams in Figs. (f) and (g) are the plots of the expressions for V and M derived in Part 1. By placing these plots directly below the sketch of the beam in Fig. (e), we establish a clear visual relationship between the diagrams and locations on the beam.

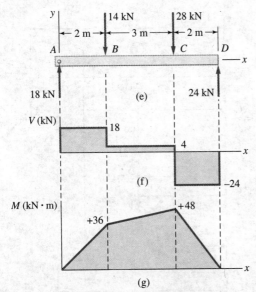

Shear force and bending moment diagrams

An inspection of the V-diagram reveals that the largest shear force in the beam is -24 kN and that it occurs at every cross section of the beam in segment CD. From the M-diagram we see that the maximum bending moment is $+48$ kN·m, which occurs under the 28-kN load at C. Note that at each concentrated force the V-diagram "jumps" by an amount equal to the force. Furthermore, there is a discontinuity in the slope of the M-diagram at each concentrated force.

Sample Problem 4.2

The simply supported beam in Fig. (a) is loaded by the clockwise couple C_0 at B. (1) Derive the shear force and bending moment equations, and (2) draw the shear force and bending moment diagrams. Neglect the weight of the beam. The support reactions A and C have been computed, and their values are shown in Fig. (a).

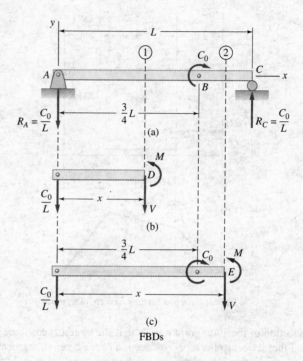

(c) FBDs

Solution

Part 1

Due to the presence of the couple C_0, we must analyze segments AB and BC separately.

Segment AB ($0 < x < 3L/4$) Figure (b) shows the FBD of the part of the beam to the left of section ① (we could also use the part to the right). Note that V and M are assumed to act in their positive directions according to the sign conventions in Fig. 4.3. The equilibrium equations for this portion of the beam yield

$$\Sigma F_y = 0 \quad +\uparrow \quad -\frac{C_0}{L} - V = 0$$

$$V = -\frac{C_0}{L} \quad \quad \text{Answer}$$

$$\Sigma M_D = 0 \quad +\circlearrowleft \quad \frac{C_0}{L}x + M = 0$$

$$M = -\frac{C_0}{L}x \quad \quad \text{Answer}$$

Segment BC ($3L/4 < x < L$) Figure (c) shows the FBD of the portion of the beam to the left of section ② (the right portion could also be used). Once again, V and M are assumed to act in their positive directions. Applying the equilibrium equations to the beam segment, we obtain

106

$$\Sigma F_y = 0 \quad +\uparrow \quad -\frac{C_0}{L} - V = 0$$

$$V = -\frac{C_0}{L} \qquad \text{Answer}$$

$$\Sigma M_E = 0 \quad +\circlearrowleft \quad \frac{C_0}{L}x - C_0 + M = 0$$

$$M = -\frac{C_0}{L}x + C_0 \qquad \text{Answer}$$

Part 2

The shear force and bending moment diagrams shown in Figs. (d) and (e) are obtained by plotting the expressions for V and M found in Part 1. From the V-diagram, we see that the shear force is the same for all cross sections of the beam. The M-diagram shows a jump of magnitude C_0 at the point of application of the couple.

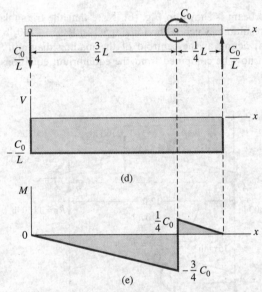

Shear force and bending moment diagrams

Sample Problem 4.3

The cantilever beam in Fig. (a) carries a triangular load, the intensity of which varies from zero at the left end to 360 lb/ft at the right end. In addition, a 1000-lb upward vertical load acts at the free end of the beam. (1) Derive the shear force and bending moment equations, and (2) draw the shear force and bending moment diagrams. Neglect the weight of the beam.

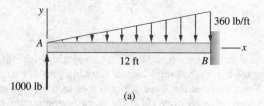

(a)

Solution

The FBD of the beam is shown in Fig. (b). Note that the triangular load has been replaced by its resultant, which is the force $0.5(12)(360) = 2160$ lb (area under the loading diagram) acting at the centroid of the loading diagram. The support reactions at B can now be computed from the equilibrium equations; the results are shown in Fig. (b).

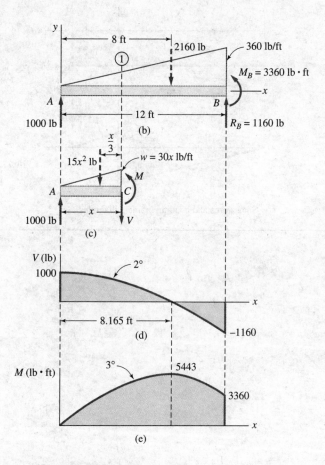

Because the loading is continuous, the beam does not have to be divided into segments. Therefore, only one expression for V and one expression for M apply to the entire beam.

Part 1

Figure (c) shows the FBD of the part of the beam that lies to the left of section ①. Letting w be the intensity of the loading at section ①, as shown in Fig. (b), we have from similar triangles, $w/x = 360/12$, or $w = 30x$ lb/ft. Now the triangular load in Fig. (c) can be replaced by its resultant force $15x^2$ lb acting at the centroid of the loading diagram, which is located at $x/3$ ft from section ①. The shear force V and bending moment M acting at section ① are shown acting in their positive directions according to the sign conventions in Fig. 4.3. Equilibrium analysis of the FBD in Fig. (c) yields

$$\Sigma F_y = 0 \quad +\uparrow \quad 1000 - 15x^2 - V = 0$$

$$V = 1000 - 15x^2 \text{ lb} \qquad \text{Answer}$$

$$\Sigma M_C = 0 \quad +\circlearrowleft \quad -1000x + 15x^2\left(\frac{x}{3}\right) + M = 0$$

$$M = 1000x - 5x^3 \text{ lb} \cdot \text{ft} \qquad \text{Answer}$$

Part 2

Plotting the expressions for V and M found in Part 1 gives the shear force and bending moment diagrams shown in Figs. (d) and (e). Observe that the shear force diagram is a parabola and the bending moment diagram is a third-degree polynomial in x.

The location of the section where the shear force is zero is found from

$$V = 1000 - 15x^2 = 0$$

which gives

$$x = 8.165 \text{ ft}$$

The maximum bending moment occurs where the slope of the M-diagram is zero—that is, where $dM/dx = 0$. Differentiating the expression for M, we obtain

$$\frac{dM}{dx} = 1000 - 15x^2 = 0$$

which again yields $x = 8.165$ ft. (In the next article, we will show that the slope of the bending moment is always zero at a section where the shear force vanishes.) Substituting this value of x into the expression for M, we find that the maximum bending moment is

$$M_{\max} = 1000(8.165) - 5(8.165)^3 = 5443 \text{ lb} \cdot \text{ft}$$

Problems

4.1–4.18 For the beam shown, derive the expressions for V and M, and draw the shear force and bending moment diagrams. Neglect the weight of the beam.

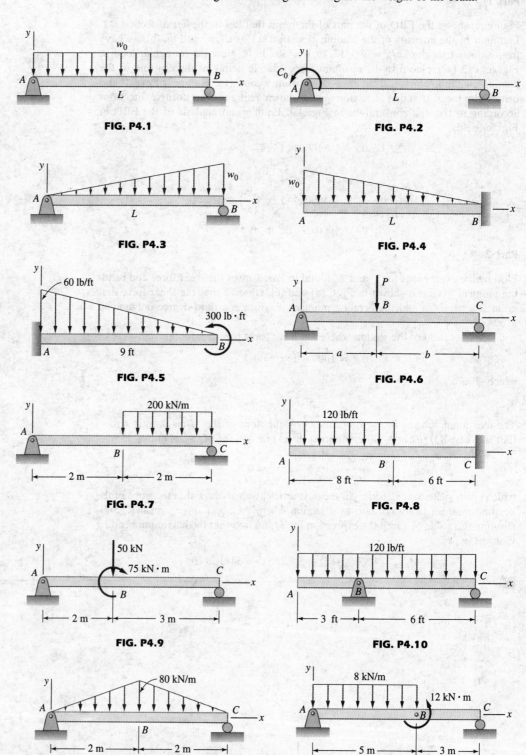

4.4 Area Method for Drawing Shear-Moment Diagrams   **111**

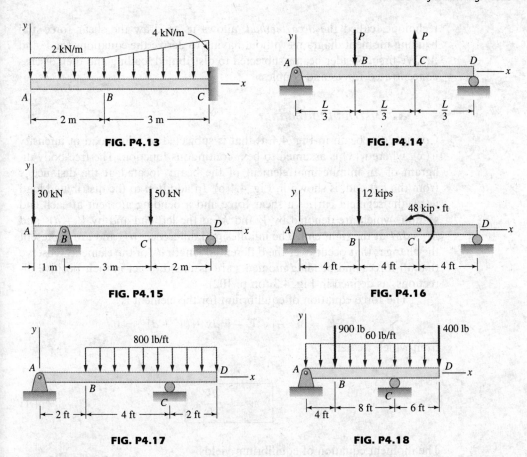

FIG. P4.13

FIG. P4.14

FIG. P4.15

FIG. P4.16

FIG. P4.17

FIG. P4.18

4.19–4.20 Derive the shear force and the bending moment as functions of the angle θ for the arch shown. Neglect the weight of the arch.

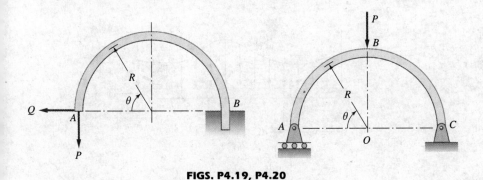

FIGS. P4.19, P4.20

4.4 Area Method for Drawing Shear-Moment Diagrams

Useful relationships between the loading, shear force, and bending moment can be derived from the equilibrium equations. These relationships enable us to plot the shear force diagram directly from the load diagram, and then construct the bending moment diagram from the shear force diagram. This

a. Distributed loading

Consider the beam in Fig. 4.4(a) that is subjected to a line load of intensity $w(x)$, where $w(x)$ is assumed to be a continuous function. The free-body diagram of an infinitesimal element of the beam, located at the distance x from the left end, is shown in Fig. 4.4(b). In addition to the distributed load $w(x)$, the segment carries a shear force and a bending moment at each end section, which are denoted by V and M at the left end and by $V + dV$ and $M + dM$ at the right end. The infinitesimal differences dV and dM represent the changes that occur over the differential length dx of the element. Observe that all forces and bending moments are assumed to act in their positive directions, as defined in Fig. 4.3 (on p. 102).

The force equation of equilibrium for the element is

$$\Sigma F_y = 0 \quad +\uparrow \quad V - w\,dx - (V + dV) = 0$$

from which we get

$$\boxed{w = -\frac{dV}{dx}} \tag{4.1}$$

The moment equation of equilibrium yields

$$\Sigma M_O = 0 \quad +\circlearrowleft \quad -M - V\,dx + (M + dM) + w\,dx\,\frac{dx}{2} = 0$$

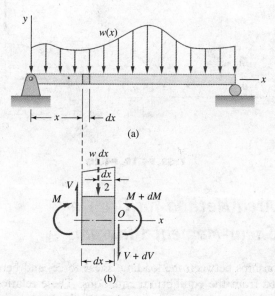

FIG. 4.4 (a) Simply supported beam carrying distributed loading; (b) free-body diagram of an infinitesimal beam segment.

After canceling M and dividing by dx, we get

$$-V + \frac{dM}{dx} + \frac{w\,dx}{2} = 0$$

Because dx is infinitesimal, the last term can be dropped (this is not an approximation), yielding

$$V = \frac{dM}{dx} \quad (4.2)$$

Equations (4.1) and (4.2) are called the *differential equations of equilibrium* for beams. The following five theorems relating the load, the shear force, and the bending moment diagrams follow from these equations.

1. The load intensity at any section of a beam is equal to the negative of the slope of the shear force diagram at the section.
   Proof—follows directly from Eq. (4.1).
2. The shear force at any section is equal to the slope of the bending moment diagram at that section.
   Proof—follows directly from Eq. (4.2).
3. The difference between the shear forces at two sections of a beam is equal to the negative of the area under the load diagram between those two sections.
   Proof—integrating Eq. (4.1) between sections A and B in Fig. 4.5, we obtain

$$\int_{x_A}^{x_B} \frac{dV}{dx}\,dx = V_B - V_A = -\int_{x_A}^{x_B} w\,dx$$

Recognizing that the integral on the right-hand side of this equation represents the area under the load diagram between A and B, we get

$$V_B - V_A = -\text{area of } w\text{-diagram}\big]_A^B \quad \text{Q.E.D.}$$

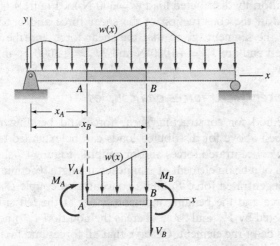

FIG. 4.5 (a) Simply supported beam carrying distributed loading; (b) free-body diagram of a finite beam segment.

For computational purposes, a more convenient form of this equation is

$$V_B = V_A - \text{area of } w\text{-diagram}\big]_A^B \qquad (4.3)$$

Note that the signs in Eq. (4.3) are correct only if $x_B > x_A$.

4. The difference between the bending moments at two sections of a beam is equal to the area of the shear force diagram between these two sections.

   Proof—integrating Eq. (4.2) between sections A and B (see Fig. 4.5), we have

   $$\int_{x_A}^{x_B} \frac{dM}{dx} dx = M_B - M_A = \int_{x_A}^{x_B} V\, dx$$

   Because the right-hand side of this equation is the area of the shear force diagram between A and B, we obtain

   $$M_B - M_A = \text{area of } V\text{-diagram}\big]_A^B \qquad \text{Q.E.D.}$$

   We find it convenient to use this equation in the form

   $$M_B = M_A + \text{area of } V\text{-diagram}\big]_A^B \qquad (4.4)$$

   The signs in Eq. (4.4) are correct only if $x_B > x_A$.

5. If the load diagram is a polynomial of degree n, then the shear force diagram is a polynomial of degree $(n+1)$, and the bending moment diagram is a polynomial of degree $(n+2)$.

   Proof—follows directly from the integration of Eqs. (4.1) and (4.2).

The area method for drawing shear force and bending moment diagrams is a direct application of the foregoing theorems. For example, consider the beam segment shown in Fig. 4.6(a), which is 2 m long and is subjected to a uniformly distributed load $w = 300$ N/m. Figure 4.6(b) shows the steps required in the construction of the shear force and bending moment diagrams for the segment, given that the shear force and the bending moment at the left end are $V_A = +1000$ N and $M_A = +3000$ N·m.

b. Concentrated forces and couples

The area method for constructing shear force and bending moment diagrams described above for distributed loads can be extended to beams that are loaded by concentrated forces and/or couples. Figure 4.7 shows the free-body diagram of a beam element of infinitesimal length dx containing a point A where a concentrated force P_A and a concentrated couple C_A are applied. The shear force and the bending moment acting at the left side of the element are denoted by V_A^- and M_A^-, whereas the notation V_A^+ and M_A^+ is used for the right side of the element. Observe that all forces and moments in Fig. 4.7 are assumed to be positive according to the sign conventions in Fig. 4.3.

The force equilibrium equation gives

$$\Sigma F_y = 0 \quad +\uparrow \quad V_A^- - P_A - V_A^+ = 0$$

4.4 Area Method for Drawing Shear-Moment Diagrams

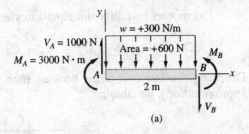

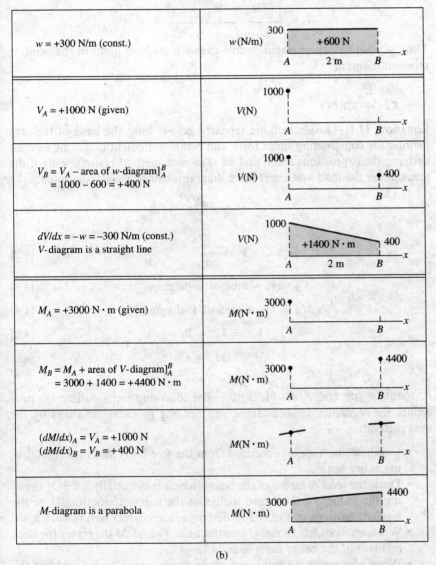

FIG. 4.6 (a) Free-body diagram of a beam segment carrying uniform loading; (b) constructing shear force and bending moment diagrams for the beam segment.

$$V_A^+ = V_A^- - P_A \qquad (4.5)$$

Equation (4.5) indicates that a positive concentrated force causes a negative jump discontinuity in the shear force diagram at A (a concentrated couple does not affect the shear force diagram).

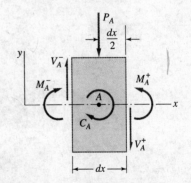

FIG. 4.7 Free-body diagram of an infinitesimal beam element carrying a concentrated force P_A and a concentrated couple C_A.

The moment equilibrium equation yields

$$\Sigma M_A = 0 \quad +\circlearrowleft \quad M_A^+ - M_A^- - C_A - V_A^+ \frac{dx}{2} - V_A^- \frac{dx}{2} = 0$$

Dropping the last two terms because they are infinitesimal (this is not an approximation), we obtain

$$M_A^+ = M_A^- + C_A \quad (4.6)$$

Thus, a positive concentrated couple causes a positive jump in the bending moment diagram.

c. Summary

Equations (4.1)–(4.6), which are repeated below, form the basis of the area method for constructing shear force and bending moment diagrams without deriving the expressions for V and M. The area method is useful only if the areas under the load and shear force diagrams can be easily computed.

$$w = -\frac{dV}{dx} \quad (4.1)$$

$$V = \frac{dM}{dx} \quad (4.2)$$

$$V_B = V_A - \text{area of } w\text{-diagram}]_A^B \quad (4.3)$$

$$M_B = M_A + \text{area of } V\text{-diagram}]_A^B \quad (4.4)$$

$$V_A^+ = V_A^- - P_A \quad (4.5)$$

$$M_A^+ = M_A^- + C_A \quad (4.6)$$

Procedure for the Area Method The following steps outline the procedure for constructing shear force and bending moment diagrams by the area method:

- Compute the support reactions from the free-body diagram (FBD) of the entire beam.
- Draw the load diagram of the beam (which is essentially a FBD) showing the values of the loads, including the support reactions. Use the sign conventions in Fig. 4.3 to determine the correct sign of each load.
- Working from left to right, construct the V- and M-diagrams for each segment of the beam using Eqs. (4.1)–(4.6).
- When you reach the right end of the beam, check to see whether the computed values of V and M are consistent with the end conditions. If they are not, you have made an error in the computations.

At first glance, using the area method may appear to be more cumbersome than plotting the shear force and bending moment equations. However, with practice you will find that the area method is not only much faster but also less susceptible to numerical errors because of the self-checking nature of the computations.

Sample Problem 4.4

The simply supported beam in Fig. (a) supports a 30-kN concentrated force at B and a 40-kN·m couple at D. Sketch the shear force and bending moment diagrams by the area method. Neglect the weight of the beam.

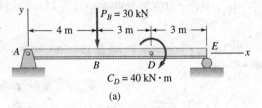

(a)

Solution

Load Diagram

The load diagram for the beam is shown in Fig. (b). The reactions at A and E are found from equilibrium analysis. The numerical value of each force (and the couple) is followed by a plus or minus sign in parentheses, indicating its sign as established by the sign conventions in Fig. 4.3.

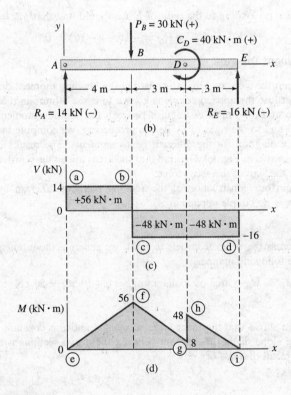

Shear Force Diagram

We now explain the steps used to construct the shear force diagram in Fig. (c). From the load diagram, we see that there are concentrated forces at A, B, and E that will cause jumps in the shear force diagram at these points. Therefore, our discussion of shear force must distinguish between sections of the beam immediately to the left and to the right of each of these points.

117

We begin by noting that $V_A^- = 0$ because no loading is applied to the left of A. We then proceed across the beam from left to right, constructing the diagram as we go:

$$V_A^+ = V_A^- - R_A = 0 - (-14) = +14 \text{ kN}$$

Plot point ⓐ.

$$V_B^- = V_A^+ - \text{area of } w\text{-diagram}]_A^B = 14 - 0 = 14 \text{ kN}$$

Plot point ⓑ.

Because $w = -dV/dx = 0$ between A and B, the slope of the V-diagram is zero between these points.

Connect ⓐ *and* ⓑ *with a horizontal straight line.*

$$V_B^+ = V_B^- - P_B = 14 - (+30) = -16 \text{ kN}$$

Plot point ⓒ.

$$V_E^- = V_B^+ - \text{area of } w\text{-diagram}]_B^E = -16 - 0 = -16 \text{ kN}$$

Plot point ⓓ.

Noting that $w = -dV/dx = 0$ between B and E, we conclude that the slope of the V-diagram is zero in segment BE.

Connect ⓒ *and* ⓓ *with a horizontal straight line.*

Because there is no loading to the right of E, we should find that $V_E^+ = 0$.

$$V_E^+ = V_E^- - R_E = -16 - (-16) = 0 \qquad \textit{Checks!}$$

Bending Moment Diagram

We now explain the steps required to construct the bending moment diagram shown in Fig. (d). Because the applied couple is known to cause a jump in the bending moment diagram at D, we must distinguish between the bending moments at sections just to the left and to the right of D. Before proceeding, we compute the areas under the shear force diagram for the different beam segments. The results of these computations are shown in Fig. (c). Observe that the areas are either positive or negative, depending on the sign of the shear force.

We begin our construction of the bending moment diagram by noting that $M_A = 0$ (there is no couple applied at A).

Plot point ⓔ.

Proceeding across the beam from left to right, we generate the moment diagram in Fig. (d) in the following manner:

$$M_B = M_A + \text{area of } V\text{-diagram}]_A^B = 0 + (+56) = 56 \text{ kN} \cdot \text{m}$$

Plot point ⓕ.

The V-diagram shows that the shear force between A and B is constant and positive. Therefore, the slope of the M-diagram between these two sections is also constant and positive (recall that $dM/dx = V$).

Connect ⓔ *and* ⓕ *with a straight line.*

$$M_D^- = M_B + \text{area of } V\text{-diagram}]_B^D = 56 + (-48) = 8 \text{ kN} \cdot \text{m}$$

Plot point ⓖ.

Because the slope of the V-diagram between B and D is negative and constant, the M-diagram has a constant, negative slope in this segment.

Connect ⓕ *and* ⓖ *with a straight line.*

$$M_D^+ = M_D^- + C_D = 8 + (+40) = 48 \text{ kN} \cdot \text{m}$$

Plot point (h).

Next, we note that $M_E = 0$ (there is no couple applied at E). Our computation based on the area of the V-diagram should verify this result.

$$M_E = M_D^+ + \text{area of } V\text{-diagram}]_D^E = 48 + (-48) = 0 \qquad \textit{Checks!}$$

Plot point (i).

The shear force between D and E is negative and constant, which means that the slope of the M-diagram for this segment is also constant and negative.

Connect (h) and (i) with a straight line.

∎

Sample Problem 4.5

The overhanging beam in Fig. (a) carries two uniformly distributed loads and a concentrated load. Using the area method, draw the shear force and bending moment diagrams for the beam.

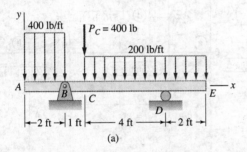

(a)

Load Diagram

The load diagram for the beam is given in Fig. (b); the reactions at B and D are determined by equilibrium analysis. Each of the numerical values is followed by a plus or minus sign in parentheses, determined by the sign conventions established in Fig. 4.3. The significance of the section labeled F will become apparent in the discussion that follows.

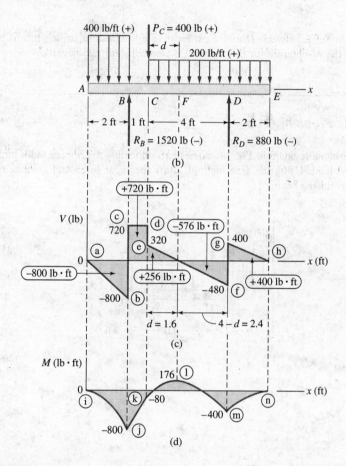

Shear Force Diagram

The steps required to construct the shear force diagram in Fig. (c) are now detailed. From the load diagram, we see that there are concentrated forces at B, C, and D, which means that there will be jumps in the shear diagram at these points. Therefore, we must differentiate between the shear force immediately to the left and to the right of each of these points.

We begin our construction of the V-diagram by observing that $V_A = 0$ because no force is applied at A.

Plot point ⓐ.

$$V_B^- = V_A - \text{area of } w\text{-diagram}]_A^B = 0 - (+400)(2) = -800 \text{ lb}$$

Plot point ⓑ.

We observe from Fig. (b) that the applied loading between A and B is constant and positive, so the slope of the shear diagram between the two cross sections is constant and negative (recall that $dV/dx = -w$).

Connect ⓐ and ⓑ with a straight line.

120

$$V_B^+ = V_B^- - R_B = -800 - (-1520) = 720 \text{ lb}$$

Plot point ⓒ.

$$V_C^- = V_B^+ - \text{area of } w\text{-diagram}]_B^C = 720 - 0 = 720 \text{ lb}$$

Plot point ⓓ.

Because $w = -dV/dx = 0$ between B and C, the slope of the V-diagram is zero in this segment.

Connect ⓒ and ⓓ with a horizontal straight line.

$$V_C^+ = V_C^- - P_C = 720 - (+400) = 320 \text{ lb}$$

Plot point ⓔ.

$$V_D^- = V_C^+ - \text{area of } w\text{-diagram}]_C^D = 320 - (+200)4 = -480 \text{ lb}$$

Plot point ⓕ.

Because the loading between C and D is constant and positive, the slope of the V-diagram between these two sections is constant and negative.

Connect ⓔ and ⓕ with a straight line.

Our computations have identified an additional point of interest—the point where the shear force is zero, labeled F on the load diagram in Fig. (b). The location of F can be found from

$$V_F = V_C^+ - \text{area of } w\text{-diagram}]_C^F = 320 - (+200)d = 0$$

which gives $d = 1.60$ ft, as shown in Fig. (c).

Continuing across the beam, we have

$$V_D^+ = V_D^- - R_D = -480 - (-880) = 400 \text{ lb}$$

Plot point ⓖ.

Next, we note that $V_E = 0$ (there is no force acting at E). The computation based on the area of the load diagram should verify this result.

$$V_E = V_D^+ - \text{area of } w\text{-diagram}]_D^E = 400 - (+200)2 = 0 \quad \text{Checks!}$$

Plot point ⓗ.

From Fig. (b), we see that the applied loading between D and E is constant and positive. Therefore, the slope of the V-diagram between these two cross sections is constant and negative.

Connect ⓖ and ⓗ with a straight line.

This completes the construction of the shear force diagram.

Bending Moment Diagram

We now explain the steps required to construct the bending moment diagram shown in Fig. (d). Because there are no applied couples, there will be no jumps in the M-diagram. The areas of the shear force diagram for the different segments of the beam are shown in Fig. (c).

We begin by noting that $M_A = 0$ because no couple is applied at A.

Plot point ⓘ.

Proceeding from left to right across the beam, we construct the bending moment diagram as follows:

$$M_B = M_A + \text{area of } V\text{-diagram}]_A^B = 0 + (-800) = -800 \text{ lb} \cdot \text{ft}$$

Plot point ⓙ.

We note from Fig. (c) that the V-diagram between A and B is a first-degree polynomial (inclined straight line). Therefore, the M-diagram between these two cross sections is a second-degree polynomial—that is, a parabola. From $dM/dx = V$, we see that the slope of the M-diagram is zero at A and -800 lb/ft at B.

Connect ⓘ and ⓙ with a parabola that has zero slope at ⓘ and negative slope at ⓙ. The parabola will be concave downward.

$$M_C = M_B + \text{area of } V\text{-diagram}]_B^C = -800 + (+720) = -80 \text{ lb} \cdot \text{ft}$$

Plot point ⓚ.

Because the V-diagram is constant and positive between B and C, the slope of the M-diagram is constant and positive between those two cross sections.

Connect ⓙ and ⓚ with a straight line.

$$M_F = M_C + \text{area of } V\text{-diagram}]_C^F = -80 + (+256) = +176 \text{ lb} \cdot \text{ft}$$

Plot point ⓛ.

Using $V = dM/dx$, we know that the slope of the M-diagram is $+320$ lb/ft at C and zero at F, and that the curve is a parabola between these two cross sections.

Connect ⓚ and ⓛ with a parabola that has positive slope at ⓚ and zero slope at ⓛ. The parabola will be concave downward.

$$M_D = M_F + \text{area of } V\text{-diagram}]_F^D = 176 + (-576) = -400 \text{ lb} \cdot \text{ft}$$

Plot point ⓜ.

The M-diagram between F and D is again a parabola, with a slope of zero at F and -480 lb/ft at D.

Connect ⓛ and ⓜ with a parabola that has zero slope at ⓛ and negative slope at ⓜ. The parabola will be concave downward.

Next, we note that $M_E = 0$ because no couple is applied at E. Our computation based on the area of the V-diagram should verify this result.

$$M_E = M_D + \text{area of } V\text{-diagram}]_D^E = -400 + (+400) = 0 \qquad \text{Checks!}$$

Plot point ⓝ.

From the familiar arguments, the M-diagram between D and E is a parabola with a slope equal to $+400$ lb/ft at D and zero at E.

Connect ⓜ and ⓝ with a parabola that has positive slope at ⓜ and zero slope at ⓝ. The parabola will be concave downward.

This completes the construction of the bending moment diagram. It is obvious in Fig. (d) that the slope of the M-diagram is discontinuous at ⓙ and ⓜ. Not so obvious is the slope discontinuity at ⓚ: From $dM/dx = V$, we see that the slope of the M-diagram to the left of ⓚ equals $+720$ lb/ft, whereas to the right of ⓚ the slope equals $+320$ lb/ft. Observe that the slope of the M-diagram is continuous at ⓛ because the shear force has the same value (zero) to the left and to the right of ⓛ.

Problems

4.21–4.40 Construct the shear force and bending moment diagrams for the beam shown by the area method.

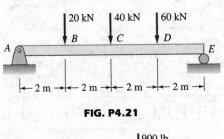

FIG. P4.21

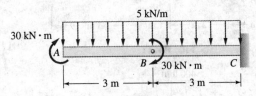

FIG. P4.22

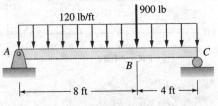

FIG. P4.23

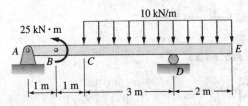

FIG. P4.24

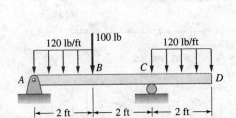

FIG. P4.25

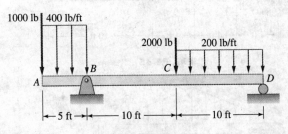

FIG. P4.26

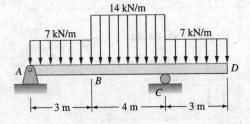

FIG. P4.27

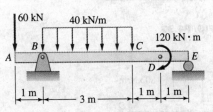

FIG. P4.28

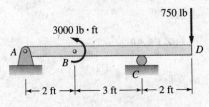

FIG. P4.29

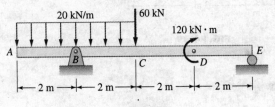

FIG. P4.30

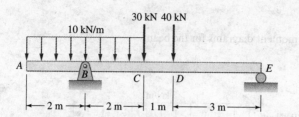

FIG. P4.31

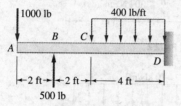

FIG. P4.33

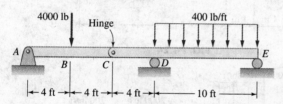

FIG. P4.35

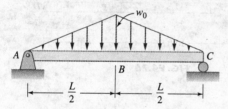

FIG. P4.37

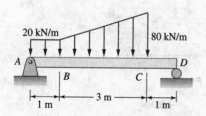

FIG. P4.39

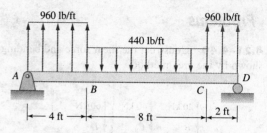

FIG. P4.32

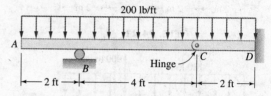

FIG. P4.34

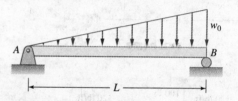

FIG. P4.36

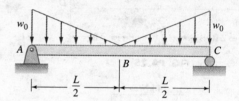

FIG. P4.38

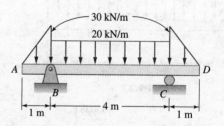

FIG. P4.40

4.41–4.45 Draw the load and the bending moment diagrams that correspond to the given shear force diagram. Assume no couples are applied to the beam.

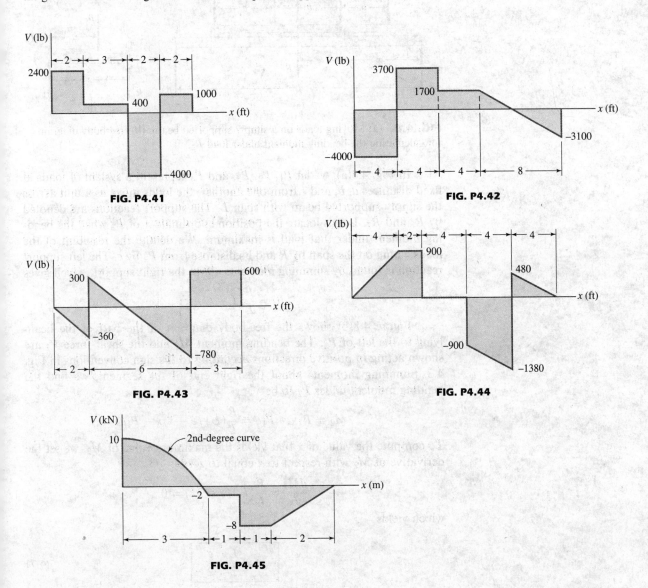

FIG. P4.41

FIG. P4.42

FIG. P4.43

FIG. P4.44

FIG. P4.45

4.5 Moving Loads

A truck or other vehicle rolling across a beam constitutes a system of moving, concentrated loads. The fixed distances between the loads are determined by the wheelbase of the vehicle. For simply supported beams carrying only concentrated loads, the maximum bending moment always occurs under one of the loads. Therefore, when designing for moving loads, we must calculate the bending moment under each load when the vehicle is in the position that results in a maximum moment under that load. The largest of these values is the maximum moment that governs the design of the beam.

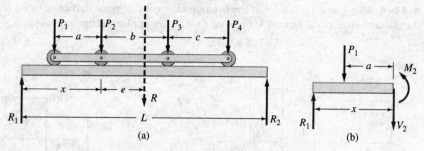

FIG. 4.8 (a) Moving loads on a simply supported beam; (b) free-body diagram for calculating the bending moment under load P_2.

In Fig. 4.8(a), we let P_1, P_2, P_3, and P_4 represent a system of loads at fixed distances a, b, and c from one another; the loads move as a unit across the simply supported beam with span L. The support reactions are denoted by R_1 and R_2. Let us locate the position coordinate x of P_2 when the bending moment under that load is maximum. We denote the resultant of the loads acting on the span by R and its distance from P_2 by e. The left support reaction is found by summing moments about the right support, which gives

$$R_1 = \frac{R}{L}(L - e - x)$$

Figure 4.8(b) shows the free-body diagram of the part of the beam lying to the left of P_2. The bending moment M_2 and the shear force V_2 are shown acting in positive directions according to the sign conventions in Fig. 4.3. Summing moments about the right end of the segment, we find the bending moment under P_2 to be

$$M_2 = R_1 x - P_1 a = \frac{R}{L}(L - e - x)x - P_1 a$$

To compute the value of x that yields the maximum value of M_2, we set the derivative of M_2 with respect to x equal to zero:

$$\frac{dM_2}{dx} = \frac{R}{L}(L - e - 2x) = 0$$

which yields

$$\boxed{x = \frac{L}{2} - \frac{e}{2}} \qquad (4.7)$$

This value of x is independent of the number of loads to the left of P_2 (these loads contribute constant terms to M_2, which vanish upon differentiation).

Equation (4.7) gives us the following rule:

The bending moment under a load P is maximized when the resultant R of all loads acting on the span is midway between P and the centerline of the beam.

Using this rule, we can locate the position of each load when the moment under that load is a maximum.

The maximum shear force occurs when one of the loads is over a support and is equal to the reaction at that support. The identity of the support and the load corresponding to the maximum shear force must be determined by trial-and-error.

Sample Problem 4.6

A truck and trailer combination with the axle loads shown in Fig. (a) rolls across the simply supported 12-m span. Compute (1) the maximum bending moment and (2) the maximum shear force.

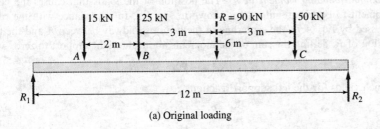

(a) Original loading

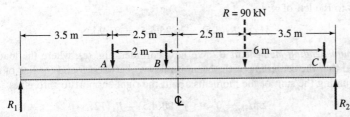

(b) Position of loads for maximum moment at A

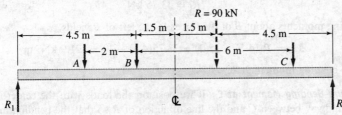

(c) Position of loads for maximum moment at B

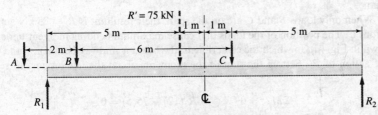

(d) Position of loads for maximum moment at C with only B and C on span

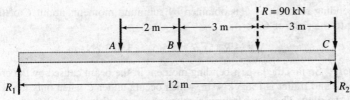

(e) Position of loads for maximum shear force

Solution

Part 1

If we assume that all three loads act on the span, their resultant is $R = 90$ kN, located as shown in Fig. (a).

Maximum Bending Moment at A The position of the loads that causes the bending moment to be maximum at A is shown in Fig. (b). In accordance with the rule expressed by Eq. (4.7), the centerline of the beam is midway between A and the line of action of R. Setting the sum of the moments about the right support to zero, we get

$$\Sigma M_{R_2} = 0 \quad +\circlearrowleft \quad 90(3.5) - R_1(12) = 0$$

which yields for the left support reaction

$$R_1 = 26.25 \text{ kN}$$

The bending moment at A is obtained by summing the moments about A of the forces to the left of A:

$$M_A = R_1(3.5) = 26.25(3.5) = 91.9 \text{ kN} \cdot \text{m}$$

Maximum Bending Moment at B We next consider Fig. (c), where the loads are so located that the centerline of the beam is midway between B and the line of action of R. Equating the sum of the moments about the right support to zero gives

$$\Sigma M_{R_2} = 0 \quad +\circlearrowleft \quad 90(4.5) - R_1(12) = 0$$

from which

$$R_1 = 33.75 \text{ kN}$$

Summing moments about B of the forces to the left of B yields

$$M_B = R_1(4.5) - 15(2) = 33.75(4.5) - 30 = 121.9 \text{ kN} \cdot \text{m}$$

for the bending moment at B.

Maximum Bending Moment at C If we position the loads with the centerline of the beam midway between C and the line of action of R so that the bending moment at C is maximized, we find that the load at A is no longer on the span. Therefore, the maximum bending moment at C may occur when only the loads at B and C are on the span.

When only loads B and C are on the span, their resultant is $R' = 75$ kN acting 2 m from C. The position of the loads that cause maximum bending moment under C is shown in Fig. (d), in which the centerline of the beam is midway between C and the line of action of R'. Setting the sum of the moments about the left support to zero, we get

$$\Sigma M_{R_1} = 0 \quad +\circlearrowleft \quad R_2(12) - 75(5) = 0$$

Therefore, the reaction at the right support is

$$R_2 = 31.25 \text{ kN}$$

The bending moment at C is obtained by summing moments about C of the forces acting to the right of C:

$$M_C = R_2(5) = 31.25(5) = 156.3 \text{ kN} \cdot \text{m}$$

Summary So far, the largest bending moment in the beam caused by the crossing of the truck and trailer is 156.3 kN·m, which occurs under C when only loads B and C are on the span. There are, however, other positions of the loads that may result in a larger bending moment: when only loads A and B are on the span, and when only the load at C is on the span. It can be shown (you should verify this) that in each case the largest bending moment is smaller than 156.3 kN. Therefore, the maximum bending moment as the truck and trailer cross the span is

$$M_{max} = 156.3 \text{ kN} \cdot \text{m} \qquad \text{Answer}$$

This moment occurs under the load at C when only B and C are on the span.

Part 2

We assume initially that the maximum shear force occurs when all the loads are on the span. The maximum reaction at the left support occurs when A is directly over the support. We note from Fig. (a) that in this case the resultant load R is 5 m from the left support. Similarly, the reaction at the right support reaches its maximum value when C is over that support, R being 3 m from the support, as shown in Fig. (e). Evidently R_2 is the higher of the two reactions because it is closer to the resultant load. To find R_2, we sum the moments of the forces in Fig. (e) about the left support and set the result equal to zero:

$$\Sigma M_{R_1} = 0 \quad +\circlearrowleft \quad R_2(12) - 90(12 - 3) = 0$$

which yields $R_2 = 67.5$ kN

We must also investigate the possibility of the maximum shearing force occurring when only loads B and C are on the span. The maximum reaction in this case is $R_1 = 50$ kN when B is over the left support. The case when only A and B are on the span need not be investigated because their resultant is 40 kN, which is less than $R_2 = 67.5$ kN computed above. Therefore, the maximum shear force in the beam is

$$V_{max} = 67.5 \text{ kN} \qquad \text{Answer}$$

occurring at the right support when all three loads are on the span.

∎

Problems

4.46 A truck with axle loads of 40 kN and 60 kN on a wheelbase of 5 m travels across a 10-m span. Compute the maximum bending moment and the maximum shear force in the span. Neglect the mass of the beam.

4.47 The trolley AB carrying a 80-kN load travels on the 15-m-long beam. Determine the maximum bending moment and the maximum shear force in the beam. Neglect the weights of the trolley and the beam.

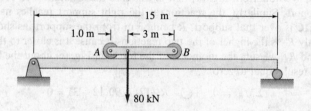

FIG. P4.47

4.48 Show that if the weight of the beam is not neglected, Eq. (4.7) becomes

$$x = \frac{L}{2} - \frac{e}{2 + (W/R)}$$

where W is the weight of the beam (assumed to be uniformly distributed) and R represents the resultant of the moving loads.

4.49 Solve Prob. 4.47 if the weight of the beam is 20 kN. Use the formula given in Prob. 4.48.

4.50 The three wheel loads roll as a unit across the 44-ft span of a simply supported beam. Find the maximum bending moment and the maximum shear force in the beam. Neglect the weight of the beam.

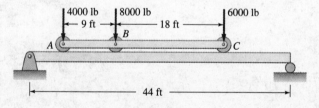

FIG. P4.50

4.51 A truck-trailer combination crossing a 12-m simply supported span has axle loads of 10 kN, 20 kN, and 30 kN separated by distances of 3 m and 5 m, respectively. Compute the maximum bending moment and the maximum shear force in the span. Neglect the weight of the beam.

Review Problems

4.52–4.62 Draw the shear force and bending moment diagrams for the beam shown. Neglect the weight of the beam.

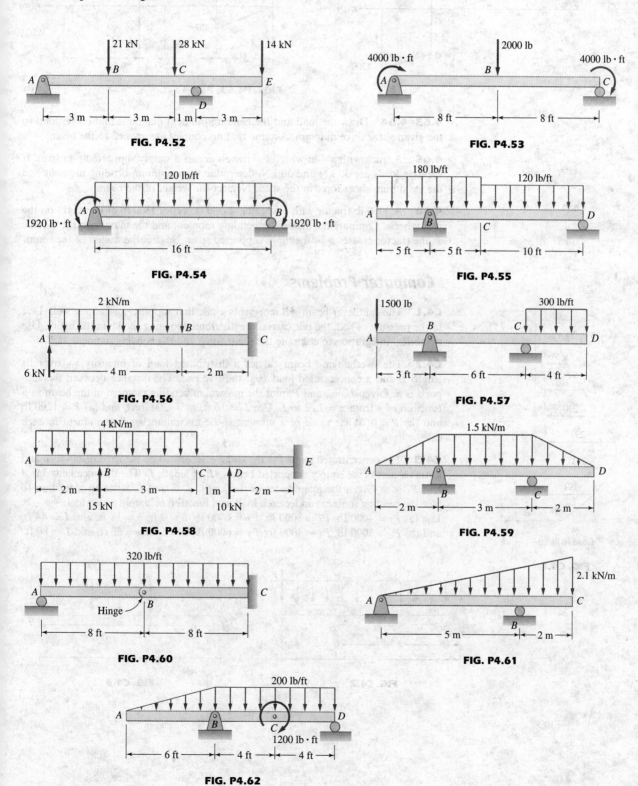

FIG. P4.52

FIG. P4.53

FIG. P4.54

FIG. P4.55

FIG. P4.56

FIG. P4.57

FIG. P4.58

FIG. P4.59

FIG. P4.60

FIG. P4.61

FIG. P4.62

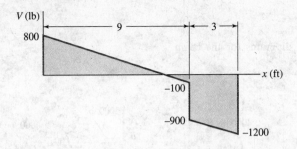
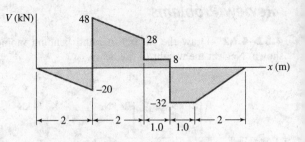

FIGS. P4.63, P4.64

4.63–4.64 Draw the load and the bending moment diagrams that correspond to the given shear force diagram. Assume that no couples are applied to the beam.

4.65 A truck with a 4-m wheelbase travels across a simply supported 8-m span. If the axle loads are 30 kN and 50 kN, determine the maximum bending moment and the maximum shear force in the span. Neglect the weight of the beam.

4.66 A 3000-lb tractor with a wheelbase of 9 ft carries 1800 lb of its weight on the rear wheels. Compute the maximum bending moment and the maximum shear force as the tractor crosses a 14-ft simply supported span. Neglect the weight of the beam.

Computer Problems

C4.1 The cantilever beam AB represents a pile that supports a retaining wall. Due to the pressure of soil, the pile carries the distributed loading shown in the figure. Use numerical integration to compute the shear force and the bending moment at B.

C4.2 The overhanging beam carries a distributed load of intensity w_0 over its length L and a concentrated load P at the free end. The distance between the supports is x. Given L, w_0, and P, plot the maximum bending moment in the beam as a function of x from $x = L/2$ to L. Use $L = 16$ ft, $w_0 = 200$ lb/ft, and (a) $P = 1200$ lb and (b) $P = 0$. What value of x minimizes the maximum bending moment in each case?

C4.3 The concentrated loads P_1, P_2, and P_3, separated by the fixed distances a and b, travel across the simply supported beam AB of length L. The distance between A and P_1 is x. Given the magnitudes of the loads, a, b, and L, write an algorithm to plot the bending moment under each load as a function of x from $x = 0$ to $L - a - b$. Use (a) $P_1 = 4000$ lb, $P_2 = 8000$ lb, $P_3 = 6000$ lb, $a = 9$ ft, $b = 18$ ft, and $L = 44$ ft; and (b) $P_1 = 8000$ lb, $P_2 = 4000$ lb, $P_3 = 6000$ lb, $a = 5$ ft, $b = 28$ ft, and $L = 80$ ft.

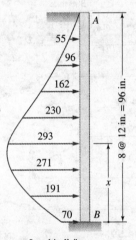

Load in lb/in.

FIG. C4.1

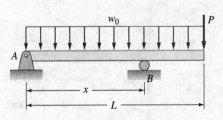

FIG. C4.2

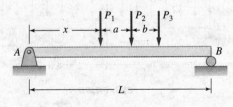

FIG. C4.3

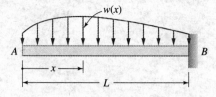

FIG. C4.4, C4.5

C4.4 The cantilever beam AB of length L carries a distributed loading w that varies with the distance x. Given L and $w(x)$, construct an algorithm to plot the shear force and bending moment diagrams. Use (a) $L = 3$ m and $w = (50 \text{ kN/m}) \sin(\pi x/2L)$, and (b) $L = 5$ m and

$$w = \begin{cases} 20 \text{ kN/m} & \text{if } x \leq 1.0 \text{ m} \\ (20 \text{ kN/m}) \dfrac{x}{1.0 \text{ m}} & \text{if } 1.0 \text{ m} \leq x \leq 4 \text{ m} \\ 0 & \text{if } x > 4 \text{ m} \end{cases}$$

C4.5 Solve Prob. C4.4 if the beam is simply supported at A and B.

5 Stresses in Beams

A stack of wide-flange steel sections at a construction site. Wide-flange sections are the most commonly used shapes in steel construction because their shape is optimized for resisting bending.

5.1 Introduction

In previous chapters, we considered stresses in bars caused by axial loading and torsion. Here we consider the third fundamental loading: bending. When deriving the relationships between the bending moment and the stresses it causes, we find it again necessary to make certain simplifying assumptions. Although these assumptions may appear to be overly restrictive, the resulting equations have served well in the design of straight, elastic beams. Furthermore, these equations can be extended to the more complicated bending problems discussed in later chapters.

We use the same steps in the analysis of bending that we used for torsion in Chapter 3:

136 CHAPTER 5 Stresses in Beams

- Make simplifying assumptions about the deformation based upon experimental evidence.
- Determine the strains that are geometrically compatible with the assumed deformations.
- Use Hooke's law to express the equations of compatibility in terms of stresses.
- Derive the equations of equilibrium. (These equations provide the relationships between the stresses and the applied loads.)

5.2 Bending Stress

a. Simplifying assumptions

The stresses caused by the bending moment are known as *bending stresses*, or *flexure stresses*. The relationship between these stresses and the bending moment is called the *flexure formula*. In deriving the flexure formula, we make the following assumptions:

- The beam has an axial plane of symmetry, which we take to be the *xy*-plane (see Fig. 5.1).
- The applied loads (such as F_1, F_2, and F_3 in Fig. 5.1) lie in the plane of symmetry and are perpendicular to the axis of the beam (the *x*-axis).
- The axis of the beam bends but does not stretch (the axis lies somewhere in the plane of symmetry; its location will be determined later).
- Plane sections of the beam remain plane (do not warp) and perpendicular to the deformed axis of the beam.
- Changes in the cross-sectional dimensions of the beam are negligible.

Because the shear stresses caused by the vertical shear force will distort (warp) an originally plane section, we are limiting our discussion here to the deformations caused by the bending moment alone. However, it can be shown that the deformations due to the vertical shear force are negligible in slender beams (the length of the beam is much greater than the cross-sectional dimensions) compared to the deformations caused by bending.

The above assumptions lead us to the following conclusion: *Each cross section of the beam rotates as a rigid entity about a line called the neutral axis of the cross section.* The *neutral axis* passes through the axis of the beam and

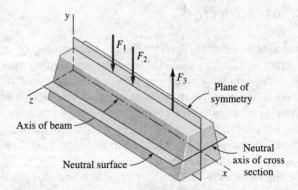

FIG. 5.1 Symmetrical beam with loads lying in the plane of symmetry.

is perpendicular to the plane of symmetry, as shown in Fig. 5.1. The xz-plane that contains the neutral axes of all the cross sections is known as the *neutral surface* of the beam.

b. Compatibility

Figure 5.2 shows a segment of the beam bounded by two cross sections that are separated by the infinitesimal distance dx. Due to the bending moment M caused by the applied loading, the cross sections rotate relative to each other by the amount $d\theta$. Note that the bending moment is assumed to be positive according to the sign conventions established in Fig. 4.3. Consistent with the assumptions made about deformation, the cross sections do not distort in any manner.

Because the cross sections are assumed to remain perpendicular to the axis of the beam, the neutral surface becomes curved upon deformation, as indicated in Fig. 5.2. The radius of curvature of the deformed surface is denoted by ρ. Note that the distance between the cross sections, measured along the neutral surface, remains unchanged at dx (it is assumed that the axis of the beam does not change length). Therefore, the longitudinal fibers lying on the neutral surface are undeformed, whereas the fibers above the surface are compressed and the fibers below are stretched.

Consider now the deformation of the longitudinal fiber ab that lies a distance y above the neutral surface, as shown in Fig. 5.2. In the deformed state, the fiber forms the arc $a'b'$ of radius $(\rho - y)$, subtended by the angle $d\theta$. Therefore, its deformed length is

$$\overline{a'b'} = (\rho - y)\,d\theta$$

The original length of this fiber is $\overline{ab} = dx = \rho\,d\theta$. The normal strain of the fiber is found by dividing the change in length by the original length, yielding

$$\epsilon = \frac{\overline{a'b'} - \overline{ab}}{\overline{ab}} = \frac{(\rho - y)\,d\theta - \rho\,d\theta}{\rho\,d\theta} = -\frac{y}{\rho}$$

Assuming that the stress is less than the proportional limit of the material, we can obtain the normal stress in fiber ab from Hooke's law:

$$\sigma = E\epsilon = -\frac{E}{\rho}y \tag{5.1}$$

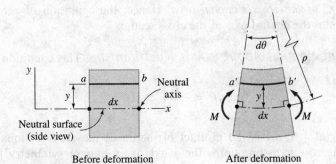

FIG. 5.2 Deformation of an infinitesimal beam segment.

Equation (5.1) shows that the normal stress of a longitudinal fiber is proportional to the distance y of the fiber from the neutral surface. The negative sign indicates that positive bending moment causes compressive stress when y is positive (fibers above the neutral surface) and tensile stress when y is negative (fibers below the neutral surface), as expected.

c. Equilibrium

To complete the derivation of the flexure formula, we must locate the neutral axis of the cross section and derive the relationship between ρ and M. Both tasks can be accomplished by applying the equilibrium conditions.

Figure 5.3 shows a typical cross section of a beam. The normal force acting on the infinitesimal area dA of the cross section is $dP = \sigma \, dA$. Substituting $\sigma = -(E/\rho)y$, we obtain

$$dP = -\frac{E}{\rho} y \, dA \tag{a}$$

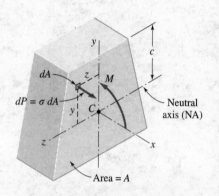

FIG. 5.3 Calculating the resultant of the normal stress acting on the cross section. Resultant is a couple equal to the internal bending moment M.

where y is the distance of dA from the neutral axis (NA). Equilibrium requires that the resultant of the normal stress distribution over the cross section must be equal to the bending moment M acting about the neutral axis (z-axis). In other words, $-\int_A y \, dP = M$, where the integral is taken over the entire cross-sectional area A (the minus sign in the expression is needed because the moment of dP and positive M have opposite sense). Moreover, the resultant axial force and the resultant bending moment about the y-axis must be zero; that is, $\int_A dP = 0$ and $\int_A z \, dP = 0$. These three equilibrium equations are developed in detail below.

Resultant Axial Force Must Vanish The condition for zero axial force is

$$\int_A dP = -\frac{E}{\rho} \int_A y \, dA = 0$$

Because $E/\rho \neq 0$, this equation can be satisfied only if

$$\int_A y \, dA = 0 \tag{b}$$

The integral in Eq. (b) is the first moment of the cross-sectional area about the neutral axis. It can be zero only if *the neutral axis passes through the centroid C of the cross-sectional area*. Hence, the condition of zero axial force locates the neutral axis of the cross section.

Resultant Moment About y-Axis Must Vanish This condition is

$$\int_A z \, dP = -\frac{E}{\rho} \int_A zy \, dA = 0 \tag{c}$$

The integral $\int_A zy \, dA$ is the product of inertia of the cross-sectional area. According to our assumptions, the y-axis is an axis of symmetry for the cross section, in which case this integral is zero and Eq. (c) is automatically satisfied.

Resultant Moment About the Neutral Axis Must Equal M

Equating the resultant moment about the z-axis to M gives us

$$-\int_A y\, dP = \frac{E}{\rho}\int_A y^2\, dA = M$$

Recognizing that $\int_A y^2\, dA = I$ is the moment of inertia[1] of the cross-sectional area about the neutral axis (the z-axis), we obtain the *moment-curvature relationship*

$$M = \frac{EI}{\rho} \tag{5.2a}$$

A convenient form of this equation is

$$\boxed{\frac{1}{\rho} = \frac{M}{EI}} \tag{5.2b}$$

d. Flexure formula; section modulus

Substituting the expression for $1/\rho$ from Eq. (5.2b) into Eq. (5.1), we get the *flexure formula*:

$$\boxed{\sigma = -\frac{My}{I}} \tag{5.3}$$

Note that a positive bending moment M causes negative (compressive) stress above the neutral axis and positive (tensile) stress below the neutral axis, as discussed previously.

The maximum value of bending stress without regard to its sign is given by

$$\boxed{\sigma_{\max} = \frac{|M_{\max}|c}{I}} \tag{5.4a}$$

where c is the distance from the neutral axis to the outermost point of the cross section, as illustrated in Fig. 5.3. Equation (5.4a) is frequently written in the form

$$\boxed{\sigma_{\max} = \frac{|M_{\max}|}{S}} \tag{5.4b}$$

where $S = I/c$ is called the *section modulus* of the beam. The dimension of S is $[L^3]$, so that its units are in.^{3}, mm^3, and so on. The formulas for the

---

[1] The moment of inertia is reviewed in Appendix A.

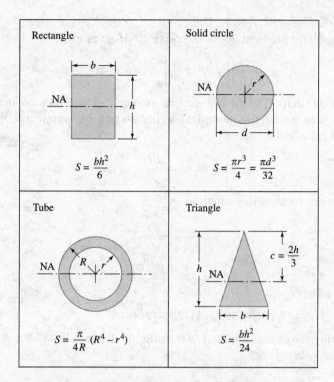

FIG. 5.4 Section moduli of simple cross-sectional shapes.

section moduli of common cross sections are given in Fig. 5.4. The section moduli of standard structural shapes are listed in various handbooks; an abbreviated list is given in Appendix B.

e. Procedures for determining bending stresses

Stress at a Given Point

- Use the method of sections to determine the bending moment M (with its correct sign) at the cross section containing the given point.
- Determine the location of the neutral axis.
- Compute the moment of inertia I of the cross-sectional area about the neutral axis. (If the beam is a standard structural shape, its cross-sectional properties are listed in Appendix B.)
- Determine the y-coordinate of the given point. Note that y is positive if the point lies above the neutral axis and negative if it lies below the neutral axis.
- Compute the bending stress from $\sigma = -My/I$. If correct signs are used for M and y, the stress will also have the correct sign (tension positive, compression negative).

Maximum Bending Stress: Symmetric Cross Section If the neutral axis is an axis of symmetry of the cross section, the maximum tensile and compressive bending stresses are equal in magnitude and occur at the section of the largest bending moment. The following procedure is recommended for determining the maximum bending stress in a prismatic beam:

- Draw the bending moment diagram by one of the methods described in Chapter 4. Identify the bending moment M_{max} that has the largest magnitude (disregard the sign).
- Compute the moment of inertia I of the cross-sectional area about the neutral axis. (If the beam is a standard structural shape, its cross-sectional properties are listed in Appendix B.)
- Calculate the maximum bending stress from $\sigma_{max} = |M_{max}|c/I$, where c is the distance from the neutral axis to the top or bottom of the cross section.

Maximum Tensile and Compressive Bending Stresses: Unsymmetrical Cross Section If the neutral axis is not an axis of symmetry of the cross section, the maximum tensile and compressive bending stresses may occur at different sections. The recommended procedure for computing these stresses in a prismatic beam follows:

- Draw the bending moment diagram by one of the methods described in Chapter 4. Identify the largest positive and negative bending moments.
- Determine the location of the neutral axis and record the distances c_{top} and c_{bot} from the neutral axis to the top and bottom of the cross section.
- Compute the moment of inertia I of the cross section about the neutral axis. (If the beam is a standard structural shape, its cross-sectional properties are listed in Appendix B.)
- Calculate the bending stresses at the top and bottom of the cross section where the largest positive bending moment occurs from $\sigma = -My/I$. At the top of the cross section, where $y = c_{top}$, we obtain $\sigma_{top} = -Mc_{top}/I$. At the bottom of the cross section, we have $y = -c_{bot}$, so that $\sigma_{bot} = Mc_{bot}/I$. Repeat the calculations for the cross section that carries the largest negative bending moment. Inspect the four stresses thus computed to determine the largest tensile (positive) and compressive (negative) bending stresses in the beam.

Note on Units Make sure that the units of the terms in the flexure formula $\sigma = -My/I$ are consistent. In the U.S. Customary system, M is often measured in pound-feet and the cross-sectional properties in inches. It is recommended that you convert M into lb·in. and compute σ in lb/in.^{2} (psi). Thus, the units in the flexure formula become

$$\sigma\,[\text{lb/in.}^2] = -\frac{M\,[\text{lb}\cdot\text{in.}]\,y\,[\text{in.}]}{I\,[\text{in.}^4]}$$

In the SI system, M is usually expressed in N·m, whereas the cross-sectional dimensions are in mm. To obtain σ in N/m^2 (Pa), the cross-sectional properties must be converted to meters, so that the units in the flexure equation are

$$\sigma\,[\text{N/m}^2] = -\frac{M\,[\text{N}\cdot\text{m}]\,y\,[\text{m}]}{I\,[\text{m}^4]}$$

Sample Problem 5.1

The simply supported beam in Fig. (a) has a rectangular cross section 120 mm wide and 200 mm high. (1) Compute the maximum bending stress in the beam. (2) Sketch the bending stress distribution over the cross section on which the maximum bending stress occurs. (3) Compute the bending stress at a point on section B that is 25 mm below the top of the beam.

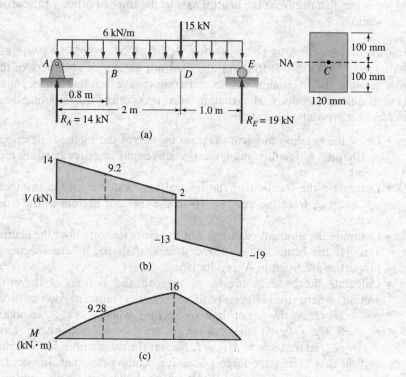

Solution

Preliminary Calculations

Before we can find the maximum bending stress in the beam, we must find the maximum bending moment. We begin by computing the external reactions at A and E; the results are shown in Fig. (a). Then we sketch the shear force and bending moment diagrams using one of the methods (for example, the area method) described in Chapter 4, obtaining the results in Figs. (b) and (c). We see that the maximum bending moment is $M_{max} = +16$ kN·m, occurring at D.

In this case, the neutral axis (NA) is an axis of symmetry of the cross section, as shown in Fig. (a). The moment of inertia of the cross section about the neutral axis is

$$I = \frac{bh^3}{12} = \frac{0.12(0.2)^3}{12} = 80.0 \times 10^{-6} \text{ m}^4$$

and the distance between the neutral axis and the top (or bottom) of the cross section is $c = 100$ mm $= 0.1$ m.

Part 1

The maximum bending stress in the beam occurs on the cross section that carries the largest bending moment, which is the section at D. Using the flexure formula,

Eq. (5.4a), we obtain for the maximum bending stress in the beam

$$\sigma_{max} = \frac{|M_{max}|c}{I} = \frac{(16 \times 10^3)(0.1)}{80.0 \times 10^{-6}} = 20.0 \times 10^6 \text{ Pa} = 20.0 \text{ MPa} \quad \textit{Answer}$$

Part 2

The stress distribution on the cross section at D is shown in Fig. (d). When drawing the figure, we were guided by the following observations: (i) the bending stress varies linearly with distance from the neutral axis; (ii) because M_{max} is positive, the top half of the cross section is in compression and the bottom half is in tension; and (iii) due to symmetry of the cross section about the neutral axis, the maximum tensile and compressive stresses are equal in magnitude.

Part 3

From Fig. (c) we see that the bending moment at section B is $M = +9.28$ kN·m. The y-coordinate of the point that lies 25 mm below the top of the beam is $y = 100 - 25 = 75$ mm $= 0.075$ m. If we substitute these values into Eq. (5.3), the bending stress at the specified location becomes

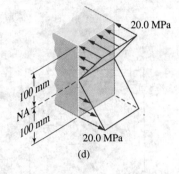

(d)

$$\sigma = -\frac{My}{I} = -\frac{(9.28 \times 10^3)(0.075)}{80.0 \times 10^{-6}} = -8.70 \times 10^6 \text{ Pa} = -8.70 \text{ MPa} \quad \textit{Answer}$$

The negative sign indicates that this bending stress is compressive, which is expected because the bending moment is positive and the point of interest lies above the neutral axis.

■

Sample Problem 5.2

The simply supported beam in Fig. (a) has the T-shaped cross section shown. Determine the values and locations of the maximum tensile and compressive bending stresses.

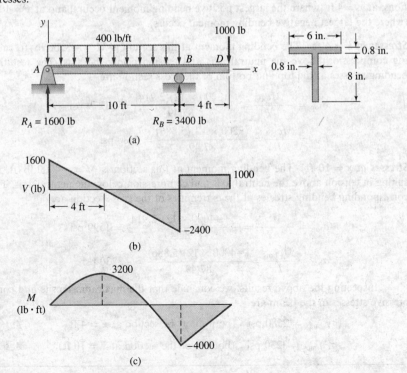

Solution

Preliminary Calculations

Before we can find the maximum tensile and compressive bending stresses, we must find the largest positive and negative bending moments. Therefore, we start by computing the external reactions at A and B, and then sketch the shear force and bending moment diagrams. The results are shown in Figs. (a)–(c). From Fig. (c), we see that the largest positive and negative bending moments are 3200 lb·ft and -4000 lb·ft, respectively.

Because the cross section does not have a horizontal axis of symmetry, we must next locate the neutral (centroidal) axis of the cross section. As shown in Fig. (d), we consider the cross section to be composed of the two rectangles with areas $A_1 = 0.8(8) = 6.4$ in.^{2} and $A_2 = 0.8(6) = 4.8$ in.^{2}. The centroidal coordinates of the areas are $\bar{y}_1 = 4$ in. and $\bar{y}_2 = 8.4$ in., measured from the bottom of the cross section. The coordinate $\bar{y}$ of the centroid C of the cross section is

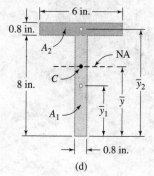

(d)

$$\bar{y} = \frac{A_1 \bar{y}_1 + A_2 \bar{y}_2}{A_1 + A_2} = \frac{6.4(4) + 4.8(8.4)}{6.4 + 4.8} = 5.886 \text{ in.}$$

We can now compute the moment of inertia I of the cross-sectional area about the neutral axis. Using the parallel-axis theorem, we have $I = \sum [\bar{I}_i + A_i(\bar{y}_i - \bar{y})^2]$, where $\bar{I}_i = b_i h_i^3 / 12$ is the moment of inertia of a rectangle about its own centroidal axis. Thus,

$$I = \left[\frac{0.8(8)^3}{12} + 6.4(4 - 5.886)^2\right] + \left[\frac{6(0.8)^3}{12} + 4.8(8.4 - 5.886)^2\right]$$

$$= 87.49 \text{ in.}^4$$

Maximum Bending Stresses

The distances from the neutral axis to the top and the bottom of the cross section are $c_{\text{top}} = 8.8 - \bar{y} = 8.8 - 5.886 = 2.914$ in. and $c_{\text{bot}} = \bar{y} = 5.886$ in., as shown in Fig. (e). Because these distances are different, we must investigate stresses at two locations: at $x = 4$ ft (where the largest positive bending moment occurs) and at $x = 10$ ft (where the largest negative bending moment occurs).

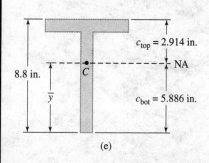

(e)

Stresses at $x = 4$ ft The bending moment at this section is $M = +3200$ lb·ft, causing compression above the neutral axis and tension below the axis. The resulting bending stresses at the top and bottom of the cross section are

$$\sigma_{\text{top}} = -\frac{Mc_{\text{top}}}{I} = -\frac{(3200 \times 12)(2.914)}{87.49} = -1279 \text{ psi}$$

$$\sigma_{\text{bot}} = \frac{Mc_{\text{bot}}}{I} = \frac{(3200 \times 12)(5.886)}{87.49} = 2580 \text{ psi}$$

Stresses at $x = 10$ ft The bending moment at this section is $M = -4000$ lb·ft, resulting in tension above the neutral axis and compression below the neutral axis. The corresponding bending stresses at the extremities of the cross section are

$$\sigma_{\text{top}} = -\frac{Mc_{\text{top}}}{I} = -\frac{(-4000 \times 12)(2.914)}{87.49} = 1599 \text{ psi}$$

$$\sigma_{\text{bot}} = \frac{Mc_{\text{bot}}}{I} = \frac{(-4000 \times 12)(5.886)}{87.49} = -3230 \text{ psi}$$

Inspecting the above results, we conclude that the maximum tensile and compressive stresses in the beam are

$(\sigma_T)_{\max} = 2580$ psi   (bottom of the section at $x = 4$ ft)   **Answer**

$(\sigma_C)_{\max} = 3230$ psi   (bottom of the section at $x = 10$ ft)   **Answer**

Sample Problem 5.3

The cantilever beam in Fig. (a) is composed of two segments with rectangular cross sections. The width of each section is 2 in., but the depths are different, as shown in the figure. Determine the maximum bending stress in the beam.

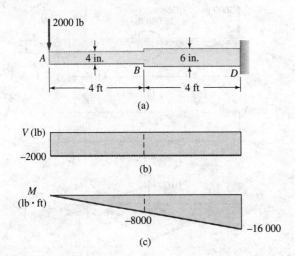

Solution

The shear force and bending moment diagrams are shown in Figs. (b) and (c). Because the cross section of the beam is not constant, the maximum stress occurs either at the section just to the left of B ($M_B = -8000$ lb·ft) or at the section at D ($M_D = -16\,000$ lb·ft). Referring to Fig. 5.4, we find that the section moduli of the two segments are

$$S_{AB} = \frac{bh_{AB}^2}{6} = \frac{(2)(4)^2}{6} = 5.333 \text{ in.}^3$$

$$S_{BD} = \frac{bh_{BD}^2}{6} = \frac{(2)(6)^2}{6} = 12.0 \text{ in.}^3$$

From Eq. (5.4b), the maximum bending stresses on the two cross sections of interest are

$$(\sigma_B)_{max} = \frac{|M_B|}{S_{AB}} = \frac{8000 \times 12}{5.333} = 18\,000 \text{ psi}$$

$$(\sigma_D)_{max} = \frac{|M_D|}{S_{BD}} = \frac{16\,000 \times 12}{12.0} = 16\,000 \text{ psi}$$

Comparing the above values, we find that the maximum bending stress in the beam is

$$\sigma_{max} = 18\,000 \text{ psi} \quad \text{(on the cross section just to the left of } B\text{)} \quad \textit{Answer}$$

This is an example where the maximum bending stress occurs on a cross section at which the bending moment is not maximum.

Sample Problem 5.4

The wide-flange section[2] W14 × 30 is used as a cantilever beam, as shown in Fig. (a). Find the maximum bending stress in the beam.

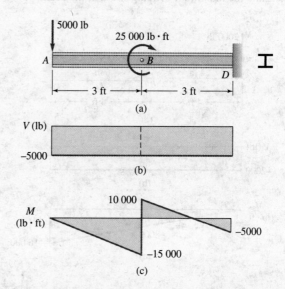

Solution

The shear force and bending moment diagrams for the beam are shown in Figs. (b) and (c). We note that the largest bending moment is $|M_{max}| = 15\,000$ lb·ft, acting just to the left of section B. From the tables in Appendix B, we find that the section modulus of a W14 × 30 section is $S = 42.0$ in.³ Therefore, the maximum bending stress in the beam is

$$\sigma_{max} = \frac{|M_{max}|}{S} = \frac{15\,000 \times 12}{42.0} = 4290 \text{ psi} \qquad \text{Answer}$$

---

[2] The designation of wide flange and other common structural shapes will be discussed in Art. 5.3. The properties of structural shapes are tabulated in Appendix B.

Problems

Unless directed otherwise, neglect the weight of the beam in the following problems. For standard structural shapes, use the properties tabulated in Appendix B.

5.1 A beam constructed from 2-in. by 8-in. boards has the cross section shown in the figure. If the maximum bending moment acting in the beam is $M = 20\,000$ lb · ft, determine the maximum bending stress in (a) board A; and (b) board B.

5.2 The magnitude of the bending moment acting on the circular cross section of a beam is $M = 30\,000$ lb · ft. Calculate the bending stresses at the following points on the cross section: (a) A; (b) B; and (c) D.

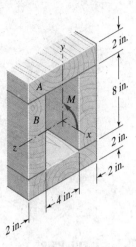

FIG. P5.1

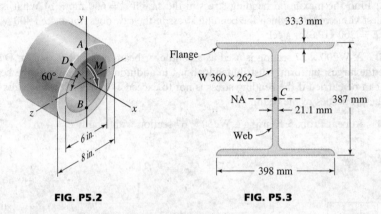

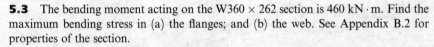

FIG. P5.2   FIG. P5.3

5.3 The bending moment acting on the W360 × 262 section is 460 kN · m. Find the maximum bending stress in (a) the flanges; and (b) the web. See Appendix B.2 for properties of the section.

5.4 The bending moment acting on the triangular cross section of a beam is $M = 3.6$ kN · m. Determine the maximum tensile and compressive bending stresses acting on the cross section.

5.5 The beam described in Sample Problem 4.3 has a rectangular cross section 8 in. high and 4 in. wide. Calculate the maximum bending stress in the beam.

5.6 The beam in Sample Problem 4.5 has a solid circular cross section of diameter d. If the bending working stress for the beam is 18 ksi, determine the smallest allowable value of d.

5.7 For the cantilever beam shown in the figure, find (a) the maximum bending stress and its location; and (b) the bending stress at a point 20 mm from the top of the beam on section B.

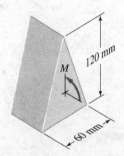

FIG. P5.4

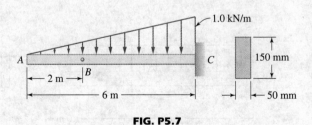

FIG. P5.7

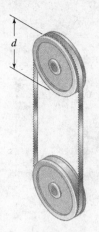

FIG. P5.9

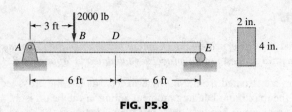

FIG. P5.8

5.8 For the beam shown, calculate (a) the maximum bending stress and (b) the bending stress at a point 0.5 in. from the top of the beam on section D.

5.9 A steel band saw, 20 mm wide and 0.8 mm thick, runs over pulleys of diameter d. (a) Find the maximum bending stress in the saw if $d = 600$ mm. (b) What is the smallest value of d for which the bending stress in the saw does not exceed 400 MPa? Use $E = 200$ GPa for steel.

5.10 A W200 × 27 section is used as a cantilever beam of length $L = 6$ m. Determine the largest uniformly distributed load w_0, in addition to the weight of the beam, that can be carried if the bending stress is not to exceed 140 MPa. See Appendix B.2 for the properties of the beam.

5.11 Repeat Prob. 5.10 using a W250 × 67 section with length $L = 4$ m.

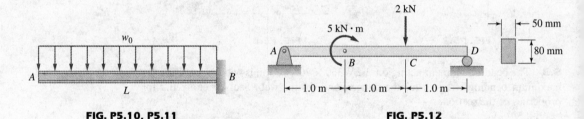

FIG. P5.10, P5.11

FIG. P5.12

5.12 The beam $ABCD$ with a rectangular cross section carries the loading shown in the figure. Determine the magnitude and location of the maximum bending stress in the beam.

5.13 An S380 × 74 section is used as a simply supported beam to carry the uniformly distributed load of magnitude $3W$ and the concentrated load W. What is the maximum allowable value of W if the working stress in bending is 120 MPa?

5.14 The simply supported beam of rectangular cross section carries a distributed load of intensity $w_0 = 3$ kN/m and a concentrated force P. Determine the largest allowable value of P if the bending stress is not to exceed 10 MPa.

5.15 Repeat Prob. 5.14 if $w_0 = 6$ kN/m.

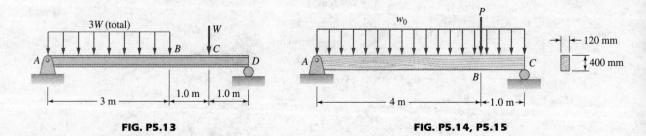

FIG. P5.13

FIG. P5.14, P5.15

5.16 The box beam is made by nailing four 2-in. by 8-in. planks together as shown. (a) Show that the moment of inertia of the cross-sectional area about the neutral axis is 981.3 in.^{4}. (b) Given that $w_0 = 300$ lb/ft, find the largest allowable force P if the bending stress is limited to 1400 psi.

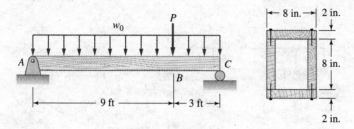

FIG. P5.16, P5.17

5.17 Solve Prob. 5.16 if $w_0 = 600$ lb/ft.

5.18 A wood beam carries the loading shown in the figure. Determine the smallest allowable width b of the beam if the working stress in bending is 10 MPa.

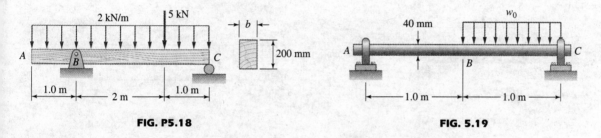

FIG. P5.18 **FIG. 5.19**

5.19 The 40-mm-diameter shaft carries a uniformly distributed load of intensity w_0 over half of its span. The self-aligning bearings at A and C act as simple supports. Find the largest allowable value of w_0 if the bending stress in the shaft is limited to 60 MPa.

5.20 The overhanging beam is made by riveting two C380 × 50 channels back-to-back as shown. The beam carries a uniformly distributed load of intensity w_0 over its entire length. Determine the largest allowable value of w_0 if the working stress in bending is 120 MPa. See Appendix B.4 for properties of the channel section.

5.21 Determine the minimum allowable height h of the beam shown in the figure if the bending stress is not to exceed 20 MPa.

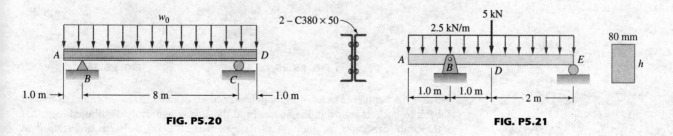

FIG. P5.20 **FIG. P5.21**

5.22 The simply supported beam consists of six tubes that are connected by thin webs. Each tube has a cross-sectional area of 0.2 in.². The beam carries a uniformly distributed load of intensity w_0. If the average bending stress in the tubes is not to exceed 10 ksi, determine the largest allowable value of w_0. Neglect the cross-sectional areas of the webs.

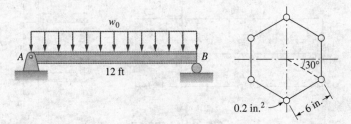

FIG. P5.22

5.23 The simply supported beam of circular cross section carries a uniformly distributed load of intensity w_0 over two-thirds of its length. What is the maximum allowable value of w_0 if the working stress in bending is 50 MPa?

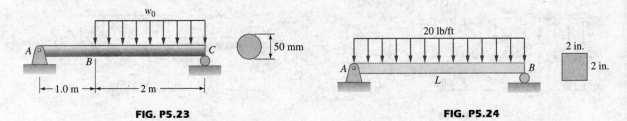

FIG. P5.23

FIG. P5.24

5.24 Find the maximum length L of the beam shown for which the bending stress will not exceed 3000 psi.

5.25 A circular bar of 1.0-in. diameter is formed into the semicircular arch. Determine the maximum bending stress at section B. Assume that the flexure formula for straight beams is applicable.

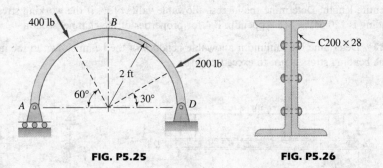

FIG. P5.25

FIG. P5.26

5.26 A cantilever beam, 4 m long, is composed of two C200 × 28 channels riveted back to back as shown in the figure. Find the largest uniformly distributed load that the beam can carry, in addition to its own weight, if (a) the webs are vertical as shown and (b) the webs are horizontal. Use 120 MPa as the working stress in bending. See Appendix B.4 for properties of the channel section.

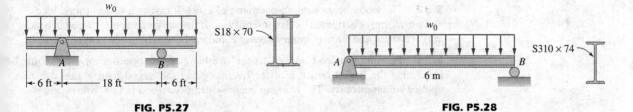

FIG. P5.27 **FIG. P5.28**

5.27 The overhanging beam is made by welding two S18 × 70 sections along their flanges as shown. The beam carries a uniformly distributed load of intensity w_0 in addition to its own weight. Calculate the maximum allowable value of w_0 if the working stress in bending is 20 ksi. See Appendix B.8 for properties of the S-section.

5.28 The S310 × 74 section is used as a simply supported beam to carry a uniformly distributed load of intensity w_0 in addition to its own weight. Determine the largest allowable value of w_0 if the working stress in bending is 120 MPa. See Appendix B.3 for properties of the S-section.

5.29 The stepped shaft carries a concentrated load P at its midspan. If the working stress in bending is 18 ksi, find the largest allowable value of P. Assume that the bearings at A and E act as simple supports.

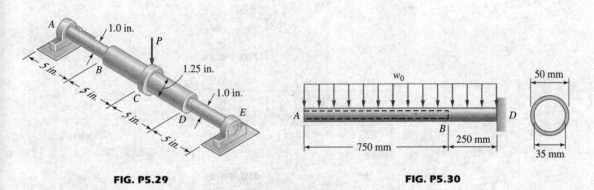

FIG. P5.29 **FIG. P5.30**

5.30 The cantilever beam has a circular cross section of 50-mm outer diameter. Portion AB of the beam is hollow, with an inner diameter of 35 mm. If the working bending stress is 120 MPa, determine the largest allowable intensity w_0 of the uniformly distributed load that can be applied to the beam.

5.31 The square timber used as a railroad tie carries two uniformly distributed loads, each totaling 48 kN. The reaction from the ground is uniformly distributed. Determine the smallest allowable dimension b of the section if the bending stress in timber is limited to 8 MPa.

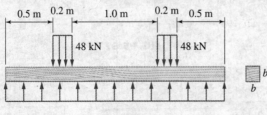

FIG. P5.31

5.32 The wood beam with an overhang of $b = 6$ ft carries a concentrated load P and a uniformly distributed load of intensity w_0. If the working stress for wood in bending is 1200 psi, find the maximum values of P and w_0 that can be applied simultaneously.

5.33 The uniform load applied to the overhang of the beam is $w_0 = 600$ lb/ft. Determine the largest length b of the overhang and the largest load P that can be applied simultaneously. The working bending stress for wood is 1200 psi.

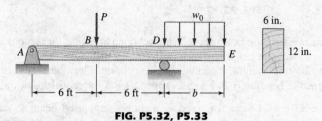

FIG. P5.32, P5.33

5.34–5.38 Determine the maximum tensile and compressive bending stresses in the beam shown.

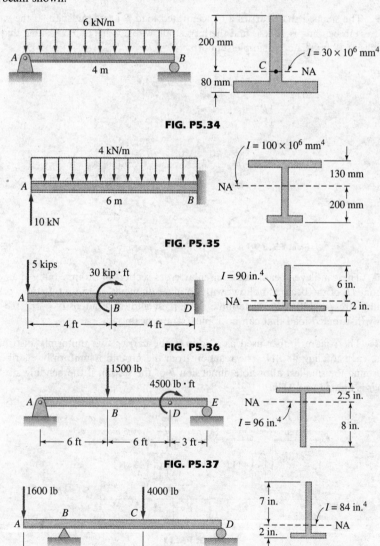

FIG. P5.34

FIG. P5.35

FIG. P5.36

FIG. P5.37

FIG. P5.38

5.39 The overhanging beam carries a uniformly distributed load totaling $8W$ and two concentrated loads of magnitude W each. Determine the maximum safe value of W if the working stresses are 3000 psi in tension and 10 000 psi in compression.

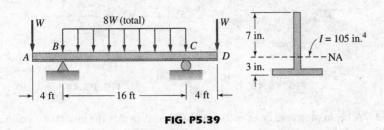

FIG. P5.39

5.40 The beam carries a concentrated load W and a uniformly distributed load that totals $4W$. Determine the largest allowable value of W if the working stresses are 60 MPa in tension and 100 MPa in compression.

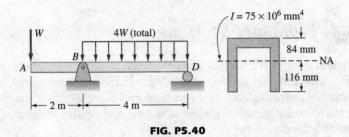

FIG. P5.40

5.41 The inverted T-beam supports three concentrated loads as shown in the figure. Find the maximum allowable value of P if the bending stresses are not to exceed 4 ksi in tension and 10 ksi in compression.

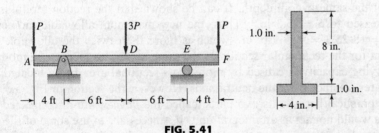

FIG. 5.41

5.42 The intensity of the triangular load carried by the T-section varies from zero at the free end to w_0 at the support. Find the maximum safe value of w_0 given that the working stresses are 4000 psi in tension and 10 000 psi in compression.

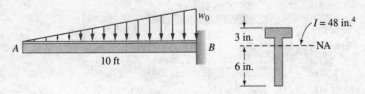

FIG. 5.42

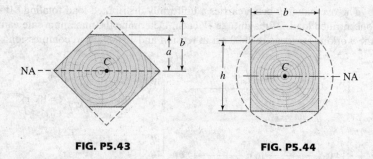

FIG. P5.43   **FIG. P5.44**

5.43 A beam of square cross section is positioned so that the neutral axis coincides with one of the diagonals. The section modulus of this beam can be increased by removing the top and bottom corners as shown. Find the ratio a/b that maximizes the section modulus.

5.44 The beam of rectangular cross section is cut from a round log. Find the ratio b/h that maximizes the section modulus of the beam.

5.3 Economic Sections

The portions of a beam located near the neutral surface are understressed compared with those at the top or bottom. Therefore, beams with certain cross-sectional shapes (including a rectangle and a circle) utilize the material inefficiently because much of the cross section contributes little to resisting the bending moment.

Consider, for example, a beam with the rectangular cross section shown in Fig. 5.5(a). The section modulus of this beam is $S = bh^2/6 = 2(6)^2/6 = 12$ in.^{3}. If the working stress is $\sigma_w = 18$ ksi, the maximum safe bending moment for the beam is $M = \sigma_w S = 18(12) = 216$ kip·in.

In Fig. 5.5(b), we have rearranged the area of the cross section but kept the same overall depth. It can be shown that the section modulus has increased to $S = 25.3$ in.^{3}. Thus, the new maximum allowable moment is $M = 18(25.3) = 455$ kip·in., which is more than twice the allowable moment for the rectangular section of the same area. This increase in moment-carrying capacity is caused by more cross-sectional area being located at a greater distance from the neutral axis. However, the section in Fig. 5.5(b) is not practical because its two parts, called the *flanges*, are disconnected and thus would not act as an integral unit. It is necessary to use some of the area to attach the flanges to each other, as in Fig. 5.5(c). The vertical connecting piece is known as the *web* of the beam. As you will learn later in this chapter, the web functions as the main shear-carrying component of the beam.

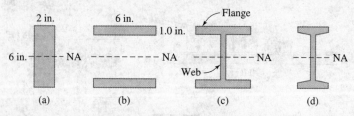

FIG. 5.5 Different ways to distribute the 12-in.^{2} cross-sectional area in (a) without changing the depth.

a. Standard structural shapes

Figure 5.5(c) is similar to a *wide-flange beam,* referred to as a W-shape. A W-shape is one of the most efficient standard structural shapes manufactured because it provides great flexural strength with minimum weight of material. Another "slimmer" version of this shape is the I-beam (referred to as an S-shape) shown in Fig. 5.5(d). The I-beam preceded the wide-flange beam, but because it is not as efficient, it has largely been replaced by the wide-flange beam.

Properties of W- and S-shapes are given in Appendix B. In SI units, a beam of either type is specified by stating its nominal depth in millimeters and its nominal mass per unit length in kilograms per meter. For example, the designation W610 × 140 indicates a wide-flange beam with a nominal depth of 610 mm and a nominal mass per unit length of 140 kg/m. The tables in Appendix B indicate that the actual depth of this beam is 617 mm and the actual mass[3] is 140.1 kg/m. In U.S. Customary units, a structural section is specified by stating its nominal depth in inches followed by its weight in pounds per linear foot. As an example, a W36 × 300 is a wide-flange beam with a nominal depth of 36 in. that weighs 300 lb/ft. The actual depth of this section is 36.74 in. Referring to Appendix B, you will see that in addition to listing the dimensions, tables of structural shapes give properties of the cross-sectional area, such as moment of inertia (I), section modulus (S), and radius of gyration (r)[4] for each principal axis of the area.

When a structural section is selected to be used as a beam, the section modulus must be equal to or greater than the section modulus determined by the flexure equation; that is,

$$S \geq \frac{|M_{\max}|}{\sigma_w} \quad (5.5)$$

This equation indicates that the section modulus of the selected beam must be equal to or greater than the ratio of the bending moment to the working stress.

If a beam is very slender (large L/r), it may fail by *lateral buckling* before the working stress is reached. Lateral buckling entails loss of resistance resulting from a combination of sideways bending and twisting. I-beams are particularly vulnerable to lateral buckling because of their low torsional rigidity and small moment of inertia about the axis parallel to the web. When lateral deflection is prevented by a floor system, or by bracing the flanges at proper intervals, the full allowable stresses may be used; otherwise, reduced stresses should be specified in design. Formulas for the reduction of the allowable stress are specified by various professional organizations, such as the American Institute of Steel Construction (AISC). In this chapter, we assume that all beams are properly braced against lateral deflection.

---

[3] Many designs are based on the nominal mass per meter. However, to illustrate the use of the tables, we will use the actual mass per meter.
[4] The use of r for radius of gyration conforms to the notation of the American Institute of Steel Construction. Be careful not to confuse this term with the r that is frequently used to indicate the radius of a circle.

b. Procedure for selecting standard shapes

A design engineer is often required to select the lightest standard structural shape (such as a W-shape) that can carry a given loading in addition to the weight of the beam. Following is an outline of the selection process:

- Neglecting the weight of the beam, draw the bending moment diagram to find the largest bending moment $M_{\max}$.
- Determine the minimum allowable section modulus from $S_{\min} = |M_{\max}|/\sigma_w$, where σ_w is the working stress.
- Choose the lightest shape from the list of structural shapes (such as in Appendix B) for which $S \geq S_{\min}$ and note its weight.
- Calculate the maximum bending stress $\sigma_{\max}$ in the selected beam caused by the prescribed loading plus the *weight of the beam*. If $\sigma_{\max} \leq \sigma_w$, the selection is finished. Otherwise, the second-lightest shape with $S \geq S_{\min}$ must be considered and the maximum bending stress recalculated. The process must be repeated until a satisfactory shape is found.

Sample Problem 5.5

What is the lightest W-shape beam that will support the 45-kN load shown in Fig. (a) without exceeding a bending stress of 120 MPa? Determine the actual bending stress in the beam.

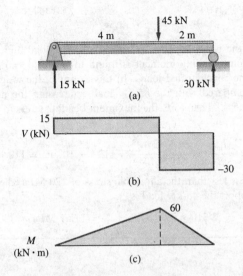

Solution

After finding the reactions shown in Fig. (a), we sketch the shear force and bending moment diagrams in Figs. (b) and (c). The maximum bending moment is $M_{max} = 60$ kN·m, occurring under the applied load. The minimum acceptable section modulus that can carry this moment is

$$S_{min} = \frac{|M_{max}|}{\sigma_w} = \frac{60 \times 10^3}{120 \times 10^6} = 500 \times 10^{-6} \text{ m}^3 = 500 \times 10^3 \text{ mm}^3$$

Referring to the table of properties of W-shapes (Appendix B) and starting at the bottom, we find that the following are the lightest beams in each size group that satisfy the requirement $S \geq S_{min}$:

Section	S (mm³)	Mass (kg/m)
W200 × 52	512 × 10³	52.3
W250 × 45	534 × 10³	44.9
W310 × 39	549 × 10³	38.7

All the beams in the remaining size groups are heavier than those listed above. Therefore, our first choice is the W310 × 39 section with $S = 549 \times 10^{-6}$ m³. (One may wonder why several sizes of beams are manufactured with approximately the same section modulus. The reason is that although the lightest beam is the cheapest on the basis of weight alone, headroom clearances frequently require a beam with less depth than the lightest one.)

The selection of the beam is not complete until a stress calculation is made that includes the weight of the beam, which for the W310 × 39 section is

$$w_0 = (38.7 \text{ kg/m}) \times (9.81 \text{ m/s}^2) = 380 \text{ N/m} = 0.380 \text{ kN/m}$$

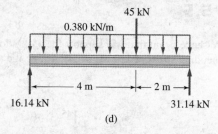

(d)

Figure (d) shows the beam supporting both the 45-kN load and the weight of the beam. The maximum bending moment is found to be $M_{max} = 61.52$ kN·m, again occurring under the concentrated load. (In this example, the weight of the beam is relatively small compared with the 45-kN load, increasing the maximum bending moment by only 2.5%.) Therefore, the maximum bending stress in the selected beam is

$$\sigma_{max} = \frac{|M_{max}|}{S} = \frac{61.52 \times 10^3}{549 \times 10^{-6}} = 112.1 \times 10^6 \text{ Pa} = 112.1 \text{ MPa}$$

Because this stress is less than the allowable stress of 120 MPa, the lightest W-shape that can safely support the 45-kN load is

$$\text{W310} \times 39 \quad (\text{with } \sigma_{max} = 112.1 \text{ MPa}) \qquad \textit{Answer}$$

Problems

5.45 A simply supported beam, 10 m long, carries a uniformly distributed load of intensity 16 kN/m over its entire span. Find the lightest W-shape for which the bending stress does not exceed 120 MPa. What is the actual maximum bending stress in the beam selected?

5.46 Solve Prob. 5.45 if the distributed load is 12 kN/m and the length of the beam is 8 m.

5.47 A simply supported beam with a 15-ft span carries a 9000-lb concentrated load at its midpoint. Select the lightest S-shape for which the bending stress does not exceed 18 ksi. What is the actual maximum bending stress in the beam selected?

5.48 The simply supported beam carries the uniformly distributed load $w_0 = 2000$ lb/ft over a part of its span. Using a working stress of 20 ksi in bending, find the lightest suitable W-shape. What is the actual maximum bending stress in the selected beam?

5.49 Solve Prob. 5.48 if $w_0 = 5000$ lb/ft.

5.50 Find the lightest W-shape for the simply supported beam if the working stress in bending is 18 ksi. What is the actual maximum bending stress in the beam selected?

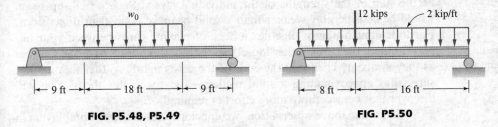

FIG. P5.48, P5.49　　　　　**FIG. P5.50**

5.51 The figure shows the cross section of a reinforced concrete floor that is supported by steel beams. The beams are simply supported, 12 ft long and spaced 8 ft apart. Headroom constraints require the height h of the beams to be no more than 12 in. The loading on the floor is 400 lb/ft^2, including the weight of the concrete. Find the lightest acceptable W-shape for the beams if the working stress is 18.5 ksi. What is the actual maximum bending stress in the beam selected? Assume that the floor loading is equally distributed among the beams.

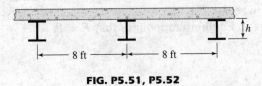

FIG. P5.51, P5.52

5.52 Solve Prob. 5.51 if the constraint on the height of the beams is 8 in.

5.53 Steel beams, spaced 6 ft apart, are driven into the ground to support the sheet piling of a coffer dam. If the working stress in bending is 12 ksi, what is the lightest S-shape that can be used for the beams? The weight of water is 62.5 lb/ft^3.

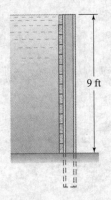

FIG. P5.53

5.54 The beams *ABD* and *DE* are joined by a hinge at *D*. Select the lightest allowable W-shape for each beam if the working stress in bending is 120 MPa. Also calculate the actual maximum bending stress in each beam.

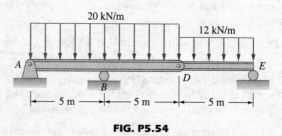

FIG. P5.54

5.4 Shear Stress in Beams

a. Analysis of flexure action

If a beam were composed of many thin layers placed on one another, bending would produce the effect shown in Fig. 5.6. The separate layers would slide past one another, and the total bending strength of the beam would be the sum of the strengths of the individual layers. Such a built-up beam would be considerably weaker than a solid beam of equivalent dimensions. For a demonstration of this, flex a deck of playing cards between your fingers, holding the cards rather loosely so that they can slide past one another as they are bent. Then grip the ends of the cards tightly so that they cannot slip—thus approximating a solid section—and try to flex them. You will discover that considerably more effort is required.

From the above observation, we conclude that the horizontal layers in a solid beam are prevented from sliding by shear stresses that act between the layers. It is this shear stress that causes the beam to act as an integral unit rather than as a stack of individual layers.

To further illustrate shear stress, consider the simply supported beam in Fig. 5.7. We isolate the shaded portion of the beam by using two cutting planes: a vertical cut along section ① and a horizontal cut located at the distance y' above the neutral axis. The isolated portion is subjected to the two horizontal forces P and F shown in the figure (vertical forces are not shown). The axial force P is due to the bending stress acting on the area A' of section ①, whereas F is the resultant of the shear stress acting on the horizontal surface. Equilibrium requires that $F = P$.

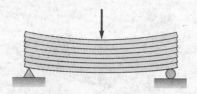

FIG. 5.6 Bending of a layered beam with no adhesive between the layers.

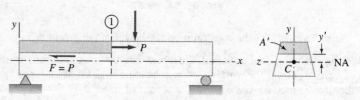

FIG. 5.7 Equilibrium of the shaded portion of the beam requires a longitudinal shear force $F = P$, where P is the resultant of the normal stress acting on area A' of section ①.

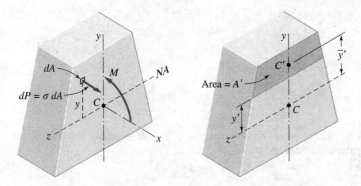

FIG. 5.8 Calculating the resultant force of the normal stress over a portion of the cross-sectional area.

We can calculate P using Fig. 5.8. The axial force acting on the area element dA of the cross section is $dP = \sigma\, dA$. If M is the bending moment acting at section ① of the beam, the bending stress is given by Eq. (5.3): $\sigma = -My/I$, where y is the distance of the element from the neutral axis, and I is the moment of inertia of the *entire cross-sectional area* of the beam about the neutral axis. Therefore,

$$dP = -\frac{My}{I}\, dA$$

Integrating over the area A', we get

$$P = \int_{A'} dP = -\frac{M}{I} \int_{A'} y\, dA = -\frac{MQ}{I} \qquad (5.6)$$

where

$$Q = \int_{A'} y\, dA \qquad (5.7\text{a})$$

is the *first moment of area A'* about the neutral axis. The negative sign in Eq. (5.6) indicates that positive M results in forces P and F that are directed opposite to those shown in Fig. 5.7. Denoting the distance between the neutral axis and the centroid C' of the area A' by $\bar{y}'$, we can write Eq. (5.7) as

$$Q = A'\bar{y}' \qquad (5.7\text{b})$$

In Eqs. (5.7), Q represents the first moment of the cross-sectional area that lies *above* y'. Because the first moment of the total cross-sectional area about the neutral axis is zero, the first moment of the area *below* y' is $-Q$. Therefore, the magnitude of Q can be computed by using the area either above or below y', whichever is more convenient. The maximum value of Q occurs at the neutral axis where $y' = 0$. It follows that the horizontal shear force F is largest on the neutral surface. The variation of Q with y' for a rectangular cross section is illustrated in Fig. 5.9.

b. Horizontal shear stress

Consider the free-body diagram (FBD) of the shaded portion of the beam in Fig. 5.10 (we show only the horizontal forces). This body is bounded by sections ① and ② that are separated by the infinitesimal distance dx, and

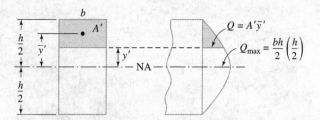

FIG. 5.9 Variation of the first moment Q of area A' about the neutral axis for a rectangular cross section.

a horizontal plane located a distance y' above the neutral axis of the cross section. If the bending moment at section ① of the beam is M, the resultant force acting on face ① of the body is given by Eq. (5.6):

$$P = -M\frac{Q}{I}$$

As explained before, Q is the first moment of the area A' (the area of face ① of the body), and I is the moment of inertia of the entire cross-sectional area of the beam. The bending moment acting at section ② is $M + dM$, where dM is the infinitesimal change in M over the distance dx. Therefore, the resultant normal force acting on face ② of the body is

$$P + dP = -(M + dM)\frac{Q}{I}$$

Because these two forces differ by

$$(P + dP) - P = -(M + dM)\frac{Q}{I} - \left(-M\frac{Q}{I}\right) = -dM\frac{Q}{I} \qquad \text{(a)}$$

equilibrium can exist only if there is an equal and opposite shear force dF acting on the horizontal surface.

If we let τ be the *average shear stress* acting on the horizontal surface, its resultant is $dF = \tau b\, dx$, where b is the width of the cross section at $y = y'$, as shown in Fig. 5.10. The equilibrium requirement for the horizontal forces is

$$\Sigma F = 0: \qquad (P + dP) - P + \tau b\, dx = 0$$

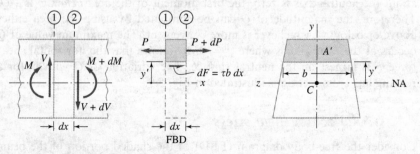

FIG. 5.10 Determining the longitudinal shear stress from the free-body diagram of a beam element.

Substituting for $(P + dP) - P$ from Eq. (a), we get

$$-dM\frac{Q}{I} + \tau b\, dx = 0$$

which gives

$$\tau = \frac{dM}{dx}\frac{Q}{Ib} \qquad \text{(b)}$$

Recalling the relationship $V = dM/dx$ between the shear force and the bending moment, we obtain for the average horizontal shear stress

$$\boxed{\tau = \frac{VQ}{Ib}} \qquad (5.8)$$

Often the shear stress is uniform over the width b of the cross section, in which case τ can be viewed as the actual shear stress.

c. Vertical shear stress

Strictly speaking, Eq. (5.8) represents the shear stress acting on a horizontal plane of the beam (a plane parallel to the neutral surface). However, we pointed out in Chapter 1 (without general proof) that a shear stress is always accompanied by a complementary shear stress of equal magnitude, the two stresses acting on mutually perpendicular planes. In a beam, the complementary stress τ' is a vertical shear stress that acts on the cross section of the beam, as illustrated in Fig. 5.11(a). Because $\tau = \tau'$, Eq. (5.8) can be used to compute the vertical as well as the horizontal shear stress at a point in a beam. The resultant of the vertical shear stress on the cross-sectional area A of the beam is, of course, the shear force V; that is, $V = \int_A \tau\, dA$.

To prove that $\tau = \tau'$, consider the infinitesimal element of the beam in Fig. 5.11(b) as a free body. For translational equilibrium of the element, the shear stress τ on the top face requires an equal, but opposite, balancing shear stress on the bottom face. Similarly, the complementary shear stress τ' on the front face must be balanced by an opposite stress on the back face. The corresponding forces on the faces of the element are obtained by multiplying each stress by the area of the face on which it acts. Thus, the horizontal and vertical forces are $\tau\, dx\, dz$ and $\tau'\, dy\, dz$, respectively. These forces form two couples of opposite sense. For rotational equilibrium, the magnitudes of the couples must be equal; that is, $(\tau\, dx\, dz)\, dy = (\tau'\, dy\, dz)\, dx$, which yields $\tau = \tau'$.

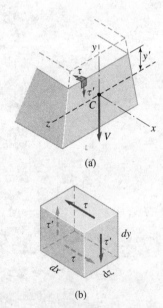

FIG. 5.11 The vertical stress τ' acting at a point on a cross section equals the longitudinal shear stress τ acting at the same point.

d. Discussion and limitations of the shear stress formula

We see that the shear stress formula $\tau = VQ/(Ib)$ predicts that the largest shear stress in a prismatic beam occurs at the cross section that carries the largest vertical shear force V. The location (the value of y') of the maximum shear stress within that section is determined by the ratio Q/b. Because Q is always maximum at $y' = 0$, the neutral axis is usually a candidate for the location of the maximum shear stress. However, if the width b at the neutral axis is larger than at other parts of the cross section, it is necessary to compute τ at two or more values of y' before its maximum value can be determined.

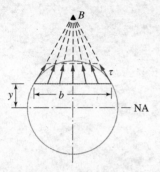

FIG. 5.12 Shear stress distribution along a horizontal line of a circular cross section.

When deriving the shear stress formula, Eq. (5.8), we stated that τ should be considered as the *average* shear stress. This restriction is necessary because the variation of shear stress across the width b of the cross section is often unknown. Equation (5.8) is sufficiently accurate for rectangular cross sections and for cross sections that are composed of rectangles, such as W- and S-shapes. For other cross-sectional shapes, however, the formula for τ must be applied with caution. Let us consider as an example the circular cross section in Fig. 5.12. It can be shown that the shear stress at the periphery of the section must be tangent to the boundary, as shown in the figure. The direction of shear stresses at interior points is unknown, except at the centerline, where the stress is vertical due to symmetry. To obtain an estimate of the maximum shear stress, the stresses are assumed to be directed toward a common center B, as shown. The vertical components of these shear stresses are assumed to be uniform across the width of the section and are computed from Eq. (5.8). Under this assumption, the shear stress at the neutral axis is $1.333 V/(\pi r^2)$. A more elaborate analysis[5] shows that the shear stress actually varies from $1.23 V/(\pi r^2)$ at the edges to $1.38 V/(\pi r^2)$ at the center.

Shear stress, like normal stress, exhibits stress concentrations near sharp corners, fillets, and holes in the cross section. The junction between the web and the flange of a W-shape is also an area of stress concentration.

e. Rectangular and wide-flange sections

We will now determine the shear stress as a function of y for a rectangular cross section of base b and height h. From Fig. 5.13, the shaded area is $A' = b[(h/2) - y]$, its centroidal coordinate being $\bar{y}' = [(h/2) + y]/2$. Thus,

$$Q = A'\bar{y}' = \left[b\left(\frac{h}{2} - y\right)\right]\left[\frac{1}{2}\left(\frac{h}{2} + y\right)\right] = \frac{b}{2}\left(\frac{h^2}{4} - y^2\right)$$

and Eq. (5.8) then becomes

$$\tau = \frac{VQ}{Ib} = \frac{V}{2I}\left(\frac{h^2}{4} - y^2\right) \tag{c}$$

We see that the shear stress is distributed parabolically across the depth of the section, as shown in Fig. 5.13. The maximum shear stress occurs at the neutral axis. If we substitute $y = 0$ and $I = bh^3/12$, Eq. (c) reduces to

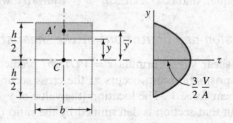

FIG. 5.13 Shear stress distribution on a rectangular cross section.

---

[5] See S. Timoshenko and J. N. Goodier, *Theory of Elasticity*, 3d ed. (New York: McGraw-Hill, 1970).

$$\tau_{max} = \frac{3}{2}\frac{V}{bh} = \frac{3}{2}\frac{V}{A} \qquad (5.9)$$

where A is the cross-sectional area. Therefore, the shear stress in a rectangular section is 50% greater than the average shear stress on the cross section.

In wide-flange sections (W-shapes), most of the bending moment is carried by the flanges, whereas the web resists the bulk of the vertical shear force. Figure 5.14 shows the shear stress distribution in the web of a typical W-shape. In this case, Q (the first moment of A' about the neutral axis) is contributed mainly by the flanges of the beam. Consequently, Q does not vary much with y, so that the shear stress in the web is almost constant. In fact, $\tau_{max} = V/A_{web}$ can be used as an approximation to the maximum shear stress in most cases, where A_{web} is the cross-sectional area of the web.

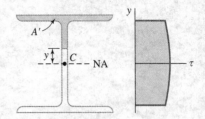

FIG. 5.14 Shear stress distribution on the web of a wide-flange beam.

f. Procedure for analysis of shear stress

The following procedure can be used to determine the shear stress at a given point in a beam:

- Use equilibrium analysis to determine the vertical shear force V acting on the cross section containing the specified point (the construction of a shear force diagram is usually a good idea).
- Locate the neutral axis and compute the moment of inertia I of the cross-sectional area about the neutral axis. (If the beam is a standard structural shape, its cross-sectional properties are listed in Appendix B.)
- Compute the first moment Q of the cross-sectional area that lies above (or below) the specified point.
- Calculate the shear stress from $\tau = VQ/(Ib)$, where b is the width of the cross section at the specified point. Note that τ is the actual shear stress only if it is uniform across b; otherwise, τ should be viewed as the average shear stress.

The maximum shear stress τ_{max} on a given cross section occurs where Q/b is largest. If the width b is constant, then τ_{max} occurs at the neutral axis because that is where Q has its maximum value. If b is not constant, it is necessary to compute the shear stress at more than one point in order to determine its maximum value.

In the U.S. Customary system, τ is commonly expressed in lb/in.^{2} (psi). Consistency of units in the shear stress formula then requires the cross-sectional properties to be in inches and V in pounds. Thus,

$$\tau\,[\text{lb/in.}^2] = \frac{V\,[\text{lb}]\,Q\,[\text{in.}^3]}{I\,[\text{in.}^4]\,b\,[\text{in.}]}$$

In the SI system, where τ is measured in N/m^2 (Pa), meters must be used for the cross-sectional dimensions and V must be in newtons, yielding

$$\tau\,[\text{N/m}^2] = \frac{V\,[\text{N}]\,Q\,[\text{m}^3]}{I\,[\text{m}^4]\,b\,[\text{m}]}$$

Sample Problem 5.6

The simply supported wood beam in Fig. (a) is fabricated by gluing together three 160-mm by 80-mm planks as shown. Calculate the maximum shear stress in (1) the glue; and (2) the wood.

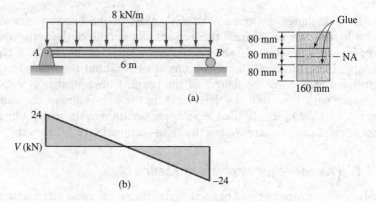

(a)

(b)

Solution

From the shear force diagram in Fig. (b) we see that the maximum shear force in the beam is $V_{max} = 24$ kN, occurring at the supports. The neutral axis is the axis of symmetry of the cross section. The moment of inertia of the cross-sectional area of the beam about the neutral axis is

$$I = \frac{bh^3}{12} = \frac{160(240)^3}{12} = 184.32 \times 10^6 \text{ mm}^4 = 184.32 \times 10^{-6} \text{ m}^4$$

Part 1

The shear stress in the glue corresponds to the horizontal shear stress discussed in Art. 5.4. Its maximum value can be computed from Eq. (5.8): $\tau_{max} = V_{max} Q/(Ib)$, where Q is the first moment of the area A' shown in Fig. (c); that is,

$$Q = A'\bar{y}' = (160 \times 80)(80) = 1.024 \times 10^6 \text{ mm}^3 = 1.024 \times 10^{-3} \text{ m}^3$$

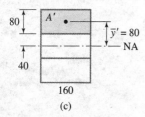

Therefore, the maximum shear stress in the glue, which occurs over either support, is

$$\tau_{max} = \frac{V_{max} Q}{Ib} = \frac{(24 \times 10^3)(1.024 \times 10^{-3})}{(184.32 \times 10^{-6})(0.160)}$$

$$= 833 \times 10^3 \text{ Pa} = 833 \text{ kPa} \qquad \textbf{Answer}$$

Part 2

Because the cross section is rectangular, the maximum shear stress in the wood can be calculated from Eq. (5.9):

$$\tau_{max} = \frac{3}{2}\frac{V_{max}}{A} = \frac{3}{2}\frac{(24 \times 10^3)}{(0.160)(0.240)} = 938 \times 10^3 \text{ Pa} = 938 \text{ kPa} \qquad \textbf{Answer}$$

The same result can be obtained from Eq. (5.8), where now A' is the area above the neutral axis, as indicated in Fig. (d). The first moment of this area about the neutral axis is

$$Q = A'\bar{y}' = (160 \times 120)(60) = 1.152 \times 10^6 \text{ mm}^3 = 1.152 \times 10^{-3} \text{ m}^3$$

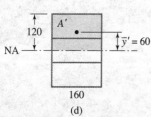

Equation (5.8) thus becomes

$$\tau_{max} = \frac{V_{max}Q}{Ib} = \frac{(24 \times 10^3)(1.152 \times 10^{-3})}{(184.32 \times 10^{-6})(0.160)}$$

$$= 938 \times 10^3 \text{ Pa} = 938 \text{ kPa}$$

which agrees with the previous result.

■

Sample Problem 5.7

The W12 × 40 section in Fig. (a) is used as a beam. If the vertical shear force acting at a certain section of the beam is 16 kips, determine the following at that section: (1) the minimum shear stress in the web; (2) the maximum shear stress in the web; and (3) the percentage of the shear force that is carried by the web.

Solution

An idealized drawing of the W12 × 40 section is shown in Fig. (b), where the dimensions were obtained from the tables in Appendix B. The drawing approximates the web and the flanges by rectangles, thereby ignoring the small fillets and rounded corners present in the actual section. The tables also list the moment of inertia of the section about the neutral axis as $I = 310$ in.^{4}.

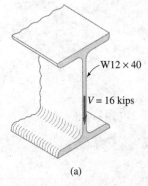

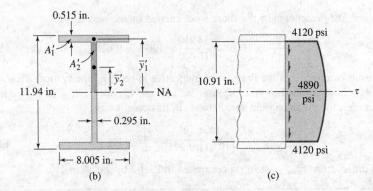

Part 1

The minimum shear stress in the web occurs at the junction with the flange, where Q/b is smallest (note that $b = 0.295$ in. is constant within the web). Therefore, in the shear stress formula $\tau = VQ/(Ib)$, Q is the first moment of the area A_1' shown in Fig. (b) about the neutral axis:

$$Q = A_1'\bar{y}_1' = (8.005 \times 0.515)\frac{11.94 - 0.515}{2} = 23.55 \text{ in.}^3$$

The minimum shear stress in the web thus becomes

$$\tau_{min} = \frac{VQ}{Ib} = \frac{(16 \times 10^3)(23.55)}{(310)(0.295)} = 4120 \text{ psi} \qquad \textit{Answer}$$

Part 2

The maximum shear stress is located at the neutral axis, where Q/b is largest. Hence, Q is the first moment of the area above (or below) the neutral axis—that is, the combined moment of areas A_1' and A_2' in Fig. (a). The moment of A_1' was calculated in Part 1. The moment of A_2' about the neutral axis is $A_2'\bar{y}_2'$, where

$$A'_2 = \left(\frac{11.94}{2} - 0.515\right)(0.295) = 1.6092 \text{ in.}^2$$

$$\bar{y}'_2 = \frac{1}{2}\cdot\left(\frac{11.94}{2} - 0.515\right) = 2.7275 \text{ in.}$$

Therefore,

$$Q = A'_1\bar{y}'_1 + A'_2\bar{y}'_2 = 23.55 + (1.6092)(2.7275) = 27.94 \text{ in.}^3$$

and the maximum shear stress in the web becomes

$$\tau_{max} = \frac{VQ}{Ib} = \frac{(16\times 10^3)(27.94)}{(310)(0.295)} = 4890 \text{ psi} \qquad \textbf{Answer}$$

Part 3

The distribution of the shear stress in the web is shown in Fig. (c). The shear force carried by the web is

$$V_{web} = (\text{cross-sectional area of web}) \times (\text{area of shear diagram})$$

We know from the discussion in Art. 5.4 that shear stress distribution is parabolic. Recalling that the area of a parabola is $(2/3)(\text{base} \times \text{height})$, we obtain

$$V_{web} = (10.91 \times 0.295)\left[4120 + \frac{2}{3}(4890 - 4120)\right] = 14\,910 \text{ lb}$$

Therefore, the percentage of the shear force carried by the web is

$$\frac{V_{web}}{V} \times 100\% = \frac{14\,910}{16\,000} \times 100\% = 93.2\% \qquad \textbf{Answer}$$

This result confirms that the flanges are ineffective in resisting the vertical shear.

It was mentioned in Art. 5.5 that we can use $\tau_{max} = V/A_{web}$ as a rough approximation for the maximum shear stress. In this case, we get

$$\frac{V}{A_{web}} = \frac{16 \times 10^3}{(10.91)(0.295)} = 4970 \text{ psi}$$

which differs from $\tau_{max} = 4890$ psi computed in Part 2 by less than 2%.

Sample Problem 5.8

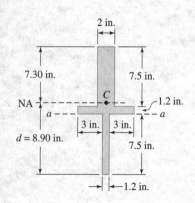

The figure shows the cross section of a beam that carries a vertical shear force $V = 12$ kips. The distance from the bottom of the section to the neutral axis is $d = 8.90$ in., and the moment of inertia of the cross-sectional area about the neutral axis is $I = 547$ in.^{4}. Determine the maximum shear stress on this cross section.

Solution

The maximum shear stress may occur at the neutral axis (where Q is largest) or at level a-a in the lower fin (where the width of the cross section is smaller than at the neutral axis). Therefore, we must calculate the shear stress at both locations.

Shear Stress at Neutral Axis We take Q to be the first moment of the rectangular area *above* the neutral axis (the area below the neutral axis could also be used). Noting that the dimensions of this area are 2 in. by 7.30 in., we have

$$Q = A'\bar{y}' = (2 \times 7.30)\frac{7.30}{2} = 53.29 \text{ in.}^3$$

and the shear stress at the neutral axis is

$$\tau = \frac{VQ}{Ib} = \frac{(12 \times 10^3)(53.29)}{(547)(2)} = 585 \text{ psi}$$

Shear Stress at a-a It is easier to compute Q by using the area *below* the line *a-a* rather than the area above the line. The dimensions of this area are $b = 1.2$ in. and $h = 7.5$ in. Consequently,

$$Q = A'\bar{y}' = (1.2 \times 7.5)\left(8.90 - \frac{7.5}{2}\right) = 46.35 \text{ in.}^3$$

and the shear stress becomes

$$\tau = \frac{VQ}{Ib} = \frac{(12 \times 10^3)(46.35)}{(547)(1.2)} = 847 \text{ psi}$$

The maximum shear stress is the larger of the two values:

$$\tau_{max} = 847 \text{ psi} \quad \text{(occurring at } a\text{-}a) \qquad\qquad\qquad \textbf{Answer}$$

Problems

5.55 The cross section of a timber beam is 80 mm wide and 160 mm high. The vertical shear force acting on the section is 40 kN. Determine the shear stress at (a) the neutral axis; and (b) 40 mm above the neutral axis.

5.56 Show that the average shear stress at the neutral axis of a circular cross section is $\tau = 4V/(3\pi r^2)$, where V is the shear force and r is the radius of the section.

5.57 Show that the average shear stress acting at the neutral axis of a beam with a thin-walled tubular cross section is $\tau_{max} = 2V/A$, where V is the shear force acting in the section and A is the cross-sectional area.

5.58 The figure shows the cross section of a simply supported beam that carries a uniformly distributed loading of intensity 200 lb/ft over its entire length L. If the working shear stress is 80 psi, determine the largest allowable value of L.

5.59 The vertical shear force acting on the cross section shown is 1800 lb. Determine the shear stress at (a) the neutral axis; and (b) 4 in. above the neutral axis.

5.60 The vertical shear force acting in a beam with the cross section shown in the figure is 20 kips. Find the maximum shear stress in the beam.

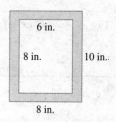

FIG. P5.58, P5.59

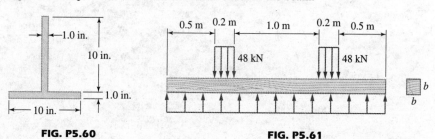

FIG. P5.60 FIG. P5.61

5.61 The square timber is used as a railroad tie. It carries two uniformly distributed loads of 48 kN each. The reaction from the ground is distributed uniformly over the length of the tie. Determine the smallest allowable dimension b if the working stress in shear is 1.0 MPa.

5.62 The simply supported beam consists of six tubes, each of cross-sectional area 0.2 in.^{2}. The tubes are connected by thin webs of thickness t. A uniformly distributed loading of intensity $w_0 = 120$ lb/ft is applied over the entire length of the beam. Calculate the smallest allowable value of t if the working shear stress for the webs is 2000 psi. Neglect the contribution of the webs when calculating I and Q.

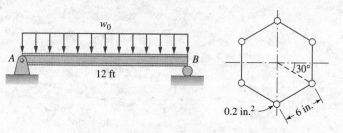

FIG. P5.62

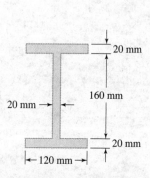

FIG. P5.63, P5.64

5.63 The vertical shear force acting on the I-section shown is 100 kN. Compute (a) the maximum shear stress acting on the section; and (b) the percentage of the shear force carried by the web.

5.64 Solve Prob. 5.63 if the height of the web is 200 mm instead of 160 mm.

5.65 The beam is built up of 1/4-in. vertical plywood strips separated by wood blocks. Determine the vertical shear force that causes a maximum shear stress of 200 psi.

5.66 The manufactured wood beam carries a uniformly distributed load of intensity w_0. Determine the largest safe value of w_0 if the maximum shear stress in the wood is limited to 300 psi.

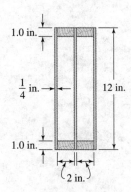

FIG. P5.65

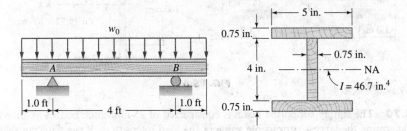

FIG. P5.66

5.67 For the beam shown in the figure, find the shear stress at a point 30 mm above the bottom of the beam at section C.

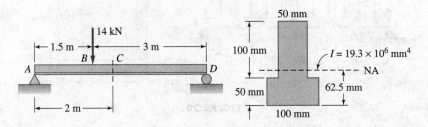

FIG. P5.67

5.68 For the beam shown, compute the shear stress at 1.0-in. vertical intervals on the cross section that carries the maximum shear force. Plot the results.

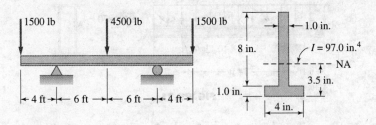

FIG. P5.68

5.69 The manufactured wood beam carries the concentrated loads shown. What is the maximum safe value of P if the working stress in shear is 6 MPa?

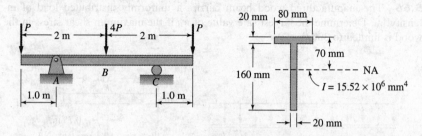

FIG. P5.69

5.70 The simply supported beam is constructed of 25-mm-thick boards as shown. Determine the largest permissible value of the load intensity w_0 if the working shear stress is 1.2 MPa.

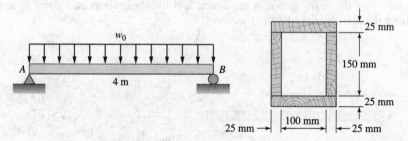

FIG. P5.70

5.71 The beam consists of two S18 × 70 sections that are welded together as shown. If the intensity of the uniformly distributed load is $w_0 = 15$ kip/ft, calculate the maximum shear stress in the beam.

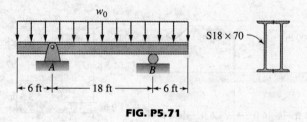

FIG. P5.71

5.72 If the maximum vertical shear force acting in a beam with the cross section shown is 30 kN, determine the maximum shear stress in the beam.

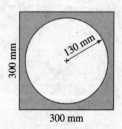

FIG. P5.72

5.73 The maximum shear force in a beam with the cross section shown in the figure is 36 kips. Determine the maximum shear stress in the beam.

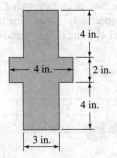

FIG. P5.73

5.74 The beam shown in cross section is fabricated by welding a fin to the back of a C250 × 45 section. The neutral axis of the cross section is located at $d = 8.78$ mm, and the moment of inertia about this axis is $I = 4.35 \times 10^6$ mm^4. What is the maximum shear stress caused by a 200-kN vertical shear force?

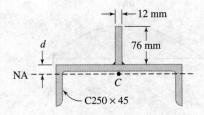

FIG. P5.74

5.5 Design for Flexure and Shear

Up to this point, we have considered bending and shear stresses in beams separately. We now explore the design of beams that satisfy the prescribed design criteria for both bending and shear. In general, bending stress governs the design of long beams, whereas shear stress is critical in short beams. We can draw this conclusion by observing that the shear force V is determined only by the magnitude of the loading, whereas the bending moment M depends on the magnitude of the loading *and* the length L of the beam. In other words, for a given loading, V_{max} is independent of L, but M_{max} increases as L is increased.

Shear stress is of concern in timber beams because of the low shear strength of wood along the grain; the typical ratio of shear strength to bending strength is 1:10. Very thin webs in metal beams can also fail in shear or by buckling caused by the shear stress.

The most direct method for satisfying both design criteria is to perform two separate computations: one based on the bending stress criterion and the other on the shear stress criterion. Examination of the results will then reveal which of the designs satisfies both criteria.

Sample Problem 5.9

The simply supported beam of rectangular cross section in Fig. (a) carries a total load W that is distributed uniformly over its length L. The working stresses in bending and shear are σ_w and τ_w, respectively. Determine the critical value of L for which the maximum shear stress and the maximum bending stress reach their working values simultaneously.

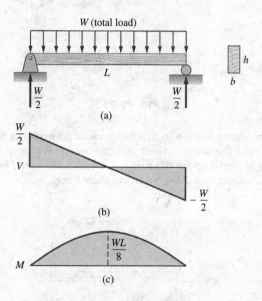

Solution

Figures (b) and (c) show the shear force and bending moment diagrams for the beam. We see that the maximum shear force $V_{max} = W/2$ occurs over the supports, and the maximum bending moment $M_{max} = WL/8$ occurs at midspan.

Design for Shear The maximum value of W that does not violate the shear stress criterion $\tau \leq \tau_w$ is obtained by setting $\tau_{max} = \tau_w$ in Eq. (5.9):

$$\tau_w = \frac{3}{2}\frac{V_{max}}{A} = \frac{3}{2}\frac{W/2}{bh}$$

which gives

$$W = \frac{4}{3}bh\tau_w \qquad (a)$$

Note that this value of W is independent of the length of the beam.

Design for Bending Letting $\sigma_{max} = \sigma_w$ in Eq. (5.4b), we get

$$\sigma_w = \frac{|M_{max}|}{S} = \frac{WL/8}{bh^2/6} = \frac{3}{4}\frac{WL}{bh^2}$$

yielding

$$W = \frac{4}{3}\frac{bh^2}{L}\sigma_w \qquad (b)$$

which is the maximum W that does not violate the bending stress criterion $\sigma \leq \sigma_w$. Observe that W decreases with increasing L.

Equating the expressions for W in Eqs. (a) and (b), we obtain

$$\frac{4}{3} bh\tau_w = \frac{4}{3} \frac{bh^2}{L} \sigma_w$$

from which

$$L = \frac{\sigma_w}{\tau_w} h \qquad \text{Answer}$$

For beams longer than this critical length, bending stress governs the design; otherwise, shear stress governs. If we assume, for example, that $\sigma_w = 10\tau_w$ (typical of timber), we obtain $L = 10h$.

Sample Problem 5.10

The box beam in Fig. (a) supports the concentrated loads $2P$ and P. Compute the maximum allowable value of P if the working stresses in bending and shear are $\sigma_w = 1000$ psi and $\tau_w = 100$ psi, respectively.

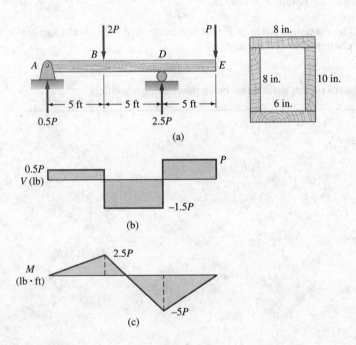

Solution

The support reactions, the shear force diagram, and the bending moment diagram, shown in Figs. (a)–(c), were obtained by equilibrium analysis (P is assumed to be measured in pounds). We see that the largest magnitude of the shear force is $|V_{\max}| = 1.5P$ lb, occurring in the segment BD. The magnitude of the maximum bending moment is $|M_{\max}| = 5P$ lb·ft $= 60P$ lb·in. at D.

The moment of inertia of the cross section in Fig. (a) about the neutral axis is the difference between the moments of inertia of the outer and inner rectangles:

$$I = \left(\frac{bh^3}{12}\right)_{\text{outer}} - \left(\frac{bh^3}{12}\right)_{\text{inner}} = \frac{8(10)^3}{12} - \frac{6(8)^3}{12} = 410.7 \text{ in.}^4$$

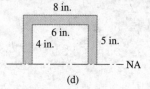

(d)

Design for Shear The maximum shear stress occurs at the neutral axis in segment BD. The first moment of the cross-sectional area above the neutral axis is computed by subtracting the first moment of the inner rectangle in Fig. (d) from the first moment of the outer rectangle:

$$Q = (A'\bar{y}')_{outer} - (A'\bar{y}')_{inner} = (8 \times 5)(2.5) - (6 \times 4)(2) = 52.0 \text{ in.}^3$$

The largest P that can be applied without exceeding the working shear stress is obtained from the shear formula:

$$\tau_w = \frac{|V_{max}|Q}{Ib} \qquad 100 = \frac{(1.5P)(52.0)}{(410.7)(2)}$$

which gives $P = 1053$ lb.

Design for Bending The flexure formula yields the largest P that will not violate the bending stress constraint. Letting $\sigma_{max} = \sigma_w$ and noting that the distance from the neutral axis to the top of the cross section is $c = 5$ in., we get

$$\sigma_w = \frac{|M_{max}|c}{I} \qquad 1000 = \frac{(60P)(5)}{410.7}$$

from which we obtain $P = 1369$ lb.

The maximum value of P that can be applied safely is the smaller of the two values computed above; namely,

$$P = 1053 \text{ lb} \qquad \textit{Answer}$$

with the maximum shear stress being the limiting condition.

Problems

Neglect the weight of the beam in the following problems.

5.75 A simply supported beam of length L has a rectangular cross section of width b and height h. The beam carries the concentrated load P in the middle of its span. If $L = 10h$, determine the ratio of σ_{max}/τ_{max}.

5.76 The laminated beam, shown in cross section, is composed of five 6-in. by 2-in. planks that are glued together. The beam carries a uniformly distributed load of intensity w_0 over its 6-ft simply supported span. If the working stresses are 90 psi for shear in glue, 120 psi for shear in wood, and 1200 psi for bending in wood, determine the maximum allowable value of w_0.

5.77 The cantilever beam of length L has a circular cross section of diameter d. The beam carries a distributed load that varies linearly as shown in the figure. Find the expression for the ratio σ_{max}/τ_{max} in terms of L and d. Assume that τ_{max} occurs at the neutral axis and that its distribution along the diameter of the cross section is uniform.

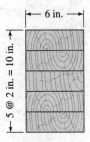

FIG. P5.76

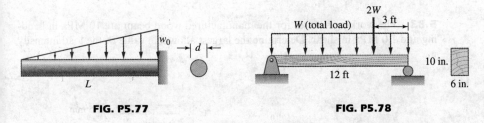

FIG. P5.77   **FIG. P5.78**

5.78 The simply supported wood beam supports a load W that is distributed uniformly over its length and a concentrated force $2W$. If the working stresses are 1500 psi in bending and 120 psi in shear, determine the maximum allowable value of W.

5.79 A W250 × 49 section (with the web vertical) is used as a cantilever beam 4 m long. The beam supports a uniformly distributed loading of intensity w_0 over its entire length. Determine the ratio σ_{max}/τ_{max}.

5.80 A simply supported timber beam with the cross section shown carries a uniformly distributed loading of intensity 6 kN/m over its length L. Find the value of L for which $\sigma_{max} = 8\tau_{max}$ and the corresponding values of σ_{max} and τ_{max}.

5.81 The uniformly distributed load is carried by a beam that has the same cross section as in Fig. P5.80. The working stresses are 10 MPa in bending and 1.0 MPa in shear. Determine the largest allowable value of the load intensity w_0.

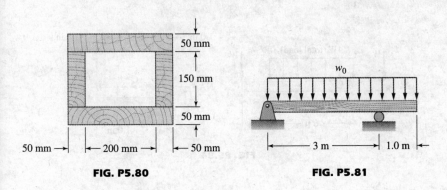

FIG. P5.80   **FIG. P5.81**

5.82 The beam supports two concentrated loads P and a triangular load of magnitude $3P$. The beam is made of three 1/4-in.-thick plywood strips separated by wood blocks as shown. Determine the maximum allowable value of P if the working stresses are 1200 psi in bending and 200 psi in shear.

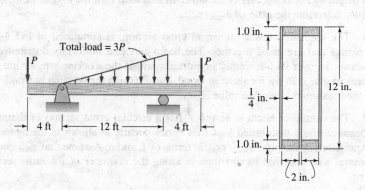

FIG. P5.82

5.83 The working stresses for the manufactured wood beam are 10 MPa in bending and 1.0 MPa in shear. Determine the largest allowable value of the load intensity w_0.

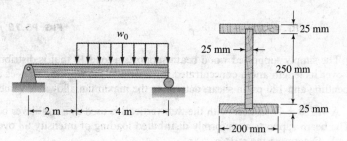

FIG. P5.83

5.84 The overhanging beam carries two concentrated loads W and a uniformly distributed load of magnitude $4W$. The working stresses are 6000 psi in tension, 10 000 psi in compression, and 8000 psi in shear. (a) Show that the neutral axis of the cross section is located at $d = 2.167$ in. and that the moment of inertia of the cross-sectional area about the neutral axis is $I = 61.55$ in.^{4}. (b) Determine the largest allowable value of W.

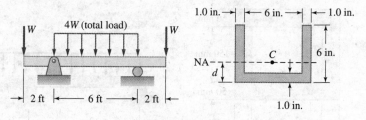

FIG. P5.84

5.85 The thin-walled tube is used as a beam to support the uniformly distributed load of intensity w_0. (a) Find the largest allowable value of w_0 based on the working stress in bending of 100 MPa. (b) Compute the corresponding maximum shear stress in the beam.

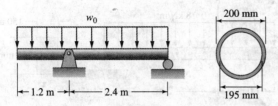

FIG. P5.85

5.86 The cast iron inverted T-section supports two concentrated loads of magnitude P. The working stresses are 48 MPa in tension, 140 MPa in compression, and 30 MPa in shear. (a) Show that the neutral axis of the cross section is located at $d = 48.75$ mm and that the moment of inertia of the cross-sectional area about this axis is $I = 11.918 \times 10^6$ mm^4. (b) Find the maximum allowable value of P.

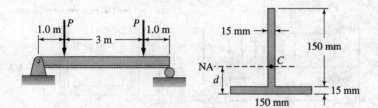

FIG. P5.86

5.87 Determine the largest safe value of the load intensity w_0 carried by the I-beam if the working stresses are 4000 psi in bending and 300 psi in shear.

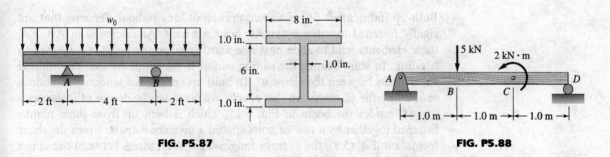

FIG. P5.87

FIG. P5.88

5.88 The wood beam has a square cross section. Find the smallest allowable cross-sectional dimensions if the working stresses are 8 MPa in bending and 1.0 MPa in shear.

5.89 The rectangular wood beam is loaded as shown in the figure. Determine the largest allowable magnitude of the load P if the working stresses are 10 MPa in bending and 1.2 MPa in shear.

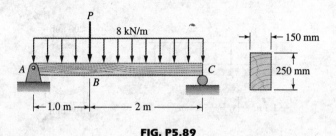

FIG. P5.89

5.90 The channel section carries a uniformly distributed load totaling $8W$ and two concentrated loads of magnitude W. (a) Verify that the neutral axis is located at $d = 50$ mm and that the moment of inertia about that axis is 15.96×10^6 mm^4. (b) Determine the maximum allowable value for W if the working stresses are 30 MPa in tension, 70 MPa in compression, and 20 MPa in shear.

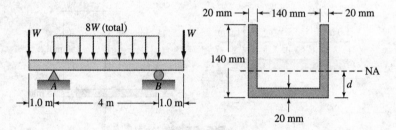

FIG. P5.90

5.6 Design of Fasteners in Built-up Beams

Built-up (fabricated) beams are composed of longitudinal elements that are rigidly fastened together by rivets, bolts, or nails. As discussed in Art. 5.4, these elements tend to slide past one another when the beam is subjected to bending. In solid beams, the sliding action is prevented by the longitudinal shear stress between the elements. In built-up beams, the tendency to slide is resisted by the fasteners. In this article, we consider the design of fasteners.

Consider the beam in Fig. 5.15, which is built up from three planks fastened together by a row of bolts spaced a distance e apart. From the shear formula in Eq. (5.8), the average longitudinal shear stress between the upper two planks is

$$\tau = \frac{VQ}{Ib}$$

where Q is the first moment of the shaded area in Fig. 5.15(c) about the neutral axis. The shear force F that must be carried by a bolt is obtained by multiplying this shear stress by the shaded area of magnitude eb in Fig. 5.15(a). Thus,

$$F = \tau e b = \frac{VQ}{Ib}(eb) = \frac{VQe}{I} \qquad (5.10a)$$

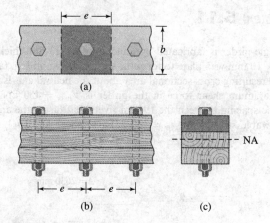

FIG. 5.15 Three planks fastened by a row of bolts.

Let us now assume that the allowable force (working force) F_w for a bolt in shear is given. The value of F_w may be governed by the shear strength of the bolt or by the bearing strength of the planks. If we neglect friction between the planks, the largest allowable spacing of the bolts is obtained by setting $F = F_w$ in Eq. (5.10a), yielding

$$e = \frac{F_w I}{VQ} \qquad (5.10b)$$

If we follow the common practice of having constant spacing of fasteners throughout the length of the beam, then V in Eq. (5.10b) represents the maximum shear force in the beam.

Sample Problem 5.11

A plate and angle girder is fabricated by attaching four 13-mm-thick angle sections to a 1100 mm × 10 mm web plate to form the section shown in Fig. (a). The moment of inertia of the resulting cross-sectional area about the neutral axis is $I = 4140 \times 10^6$ mm^4. If the maximum shear force in the girder is $V_{max} = 450$ kN, determine the largest allowable spacing between the 19-mm rivets that fasten the angles to the web plate. The allowable stresses are $\tau_w = 100$ MPa in shear and $(\sigma_b)_w = 280$ MPa in bearing.

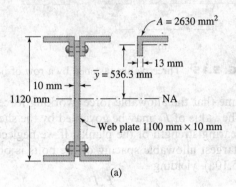

(a)

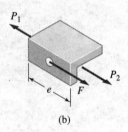

(b)

Solution

The rivets provide the shear connection between the angle sections and the web plate. Figure (b) shows a segment of an angle section of length e, where e is the spacing of the rivets. The shear force F in the rivet resists the difference of the longitudinal forces P_1 and P_2 (caused by the bending stresses) that act at the two ends of the segment. The value of F can be obtained from Eq. (5.10a): $F = VQe/I$, where Q is the first moment of the cross-sectional area of the angle section about the neutral axis of the beam. Referring to the data in Fig. (a), we obtain

$$Q = A\bar{y} = (2630)(536.3) = 1.4105 \times 10^6 \text{ mm}^3 = 1.4105 \times 10^{-3} \text{ m}^3$$

Before we can determine the spacing of the rivets, we must calculate the allowable force F_w that can be transmitted by a rivet. Assuming that F_w is governed by the working *shear stress* in the rivet, we have

$$F_w = A_{rivet}\tau_w = \frac{\pi(0.019)^2}{4}(100 \times 10^6) = 28.35 \times 10^3 \text{ N}$$

Because the rivets are in double shear, the bearing force between the rivet and the web plate is $2F$. Thus, the allowable value of F_w, determined by the *bearing stress* of the web plate, is given by

$$2F_w = (d_{rivet}\,t_{web})(\sigma_b)_w = (0.019)(0.010)(280 \times 10^6)$$

which yields $F_w = 26.60 \times 10^3$ N. There is no need to consider the bearing stress between the rivet and the angle sections because the bearing force is F, which is only one-half of the force that acts between the rivet and the web plate. We conclude that the allowable force transmitted by a rivet is governed by bearing stress between the rivet and the web plate, its value being

$$F_w = 26.60 \times 10^3 \text{ N}$$

The largest allowable spacing of the rivets can now be calculated from Eq. (5.10b):

$$e = \frac{F_w I}{VQ} = \frac{(26.60 \times 10^3)(4140 \times 10^{-6})}{(450 \times 10^3)(1.4105 \times 10^{-3})} = 0.1735 \text{ m} = 173.5 \text{ mm} \qquad \text{Answer}$$

Problems

5.91 The beam shown in cross section is fabricated by bolting three 80-mm by 200-mm wood planks together. The beam is loaded so that the maximum shear stress in the wood is 1.4 MPa. If the maximum allowable shear force in a bolt is $F_w = 8$ kN, determine the largest permissible spacing of the bolts.

5.92 The box beam in the figure is secured by screws spaced 5 in. apart along the length of the beam. The beam carries a concentrated force P at the third point of its 12-ft simply supported span. (a) Determine the maximum allowable value of P if the working shear stress in the beam is $\tau_w = 120$ psi and the shear force in each screw is limited to $F_w = 300$ lb. (b) What is the corresponding maximum bending stress in the beam?

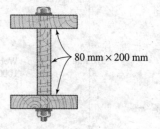

FIG. 5.91

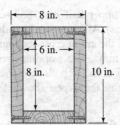

FIG. P5.92, P5.93

5.93 A uniformly distributed load of intensity w_0 is applied over the middle 6 ft of a simply supported beam 12 ft long. The cross section of the beam is as shown in the figure, but it is turned by 90°, so that the 10-in. dimension is horizontal. The working stresses for the beam are $\sigma_w = 1200$ psi in bending and $\tau_w = 120$ psi in shear. The screws are spaced 2 in. apart, the allowable shear force in each screw being 200 lb. Determine the maximum allowable value of w_0.

5.94 The 12-ft-long walkway of a scaffold is made by screwing two 12-in. by 1/2-in. sheets of plywood to 1.5-in. by 3.5-in. timbers as shown. The screws have a 5-in. spacing along the length of the walkway. The working stress in bending is $\sigma_w = 850$ psi for the plywood and the timbers, and the allowable shear force in each screw is $F_w = 250$ lb. What limit should be placed on the weight W of a person who walks across the plank?

5.95 A simply supported beam, 12 ft long, consists of three 4-in. by 6-in. planks that are secured by bolts spaced 12 in. apart. The bolts are tightened to a tensile stress of 20 ksi. The beam carries a concentrated load at its midspan that causes a maximum bending stress of 1200 psi. If the coefficient of friction between the planks is 0.4, determine the bolt diameters so that the shear between planks can be transmitted by friction only.

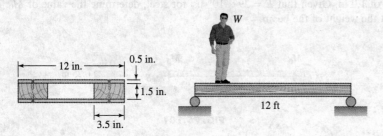

FIG. P5.94

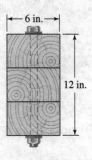

FIG. P5.95

184 CHAPTER 5 Stresses in Beams

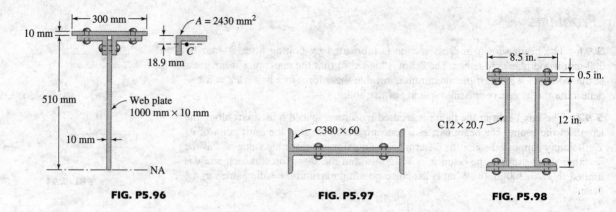

FIG. P5.96   FIG. P5.97   FIG. P5.98

5.96 The figure shows the upper half of a built-up girder (the cross section is symmetric about the neutral axis). All rivets used in fabrication have a diameter of 22 mm. The moment of inertia of the entire cross-sectional area of the girder about the neutral axis is $I = 4770 \times 10^6$ mm^4. The working stresses are 100 MPa for rivets in shear and 280 MPa for bearing of the web plate. If the maximum shear force carried by the girder is 450 kN, determine the largest allowable spacing of rivets that join the angles to the web plate.

5.97 Two C380 × 60 channels are riveted back-to-back and used as a beam with the web horizontal. The 19-mm rivets are spaced 200 mm apart along the length of the beam. What is the largest allowable shear force in the beam if the allowable stresses are 100 MPa for rivets in shear and 220 MPa for the channels in bearing?

5.98 Two C12 × 20.7 channels are joined to 8.5-in. by 0.5-in. plates with 3/4-in. rivets to form a beam with the cross section shown in the figure. The maximum shear force in the beam is 20 kips. (a) Determine the maximum allowable spacing of the rivets using 5000 psi for the working shear stress. (b) Compute the corresponding maximum bearing stress in the channels.

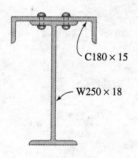

FIG. P5.99

5.99 The beam is fabricated by attaching a C180 × 15 channel to a W250 × 18 shape with 15-mm rivets as shown. The maximum shear force in the beam is 65 kN. (a) Find the maximum allowable spacing of the rivets if the working stress is 100 MPa in shear. (b) What is the corresponding maximum bearing stress exerted by the rivets?

Review Problems

5.100 The bending moment acting on the cross section of the beam is $M = 1.8$ kN·m. Find the maximum tensile and compressive bending stresses acting on the cross section. Neglect the weight of the beam.

5.101 The steel beam is bent by the end couples M_0 that cause a midspan deflection of 1.0 in. Given that $E = 29 \times 10^6$ psi for steel, determine the value of M_0. Neglect the weight of the beam.

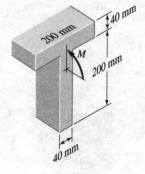

FIG. P5.100

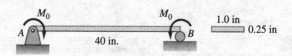

FIG. P5.101

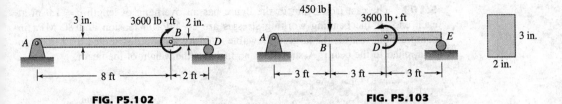

FIG. P5.102 **FIG. P5.103**

5.102 The stepped beam has a rectangular cross section 2 in. wide. Determine the maximum bending stress in the beam due to the 3600-lb·ft couple. Neglect the weight of the beam.

5.103 Determine the magnitude and location of the maximum bending stress for the beam. Neglect the weight of the beam.

5.104 Determine the maximum tensile and compressive bending stresses in the beam. Neglect the weight of the beam.

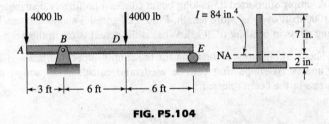

FIG. P5.104

5.105 The overhanging beam carries concentrated loads of magnitudes P and $2P$. If the bending working stresses are 15 ksi in tension and 18 ksi in compression, determine the largest allowable value of P. Neglect the weight of the beam.

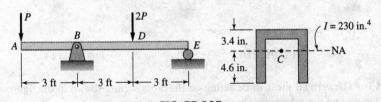

FIG. P5.105

5.106 The S380 × 74 section carries a uniformly distributed load totaling $3W$ and a concentrated load W. Determine the largest value of W if the working stress in bending is 120 MPa. Neglect the weight of the beam.

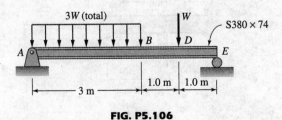

FIG. P5.106

5.107 The cast iron beam in the figure has an overhang of length $b = 1.0$ m at each end. If the bending working stresses are 20 MPa in tension and 80 MPa in compression, what is the largest allowable intensity w_0 of a distributed load that can be applied to the beam? Assume that w_0 includes the weight of the beam.

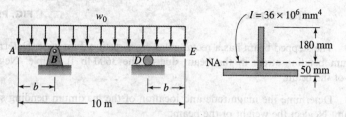

FIG. P5.107, P5.108

5.108 Solve Prob. 5.107 using $b = 3$ m, all other data remaining unchanged.

5.109 A simply supported 25-ft-long beam carries a uniformly distributed load of intensity 1000 lb/ft over its entire length. Find the lightest S-shape that can be used if the working stress in bending is 20 ksi. What is the actual stress in the beam selected?

5.110 The working stress in bending for the simply supported beam is 120 MPa. Find the lightest W-shape that can be used, and calculate the actual maximum bending stress in the beam selected.

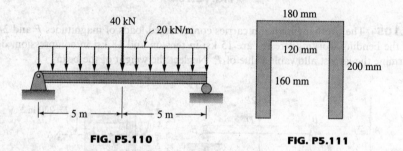

FIG. P5.110   FIG. P5.111

5.111 The vertical shear force acting on the cross section shown in the figure is 60 kN. Determine the maximum shear stress on the section.

5.112 The cross section of a beam is formed by gluing two pieces of wood together as shown. If the vertical shear force acting on the section is 60 kN, determine the shear stress (a) at the neutral axis; and (b) on the glued joint.

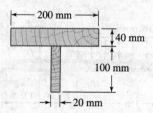

FIG. P5.112

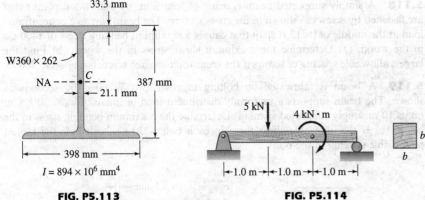

FIG. P5.113 **FIG. P5.114**

5.113 The W360 × 262 section carries a vertical shear force of 650 kN. For this section, calculate; (a) the minimum shear stress in the web; (b) the maximum shear stress in the web, and (c) the percentage of the vertical shear force carried by the web.

5.114 The simply supported timber beam has a square cross section. Find the smallest allowable value of the dimension b if the working stresses are 8 MPa in bending and 1.0 MPa in shear.

5.115 A simply supported beam with the cross section shown supports a uniformly distributed load of intensity w_0 over its length L. Determine the ratio σ_{max}/τ_{max} for this beam.

5.116 The wood beam carries a concentrated load W and a distributed load totaling $0.7W$. Find the largest allowable value of W if the working stresses are 1200 psi in bending and 150 psi in shear.

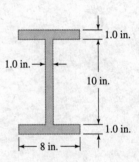

FIG. P5.115

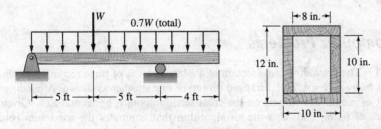

FIG. P5.116

5.117 The weight W travels across the span of the wood beam. Determine the maximum allowable value of W if the working stresses are 1200 psi in bending and 120 psi in shear.

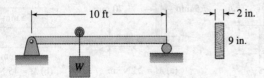

FIG. P5.117

FIG. P5.118

5.118 A simply supported beam is made of four 2-in. by 6-in. wood planks that are fastened by screws as shown in the cross section. The beam carries a concentrated load at the middle of its 12-ft span that causes a maximum bending stress of 1400 psi in the wood. (a) Determine the maximum shear stress in the wood. (b) Find the largest allowable spacing of screws if the shear force in each screw is limited to 200 lb.

5.119 A beam is fabricated by bolting together two W200 × 100 sections as shown. The beam supports a uniformly distributed load of intensity $w_0 = 30$ kN/m on its 10-m simply supported span. (a) Determine the maximum bending stress in the beam. (b) If the allowable shear force in each bolt is 30 kN, calculate the largest permissible spacing of the bolts.

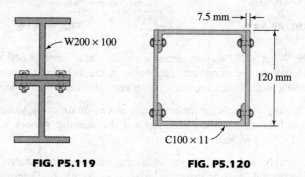

FIG. P5.119   **FIG. P5.120**

5.120 Two C100 × 11 channels are joined to 120-mm by 7.5-mm plates with 10-mm rivets to form the beam shown in the figure. The beam carries a uniformly distributed loading of intensity w_0 over its 4-m simply supported span. (a) If the working bending stress is 120 MPa, find the largest allowable value of w_0. (b) Determine the largest allowable spacing of rivets using 80 MPa as the working stress for rivets in shear.

Computer Problems

C5.1 The symmetric cross section of a beam consists of three rectangles of dimensions b_i by h_i ($i = 1, 2, 3$), arranged on top of one another as shown. A bending moment of magnitude M acts on the cross section about a horizontal axis. Given the values of b_i, h_i, and M, write an algorithm that computes the maximum bending stress acting on the cross section. Apply the algorithm to the cross sections and moments shown in parts (a) and (b) of the figure.

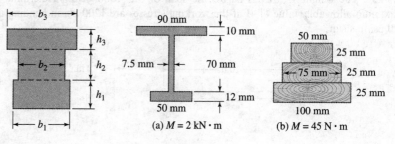

FIG. C5.1

C5.2 The cantilever beam of length L has a rectangular cross section of constant width b. The height h of the beam varies as $h = h_1 + (h_2 - h_1)(x/L)^2$. The magnitude of the uniformly distributed load is w_0. Given L, b, h_1, h_2, and w_0, construct an algorithm to plot the maximum normal stress acting on the cross section as a function of x. (a) Run the algorithm with $L = 2$ m, $b = 25$ mm; $h_1 = 30$ mm, $h_2 = 120$ mm, and $w_0 = 2$ kN/m. (b) Find the combination of h_1 and h_2 that minimizes the maximum normal stress in the beam while maintaining the 75-mm average height of the beam in part (a).

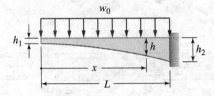

FIG. C5.2

C5.3 The simply supported beam of rectangular cross section has a constant width b, but its height h varies as $h = h_1 + (h_2 - h_1) \sin(\pi x/L)$, where L is the length of the beam. A concentrated load P acts at the distance x from the left support. Given L, b, h_1, h_2, and P, write an algorithm to plot the maximum bending stress under the load as a function of x. Run the algorithm with $P = 100$ kips, $L = 36$ ft, $b = 4$ in. and (a) $h_1 = 18$ in., $h_2 = 30$ in.; and (b) $h_1 = h_2 = 26.55$ in. (The two beams have the same volume.)

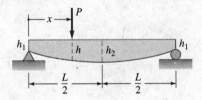

FIG. C5.3

C5.4 The cantilever beam of length L and constant flange width b is fabricated from a plate with thickness t. The height h of the web varies linearly from h_1 to h_2. The beam carries a concentrated load P at the free end. Given L, b, t, h_1, h_2, and P, construct an algorithm to plot the maximum normal and shear stresses on the cross section as functions of the distance x. Run the algorithm with $L = 8$ ft, $b = 4$ in., $t = 0.5$ in., $h_1 = 2$ in., $h_2 = 18$ in., and $P = 4800$ lb.

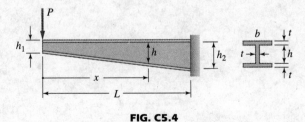

FIG. C5.4

C5.5 The simply supported I-beam of length L carries a concentrated load P at its midspan. The overall height of the cross section is h, the flange width is b, and the wall thickness is t. The working stresses are σ_w in bending and τ_w in shear. Given P, L, h, t, σ_w, and τ_w, construct an algorithm to plot σ_{max}/σ_w and τ_{max}/τ_w as functions of b in the range $b = 0.25h$ to $1.25h$, where σ_{max} and τ_{max} are the maximum normal and shear stresses in the beam. (a) Run the algorithm with $P = 60$ kN, $L = 1.5$ m, $h = 175$ mm, $t = 12.5$ mm, $\sigma_w = 80$ MPa, and $\tau_w = 30$ MPa. (b) Experiment with different values of h to determine the smallest allowable h for the beam in part (a) and the corresponding value of b.

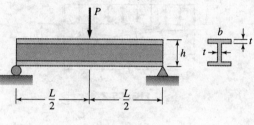

FIG. C5.5

6 Deflection of Beams

Wing of a commercial passenger plane. Wings are essentially cantilever beams subjected to aerodynamic forces, such as lift. Since excessive deformation can destroy the aerodynamic integrity, the deflection of a wing is as important as its strength.

6.1 Introduction

In this chapter, we consider the deflection of statically determinate beams. Because the design of beams is frequently governed by rigidity rather than strength, the computation of deflections is an integral component of beam analysis. For example, building codes specify limits on deflections as well as stresses. Excessive deflection of a beam not only is visually disturbing but also may cause damage to other parts of the building. For this reason, building codes limit the maximum deflection of a beam to about 1/360th of its span. Deflections can also govern the design of machinery, cars, and aircraft. In the design of a lathe, for example, the deflections must be kept below the dimensional tolerances of the parts being machined. Cars and aircraft must have sufficient rigidity to control structural vibrations.

Deflections also play a crucial role in the analysis of statically indeterminate beams. They form the bases for the compatibility equations that are needed to supplement the equations of equilibrium, as we discovered in our analysis of axially loaded bars and torsion of shafts.

A number of analytical methods are available for determining the deflections of beams. Their common basis is the differential equation that relates the deflection to the bending moment. The solution of this equation is complicated because the bending moment is usually a discontinuous function, so that the equations must be integrated in a piecewise fashion. The various methods of deflection analysis are essentially different techniques for solving this differential equation. We consider two such methods in this text:

Method of double integration This method is fairly straightforward in its application, but it often involves considerable algebraic manipulation. We also present a variation of the method that simplifies the algebra by the use of *discontinuity functions*. The primary advantage of the double-integration method is that it produces the equation for the deflection everywhere along the beam.

Moment-area method The moment-area method is a semigraphical procedure that utilizes the properties of the area under the bending moment diagram. It is the quickest way to compute the deflection at a specific location if the bending moment diagram has a simple shape. The method is not suited for deriving the deflection as a function of distance along the beam without using a computer program.

In this chapter, we also discuss the *method of superposition*, in which the applied loading is represented as a series of simple loads for which deflection formulas are available. Then the desired deflection is computed by adding the contributions of the component loads (principle of superposition).

6.2 Double-Integration Method

Figure 6.1(a) illustrates the bending deformation of a beam. Recall that in a real beam, the displacements and slopes are very small if the stresses are below the elastic limit, so that the deformation shown in the figure is greatly exaggerated. The deformed axis of the beam is called its *elastic curve*. In this article, we derive the differential equation for the elastic curve and describe a method for its solution.

a. Differential equation of the elastic curve

As shown in Fig. 6.1(a), we let x be the horizontal coordinate of an arbitrary point A on the axis of the beam, measured from the fixed origin O. As the beam deforms, its axis becomes curved and A is displaced to the position A'. The vertical deflection of A, denoted by v, is considered to be positive if directed in the positive direction of the y-axis—that is, upward in Fig. 6.1(a). Because the axis of the beam lies on the neutral surface, its length does not change. Therefore, the distance $\overline{OA'}$, measured along the elastic curve, is also x. It follows that the horizontal deflection of A is negligible provided the slope of the elastic curve remains small.

Consider next the deformation of an infinitesimal segment AB of the beam axis, as shown in Fig. 6.1(b). The elastic curve $A'B'$ of the segment has

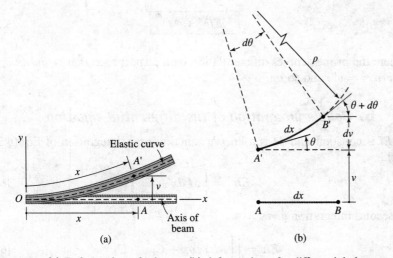

FIG. 6.1 (a) Deformation of a beam; (b) deformation of a differential element of beam axis.

the same length dx as the undeformed segment. If we let v be the deflection of A, then the deflection of B is $v + dv$, with dv being the infinitesimal change in the deflection over the length dx. Similarly, the slope angles at the ends of the deformed segment are denoted by θ and $\theta + d\theta$. From the geometry of the figure, we obtain

$$\frac{dv}{dx} = \sin\theta \approx \theta \quad (6.1)$$

The approximation is justified because θ is small. From Fig. 6.1(b), we also see that

$$dx = \rho\, d\theta \quad \text{(a)}$$

where ρ is the radius of curvature of the deformed segment. Rewriting Eq. (a) as $1/\rho = d\theta/dx$ and substituting θ from Eq. (6.1), we obtain

$$\frac{1}{\rho} = \frac{d^2v}{dx^2} \quad (6.2)$$

When deriving the flexure formula in Art. 5.2, we obtained the *moment-curvature relationship*

$$\frac{1}{\rho} = \frac{M}{EI} \quad \text{(5.2b, repeated)}$$

where M is the bending moment acting on the segment, E is the modulus of elasticity of the beam material, and I represents the moment of inertia of the cross-sectional area about the neutral (centroidal) axis. Substitution of Eq. (5.2b) into Eq. (6.2) yields

$$\frac{d^2v}{dx^2} = \frac{M}{EI} \quad (6.3a)$$

which is the *differential equation of the elastic curve*. The product EI, called the *flexural rigidity* of the beam, is usually constant along the beam. It is convenient to write Eq. (6.3a) in the form

$$EIv'' = M \qquad (6.3b)$$

where the prime denotes differentiation with respect to x; that is, $dv/dx = v'$, $d^2v/dx^2 = v''$, and so on.

b. Double integration of the differential equation

If EI is constant and M is a known function of x, integration of Eq. (6.3b) yields

$$EIv' = \int M\,dx + C_1 \qquad (6.4)$$

A second integration gives

$$EIv = \iint M\,dx\,dx + C_1 x + C_2 \qquad (6.5)$$

where C_1 and C_2 are constants of integration to be determined from the prescribed constraints (for example, the boundary conditions) on the deformation of the beam. Because Eq. (6.5) gives the deflection v as a function of x, it is called the *equation of the elastic curve*. The analysis described above is known as the *double-integration method* for calculating beam deflections.

In Eq. (6.5), the term $\iint M\,dx\,dx$ gives the shape of the elastic curve. The position of the curve relative to the coordinate axes is determined by the constants of integration: C_1 represents a rigid-body rotation about the origin and C_2 is a rigid-body displacement in the y-direction. Hence, the computation of the constants is equivalent to adjusting the position of the elastic curve so that it fits properly on the supports.

If the bending moment or flexural rigidity is not a smooth[1] function of x, a separate differential equation must be written for each beam segment between the discontinuities. This means that if there are n such segments, two integrations will produce $2n$ constants of integration (two per segment). We now show that there are also $2n$ equations available for finding the constants. We first recognize that the elastic curve must not contain gaps or kinks. In other words, the slopes and deflections must be continuous at the junctions where the segments meet. Because there are $n-1$ junctions between the n segments, these *continuity conditions* give us $2(n-1)$ equations. Two additional equations are provided by the *boundary conditions* imposed by the supports, so that there are a total of $2(n-1) + 2 = 2n$ equations. As you can see, the evaluation of the constants of integration can be tedious if the beam contains more than two segments.

c. Procedure for double integration

The following procedure assumes that EI is constant in each segment of the beam:

- Sketch the elastic curve of the beam, taking into account the boundary conditions: zero displacement at pin supports as well as zero displacement and zero slope at built-in (cantilever) supports, for example.

---

[1] The term *smooth* here means that the function and its derivatives are continuous.

- Use the method of sections to determine the bending moment M at an arbitrary distance x from the origin. Always show M acting in the *positive direction* on the free-body diagram (this assures that the equilibrium equations yield the correct sign for the bending moment). If the loading has discontinuities, a separate expression for M must be obtained for each segment between the discontinuities.
- By integrating the expressions for M twice, obtain an expression for EIv in each segment. Do not forget to include the constants of integration.
- Evaluate the constants of integration from the boundary conditions and the continuity conditions on slope and deflection between segments.

Frequently only the magnitude of the deflection, called the *displacement*, is required. We denote the displacement by δ; that is, $\delta = |v|$.

Sample Problem 6.1

The cantilever beam AB of length L shown in Fig. (a) carries a uniformly distributed load of intensity w_0, which includes the weight of the beam. (1) Derive the equation of the elastic curve. (2) Compute the maximum displacement if the beam is a W12 × 35 section using $L = 8$ ft, $w_0 = 400$ lb/ft, and $E = 29 \times 10^6$ psi.

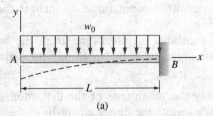

(a)

Solution

Part 1

The dashed line in Fig. (a) represents the elastic curve of the beam. The bending moment acting at the distance x from the left end can be obtained from the free-body diagram in Fig. (b) (note that V and M are shown acting in their positive directions):

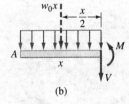

(b)

$$M = -w_0 x \left(\frac{x}{2}\right) = -\frac{w_0 x^2}{2}$$

Substituting the expression for M into the differential equation $EIv'' = M$, we get

$$EIv'' = -\frac{w_0 x^2}{2}$$

Successive integrations yield

$$EIv' = -\frac{w_0 x^3}{6} + C_1 \qquad \text{(a)}$$

$$EIv = -\frac{w_0 x^4}{24} + C_1 x + C_2 \qquad \text{(b)}$$

The constants C_1 and C_2 are obtained from the boundary conditions at the built-in end B, which are:

1. $v'|_{x=L} = 0$ (support prevents rotation at B). Substituting $v' = 0$ and $x = L$ into Eq. (a), we get

$$C_1 = \frac{w_0 L^3}{6}$$

2. $v|_{x=L} = 0$ (support prevents deflection at B). With $v = 0$ and $x = L$, Eq. (b) becomes

$$0 = -\frac{w_0 L^4}{24} + \left(\frac{w_0 L^3}{6}\right) L + C_2 \qquad C_2 = -\frac{w_0 L^4}{8}$$

If we substitute C_1 and C_2 into Eq. (b), the equation of the elastic curve is

$$EIv = -\frac{w_0 x^4}{24} + \frac{w_0 L^3}{6} x - \frac{w_0 L^4}{8}$$

$$EIv = \frac{w_0}{24}(-x^4 + 4L^3 x - 3L^4) \qquad \textit{Answer}$$

Part 2

From Table B.7 in Appendix B, the properties of a W12 × 35 shape are $I = 285$ in.^{4} and $S = 45.6$ in.^{3} (section modulus). From the result of Part 1, the maximum displacement of the beam is (converting feet to inches)

$$\delta_{max} = |v|_{x=0} = \frac{w_0 L^4}{8EI} = \frac{(400/12)(8 \times 12)^4}{8(29 \times 10^6)(285)} = 0.0428 \text{ in.} \qquad \text{Answer}$$

To get a better appreciation of the magnitude of the displacement, let us compute the maximum bending stress in the beam. The magnitude of the maximum bending moment, which occurs at B, is $M_{max} = w_0 L^2/2$. Therefore, the maximum bending stress is

$$\sigma_{max} = \frac{M_{max}}{S} = \frac{w_0 L^2}{2S} = \frac{(400/12)(8 \times 12)^2}{2(45.6)} = 33\,700 \text{ psi}$$

which is close to the proportional limit of 35 000 psi for structural steel. We see that the maximum displacement is very small compared to the length of the beam even when the material is stressed to its proportional limit.

∎

Sample Problem 6.2

The simply supported beam ABC in Fig. (a) carries a distributed load of maximum intensity w_0 over its span of length L. Determine the maximum displacement of the beam.

Solution

The bending moment and the elastic curve (the dashed line in Fig. (a)) are symmetric about the midspan. Therefore, we will analyze only the left half of the beam (segment AB).

Because of the symmetry, each support carries half of the total load, so that the reactions are $R_A = R_C = w_0 L/4$. The bending moment in AB can be obtained from the free-body diagram in Fig. (b), yielding

$$M = \frac{w_0 L}{4} x - \frac{w_0 x^2}{L}\left(\frac{x}{3}\right) = \frac{w_0}{12L}(3L^2 x - 4x^3)$$

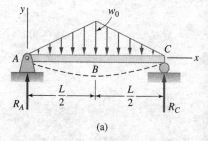

(a)

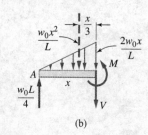

(b)

Substituting M into the differential equation of the elastic curve, Eq. (6.3b), and integrating twice, we obtain

$$EIv'' = \frac{w_0}{12L}(3L^2 x - 4x^3)$$

$$EIv' = \frac{w_0}{12L}\left(\frac{3L^2 x^2}{2} - x^4\right) + C_1 \qquad \text{(a)}$$

$$EIv = \frac{w_0}{12L}\left(\frac{L^2 x^3}{2} - \frac{x^5}{5}\right) + C_1 x + C_2 \qquad \text{(b)}$$

The two constants of integration can be evaluated from the following two conditions on the elastic curve of segment AB:

1. $v|_{x=0} = 0$ (no deflection at A due to the simple support). Substituting $x = v = 0$ in Eq. (b), we obtain

$$C_2 = 0$$

2. $v'|_{x=L/2} = 0$ (due to symmetry, the slope at midspan is zero). With $x = L/2$ and $v' = 0$, Eq. (a) becomes

$$0 = \frac{w_0}{12L}\left(\frac{3L^4}{8} - \frac{L^4}{16}\right) + C_1$$

$$C_1 = -\frac{5w_0 L^3}{192}$$

Substitution of the constants into Eq. (b) yields the equation of the elastic curve for segment AB:

$$EIv = \frac{w_0}{12L}\left(\frac{L^2 x^3}{2} - \frac{x^5}{5}\right) - \frac{5w_0 L^3}{192}x$$

$$EIv = -\frac{w_0 x}{960L}(25L^4 - 40L^2 x^2 + 16x^4) \tag{c}$$

By symmetry, the maximum displacement occurs at midspan. Evaluating Eq. (c) at $x = L/2$, we get

$$EIv|_{x=L/2} = -\frac{w_0}{960L}\left(\frac{L}{2}\right)\left[25L^4 - 40L^2\left(\frac{L}{2}\right)^2 + 16\left(\frac{L}{2}\right)^4\right] = -\frac{w_0 L^4}{120}$$

The negative sign indicates that the deflection is downward, as expected. Therefore, the maximum displacement is

$$\delta_{max} = |v|_{x=L/2} = \frac{w_0 L^4}{120EI} \quad \downarrow \qquad \text{Answer}$$

Sample Problem **6.3**

The simply supported wood beam ABC in Fig. (a) has the rectangular cross section shown. The beam supports a concentrated load of 300 N located 2 m from the left support. Determine the maximum displacement and the maximum slope angle of the beam. Use $E = 12$ GPa for the modulus of elasticity. Neglect the weight of the beam.

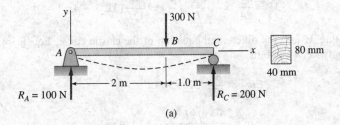

(a)

Solution
The moment of inertia of the cross-sectional area is

$$I = \frac{bh^3}{12} = \frac{40(80)^3}{12} = 1.7067 \times 10^6 \text{ mm}^4 = 1.7067 \times 10^{-6} \text{ m}^4$$

Therefore, the flexural rigidity of the beam is

$$EI = (12 \times 10^9)(1.7067 \times 10^{-6}) = 20.48 \times 10^3 \text{ N} \cdot \text{m}^2$$

The elastic curve is shown by the dashed line in Fig. (a). Because the loading is discontinuous at B, the beam must be divided into two segments: AB and BC. The bending moments in the two segments of the beam can be derived from the free-body diagrams in Fig. (b). The results are[2]

$$M = \begin{cases} 100x \text{ N} \cdot \text{m} & \text{in } AB \ (0 \le x \le 2 \text{ m}) \\ 100x - 300(x-2) \text{ N} \cdot \text{m} & \text{in } BC \ (2 \text{ m} \le x \le 3 \text{ m}) \end{cases}$$

Because the expressions for bending moments in segments AB and BC are different, they must be treated separately during double integration. Substituting the bending moments into Eq. (6.3b) and integrating twice, we get the following computations:

Segment AB

$$EIv'' = 100x \text{ N} \cdot \text{m}$$

$$EIv' = 50x^2 + C_1 \text{ N} \cdot \text{m}^2 \quad \text{(a)}$$

$$EIv = \frac{50}{3}x^3 + C_1 x + C_2 \text{ N} \cdot \text{m}^3 \quad \text{(b)}$$

Segment BC

$$EIv'' = 100x - 300(x-2) \text{ N} \cdot \text{m}$$

$$EIv' = 50x^2 - 150(x-2)^2 + C_3 \text{ N} \cdot \text{m}^2 \quad \text{(c)}$$

$$EIv = \frac{50}{3}x^3 - 50(x-2)^3 + C_3 x + C_4 \text{ N} \cdot \text{m}^3 \quad \text{(d)}$$

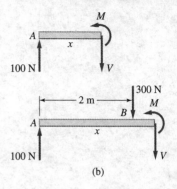

(b)

The four constants of integration, C_1 to C_4, can be found from the following boundary and continuity conditions:

1. $v|_{x=0} = 0$ (no deflection at A due to the support). Substituting $v = x = 0$ into Eq. (b), we get

$$C_2 = 0 \quad \text{(e)}$$

2. $v|_{x=3\text{m}} = 0$ (no deflection at C due to the support). Letting $x = 3$ m and $v = 0$ in Eq. (d) yields

$$0 = \frac{50}{3}(3)^3 - 50(3-2)^3 + C_3(3) + C_4$$

$$3C_3 + C_4 = -400 \text{ N} \cdot \text{m}^3 \quad \text{(f)}$$

3. $v'|_{x=2\text{m}^-} = v'|_{x=2\text{m}^+}$ (the slope at B is continuous). Equating Eqs. (a) and (c) at $x = 2$ m, we obtain

$$50(2)^2 + C_1 = 50(2)^2 + C_3$$

$$C_1 = C_3 \quad \text{(g)}$$

4. $v|_{x=2\text{m}^-} = v|_{x=2\text{m}^+}$ (the deflection at B is continuous). Substituting $x = 2$ m into Eqs. (b) and (d) and equating the results give

$$\frac{50}{3}(2)^3 + C_1(2) + C_2 = \frac{50}{3}(2)^3 + C_3(2) + C_4$$

$$2C_1 + C_2 = 2C_3 + C_4 \quad \text{(h)}$$

The solution of Eqs. (e)–(h) is

$$C_1 = C_3 = -\frac{400}{3} \text{ N} \cdot \text{m}^2 \qquad C_2 = C_4 = 0$$

---

[2] The bending moment in BC could be simplified as $M = -200x + 600$ N · m, but no advantage is gained from this simplification. In fact, the computation of the constants of integration is somewhat easier if we do not simplify.

Substituting the values of the constants and EI into Eqs. (a)–(d), we obtain the following results:

Segment AB

$$v' = \frac{50x^2 - (400/3)}{20.48 \times 10^3} = (2.441x^2 - 6.510) \times 10^{-3}$$

$$v = \frac{(50/3)x^3 - (400/3)x}{20.48 \times 10^3} = (0.8138x^3 - 6.510x) \times 10^{-3} \text{ m}$$

Segment BC

$$v' = \frac{50x^2 - 150(x-2)^2 - (400/3)}{20.48 \times 10^3}$$

$$= [2.441x^2 - 7.324(x-2)^2 - 6.150] \times 10^{-3}$$

$$v = \frac{(50/3)x^3 - 50(x-2)^3 - (400/3)x}{20.48 \times 10^3}$$

$$= [0.8138x^3 - 2.441(x-2)^3 - 6.150x] \times 10^{-3} \text{ m}$$

The maximum displacement occurs where the slope of the elastic curve is zero. This point is in the longer of the two segments—namely, in AB. Setting $v' = 0$ in segment AB, we get

$$2.441x^2 - 6.510 = 0 \qquad x = 1.6331 \text{ m}$$

The corresponding deflection is

$$v|_{x=1.6331\,\text{m}} = [(0.8138(1.6331)^3 - 6.510(1.6331)] \times 10^{-3}$$

$$= -7.09 \times 10^{-3} \text{ m} = -7.09 \text{ mm}$$

The negative sign indicates that the deflection is downward, as expected. Thus, the maximum displacement is

$$\delta_{max} = |v|_{x=1.6331\,\text{m}} = 7.09 \text{ mm} \downarrow \qquad \text{Answer}$$

By inspection of the elastic curve in Fig. (a), the largest slope occurs at C. Its value is

$$v'|_{x=3\,\text{m}} = [2.441(3)^2 - 7.324(3-2)^2 - 6.150] \times 10^{-3} = 8.50 \times 10^{-3}$$

According to the sign conventions for slopes, the positive value for v' means that the beam rotates counterclockwise at C (this is consistent with the sketch of the elastic curve in the figure). Therefore, the maximum slope angle of the beam is

$$\theta_{max} = |v'|_{x=3\,\text{m}} = 8.50 \times 10^{-3} \text{ rad} = 0.487° \circlearrowleft \qquad \text{Answer}$$

Sample Problem 6.4

The cantilever beam ABC in Fig. (a) consists of two segments with different moments of inertia: I_0 for segment AB and $2I_0$ for segment BC. Segment AB carries a uniformly distributed load of intensity 200 lb/ft. Using $E = 10 \times 10^6$ psi and $I_0 = 40$ in.^{4}, determine the maximum displacement of the beam.

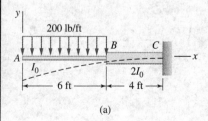

(a)

Solution

The dashed line in Fig. (a) represents the elastic curve of the beam. The bending moments in the two segments, obtained from the free-body diagrams in Fig. (b), are

$$M = \begin{cases} -100x^2 \text{ lb} \cdot \text{ft} & \text{in } AB \ (0 \le x \le 6 \text{ ft}) \\ -1200(x-3) \text{ lb} \cdot \text{ft} & \text{in } BC \ (6 \text{ ft} \le x \le 10 \text{ ft}) \end{cases}$$

Substituting the expressions for M into Eq. (6.3b) and integrating twice yield the following results:

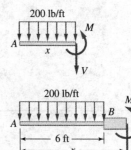

Segment AB ($I = I_0$)

$$EI_0 v'' = -100x^2 \text{ lb} \cdot \text{ft}$$

$$EI_0 v' = -\frac{100}{3}x^3 + C_1 \text{ lb} \cdot \text{ft}^2 \qquad \text{(a)}$$

$$EI_0 v = -\frac{25}{3}x^4 + C_1 x + C_2 \text{ lb} \cdot \text{ft}^3 \qquad \text{(b)}$$

Segment BC ($I = 2I_0$)

$$E(2I_0) v'' = -1200(x-3) \text{ lb} \cdot \text{ft} \quad \text{or} \quad EI_0 v'' = -600(x-3) \text{ lb} \cdot \text{ft}$$

$$EI_0 v' = -300(x-3)^2 + C_3 \text{ lb} \cdot \text{ft}^2 \qquad \text{(c)}$$

$$EI_0 v = -100(x-3)^3 + C_3 x + C_4 \text{ lb} \cdot \text{ft}^3 \qquad \text{(d)}$$

The conditions for evaluating the four constants of integration follow:

1. $v'|_{x=10\text{ft}} = 0$ (no rotation at C due to the built-in support). With $v' = 0$ and $x = 10$ ft, Eq. (c) yields

$$0 = -300(10-3)^2 + C_3$$

$$C_3 = 14.70 \times 10^3 \text{ lb} \cdot \text{ft}^2$$

2. $v|_{x=10\text{ft}} = 0$ (no deflection at C due to the built-in support). Substituting $v = 0$, $x = 10$ ft, and the value of C_3 into Eq. (d), we get

$$0 = -100(10-3)^3 + (14.70 \times 10^3)(10) + C_4$$

$$C_4 = -112.7 \times 10^3 \text{ lb} \cdot \text{ft}^3$$

3. $v'|_{x=6\text{ft}^-} = v'|_{x=6\text{ft}^+}$ (the slope at B is continuous). Equating Eqs. (a) and (c) after substituting $x = 6$ ft and the value of C_3, we obtain

$$-\frac{100}{3}(6^3) + C_1 = -300(6-3)^2 + (14.70 \times 10^3)$$

$$C_1 = 19.20 \times 10^3 \text{ lb} \cdot \text{ft}^2$$

4. $v|_{x=6\text{ft}^-} = v|_{x=6\text{ft}^+}$ (the displacement at B is continuous). Using $x = 6$ ft and the previously computed values of the constants of integration in Eqs. (b) and (d) gives

$$-\frac{25}{3}(6)^4 + (19.20 \times 10^3)(6) + C_2 = -100(6-3)^3$$

$$+ (14.70 \times 10^3)(6) - (112.7 \times 10^3)$$

$$C_2 = -131.6 \times 10^3 \text{ lb} \cdot \text{ft}^3$$

The maximum deflection of the beam occurs at A—that is, at $x = 0$. From Eq. (b), we get

$$EI_0 v|_{x=0} = C_2 = -131.6 \times 10^3 \text{ lb} \cdot \text{ft}^3 = -227.4 \times 10^6 \text{ lb} \cdot \text{in.}^3$$

The negative sign indicates that the deflection of A is downward, as anticipated. Therefore, the maximum displacement is

$$\delta_{max} = |v|_{x=0} = \frac{227.4 \times 10^6}{EI_0} = \frac{227.4 \times 10^6}{(10 \times 10^6)(40)} = 0.569 \text{ in.} \downarrow \qquad \textit{Answer}$$

Problems

6.1 For the simply supported beam carrying the concentrated load P at its midspan, determine (a) the equation of the elastic curve; and (b) the maximum displacement.

6.2 The simply supported beam carries a uniformly distributed load of intensity w_0. Determine (a) the equation of the elastic curve; and (b) the maximum displacement.

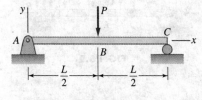

FIG. P6.1

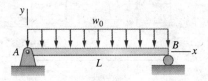

FIG. P6.2

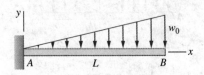

FIG. P6.3

6.3 The intensity of the distributed load on the cantilever beam varies linearly from zero to w_0. Derive the equation of the elastic curve.

6.4 The simply supported beam carries two end couples, each of magnitude M_0 but oppositely directed. Find the location and magnitude of the maximum deflection.

6.5 Solve Prob. 6.4 if the couple M_0 acting at the left support is removed.

6.6 Compute the location and maximum value of $EI\delta$ for the simply supported beam carrying the couple M_0 at the midspan. (*Hint*: By skew-symmetry, the deflection at midspan is zero.)

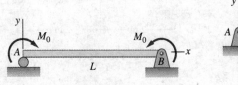

FIG. P6.4, P6.5

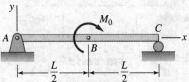

FIG. P6.6

6.7 Determine the value of $EI\delta$ at midspan of the simply supported beam. Is the deflection up or down?

6.8 Derive the equation of the elastic curve for the cantilever beam. The load intensity varies linearly from zero at the free end to w_0 at the support.

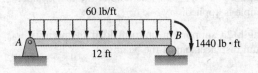

FIG. P6.7

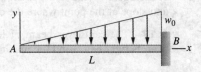

FIG. P6.8

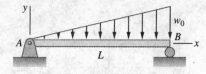

FIG. P6.9

6.9 The intensity of the distributed load on the simply supported beam varies linearly from zero to w_0. (a) Derive the equation of the elastic curve. (b) Find the location of the maximum deflection.

6.10 Determine the maximum displacement of the simply supported beam due to the distributed loading shown in the figure. (*Hint:* Utilize symmetry and analyze the right half of the beam only.)

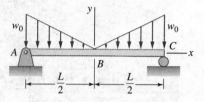

FIG. P6.10   **FIG. P6.11**

6.11 Two concentrated loads are placed symmetrically on the simply supported beam. Calculate the maximum displacement of the beam.

6.12 Determine the maximum displacement of the cantilever beam caused by the concentrated load P.

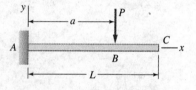

FIG. P6.12

6.13 The uniformly distributed load of intensity w_0 acts on the central portion of the simply supported beam. Find the maximum value of $EI\delta$ for the beam.

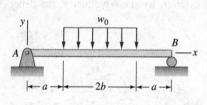   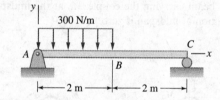

FIG. P6.13   **FIG. P6.14**

6.14 The left half of the simply supported beam carries a uniformly distributed load of intensity 300 N/m. (a) Compute the value of $EI\delta$ at midspan. (b) If $E = 10$ GPa, determine the smallest value of I that limits the midspan displacement to 1/360th of the span.

6.15 For the overhanging beam, compute (a) the displacement under the load P; and (b) the maximum displacement between the supports.

6.16 The simply supported steel beam is loaded by the 30-kN·m couple as shown in the figure. Using $E = 200$ GPa and $I = 8 \times 10^{-6}$ m^4, determine the displacement and slope at the point where the couple is applied.

6.17 The cantilever beam of length $2a$ supports a uniform load of intensity w_0 over its right half. Find the maximum displacement of the beam.

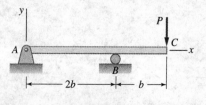   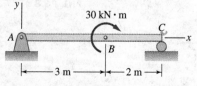   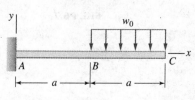

FIG. P6.15   **FIG. P6.16**   **FIG. P6.17**

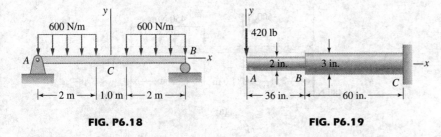

FIG. P6.18

FIG. P6.19

6.18 Two uniformly distributed loads are placed symmetrically on the simply supported beam. Calculate the maximum value of $EI\delta$ for the beam. (*Hint*: Utilize symmetry and analyze the right half of the beam only).

6.19 The steel cantilever beam consists of two cylindrical segments with the diameters shown. Determine the maximum displacement of the beam due to the 420-lb concentrated load. Use $E = 29 \times 10^6$ psi for steel.

6.20 The stepped beam of length $4a$ carries a distributed load of intensity w_0 over its middle half. The moments of inertia are $1.5I_0$ for the middle half and I_0 for the rest of the beam. Find the displacement of the beam at its midspan. (*Hint*: Utilize symmetry and analyze the right half of the beam only).

6.21 The moment of inertia of the cantilever beam varies linearly from zero at the free end to I_0 at the fixed end. Find the displacement at the free end caused by the concentrated load P.

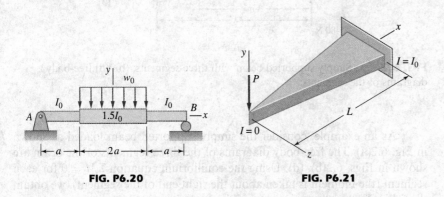

FIG. P6.20

FIG. P6.21

6.3 Double Integration Using Bracket Functions

Evaluating the constants of integration that arise in the double-integration method can become very involved if more than two beam segments must be analyzed. We can simplify the calculations by expressing the bending moment in terms of discontinuity functions, also known as *Macaulay bracket functions*. Discontinuity functions enable us to write a single expression for the bending moment that is valid for the entire length of the beam, even if the loading is discontinuous. By integrating a single, continuous expression for the bending moment, we obtain equations for slopes and deflections that are also continuous everywhere.

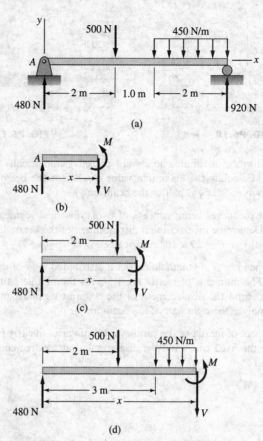

FIG. 6.2 (a) Simply supported beam with three segments; (b)–(d) free-body diagrams of the segments.

As an example, consider the simply supported beam loaded as shown in Fig. 6.2(a). The free-body diagrams of the three segments of the beam are shown in Figs. 6.2(b)–(d). Using the equilibrium equation $\Sigma M = 0$ for each segment (the moment is taken about the right end of the segment), we obtain the following bending moments:

Segment	M (N·m)
$0 \leq x \leq 2$ m	$480x$
2 m $\leq x \leq 3$ m	$480x - 500(x-2)$
3 m $\leq x \leq 5$ m	$480x - 500(x-2) - \frac{450}{2}(x-3)^2$

Note that in each successive segment an extra term is added to M, while the rest of the expression remains unchanged. This pattern suggests using the expression

$$M = 480x - 500(x-2) - \frac{450}{2}(x-3)^2 \text{ N} \cdot \text{m}$$

for the entire beam, with the understanding that the term $(x-2)$ disappears when $x \leq 2$, and $(x-3)^2$ disappears when $x \leq 3$. This idea is formalized by using the Macaulay bracket functions described below.

A Macaulay bracket function, often referred to as a "bracket function," is defined as

$$\langle x - a \rangle^n = \begin{cases} 0 & \text{if } x \leq a \\ (x - a)^n & \text{if } x \geq a \end{cases} \quad (6.6)$$

where n is a nonnegative integer.[3] The brackets $\langle \cdots \rangle$ identify the expression as a bracket function. Note that a bracket function is zero by definition if the expression in the brackets—namely, $(x - a)$—is negative; otherwise, it is evaluated as written. A bracket function can be integrated by the same rule as an ordinary function—namely,

$$\int \langle x - a \rangle^n \, dx = \frac{\langle x - a \rangle^{n+1}}{n+1} + C \quad (6.7)$$

where C is a constant of integration.

With bracket functions, the bending moment for the beam in Fig. 6.2 can be written as

$$M = 480x - 500\langle x - 2 \rangle - \frac{450}{2}\langle x - 3 \rangle^2 \ \text{N} \cdot \text{m} \quad (a)$$

This expression, valid over the entire length of the beam, is called the *global* bending moment equation for the beam. Its integrals, representing the slope and deflection of the beam, are continuous functions. Thus, double integration of Eq. (a) automatically assures continuity of deformation.

Observe that the global bending moment equation in Eq. (a) can be obtained by writing the bending moment equation for the rightmost beam segment, using the free-body diagram of the beam that lies to the *left* of the cutting plane, as in Fig. 6.2(d). Then the brackets are inserted in appropriate locations.

Referring to Fig. 6.2, we see that the bracket function $\langle x - 3 \rangle^2$ is caused by the distributed load that starts at $x = 3$ m and continues to the right end of the beam. Now suppose that the distributed load were to end at $x = 4$ m, as shown in Fig. 6.3(a). The problem is how to handle the termination in the expression for M. We can do this by letting the distributed load run to the end of the beam, as in Fig. 6.1(a), but canceling out the unwanted portion by introducing an equal but oppositely directed load between $x = 4$ m and the right end of the beam. This technique is shown in Fig. 6.3(b). The corresponding global expression for the bending moment, obtained from the free-body diagram in Fig. 6.3(c), is

$$M = 435x - 500\langle x - 2 \rangle - \frac{450}{2}\langle x - 3 \rangle^2 + \frac{450}{2}\langle x - 4 \rangle^2 \ \text{N} \cdot \text{m}$$

---

[3] Negative values of n result in a class of functions called *singularity functions*, which we do not need in our analysis.

CHAPTER 6 Deflection of Beams

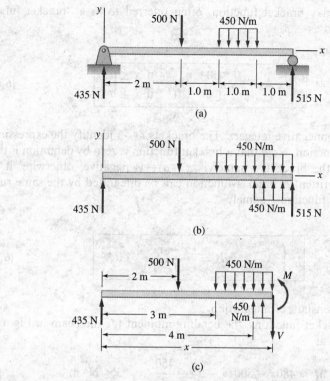

FIG. 6.3 (a) Simply supported beam; (b) same beam with equivalent loading; (c) free-body diagram for determining the bending moment M in the rightmost segment.

After the global bending moment equation has been written, it can be integrated to obtain the slope and the deflection equations for the entire beam. The two constants of integration that arise can then be computed from the boundary conditions. As mentioned before, continuity of slope and deflection at the junctions between the segments is automatically satisfied when bracket functions are used.

Sample Problem 6.5

The simply supported beam ABC in Fig. (a) carries a concentrated load of 300 N as shown. Determine the equations for the slope and deflection of the beam using $EI = 20.48 \times 10^3$ N·m². (*Note:* The same beam was analyzed in Sample Problem 6.3; see p. 198.)

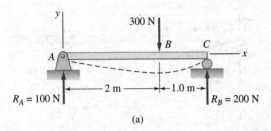

(a)

Solution

The dashed line in Fig. (a) represents the elastic curve of the beam. Using the free-body diagram in Fig. (b), we obtain the following global bending moment equation:

$$M = 100x - 300\langle x-2 \rangle \text{ N·m} \quad (a)$$

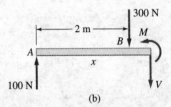

(b)

Note that for segment AB $(0 \le x \le 2$ m$)$, the last term is zero by definition of the bracket function, so that $M = 100x$ N·m. For segment BC $(2$ m $\le x \le 3$ m$)$, the bending moment is $M = 100x - 300(x-2)$ N·m. Substituting Eq. (a) into the differential equation of the elastic curve and integrating twice, we obtain

$$EIv'' = 100x - 300\langle x-2 \rangle \text{ N·m}$$

$$EIv' = 50x^2 - 150\langle x-2 \rangle^2 + C_1 \text{ N·m}^2 \quad (b)$$

$$EIv = \frac{50}{3}x^3 - 50\langle x-2 \rangle^3 + C_1 x + C_2 \text{ N·m}^3 \quad (c)$$

To evaluate C_1 and C_2, we apply the following boundary conditions:

1. $v|_{x=0} = 0$ (no deflection at A due to the simple support). Substituting $v = x = 0$ into Eq. (c) and recalling that $\langle 0-2 \rangle^3 = 0$, we get

$$C_2 = 0$$

2. $v|_{x=3\text{ m}} = 0$ (no deflection at C due to the simple support). Substituting $v = 0$, $x = 3$ m, and $C_2 = 0$ into Eq. (c) and noting that $\langle 3-2 \rangle^3 = (3-2)^3$, we obtain

$$\frac{50}{3}(3)^3 - 50(3-2)^3 + C_1(3) = 0 \qquad C_1 = -\frac{400}{3} \text{ N·m}^2$$

Substituting the values of EI and the constants of integration into Eqs. (b) and (c) yields the following global expressions for the slope and the deflection:

$$v' = \frac{50x^2 - 150\langle x-2 \rangle^2 - (400/3)}{20.48 \times 10^3}$$

$$= [2.441x^2 - 7.324\langle x-2 \rangle^2 - 6.150] \times 10^{-3} \qquad \text{Answer}$$

$$v = \frac{(50/3)x^3 - 50\langle x-2 \rangle^3 - (400/3)x}{20.48 \times 10^3}$$

$$= [0.8138x^3 - 2.441\langle x-2 \rangle^3 - 6.150x] \times 10^{-3} \text{ m} \qquad \text{Answer}$$

Sample Problem 6.6

For the overhanging beam in Fig. (a), determine (1) the equation for the elastic curve; and (2) the values of $EI\delta$ midway between the supports and at point E (indicate whether each δ is up or down).

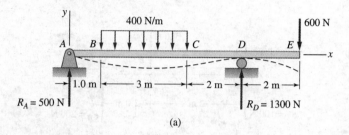

(a)

Solution

Part 1

The dashed line in Fig. (a) represents the elastic curve of the beam. Figure (b) shows the equivalent loading that is used to determine the bending moment in the beam. Recall that the use of bracket functions in the expression for the bending moment requires each distributed load to extend to the right end of the beam. We must, therefore, extend the 400-N/m loading to point E and cancel the unwanted portion by applying an equal and opposite loading to CE. The global expression for the bending moment can now be derived from the free-body diagram in Fig. (c), the result being

$$M = 500x - \frac{400}{2}\langle x - 1 \rangle^2 + \frac{400}{2}\langle x - 4 \rangle^2 + 1300\langle x - 6 \rangle \text{ N} \cdot \text{m}$$

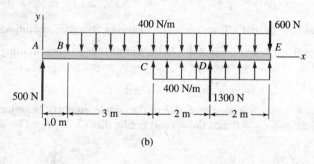

(b)

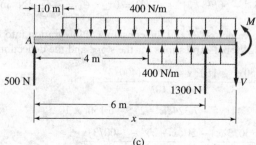

(c)

Substituting M into the differential equation for the elastic curve and integrating twice, we get

$$EIv'' = 500x - 200\langle x-1\rangle^2 + 200\langle x-4\rangle^2 + 1300\langle x-6\rangle \text{ N}\cdot\text{m}$$

$$EIv' = 250x^2 - \frac{200}{3}\langle x-1\rangle^3 + \frac{200}{3}\langle x-4\rangle^3 + 650\langle x-6\rangle^2 + C_1 \text{ N}\cdot\text{m}^2$$

$$EIv = \frac{250}{3}x^3 - \frac{50}{3}\langle x-1\rangle^4 + \frac{50}{3}\langle x-4\rangle^4 + \frac{650}{3}\langle x-6\rangle^3$$
$$+ C_1 x + C_2 \text{ N}\cdot\text{m}^3 \tag{a}$$

The boundary conditions follow:

1. $v|_{x=0} = 0$ (deflection at A is prevented by the simple support). Because all the bracket functions in Eq. (a) are zero at $x = 0$, we get

$$C_2 = 0$$

2. $v|_{x=6\text{m}} = 0$ (deflection at D is prevented by the simple support). Equation (a) now gives

$$0 = \frac{250}{3}(6)^3 - \frac{50}{3}(6-1)^4 + \frac{50}{3}(6-4)^4 + C_1(6)$$

$$C_1 = -\frac{3925}{3} \text{ N}\cdot\text{m}^2$$

When we substitute the values for C_1 and C_2 into Eq. (a), the equation for the elastic curve becomes

$$EIv = \frac{250}{3}x^3 - \frac{50}{3}\langle x-1\rangle^4 + \frac{50}{3}\langle x-4\rangle^4$$
$$+ \frac{650}{3}\langle x-6\rangle^3 - \frac{3925}{3}x \text{ N}\cdot\text{m}^3 \qquad \text{Answer}$$

Part 2

The deflection midway between the supports is obtained by substituting $x = 3$ m into the expression for EIv. Noting that $\langle 3-4\rangle^4 = 0$ and $\langle 3-6\rangle^3 = 0$, we obtain

$$EIv|_{x=3\text{m}} = \frac{250}{3}(3)^3 - \frac{50}{3}(3-1)^4 - \frac{3925}{3}(3) = -1942 \text{ N}\cdot\text{m}^3$$

The negative sign shows that the deflection is down, so that the value of $EI\delta$ at midspan is

$$EI\delta_{\text{mid}} = 1942 \text{ N}\cdot\text{m}^3 \downarrow \qquad \text{Answer}$$

At point E, we have

$$EIv|_{x=8\text{m}} = \frac{250}{3}(8)^3 - \frac{50}{3}(8-1)^4 + \frac{50}{3}(8-4)^4 + \frac{650}{3}(8-6)^3 - \frac{3925}{3}(8)$$

$$= -1817 \text{ N}\cdot\text{m}^3$$

Again, the minus sign indicates a downward deflection. Therefore,

$$EI\delta_E = 1817 \text{ N}\cdot\text{m}^3 \downarrow \qquad \text{Answer}$$

∎

Sample Problem 6.7

A couple M_0 is applied at the midpoint of the cantilever beam of length L, as shown in Fig. (a). Find the magnitude of the vertical force P for which the deflection at end C is zero.

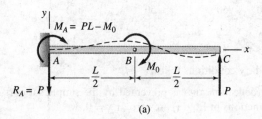

(a)

Solution

The elastic curve of the beam is shown in Fig. (a) by the dashed line. From the free-body diagram in Fig. (b), we get for the bending moment

$$M = (PL - M_0) - Px + M_0 \left\langle x - \frac{L}{2} \right\rangle^0$$

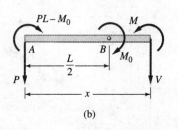

(b)

Note that $\langle x - L/2 \rangle^0 = 1$ for $x \geq L/2$. Substituting this expression into the differential equation for the elastic curve, and integrating twice, we obtain

$$EIv'' = (PL - M_0) - Px + M_0 \left\langle x - \frac{L}{2} \right\rangle^0$$

$$EIv' = (PL - M_0)x - \frac{P}{2}x^2 + M_0 \left\langle x - \frac{L}{2} \right\rangle + C_1 \quad \text{(a)}$$

$$EIv = (PL - M_0)\frac{x^2}{2} - \frac{P}{6}x^3 + \frac{M_0}{2}\left\langle x - \frac{L}{2} \right\rangle^2 + C_1 x + C_2 \quad \text{(b)}$$

The boundary conditions are:

1. $v'|_{x=0} = 0$ (the fixed support at A prevents rotation). Substituting $x = v' = 0$ into Eq. (a) yields $C_1 = 0$.
2. $v|_{x=0} = 0$ (the fixed support at A prevents deflection). Setting $x = v = 0$ in Eq. (b), we get $C_2 = 0$.

Therefore, the equation of the elastic curve is

$$EIv = (PL - M_0)\frac{x^2}{2} - P\frac{x^3}{6} + \frac{M_0}{2}\left\langle x - \frac{L}{2} \right\rangle^2$$

At end C, we have

$$EIv|_{x=L} = (PL - M_0)\frac{L^2}{2} - \frac{PL^3}{6} + \frac{M_0}{2}\left(L - \frac{L}{2}\right)^2 = \left(\frac{PL}{3} - \frac{3M_0}{8}\right)L^2$$

To find the force P that results in zero displacement at C, we set $v|_{x=L} = 0$ and solve for P. The result is

$$P = \frac{9}{8}\frac{M_0}{L} \qquad \text{Answer}$$

Problems

6.22 The cantilever beam has a rectangular cross section 50 mm wide and h mm high. Find the smallest allowable value of h if the maximum displacement of the beam is not to exceed 10 mm. Use $E = 10$ GPa.

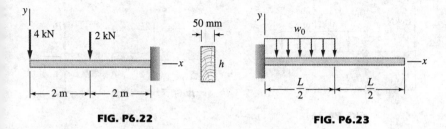

FIG. P6.22

FIG. P6.23

6.23 Find the value of $EI\theta$ at the free end of the cantilever beam.

6.24 Determine the value of $EI\delta$ at midspan for the beam loaded by two concentrated forces.

6.25 Compute the midspan value of $EI\delta$ for the simply supported beam carrying a uniformly distributed load over part of its span.

6.26 Find the value of $EI\delta$ at the free end of the overhanging beam. Is this displacement up or down?

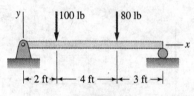

FIG. P6.24

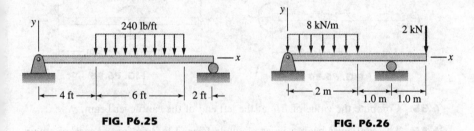

FIG. P6.25

FIG. P6.26

6.27 For the overhanging beam shown, (a) derive the equation of the elastic curve; and (b) compute the value of $EI\delta$ at the right end.

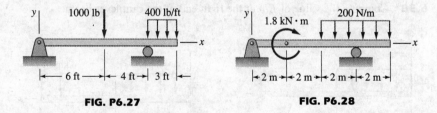

FIG. P6.27

FIG. P6.28

6.28 (a) Determine the equation of the elastic curve for the overhanging beam; and (b) calculate the value of $EI\delta$ midway between the supports.

6.29 The overhanging beam carries a uniformly distributed load over its entire length. Determine the ratio b/a for which the deflection midway between the supports is zero. (*Hint*: Use symmetry and analyze the right half of the beam only.)

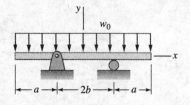

FIG. P6.29

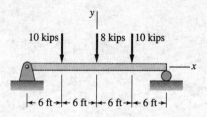

FIG. P6.30

6.30 The simply supported beam carries three concentrated loads as shown in the figure. Determine (a) the equation of the elastic curve for the right half of the beam; and (b) the value of $EI\delta$ under the 8-kip load.

6.31 For the overhanging beam, compute the value of $EI\delta$ under the 15-kN load.

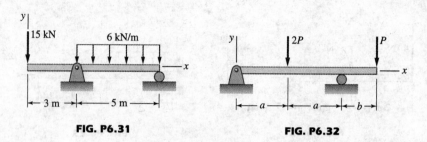

FIG. P6.31   FIG. P6.32

6.32 Determine the displacement midway between the supports for the overhanging beam.

6.33 For the overhanging beam, find the displacement at the left end.

6.34 For the overhanging beam, determine (a) the value of $EI\delta$ under the 24-kN load; and (b) the maximum value of $EI\delta$ between the supports.

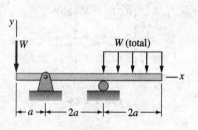

FIG. P6.33

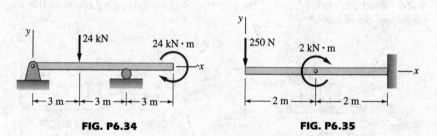

FIG. P6.34   FIG. P6.35

6.35 Compute the value of $EI\delta$ at the left end of the cantilever beam.

6.36 The cantilever beam carries a couple formed by two forces, each of magnitude $P = 2000$ lb. Determine the force R that must be applied as shown to prevent displacement of point A.

6.37 Find the maximum displacement of the cantilever beam.

6.38 Compute the value of $EI\delta$ at the right end of the cantilever beam.

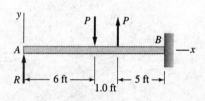

FIG. P6.36

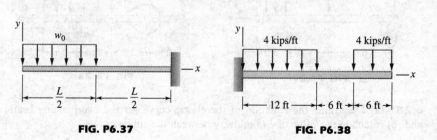

FIG. P6.37   FIG. P6.38

6.39 Determine the value of $EI\theta$ at each end of the overhanging beam.

6.40 For the simply supported beam, compute the value of $EI\delta$ at midspan.

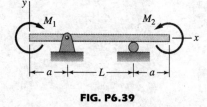

FIG. P6.39

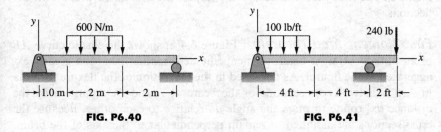

FIG. P6.40   FIG. P6.41

6.41 Calculate the value of $EI\theta$ at the right support of the overhanging beam.

6.42 Determine the maximum deflection of the cantilever beam.

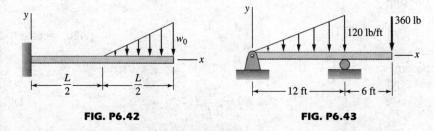

FIG. P6.42   FIG. P6.43

6.43 Compute the value of $EI\delta$ at the right end of the overhanging beam.

6.44 Derive the equation for the elastic curve for the simply supported beam that carries the distributed load shown in the figure.

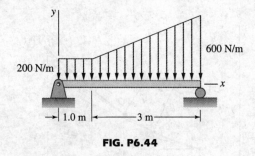

FIG. P6.44

*6.4 Moment-Area Method

The *moment-area method* is useful for determining the slope or deflection of a beam at a specified location. It is a semigraphical method in which the integration of the bending moment is carried out indirectly, using the geometric properties of the area under the bending moment diagram. As in the method of double integration, we assume that the deformation is within the elastic range, resulting in small slopes and small displacements.

a. Moment-area theorems

We will now derive two theorems that are the bases of the moment-area method. The first theorem deals with slopes; the second theorem with deflections.

First Moment-Area Theorem Figure 6.4(a) shows the elastic curve AB of an initially straight beam segment (the deformation has been greatly exaggerated in the figure). As discussed in the derivation of the flexure formula in Art. 5.2, two cross sections of the beam at P and Q, separated by the distance dx, rotate through the angle $d\theta$ relative to each other. Because the cross sections are assumed to remain perpendicular to the axis of the beam, $d\theta$ is also the difference in the slope of the elastic curve between P and Q, as shown in Fig. 6.4(a). From the geometry of the figure, we see that $dx = \rho\, d\theta$, where ρ is the radius of curvature of the elastic curve of the deformed element. Therefore, $d\theta = dx/\rho$, which upon using the moment-curvature relationship

$$\frac{1}{\rho} = \frac{M}{EI} \qquad \text{(5.2b, repeated)}$$

becomes

$$d\theta = \frac{M}{EI}\, dx \qquad \text{(a)}$$

Integrating Eq. (a) over the segment AB yields

$$\int_A^B d\theta = \int_A^B \frac{M}{EI}\, dx \qquad \text{(b)}$$

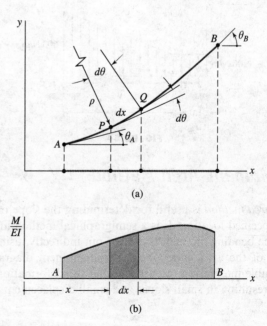

FIG. 6.4 (a) Elastic curve of a beam segment; (b) bending moment diagram for the segment.

The left side of Eq. (b) is $\theta_B - \theta_A$, which is the change in the slope between A and B. The right-hand side represents the area under the $M/(EI)$ diagram between A and B, shown as the shaded area in Fig. 6.4(b). If we introduce the notation $\theta_{B/A} = \theta_B - \theta_A$, Eq. (b) can be expressed in the form

$$\theta_{B/A} = \text{area of } \frac{M}{EI} \text{ diagram} \Big]_A^B \qquad (6.8)$$

which is the *first moment-area theorem*.

Second Moment-Area Theorem Referring to the elastic curve AB in Fig. 6.5(a), we let $t_{B/A}$ be the vertical distance of point B from the tangent to the elastic curve at A. This distance is called the *tangential deviation* of B with respect to A. To calculate the tangential deviation, we first determine the contribution dt of the infinitesimal element PQ and then use $t_{B/A} = \int_A^B dt$ to add the contributions of all the elements between A and B. As shown in the figure, dt is the vertical distance at B between the tangents drawn to the elastic curve at P and Q. Recalling that the slopes are very small, we obtain from geometry

$$dt = x' \, d\theta$$

where x' is the horizontal distance of the element from B.

Therefore, the tangential deviation is

$$t_{B/A} = \int_A^B dt = \int_A^B x' \, d\theta$$

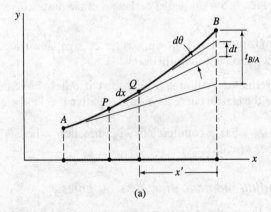

(a)

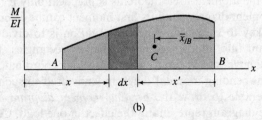

(b)

FIG. 6.5 (a) Elastic curve of a beam segment; (b) bending moment diagram for the segment.

Substituting $d\theta$ from Eq. (a), we obtain

$$t_{B/A} = \int_A^B \frac{M}{EI} x' \, dx \tag{c}$$

The right-hand side of Eq. (c) represents the first moment of the shaded area of the $M/(EI)$ diagram in Fig. 6.5(b) about point B. Denoting the distance between B and the centroid C of this area by $\bar{x}_{/B}$ (read $/B$ as "relative to B"), we can write Eq. (c) as

$$\boxed{t_{B/A} = \text{area of } \frac{M}{EI} \text{ diagram}\Big]_A^B \cdot \bar{x}_{/B}} \tag{6.9}$$

This is the *second moment-area theorem*. Note that the first moment of area, represented by the right-hand side of Eq. (6.9), is always taken about the point at which the deviation is being computed.

Do not confuse $t_{B/A}$ (the tangential deviation of B with respect to A) with $t_{A/B}$ (the tangential deviation of A with respect to B). In general, these two distances are not equal, as illustrated in Fig. 6.6.

(a) Positive deviation; B located above reference tangent

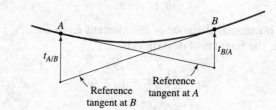

FIG. 6.6 Tangential deviations of the elastic curve.

(b) Negative deviation; B located below reference tangent

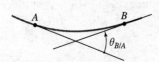

(c) Positive change of slope is counterclockwise from left tangent

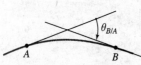

(d) Negative change of slope is clockwise from left tangent

FIG. 6.7 Sign conventions for tangential deviation and change of slope.

Sign Conventions The following rules of sign, illustrated in Fig. 6.7, apply to the two moment-area theorems:

- The tangential deviation $t_{B/A}$ is positive if B lies above the tangent line drawn to the elastic curve at A, and negative if B lies below the tangent line.
- Positive $\theta_{B/A}$ has a counterclockwise direction, whereas negative $\theta_{B/A}$ has a clockwise direction.

b. Bending moment diagrams by parts

Application of the moment-area theorems is practical only if the area under the bending moment diagram and its first moment can be calculated without difficulty. The key to simplifying the computation is to divide the bending moment diagram into simple geometric shapes (rectangles, triangles, and parabolas) that have known areas and centroidal coordinates. Sometimes the conventional bending moment diagram lends itself to such division, but often it is preferable to draw the *bending moment diagram by parts*, with each part of the diagram representing the effect of one load. Construction of the bending moment diagram by parts for *simply supported* beams proceeds as follows:

- Calculate the simple support reactions and consider them to be applied loads.
- Introduce a fixed support at a convenient location. A simple support of the original beam is usually a good choice, but sometimes another point is more convenient. The beam is now cantilevered from this support.
- Draw a bending moment diagram for each load (including the support reactions of the original beam). If all the diagrams can be fitted on a single plot, do so, drawing the positive moments above the *x*-axis and the negative moments below the *x*-axis.

Only the last step of the procedure is needed for a cantilever beam because a fixed support is already present.

As an illustration, consider the simply supported beam *ABC* in Fig. 6.8(a). We start by computing the support reactions; the results are shown in the figure. In Fig. 6.8(b), we introduce a fixed support at *C* and show the reaction at *A* as an applied load. The result is a cantilever beam that is *statically equivalent* to the original beam; that is, the cantilever beam has the same conventional bending moment diagram as the beam in Fig. 6.8(a). We now draw a bending moment diagram for each of the two loads, as shown in Fig. 6.8(c). The moment M_1 due to R_A is positive, whereas the distributed

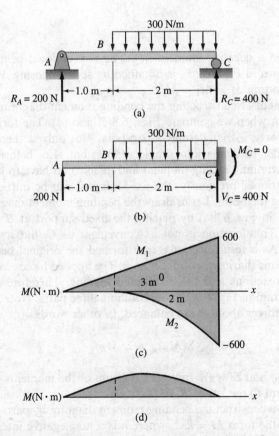

FIG. 6.8 (a) Simply supported beam; (b) equivalent beam with fixed support at *C*; (c) bending moment diagram by parts; (d) conventional bending moment diagram.

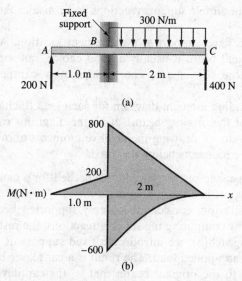

FIG. 6.9 (a) Beam with fixed support at B that is equivalent to the simply supported beam in Fig. 6.8(a); (b) bending moment diagram by parts.

load results in a negative moment M_2. The conventional bending moment diagram, shown in Fig. 6.8(d), is obtained by superimposing M_1 and M_2—that is, by plotting $M = M_1 + M_2$.

The benefit of constructing the bending moment diagram by parts becomes evident when we compare Figs. 6.8(c) and (d). The former contains two simple parts: a triangle and a parabola. Not only is the conventional diagram harder to divide into simple shapes, but also, before this can be done, the maximum bending moment and its location have to be found.

As mentioned previously, the fixed support can be introduced at any location along the beam. Let us draw the bending moment diagram by parts for the beam in Fig. 6.8(a) by placing the fixed support at B, as shown in Fig. 6.9(a). (This location is not as convenient as C, but it serves as an illustration.) As a result, we have transformed the original beam into two cantilever beams sharing the support at B. The applied forces consist of both the original reactions and the distributed loading. Therefore, the bending moment diagram in Fig. 6.9(b) now contains three parts. Note that the moments of the forces about B are balanced. In other words,

$$\Sigma(M_B)_R = \Sigma(M_B)_L$$

where $\Sigma(M_B)_R$ and $\Sigma(M_B)_L$ represent the sum of the moments of the forces to the right and to the left of B, respectively.

When we construct the bending moment diagram by parts, each part is invariably of the form $M = kx^n$, where n is a nonnegative integer that represents the degree of the moment equation. Table 6.1 shows the properties of areas under the M-diagram for $n = 0, 1, 2,$ and 3. This table is useful in computations required by the moment-area method.

6.4 Moment-Area Method

n	Plot of $M = kx^n$	Area	$\bar{x}$
0	rectangle, base b, height h	bh	$\dfrac{1}{2}b$
1	triangle, base b, height h	$\dfrac{1}{2}bh$	$\dfrac{2}{3}b$
2	parabolic spandrel, base b, height h	$\dfrac{1}{3}bh$	$\dfrac{3}{4}b$
3	cubic spandrel, base b, height h	$\dfrac{1}{4}bh$	$\dfrac{4}{5}b$

TABLE 6.1 *Properties of Areas Bounded by $M = kx^n$*

c. Application of the moment-area method

Cantilever Beams Consider the deflection of the cantilever beam shown in Fig. 6.10. Because the support at A is fixed, the tangent drawn to the elastic curve at A is horizontal. Therefore, $t_{B/A}$ (the tangential deviation of B with respect to A) has the same magnitude as the displacement of B. In other words, $\delta_B = |t_{B/A}|$, where

$$t_{B/A} = \text{area of } \frac{M}{EI} \text{ diagram}\Big]_A^B \cdot \bar{x}_{/B}$$

Simply Supported Beams The elastic curve of a simply supported beam is shown in Fig. 6.11. The problem is to compute the displacement δ_B of a point B located a distance x from A. Because the point at which a tangent to the elastic curve is horizontal is usually unknown, this computation is more involved than that for a cantilever beam. If a tangent is drawn to the elastic curve at A, the tangential deviation $t_{B/A}$ is evidently *not* the displacement δ_B. However, from the figure, we see that $\delta_B = \theta_A x - t_{B/A}$. Therefore, we must compute the slope angle θ_A as well as $t_{B/A}$.

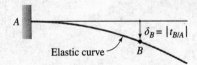

FIG. 6.10 The displacement equals the magnitude of the tangential deviation for point B on the cantilever beam.

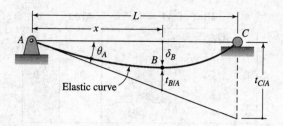

FIG. 6.11 Procedure for calculating δ_B, the displacement of point B on the simply supported beam.

The procedure for computing δ_B thus consists of the following steps:

- Compute $t_{C/A}$ from

$$t_{C/A} = \text{area of } \frac{M}{EI} \text{ diagram}\Big]_A^C \cdot \bar{x}_{/C}$$

- Determine θ_A from the geometric relationship

$$\theta_A = \frac{t_{C/A}}{L}$$

- Compute $t_{B/A}$ using

$$t_{B/A} = \text{area of } \frac{M}{EI} \text{ diagram}\Big]_A^B \cdot \bar{x}_{/B}$$

- Calculate δ_B from

$$\delta_B = \theta_A x - t_{B/A}$$

This procedure may appear to be involved, but it can be executed rapidly, especially if the bending moment diagram is drawn by parts. It must be emphasized that an accurate sketch of the elastic curve, similar to that shown in Fig. 6.11, is the basis of the procedure. Such a sketch should be the starting point of every analysis.

Sample Problem 6.8

A 600-lb/ft uniformly distributed load is applied to the left half of the cantilever beam ABC in Fig. (a). Determine the magnitude of force P that must be applied as shown so that the displacement at A is zero.

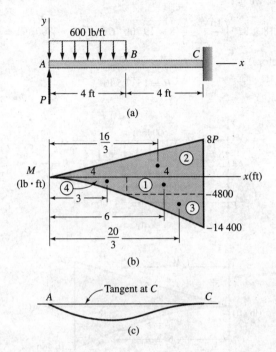

Solution

The bending moment diagram, drawn by parts, is shown in Fig. (b). The upper portion is the moment caused by P; the lower part is due to the distributed load. The area under the diagram can be divided into the four simple shapes shown: the rectangle ①, the triangles ② and ③, and the parabola ④.

The sketch of the elastic curve in Fig. (c) is drawn so that it satisfies the boundary conditions ($\delta_C = \theta_C = 0$) and the requirement that $\delta_A = 0$. Because the slope of the elastic curve at C is zero, we see that $t_{A/C}$ (the tangential deviation of A relative to C) is zero. Therefore, from the second moment-area theorem, we obtain

$$t_{A/C} = \text{area of } \frac{M}{EI} \text{ diagram} \Big]_C^A \cdot \bar{x}_{/A} = 0$$

Using the four subareas shown in Fig. (b) to compute the first moment of the bending moment diagram about A, we get (the constant EI cancels)

$$\frac{1}{2}(8 \times 8P)\left(\frac{16}{3}\right) - \frac{1}{3}(4 \times 4800)(3) - (4 \times 4800)(6) - \frac{1}{2}(4 \times 9600)\left(\frac{20}{3}\right) = 0$$

which yields

$$P = 1537.5 \text{ lb} \qquad \textbf{Answer}$$

Alternative Solution

There are other ways of drawing the bending moment diagram by parts. We could, for example, replace the distributed loading with the equivalent loading shown in Fig. (d). The resulting bending moment diagram by parts in Fig. (e) has only three parts: two parabolas and a triangle. Setting the first moment of the bending diagram about A to zero, we get

$$\frac{1}{2}(8 \times 8P)\left(\frac{16}{3}\right) - \frac{1}{3}(8 \times 19\,200)(6) + \frac{1}{3}(4 \times 4800)(7) = 0$$

giving us, as before,

$$P = 1537.5 \text{ lb} \qquad \textit{Answer}$$

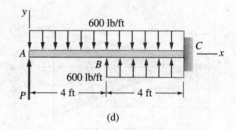

(d)

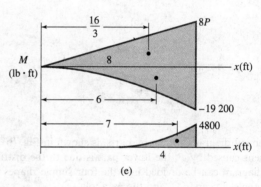

(e)

Sample Problem 6.9

The simply supported beam in Fig. (a) supports a concentrated load of 300 N as shown. Using $EI = 20.48 \times 10^3$ N·m², determine (1) the slope angle of the elastic curve at A; and (2) the displacement at D. (*Note:* This beam was analyzed in Sample Problems 6.3 and 6.5.)

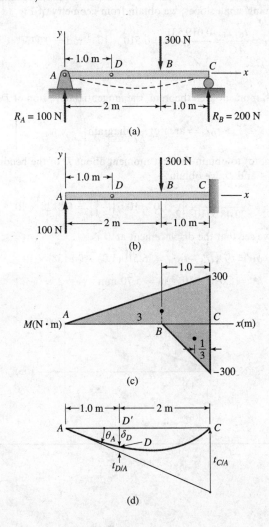

Solution

To obtain the bending moment diagram by parts, we introduce a fixed support at C and consider the reaction at A to be an applied load, as shown in Fig. (b). The resulting bending moment diagram is shown in Fig. (c). The sketch of the elastic curve of the original beam in Fig. (d) identifies the slope angle θ_A and the displacement δ_D, which are to be found, together with the tangential deviations $t_{C/A}$ and $t_{D/A}$.

Part 1

The tangential deviation $t_{C/A}$ can be found from the second moment-area theorem:

$$t_{C/A} = \text{area of } \frac{M}{EI} \text{ diagram} \bigg]_C^A \cdot \bar{x}_{/C}$$

Substituting the given value of EI and computing the first moment of the bending moment diagram about C with the aid of Fig. (c), we get

$$t_{C/A} = \frac{1}{20.48 \times 10^3} \left[\frac{1}{2}(3 \times 300)(1.0) - \frac{1}{2}(1.0 \times 300)\left(\frac{1}{3}\right) \right] = 0.019\,531 \text{ m}$$

Note that $t_{C/A}$ is positive, which means that C is above the reference tangent at A, as expected. Assuming small slopes, we obtain from geometry of Fig. (d)

$$\theta_A = \frac{t_{C/A}}{AC} = \frac{0.019\,531}{3} = 6.510 \times 10^{-3} \text{ rad} = 0.373° \;\circlearrowleft \qquad \text{Answer}$$

Part 2

From the second moment-area theorem, the tangential deviation of D relative to A is

$$t_{D/A} = \text{area of } \frac{M}{EI} \text{ diagram} \bigg]_A^D \cdot \bar{x}_{/D}$$

Referring to Fig. (e) to obtain the first moment about D of the bending moment diagram between A and D, we obtain

$$t_{D/A} = \frac{1}{20.48 \times 10^3} \left[\frac{1}{2}(1.0 \times 100)\left(\frac{1}{3}\right) \right] = 0.8138 \times 10^{-3} \text{ m}$$

From Fig. (d), we see that the displacement at D is

$$\delta_D = \theta_A \overline{AD'} - t_{D/A} = [6.510(1.0) - 0.8138] \times 10^{-3}$$
$$= 5.696 \times 10^{-3} \text{ m} = 5.70 \text{ mm} \downarrow \qquad \text{Answer}$$

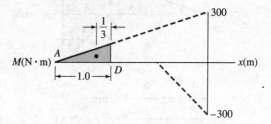

Sample Problem 6.10

Determine the value of $EI\delta$ at end D of the overhanging beam in Fig. (a).

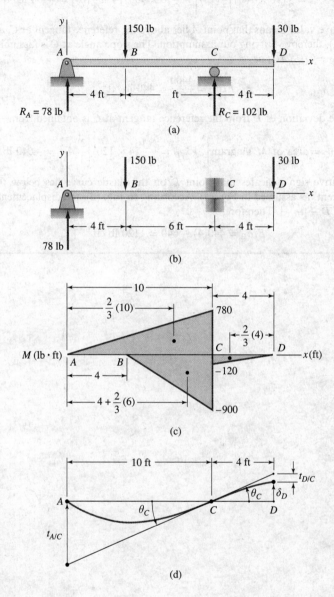

Solution

The statically equivalent beam used to draw the bending moment diagram by parts is shown in Fig. (b). We introduced a built-in support at C and show the reaction at A as an applied load. The result is, in effect, two beams that are cantilevered from C. The bending moment diagrams by parts for these beams are shown in Fig. (c).

The elastic curve of the original beam in Fig. (d) was drawn assuming that the beam rotates counterclockwise at C. The correct direction is determined from the sign of the tangential deviation $t_{A/C}$. Using the second moment-area theorem and recognizing that EI is a constant, we have

227

$$EIt_{A/C} = \text{area of } M \text{ diagram} \Big]_A^C \cdot \bar{x}_{/A}$$

$$= \frac{1}{2}(10 \times 780)\left[\frac{2}{3}(10)\right] - \frac{1}{2}(6 \times 900)\left[4 + \frac{2}{3}(6)\right] = 4400 \text{ lb} \cdot \text{ft}^3$$

The positive value means that point A lies above the reference tangent at C, as shown in Fig. (d), thereby verifying our assumption. The slope angle at C is (assuming small slopes) $\theta_C = t_{A/C}/\overline{AC}$, or

$$EI\theta_C = \frac{4400}{10} = 440 \text{ lb} \cdot \text{ft}^2$$

The deviation of D from the reference tangent at C is obtained from

$$EIt_{D/C} = \text{area of } M \text{ diagram}\Big]_C^D \cdot \bar{x}_{/D} = -\frac{1}{2}(4 \times 120)\left[\frac{2}{3}(4)\right] = -640 \text{ lb} \cdot \text{ft}^3$$

The negative sign indicates that point D on the elastic curve lies below the reference tangent, as assumed in Fig. (d). According to Fig. (d), the displacement of D is $\delta_D = \theta_C \overline{CD} - |t_{D/C}|$. Therefore,

$$EI\delta_D = 440(4) - 640 = 1120 \text{ lb} \cdot \text{ft}^3 \uparrow \qquad \text{Answer}$$

Problems

6.45 Solve Sample Problem 6.10 by introducing a built-in support at A rather than at C.

6.46 For the cantilever beam ABC, compute the value of $EI\delta$ at end C.

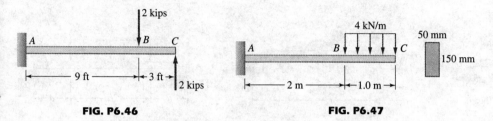

FIG. P6.46

FIG. P6.47

6.47 The cantilever beam ABC has the rectangular cross section shown in the figure. Using $E = 69$ GPa, determine the maximum displacement of the beam.

6.48 The properties of the timber cantilever beam ABC are $I = 60$ in.^{2} and $E = 1.5 \times 10^6$ psi. Determine the displacement of the free end A.

6.49 For the beam described in Prob. 6.48, compute the displacement of point B.

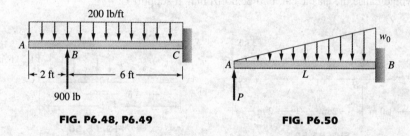

FIG. P6.48, P6.49

FIG. P6.50

6.50 The cantilever beam AB supporting a linearly distributed load of maximum intensity w_0 is propped at end A by the force P. (a) Find the value of P for which the deflection of A is zero. (b) Compute the corresponding value of $EI\theta$ at A.

6.51 Determine the displacement of end A of the cantilever beam ABC.

6.52 Compute the value of $EI\delta$ at point B for the simply supported beam ABC.

6.53 For the simply supported beam $ABCD$, determine the values of $EI\delta$ at (a) point B; and (b) point C.

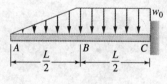

FIG. P6.51

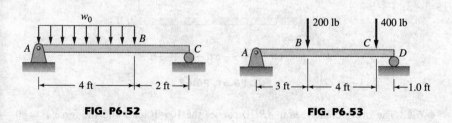

FIG. P6.52

FIG. P6.53

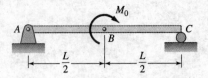

FIG. P6.54

6.54 Find the maximum displacement of the simply supported beam ABC that is loaded by a couple M_0 at its midspan. (*Hint*: By symmetry, the deflection is zero at point B, the point of application of the couple.)

6.55 Determine the value of $EI\delta$ at point C for the simply supported beam $ABCD$.

6.56 For the simply supported beam ABC, determine the values of (a) $EI\delta$ at point B; and (b) $EI\theta$ at point B.

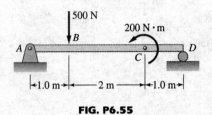

FIG. P6.55

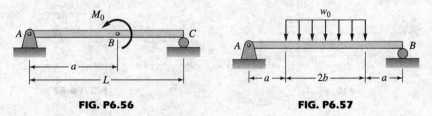

FIG. P6.56   FIG. P6.57

6.57 Find the maximum value of $EI\delta$ for the symmetrically loaded, simply supported beam AB.

6.58 Determine the value of $EI\delta$ at end A of the overhanging beam $ABCD$. (*Hint*: By symmetry, the elastic curve midway between the supports is horizontal.)

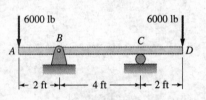

FIG. P6.58

6.59 For the overhanging beam $ABCD$, compute the magnitude of the load P that would cause the elastic curve to be horizontal at support C.

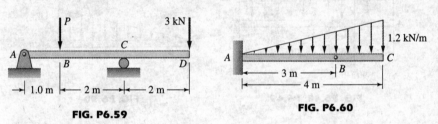

FIG. P6.59   FIG. P6.60

6.60 Determine the displacement at point B of the cantilever beam ABC. Use $E = 10$ GPa and $I = 30 \times 10^6$ mm^4.

6.61 For the overhanging beam $ABCD$, compute (a) the value of the force P for which the slope of the elastic curve at C is zero, and (b) the corresponding value of $EI\delta$ at B.

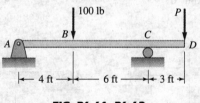

FIG. P6.61, P6.62

6.62 The overhanging beam $ABCD$ carries the 100-lb load and the force $P = 80$ lb. Compute the value of $EI\delta$ at point D.

6.63 The overhanging beam $ABCD$ carries the uniformly distributed load of intensity 120 lb/ft over the segments AB and CD. Find the value of $EI\delta$ at point B.

6.64 Determine the value of $EI\delta$ at point A of the overhanging beam ABC.

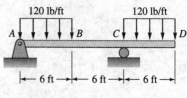

FIG. P6.63

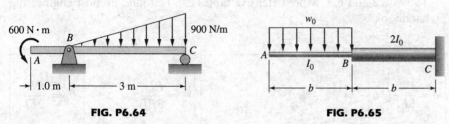

FIG. P6.64 **FIG. P6.65**

6.65 The two segments of the cantilever beam ABC have different cross sections with the moments of inertia shown in the figure. Determine the expression for the maximum displacement of the beam.

6.66 The simply supported beam ABC contains two segments. The moment of inertia of the cross-sectional area for segment AB is three times larger than the moment of inertia for segment BC. Find the expression for the displacement for point B.

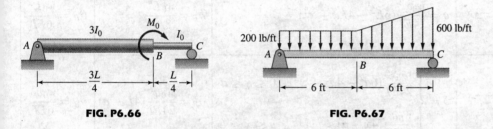

FIG. P6.66 **FIG. P6.67**

6.67 Calculate the value of $EI\delta$ at point B of the simply supported beam ABC.

6.5 Method of Superposition

The *method of superposition*, a popular method for finding slopes and deflections, is based on the *principle of superposition*:

> If the response of a structure is linear, then the effect of several loads acting simultaneously can be obtained by superimposing (adding) the effects of the individual loads.

By "linear response" we mean that the relationship between the cause (loading) and the effect (internal forces and deformations) is linear. The two requirements for linear response are (1) the material must obey Hooke's law; and (2) the deformations must be sufficiently small so that their effect on the geometry is negligible.

The method of superposition permits us to use the known displacements and slopes for simple loads to obtain the deformations for more complicated loadings. To use the method effectively requires access to tables that list the formulas for slopes and deflections for various loadings, such as Tables 6.2 and 6.3. More extensive tables can be found in most engineering handbooks.

Beam	Deflection δ	δ_B	θ_B
Cantilever with point load P at B	$\delta = \dfrac{Px^2}{6EI}(3L - x)$	$\delta_B = \dfrac{PL^3}{3EI}$	$\theta_B = \dfrac{PL^2}{2EI}$
Cantilever with point load P at distance a	$\delta = \begin{cases} \dfrac{Px^2}{6EI}(3a - x) & 0 \le x \le a \\ \dfrac{Pa^2}{6EI}(3x - a) & a \le x \le L \end{cases}$	$\delta_B = \dfrac{Pa^2}{6EI}(3L - a)$	$\theta_B = \dfrac{Pa^2}{2EI}$
Cantilever with uniform load w_0 over full length	$\delta = \dfrac{w_0 x^2}{24EI}(6L^2 - 4Lx + x^2)$	$\delta_B = \dfrac{w_0 L^4}{8EI}$	$\theta_B = \dfrac{w_0 L^3}{6EI}$
Cantilever with uniform load w_0 over length a	$\delta = \begin{cases} \dfrac{w_0 x^2}{24EI}(6a^2 - 4ax + x^2) & 0 \le x \le a \\ \dfrac{w_0 a^3}{24EI}(4x - a) & a \le x \le L \end{cases}$	$\delta_B = \dfrac{w_0 a^3}{24EI}(4L - a)$	$\theta_B = \dfrac{w_0 a^3}{6EI}$
Cantilever with triangular load	$\delta = \dfrac{w_0 x^2}{120L\,EI}(10L^3 - 10L^2 x + 5Lx^2 - x^3)$	$\delta_B = \dfrac{w_0 L^4}{30EI}$	$\theta_B = \dfrac{w_0 L^3}{24EI}$
Cantilever with moment M_0 at B	$\delta = \dfrac{M_0 x^2}{2EI}$	$\delta_B = \dfrac{M_0 L^2}{2EI}$	$\theta_B = \dfrac{M_0 L}{EI}$

TABLE 6.2 *Deflection Formulas for Cantilever Beams*

Beam	Deflection δ	Maximum / Center Deflection	Slope
Simply supported, point load P at center (span $L/2$, $L/2$)	$\delta = \dfrac{Px}{48EI}(3L^2 - 4x^2) \quad 0 \le x \le \dfrac{L}{2}$	$\delta_{\max} = \dfrac{PL^3}{48EI}$	$\theta_A = \theta_B = \dfrac{PL^2}{16EI}$
Simply supported, point load P at distance a, b	$\delta = \begin{cases} \dfrac{Pbx}{6LEI}(L^2 - x^2 - b^2) & 0 \le x \le a \\[4pt] \dfrac{Pb}{6LEI}\left[\dfrac{L}{b}(x-a)^3 + (L^2 - b^2)x - x^3\right] & a \le x \le L \end{cases}$	$\delta_{\max} = \dfrac{Pb(L^2 - b^2)^{3/2}}{9\sqrt{3}\,LEI}$ at $x = \sqrt{\dfrac{L^2 - b^2}{3}}$ if $a > b$ $\delta_{\text{center}} = \dfrac{Pb}{48EI}(3L^2 - 4b^2)$	$\theta_A = \dfrac{Pab}{6LEI}(2L - a)$ $\theta_B = \dfrac{Pab}{6LEI}(2L - b)$
Simply supported, uniform load w_0	$\delta = \dfrac{w_0 x}{24EI}(L^3 - 2Lx^2 + x^3)$	$\delta_{\max} = \dfrac{5w_0 L^4}{384EI}$	$\theta_A = \theta_B = \dfrac{w_0 L^3}{24EI}$
Simply supported, partial uniform load w_0 over a, b	$\delta = \begin{cases} \dfrac{w_0 x}{24LEI}[a^2(a - 2L)^2 + 2a(a - 2L)x^2 + Lx^3] & 0 \le x \le a \\[4pt] \dfrac{w_0 a^2}{24LEI}(-L + x)(a^2 - 4Lx + 2x^2) & a \le x \le L \end{cases}$	$\delta_{\text{center}} = \dfrac{w_0}{384EI}(5L^4 - 12L^2 b^2 + 8b^4)$ if $a \ge b$ $\delta_{\text{center}} = \dfrac{w_0 a^2}{96EI}(3L^2 - 2a^2)$ if $a \le b$	$\theta_A = \dfrac{w_0 a^2}{24LEI}(a - 2L)^2$ $\theta_B = \dfrac{w_0 a^2}{24LEI}(2L^2 - a^2)$
Simply supported, triangular load w_0	$\delta = \dfrac{w_0 x}{360LEI}(7L^4 - 10L^2 x^2 + 3x^4)$	$\delta_{\max} = 0.006522\,\dfrac{w_0 L^4}{EI}$ at $x = 0.5193L$ $\delta_{\text{center}} = 0.006510\,\dfrac{w_0 L^4}{EI}$	$\theta_A = \dfrac{7w_0 L^3}{360EI}$ $\theta_B = \dfrac{w_0 L^3}{45EI}$
Simply supported, end moment M_0	$\delta = \dfrac{M_0 x}{6LEI}(L^2 - x^2)$	$\delta_{\max} = \dfrac{M_0 L^2}{9\sqrt{3}\,EI}$ at $x = \dfrac{L}{\sqrt{3}}$ $\delta_{\text{center}} = \dfrac{M_0 L^2}{16EI}$	$\theta_A = \dfrac{M_0 L}{6EI}$ $\theta_B = \dfrac{M_0 L}{3EI}$

TABLE 6.3 Deflection Formulas for Simply Supported Beams

Sample Problem 6.11

Compute the midspan value of $EI\delta$ for the simply supported beam shown in Fig. (a) that is carrying two concentrated loads.

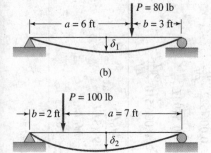

Solution

The loading on the beam can be considered to be the superposition of the loads shown in Figs. (b) and (c). According to Table 6.3, the displacement at the center of a simply supported beam is given by

$$EI\delta_{center} = \frac{Pb}{48}(3L^2 - 4b^2) \quad \text{where } a > b$$

We can use this formula to obtain the midspan displacements δ_1 and δ_2 of the beams in Figs. (b) and (c), provided that we choose for a and b the dimensions shown in the figures (note that a must be larger than b). We obtain

$$EI\delta_1 = \frac{(80)(3)}{48}[3(9)^2 - 4(3)^2] = 1035 \text{ lb} \cdot \text{ft}^3 \downarrow$$

$$EI\delta_2 = \frac{(100)(2)}{48}[3(9)^2 - 4(2)^2] = 946 \text{ lb} \cdot \text{ft}^3 \downarrow$$

The midspan deflection of the original beam is obtained by superposition:

$$EI\delta = EI\delta_1 + EI\delta_2 = 1035 + 946 = 1981 \text{ lb} \cdot \text{ft}^3 \downarrow \quad \text{Answer}$$

Sample Problem 6.12

The simply supported beam in Fig. (a) carries a uniformly distributed load over part of its length. Compute the midspan displacement.

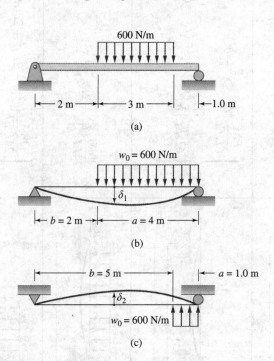

234

Solution

The given loading can be analyzed as the superposition of the two loadings shown in Figs. (b) and (c). From Table 6.3, the midspan value of $EI\delta$ for the beam in Fig. (b) is

$$EI\delta_1 = \frac{w_0}{384}(5L^4 - 12L^2b^2 + 8b^4)$$

$$= \frac{600}{384}[5(6)^4 - 12(6)^2(2)^2 + 8(2)^4] = 7625 \text{ N} \cdot \text{m}^3 \downarrow$$

Similarly, the midspan displacement of the beam in Fig. (c) is

$$EI\delta_2 = \frac{w_0 a^2}{96}(3L^2 - 2a^2) = \frac{(600)(1)^2}{96}[3(6)^2 - 2(1)^2] = 662.5 \text{ N} \cdot \text{m}^3 \uparrow$$

The midspan displacement of the original beam is obtained by superposition:

$$EI\delta = EI\delta_1 - EI\delta_2 = 7625 - 662.5 = 6960 \text{ N} \cdot \text{m}^3 \downarrow \qquad \textbf{Answer}$$

Alternative Solution

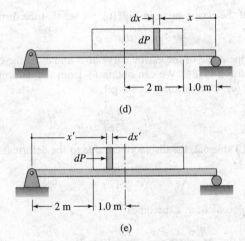

(d)

(e)

Another way to apply superposition is to consider an infinitesimal load element $dP = w\,dx = 600\,dx$ N. If the load element lies to the right of the midspan, as shown in Fig. (d), we see from Table 6.3 that its contribution to the midspan displacement is (we replace P by dP and b by x in the formula for δ_{center})

$$d\delta = \frac{dP}{48EI}x(3L^2 - 4x^2) = \frac{600\,dx}{48EI}x[3(6)^2 - 4x^2]$$

$$= \frac{600}{48EI}(108x - 4x^3)\,dx$$

If the element is on the left half of the span, we must replace x by $x' = L - x$ as indicated in Fig. (e). By adding the contributions of all the load elements, we get for the midspan displacement

$$\delta = \frac{600}{48EI}\left[\int_{1\,\text{m}}^{3\,\text{m}}(108x - 4x^3)\,dx + \int_{2\,\text{m}}^{3\,\text{m}}(108x' - 4x'^3)\,dx'\right]$$

$$= \frac{600}{48EI}(352 + 205) = \frac{6960}{EI} \text{ N} \cdot \text{m}^3 \downarrow \qquad \textbf{Answer}$$

As you can see, the formula for the displacement caused by a concentrated load can be used to find the displacement due to any distributed load by integration. If the integral is difficult, it can always be evaluated numerically.

Sample Problem 6.13

The overhanging beam ABC in Fig. (a) carries a concentrated load P at end C. Determine the displacement of the beam at C.

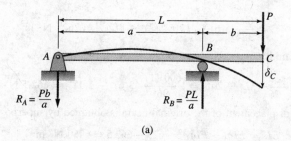

(a)

Solution

From the sketch of the elastic curve in Fig. (b), we see that the displacement of C is

$$\delta_C = \theta_B b + \delta'_C$$

where θ_B is the slope angle of the elastic curve at B and δ'_C is the displacement at C due to the deformation of BC. We can obtain θ_B from the deformation of segment AB, shown in Fig. (c). Using Table 6.3, we get

$$\theta_B = \frac{(Pb)a}{3EI}$$

From Fig. (d) and Table 6.2, the displacement due to the deformation of BC is

$$\delta'_C = \frac{Pb^3}{3EI}$$

Therefore, the displacement at C becomes

$$\delta_C = \frac{Pba}{3EI}b + \frac{Pb^3}{3EI} = \frac{Pb^2}{3EI}(a+b) = \frac{Pb^2 L}{3EI} \downarrow \qquad \text{Answer}$$

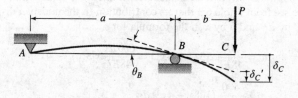

(b)

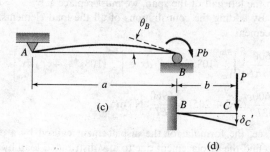

(c)

(d)

Problems

6.68 For the beam in Sample Problem 6.11, find the values of $EI\delta$ under the concentrated loads at B and C.

6.69 Determine the value of $EI\delta$ at midspan of the simply supported beam.

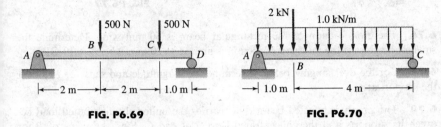

FIG. P6.69

FIG. P6.70

6.70 Find the midspan displacement of the simply supported beam using $E = 10$ GPa and $I = 20 \times 10^6$ mm^4.

6.71 Determine the midspan displacement for the simply supported beam.

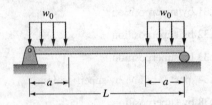

FIG. P6.71

6.72 Compute the value of $EI\delta$ at the overhanging end A of the beam.

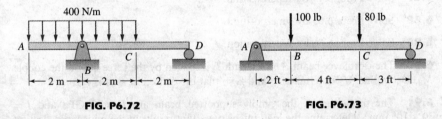

FIG. P6.72

FIG. P6.73

6.73 Determine the value of $EI\delta$ at midspan for the beam loaded by two concentrated forces.

6.74 The cross section of the wood beam is 4 in. by 8 in. Find the value of P for which the downward deflection at the midspan is 0.5 in. Use $E = 1.5 \times 10^6$ psi.

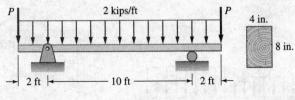

FIG. P6.74

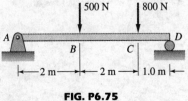

FIG. P6.75

6.75 Determine the value of $EI\delta$ under each of the concentrated loads that are applied to the simply supported beam.

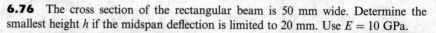
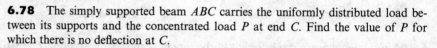

FIG. P6.76   FIG. P6.77

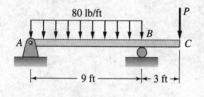

FIG. P6.78, P6.79

6.76 The cross section of the rectangular beam is 50 mm wide. Determine the smallest height h if the midspan deflection is limited to 20 mm. Use $E = 10$ GPa.

6.77 For the overhanging beam, determine the magnitude and sense of $EI\theta$ over the support at C.

6.78 The simply supported beam ABC carries the uniformly distributed load between its supports and the concentrated load P at end C. Find the value of P for which there is no deflection at C.

6.79 The simply supported beam ABC carries the uniformly distributed load between its supports and the concentrated load P at end C. Find the value of P for which the deflection curve is horizontal at B.

6.80 Solve Prob. 6.31 by superposition.

6.81 Solve Prob. 6.32 by superposition.

6.82 Solve Prob. 6.33 by superposition.

6.83 Solve Prob. 6.34, part (a) only, by superposition.

6.84 Solve Prob. 6.38 by superposition.

6.85 Solve Prob. 6.39 by superposition.

6.86 Solve Prob. 6.42 by superposition.

6.87 Solve Prob. 6.43 by superposition.

6.88 Solve Prob. 6.65 by superposition.

6.89 Solve Prob. 6.66 by superposition.

6.90 The cantilever beam AB of length L is loaded by the force P and the couple M_0. Determine M_0 in terms of P and L so that the deflection at A is zero.

6.91 The properties of the simply supported beam are $E = 70$ GPa and $I = 30 \times 10^6$ mm^4. Determine the load intensity w_0 that results in the midspan deflection being equal to 1/360th of the span.

6.92 Determine the vertical displacement of point C of the frame ABC caused by the applied couple M_0. Assume that EI is constant throughout the frame.

6.93 Solve Prob. 6.92 if the couple M_0 is replaced by a downward vertical load P.

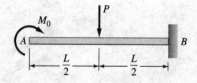

FIG. P6.90

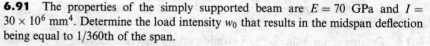
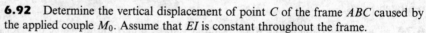

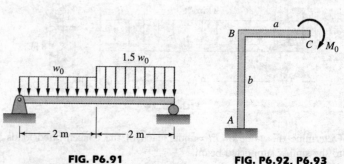

FIG. P6.91   FIG. P6.92, P6.93

6.94 Find the vertical displacement of point C of the frame ABC. The cross-sectional moments of inertia are $2I_0$ for segment AB and I_0 for segment BC.

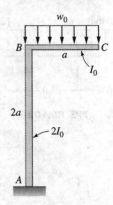

FIG. P6.94

Review Problems

6.95 (a) Determine the equation of the elastic curve for the cantilever beam. (b) Using the result of part (a), compare the displacement at the free end with the corresponding expression given in Table 6.2.

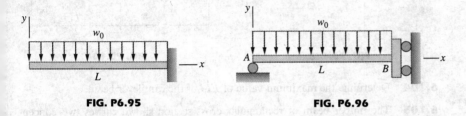

FIG. P6.95

FIG. P6.96

6.96 Derive the equation of the elastic curve for the beam AB. The support at B is free to move vertically but does not allow rotation.

6.97 Find the equation of the elastic curve for the simply supported beam that carries a distributed load of intensity $w = w_0 x^2/L^2$.

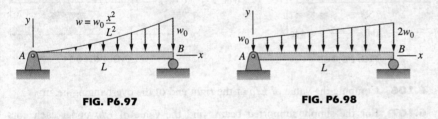

FIG. P6.97

FIG. P6.98

6.98 The intensity of the distributed loading acting on the simply supported beam varies linearly from w_0 at A to $2w_0$ at B. Determine the equation of the elastic curve of the beam.

6.99 Derive the equations of the elastic curve for the two segments of the overhanging beam ABC.

6.100 Find the equation of the elastic curve for segments AB and BC of the simply supported beam.

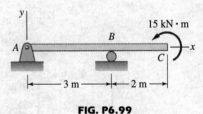

FIG. P6.99

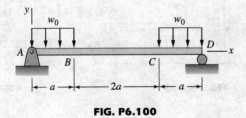

FIG. P6.100

6.101 Compute the value of $EI\theta$ at support B of the overhanging beam shown in the figure.

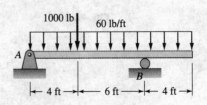

FIG. P6.101

CHAPTER 6 Deflection of Beams

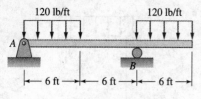

FIG. P6.102

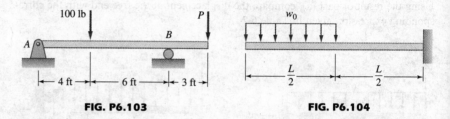

FIG. P6.103

FIG. P6.104

6.102 Determine the value of $EI\delta$ midway between the supports for the overhanging beam.

6.103 The overhanging beam carries concentrated loads of magnitudes 100 lb and P. (a) Determine P for which the slope of the elastic curve at B is zero. (b) Compute the corresponding value of $EI\delta$ under the 100-lb load.

6.104 Determine the maximum value of $EI\delta$ of the cantilever beam.

6.105 The timber beam of rectangular cross section shown carries two concentrated loads, each of magnitude P. Find the maximum allowable value of P if the midspan displacement of the beam is limited to 0.5 in. Use $E = 1.5 \times 10^6$ psi.

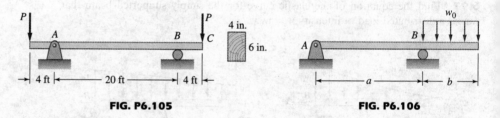

FIG. P6.105

FIG. P6.106

6.106 Compute the value of $EI\delta$ at the right end of the overhanging beam.

6.107 For the simply supported beam, find the value of $EI\delta$ under each concentrated load.

6.108 Determine the value of $EI\delta$ at midspan of the simply supported beam.

6.109 The segments AB and BC of the bent bar have the same flexural rigidity EI. Find the horizontal component of $EI\delta$ at end C.

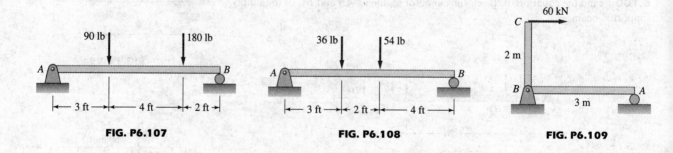

FIG. P6.107

FIG. P6.108

FIG. P6.109

6.110 Determine the midspan displacement of the simply supported beam ABC. The beam contains three segments with the cross-sectional moments of inertia shown in the figure.

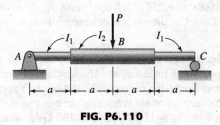

FIG. P6.110

Computer Problems

Neglect the weight of the beam in the problems.

C6.1 The uniform cantilever beam of length L carries a distributed load w that varies with the distance x. Given L, $w(x)$, E, and I, write a program to plot the deflection of the beam against x. Apply the program to the steel ($E = 29 \times 10^6$ psi) beams shown in Figs. (a) and (b). (*Hint*: Use superposition by applying the deflection formulas for the beam with concentrated load in Table 6.2 to the load element $w\,dx$ and integrating the result from $x = 0$ to L—see Sample Problem 6.11.)

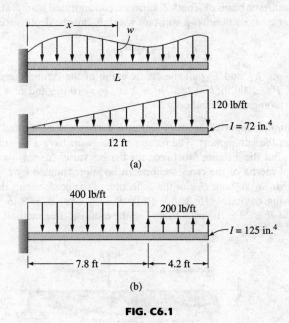

FIG. C6.1

C6.2 Solve Prob. C6.1 assuming the beam to be simply supported at each end.

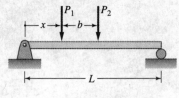

FIG. C6.3

C6.3 The concentrated loads P_1 and P_2 travel across the simply supported beam of length L and constant cross section. The loads are separated by the constant distance b. Given P_1, P_2, L, b, E, and I, plot the deflection under P_1 as a function of the distance x from $x = 0$ to L. Use the following data: (a) $P_1 = 12$ kN, $P_2 = 6$ kN, $L = 10$ m, $b = 3$ m, $E = 70$ GPa, $I = 250 \times 10^6$ mm^4; and (b) $P_2 = -6$ kN, other data the same as in part (a). (*Hint*: Use the method of superposition in conjunction with Table 6.3.)

C6.4 The overhanging beam of length L and constant cross section carries a uniformly distributed loading of intensity w_0. The distance between the supports is b. Given L, b, w_0, E, and I, plot the deflection of the beam. Experiment with the program to determine the value of b that minimizes the maximum displacement. Use the following data: $L = 6$ m, $w_0 = 12$ kN/m, $E = 200$ GPa, and $I = 95 \times 10^6$ mm^4. (*Hint*: Use superposition in conjunction with Tables 6.2 and 6.3.)

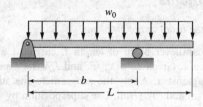

FIG. C6.4

C6.5 The cantilever beam of length L carries a concentrated load P at the free end. The rectangular cross section has a constant width b, but the depth varies as

$$h = h_1 + (h_2 - h_1)\frac{x}{L}$$

Given L, P, b, h_1, h_2, and E, plot the elastic curve of the beam. Use the following data: $L = 6$ ft, $P = 2000$ lb, $b = 2$ in., $h_1 = 2$ in., $h_2 = 10$ in., and $E = 29 \times 10^6$ psi. (*Hint*: Use the moment-area method.)

C6.6 The simply supported beam of length L carries a concentrated force P at a distance b from the left support. The flanges of the beam have a constant cross-sectional area A_f, but the distance h between the flanges varies from h_1 to h_2 as shown. The moment of inertia of the cross section can be approximated by $I = 2A_f(h/2)^2$. Given L, b, P, A_f, h_1, h_2, and E, plot the deflection of the beam versus the distance x. Use the following data: $L = 16$ ft, $b = 10$ ft, $P = 30$ kips, $A_f = 8$ in.^{2}, $h_1 = 8$ in., $h_2 = 20$ in., and $E = 29 \times 10^6$ psi. (*Hint*: Use the moment-area method.)

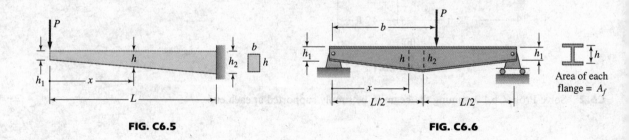

FIG. C6.5   **FIG. C6.6**

7 Statically Indeterminate Beams

Elevated concrete roadways. Such roadways can be treated as continuous beams resting on numerous supports. Continuous beams are statically indeterminate; they are analyzed by considering deflections in addition to equilibrium.

7.1 Introduction

A beam is statically indeterminate if the number of support reactions exceeds the number of independent equilibrium equations. In general, two equilibrium equations are available for a beam supporting lateral loads ($\Sigma F_y = 0$ and $\Sigma M_A = 0$, A being an arbitrary point).[1] Hence, a statically determinate beam has two support reactions, which is the minimum number needed to keep the beam in equilibrium. Additional reactions, being nonessential for equilibrium, are known as *redundant reactions*. The number of redundant reactions is called the *degree of indeterminacy* of the beam.

---

[1] We assume that the axial force in the beam is zero, so that ΣF_x is automatically satisfied.

In our study of axial and torsional loading, we found that the solution of statically indeterminate problems requires the analysis of compatibility of deformation as well as equilibrium. For beams, the compatibility equations are derived from the constraints imposed on the elastic curve by the supports. The procedure for determining the support reactions of a statically indeterminate beam depends on which of the methods described in Chapter 6 is used to analyze the deformation.

We note that each support reaction corresponds to a constraint imposed by the support. For example, a simple support provides a force that imposes the deflection constraint. A built-in support provides two reactions: a force imposing the constraint on deflection, and a couple imposing the rotational constraint. Thus, *the number of support constraints and the number of reactions are always equal.*

7.2 Double-Integration Method

Recall that in the method of double integration, we derived the equation for the elastic curve of the beam by integrating the differential equation $EIv'' = M$ two times, resulting in

$$EIv = \iint M \, dx \, dx + C_1 x + C_2 \quad \text{(6.5, repeated)}$$

If the beam is statically determinate, it has two support reactions and thus two constraints on its elastic curve. Because the reactions can be computed from the equilibrium equations, the conditions of constraint are available to compute the two constants of integration. In a statically indeterminate beam, each redundant reaction represents an additional unknown. However, there is also an additional constraint associated with each redundancy, which, when substituted into Eq. (6.5), provides an extra equation.

For example, the simply supported beam in Fig. 7.1(a) is statically determinate. It has two deflection constraints ($v_A = 0$ and $v_B = 0$) and two support reactions (R_A and R_B), as shown in the figure. The reactions can be

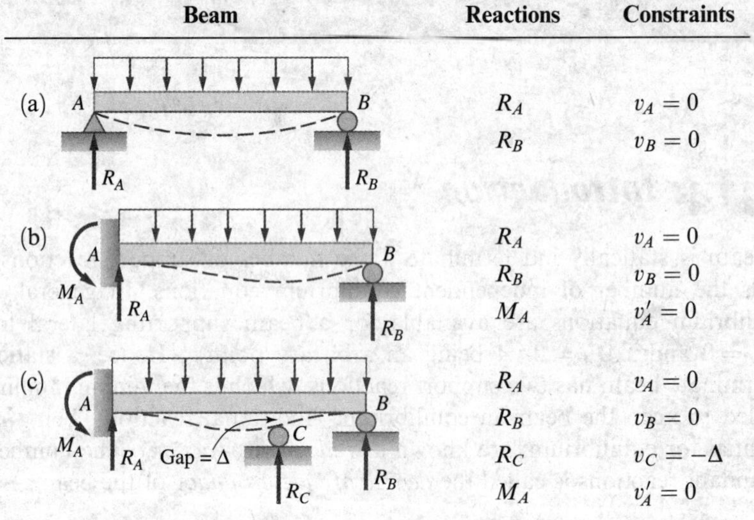

Fig. 7.1 Examples of statically determinate and indeterminate beams.

determined from the equilibrium equations, so that the constraints can be used to compute the constants C_1 and C_2 in Eq. (6.5).

By building in the support at A, as shown in Fig. 7.1(b), we introduce the additional constraint $v'_A = 0$ and the reactive couple M_A (a redundant reaction). Therefore, the beam is statically indeterminate of degree one. The number of unknowns is now five: three support reactions (R_A, R_B, and M_A) and two constants of integration (C_1 and C_2). The number of available equations is also five: two equilibrium equations and three constraints shown in the figure.

In Fig 7.1(c), we have added another support at C that has a small initial gap Δ. Assuming that the beam makes contact with the support at C when the loading is applied, we see that the support introduces another redundant reaction R_C and the corresponding constraint $v_C = -\Delta$. Since there are now two redundant support reactions, the degree of static indeterminacy of the beam is two. The number of available equations for determining the six unknowns (R_A, R_B, R_C, M_A, C_1, and C_2) is also six: two equations of equilibrium and the four constraint conditions shown in the figure.

The above discussion assumes that M is the global expression for the bending moment (applicable to the entire beam). If the beam is divided into two or more segments with different expressions for M, double integration will result in additional constants of integration. However, there will be an equal number of new constraints in the form of continuity conditions (deflections and slopes must be continuous across the junctions between the segments). Clearly, the method of double integration can become tedious for statically indeterminate beams with multiple segments, unless M is expressed in terms of bracket functions.

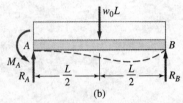

(a)

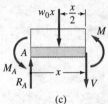

(b)

(c)

Sample Problem 7.1

Determine all the support reactions for the propped cantilever beam in Fig. (a).

Solution

Equilibrium The free-body diagram of the beam, shown in Fig. (b), yields the equilibrium equations

$$\Sigma F_y = 0 \quad +\uparrow \quad R_A + R_B - w_0 L = 0 \tag{a}$$

$$\Sigma M_A = 0 \quad +\circlearrowleft \quad M_A + R_B L - (w_0 L)\frac{L}{2} = 0 \tag{b}$$

Because there are three support reactions (R_A, R_B, and M_A) but only two independent equilibrium equations, the degree of static indeterminacy is one.

Compatibility A third equation containing the support reactions is obtained by analyzing the deformation of the beam. We start with the expression for the bending moment, obtainable from the free-body diagram in Fig. (c):

$$M = -M_A + R_A x - \frac{w_0 x^2}{2}$$

Substituting M into the differential equation for the elastic curve and integrating twice, we get

$$EIv'' = -M_A + R_A x - \frac{w_0 x^2}{2}$$

$$EIv' = -M_A x + R_A \frac{x^2}{2} - \frac{w_0 x^3}{6} + C_1$$

$$EIv = -M_A \frac{x^2}{2} + R_A \frac{x^3}{6} - \frac{w_0 x^4}{24} + C_1 x + C_2$$

Since there are three support reactions, we also have three support constraints. Applying these constraints to the elastic curve, shown by the dashed line in Fig. (b), we get

1. $v'|_{x=0} = 0$ (no rotation at A) $C_1 = 0$
2. $v|_{x=0} = 0$ (no deflection at A) $C_2 = 0$
3. $v|_{x=L} = 0$ (no deflection at B)

$$-M_A \frac{L^2}{2} + R_A \frac{L^3}{6} - \frac{w_0 L^4}{24} = 0 \tag{c}$$

The solution of Eqs. (a)–(c) is

$$M_A = \frac{w_0 L^2}{8} \quad R_A = \frac{5w_0 L}{8} \quad R_B = \frac{3w_0 L}{8} \quad \text{Answer}$$

Because the results are positive, the reactions are directed as shown in Fig. (b).

Sample Problem 7.2

The beam in Fig. (a) has built-in supports at both ends. Determine all the support reactions.

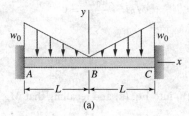

(a)

Solution

The free-body diagram of the beam in Fig. (b) shows four support reactions: the forces R_A and R_C and the couples M_A and M_C. Because there are two independent equilibrium equations, the degree of static indeterminacy is two.

We can simplify the analysis considerably by taking advantage of symmetries about the midpoint B. The symmetry of loading implies that $M_A = M_C$ and

$$R_A = R_C = \frac{w_0 L}{2}$$ **Answer**

Also, the symmetry of deformation requires that the elastic curve, shown by the dashed line in Fig. (b), has zero slope at the midpoint B.

Because of the above symmetries, we need to analyze only half of the beam, such as the segment BC shown in Fig. (c).

Equilibrium From the free-body diagram of segment BC in Fig. (c), we get

$$\Sigma F_y = 0 \quad +\uparrow \quad V_B + R_C - \frac{w_0 L}{2} = 0$$

$$\Sigma M_C = 0 \quad +\circlearrowleft \quad -M_B - M_C - V_B L + \frac{w_0 L}{2}\left(\frac{L}{3}\right) = 0$$

Substituting $R_C = w_0 L/2$ and solving yield $V_B = 0$ and

$$M_C = -M_B + \frac{w_0 L^2}{6} \quad\quad (a)$$

Compatibility From the free-body diagram in Fig. (d), the bending moment in segment BC is

$$M = M_B - \frac{w_0 x^2}{2L}\left(\frac{x}{3}\right) = M_B - \frac{w_0 x^3}{6L}$$

Substituting M into the differential equation of the elastic curve and integrating twice, we obtain

$$EIv'' = M_B - \frac{w_0 x^3}{6L}$$

$$EIv' = M_B x - \frac{w_0 x^4}{24L} + C_1$$

$$EIv = M_B \frac{x^2}{2} - \frac{w_0 x^5}{120L} + C_1 x + C_2$$

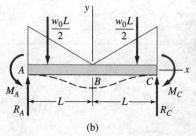

(b)

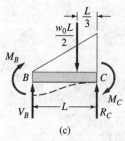

(c)

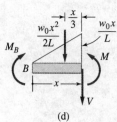

(d)

The elastic curve of segment BC is shown with the dashed line in Fig. (c). Applying the zero slope constraints on this elastic curve at B and C, we obtain

1. $v'|_{x=0} = 0 \qquad C_1 = 0$
2. $v'|_{x=L} = 0$

$$M_B L - \frac{w_0 L^3}{24} = 0$$

which yields

$$M_B = \frac{w_0 L^2}{24} \qquad \text{(b)}$$

Substituting Eq. (b) into Eq. (a) and recalling that $M_A = M_C$, we get

$$M_A = M_C = -\frac{w_0 L^2}{24} + \frac{w_0 L^2}{6} = \frac{w_0 L^2}{8} \qquad \textit{Answer}$$

Note that we did not use the deflection constraint $v|_{x=L} = 0$ because the constant C_2 was not needed in this problem.

Problems

7.1 Find all the support reactions for the propped cantilever beam that carries the couple M_0 at the propped end.

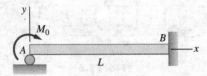

FIG. P7.1

7.2 Determine all the support reactions for the propped cantilever beam due to the parabolic loading shown in the figure.

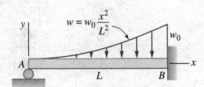

FIG. P7.2

7.3 The beam carrying the couple M_0 at its midpoint is built in at both ends. Find all the support reactions. (*Hint*: Utilize the skew-symmetry of deformation about the midpoint.)

7.4 A concentrated load is applied to the beam with built-in ends. (a) Find all the support reactions; and (b) draw the bending moment diagram. (*Hint*: Use symmetry.)

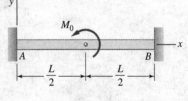

FIG. P7.3

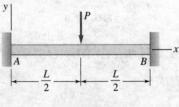

FIG. P7.4

7.5 The beam with three supports carries a uniformly distributed load. Determine all the support reactions. (*Hint*: Use symmetry.)

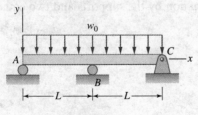

FIG. P7.5

7.6 A uniformly distributed load is applied to the beam with built-in supports. (a) Find all the support reactions; and (b) draw the bending moment diagram.

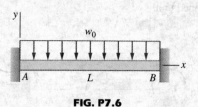

FIG. P7.6

7.7 The beam is built in at A and supported by vertical rollers at B (the rollers allow vertical deflection but prevent rotation). If the concentrated load P is applied at B, determine all the support reactions.

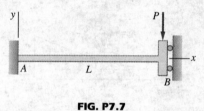

FIG. P7.7

7.8 A triangular load is applied to the beam with built-in ends. Find all the support reactions.

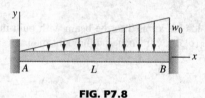

FIG. P7.8

7.3 Double Integration Using Bracket Functions

Because bracket functions enable us to write a global expression for the bending moment M, they eliminate the need to segment a beam if the loading is discontinuous. Therefore, the number of unknowns is always $n + 2$: n support reactions and two constants arising from double integration. The number of available equations is also $n + 2$: n equations of constraint imposed on the deformation by the supports and two equations of equilibrium.

Sample Problem 7.3

Before the 5000-N load is applied to the beam in Fig. (a), there is a small gap $\delta_0 = 30$ mm between the beam and the support under B. Find all the support reactions after the load is applied. Use $E = 10$ GPa and $I = 20 \times 10^6$ mm^4.

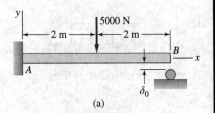

(a)

Solution

The free-body diagram of the beam is shown in Fig. (b). There are three support reactions: R_A, R_B, and M_A. By including R_B, we have assumed that the beam deflects sufficiently to make contact with the support at B (if the solution yields a positive value for R_B, we will know that this assumption is correct). The number of unknowns in this problem is five: the three support reactions and the two integration constants resulting from double integration. There are also five equations: two equations of equilibrium and three equations of constraint (the deflections at A and B, and the slope at A are known).

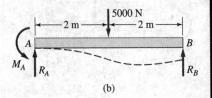

(b)

Equilibrium Referring to the free-body diagram of the beam in Fig. (b), we obtain the following two independent equilibrium equations:

$$\Sigma F_y = 0 \quad +\uparrow \quad R_A + R_B - 5000 = 0 \quad \text{(a)}$$

$$\Sigma M_B = 0 \quad +\circlearrowleft \quad M_A - R_A(4) + 5000(2) = 0 \quad \text{(b)}$$

Compatibility The free-body diagram in Fig. (c) yields the following global bending moment equation for the beam:

$$M = -M_A + R_A x - 5000\langle x - 2 \rangle \; \text{N} \cdot \text{m}$$

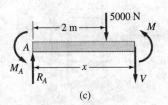

(c)

Substituting M into the differential equation of the elastic curve and integrating twice, we get

$$EIv'' = -M_A + R_A x - 5000\langle x - 2 \rangle \; \text{N} \cdot \text{m}$$

$$EIv' = -M_A x + R_A \frac{x^2}{2} - 2500\langle x - 2 \rangle^2 + C_1 \; \text{N} \cdot \text{m}^2$$

$$EIv = -M_A \frac{x^2}{2} + R_A \frac{x^3}{6} - \frac{2500}{3} \langle x - 2 \rangle^3 + C_1 x + C_2 \; \text{N} \cdot \text{m}^3$$

The constraints imposed by the supports on the elastic curve, shown as the dashed line in Fig. (b), yield

1. $v'|_{x=0} = 0$ (no rotation at A) $C_1 = 0$
2. $v|_{x=0} = 0$ (no deflection at A) $C_2 = 0$
3. $v|_{x=4\,\text{m}} = -\delta_0 = -0.03$ m (downward deflection at B equals δ_0)

$$(10 \times 10^9)(20 \times 10^{-6})(-0.03) = -M_A \frac{(4)^2}{2} + R_A \frac{(4)^3}{6} - \frac{2500}{3}(2)^3$$

Note that I was converted from mm^4 to m^4. After simplification, we get

$$-24M_A + 32R_A = 2000 \quad \text{(c)}$$

The solution of Eqs. (a)–(c) is

$$R_A = 3109 \; \text{N} \qquad R_B = 1891 \; \text{N} \qquad M_A = 2438 \; \text{N} \cdot \text{m} \qquad \textbf{Answer}$$

Because all reactions are positive, their directions shown in Fig. (b) are correct. Positive R_B indicates that the beam does make contact with the support at B, as we had assumed.

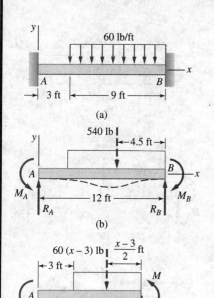

Sample Problem 7.4

The beam in Fig. (a) is built in at both ends and carries a uniformly distributed load over part of its length. Determine all the support reactions.

Solution

The free-body diagram of the entire beam in Fig. (b) contains four support reactions: the forces R_A and R_B and the couples M_A and M_B. Therefore, the total number of unknowns is six: the four reactions and two constants of integration arising from double integration. The number of available equations is also six: two equations of equilibrium and four conditions of constraint at the supports (the deflection and the slope at each support must be zero).

Equilibrium From the free-body diagram of the beam in Fig. (b), we obtain the equilibrium equations

$$\Sigma F_y = 0 \quad +\uparrow \quad R_A + R_B - 540 = 0 \tag{a}$$

$$\Sigma M_B = 0 \quad +\circlearrowleft \quad -M_A - R_A(12) + 540(4.5) - M_B = 0 \tag{b}$$

Compatibility From the free-body diagram in Fig. (c), the global expression for the bending moment is

$$M = -M_A + R_A x - \frac{60}{2}\langle x-3\rangle^2 \text{ lb}\cdot\text{ft}$$

Substituting this expression for M into the differential equation of the elastic curve and integrating twice, we obtain

$$EIv'' = -M_A + R_A x - 30\langle x-3\rangle^2 \text{ lb}\cdot\text{ft}$$

$$EIv' = -M_A x + R_A \frac{x^2}{2} - 10\langle x-3\rangle^3 + C_1 \text{ lb}\cdot\text{ft}^2 \tag{c}$$

$$EIv = -M_A \frac{x^2}{2} + R_A \frac{x^3}{6} - 2.5\langle x-3\rangle^4 + C_1 x + C_2 \text{ lb}\cdot\text{ft}^3 \tag{d}$$

The elastic curve of the beam is shown by the dashed line in Fig. (b). The constraints imposed by the supports yield

1. $v'|_{x=0} = 0$ (slope at A is zero) $\quad C_1 = 0$
2. $v|_{x=0} = 0$ (deflection at A is zero) $\quad C_2 = 0$
3. $v'|_{x=12\text{ft}} = 0$ (slope at B is zero)

$$-M_A(12) + R_A \frac{(12)^2}{2} - 10(9)^3 = 0 \tag{e}$$

4. $v|_{x=12\text{ft}} = 0$ (deflection at B is zero)

$$-M_A \frac{(12)^2}{2} + R_A \frac{(12)^3}{6} - 2.5(9)^4 = 0 \tag{f}$$

The solution of Eqs. (e) and (f) is

$$R_A = 189.8 \text{ lb} \qquad M_A = 532 \text{ lb}\cdot\text{ft} \qquad \textit{Answer}$$

The equilibrium equations, Eqs. (a) and (b), then yield

$$R_B = 350 \text{ lb} \qquad M_B = 683 \text{ lb}\cdot\text{ft} \qquad \textit{Answer}$$

All the reactions are positive, indicating that the directions assumed on the free-body diagram in Fig. (b) are correct.

Problems

7.9 Determine all the support reactions for the propped cantilever beam shown in the figure.

7.10 For the beam with built-in ends, determine (a) all the support reactions; and (b) the displacement at midspan.

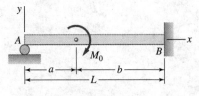

FIG. P7.9

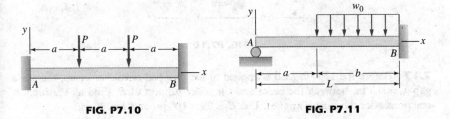

FIG. P7.10 **FIG. P7.11**

7.11 Find the support reaction at A for the propped cantilever beam.

7.12 Determine all the support reactions for the beam with built-in ends.

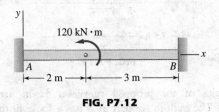

FIG. P7.12

7.13 For the beam with built-in ends, determine (a) all the support reactions; and (b) the displacement at the midpoint C. (*Hint*: Use symmetry.)

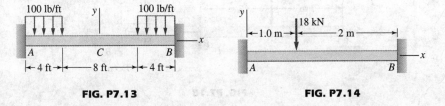

FIG. P7.13 **FIG. P7.14**

7.14 Determine all the support reactions for the beam with built-in ends.

7.15 Find all the support reactions for the beam shown in the figure.

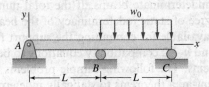

FIG. P7.15

7.16 The beam ABC has a built-in support at A and roller supports at B and C. Find all the support reactions.

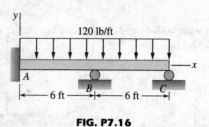

FIG. P7.16

7.17 Before the 1200-lb load is applied to the propped cantilever beam, there is a gap $\delta_0 = 0.3$ in. between the beam and the roller support at B. Find all the support reactions after the load is applied. Use $E = 29 \times 10^6$ psi and $I = 36$ in.^{4}.

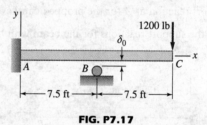

FIG. P7.17

7.18 The properties of the propped cantilever beam are $E = 72$ GPa and $I = 126 \times 10^6$ mm^4. The built-in support at B has a loose fit that allows the end of the beam to rotate through the angle $\theta_0 = 0.75°$ when the load is applied, as shown in the detail. Determine all the support reactions.

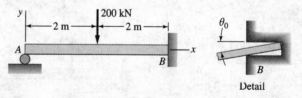

FIG. P7.18

*7.4 Moment-Area Method

The moment-area method is well suited for deriving the compatibility equations for statically indeterminate beams. If the total number of support reactions is n, the degree of static indeterminacy of the beam is $n - 2$. A total of n equations are available for computing the support reactions: two equilibrium equations and $n - 2$ compatibility equations to be obtained from the moment-area theorems. The following sample problems illustrate the application of the moment-area theorems to statically indeterminate beams.

Sample Problem 7.5

The propped cantilever beam AB in Fig. (a) carries a uniformly distributed load of intensity w_0 along its entire length L. Determine all the support reactions acting upon the beam.

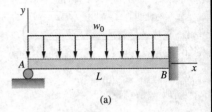

(a)

Solution

From the free-body diagram in Fig. (b), we see that there are three reactions (R_A, R_B, and M_B). Because there are only two independent equilibrium equations, the beam is statically indeterminate of degree one. Therefore, one compatibility equation is required.

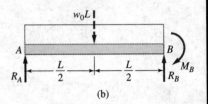

(b)

Equilibrium The following two independent equilibrium equations can be derived from the free-body diagram in Fig. (b):

$$\Sigma F_y = 0 \quad +\uparrow \quad R_A + R_B - w_0 L = 0 \tag{a}$$

$$\Sigma M_B = 0 \quad +\circlearrowleft \quad w_0 L\left(\frac{L}{2}\right) - R_A L - M_B = 0 \tag{b}$$

Compatibility Referring to the elastic curve in Fig. (c), we see that the tangent to the elastic curve at B is horizontal (rotation is prevented by the built-in support). Therefore, the tangential deviation of A with respect to B is zero. From the bending moment diagram drawn by parts in Fig. (d), the second moment-area theorem yields the compatibility equation

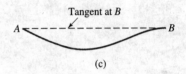

(c)

$$EI t_{A/B} = \text{area of } M\text{-diagram}]_B^A \cdot \bar{x}_{/A}$$

$$= \frac{1}{2}(L)(R_A L)\left(\frac{2L}{3}\right) - \frac{1}{3}(L)\left(\frac{w_0 L^2}{2}\right)\left(\frac{3L}{4}\right) = 0$$

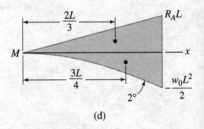

(d)

which gives

$$R_A = \frac{3}{8} w_0 L \qquad \text{Answer}$$

Substituting this value for R_A into Eqs. (a) and (b), we find the remaining two reactions:

$$R_B = w_0 L - R_A = w_0 L - \frac{3}{8} w_0 L = \frac{5}{8} w_0 L \qquad \text{Answer}$$

$$M_B = w_0 L\left(\frac{L}{2}\right) - R_A L = \frac{1}{2} w_0 L^2 - \frac{3}{8} w_0 L^2 = \frac{1}{8} w_0 L^2 \qquad \text{Answer}$$

Sample Problem 7.6

The beam AB in Fig. (a) is built in at both ends and carries a uniformly distributed load over part of its length. Compute all the support reactions acting on the beam. (*Note*: This problem was solved by double integration in Sample Problem 7.4.)

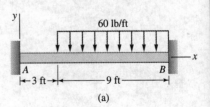

(a)

Solution

The free-body diagram of the entire beam in Fig. (b) contains four unknown end reactions: the forces R_A and R_B and the couples M_A and M_B. Because there are only two independent equilibrium equations, the beam is statically indeterminate of degree two. It follows that two compatibility equations are required for the solution.

Equilibrium From the free-body diagram in Fig. (b), the two independent equilibrium equations are

$$\Sigma F_y = 0 \quad +\uparrow \quad R_A + R_B - 540 = 0 \quad \text{(a)}$$

$$\Sigma M_B = 0 \quad +\circlearrowleft \quad M_A - R_A(12) + 540(4.5) - M_B = 0 \quad \text{(b)}$$

where the forces are in pounds and the couples are in pound-feet.

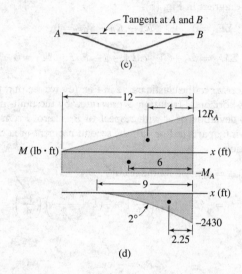

Compatibility The elastic curve of the beam is shown in Fig. (c). Because the slope at each end is horizontal due to the built-in supports, we conclude that the change in the slope between A and B is zero. From the bending moment diagram by parts in Fig. (d), the first moment-area theorem gives us

$$EI\theta_{B/A} - \text{area of } M\text{-diagram}]_A^B = \frac{1}{2}(12)(12R_A) - 12M_A - \frac{1}{3}(9)(2430) = 0 \quad \text{(c)}$$

A second compatibility equation is obtained by noting that the tangential deviation $t_{B/A}$ of B with respect to A is zero (we could also have used the condition $t_{A/B} = 0$). Applying the second moment-area theorem using the bending moment diagram in Fig. (d), we obtain

$$EIt_{B/A} = \text{area of } M\text{-diagram}]_A^B \cdot \bar{x}_{/B}$$

$$= \frac{1}{2}(12)(12R_A)(4) - 12M_A(6) - \frac{1}{3}(9)(2430)(2.25) = 0 \quad \text{(d)}$$

Solving Eqs. (a)–(d) gives

$$R_A = 189.8 \text{ lb} \quad R_B = 350 \text{ lb} \quad M_A = 532 \text{ lb} \cdot \text{ft} \quad M_B = 683 \text{ lb} \cdot \text{ft} \quad \textit{Answer}$$

These results agree with the answers obtained by the double-integration method with bracket functions in Sample Problem 7.4.

Sample Problem 7.7

The beam in Fig. (a) has three supports. Calculate all the support reactions due to the 6000-lb force.

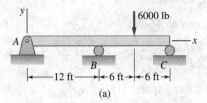

(a)

Solution

The free-body diagram in Fig. (b) shows that there are three vertical reactions: R_A, R_B, and R_C. Because there are only two independent equilibrium equations, the beam is statically indeterminate of degree one. Therefore, the computation of the reactions requires one compatibility equation in addition to the two equations of equilibrium.

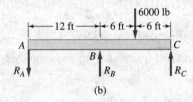

(b)

Equilibrium Using the free-body diagram in Fig. (b), we obtain the equilibrium equations

$$\Sigma F_y = 0 \quad +\uparrow \quad -R_A + R_B + R_C - 6000 = 0 \quad \text{(a)}$$

$$\Sigma M_B = 0 \quad +\circlearrowleft \quad R_A(12) + R_C(12) - 6000(6) = 0 \quad \text{(b)}$$

Compatibility From the elastic curve in Fig. (c), we see that the tangential deviations $t_{A/C}$ and $t_{B/C}$ are related by

$$t_{A/C} = 2t_{B/C} \tag{c}$$

Using the bending moment diagram drawn by parts in Fig. (d) and the second moment-area theorem, we obtain

$$EIt_{A/C} = \text{area of } M\text{-diagram}]_C^A \cdot \bar{x}_{/A}$$

$$= \frac{1}{2}(12)(12R_B)\left[12 + \frac{2}{3}(12)\right] - \frac{1}{2}(24)(24R_A)\left[\frac{2}{3}(24)\right]$$

$$- \frac{1}{2}(6)(36\,000)\left[18 + \frac{2}{3}(6)\right]$$

$$= 1440R_B - 4608R_A - (2.376 \times 10^6) \text{ lb} \cdot \text{ft}^3 \tag{d}$$

and

$$EIt_{B/C} = \text{area of } M\text{-diagram}]_C^B \cdot \bar{x}_{/B}$$

$$= \frac{1}{2}(12)(12R_B)\left[\frac{2}{3}(12)\right] - 12(12R_A)\left[\frac{1}{2}(12)\right]$$

$$- \frac{1}{2}(12)(12R_A)\left[\frac{2}{3}(12)\right] - \frac{1}{2}(6)(36\,000)\left[6 + \frac{2}{3}(6)\right]$$

$$= 576R_B - 1440R_A - (1.080 \times 10^6) \text{ lb} \cdot \text{ft}^3 \tag{e}$$

Substituting Eqs. (d) and (e) into Eq. (c) and simplifying yield

$$R_B = 750 + 6R_A \tag{f}$$

Solving Eqs. (a), (b), and (f) for the reactions, we obtain

$$R_A = 563 \text{ lb} \qquad R_B = 4130 \text{ lb} \qquad R_C = 2440 \text{ lb} \qquad \textit{Answer}$$

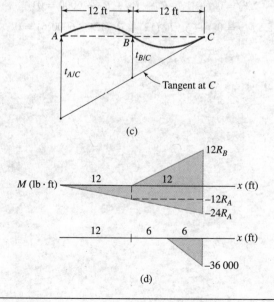

Problems

7.19 Determine the support reaction at A for the propped cantilever beam.

7.20 The propped cantilever beam carries two concentrated loads, each of magnitude P. Find the support reaction at A.

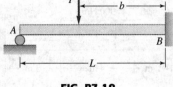

FIG. P7.19

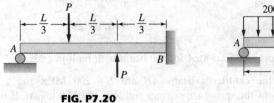

FIG. P7.20

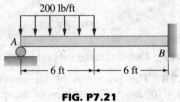

FIG. P7.21

7.21 Find the reactive couple acting on the propped cantilever beam at B.

7.22 The beam AB has a built-in support at A. The roller support at B allows vertical deflection but prevents rotation. Determine all the support reactions.

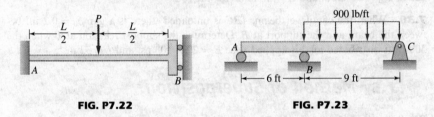

FIG. P7.22   **FIG. P7.23**

7.23 The beam ABC rests on three supports. Determine all the support reactions.

7.24 The load acting on the beam ABC has a triangular distribution. Find the reactions at all three supports.

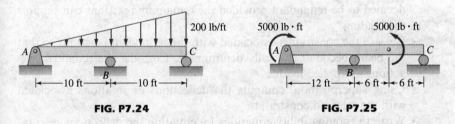

FIG. P7.24   **FIG. P7.25**

7.25 The beam ABC has three supports and carries two equal but opposite couples. Determine all the support reactions.

7.26 The beam AB has a built-in support at each end. Determine (a) the reactive couples acting on the beam at A and B; and (b) the value of $EI\delta$ at midspan. (*Hint*: Use symmetry.)

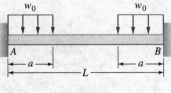

FIG. P7.26

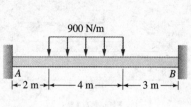

FIG. P7.27

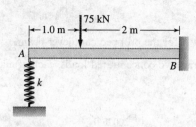

FIG. P7.28

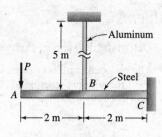

FIG. P7.29

7.27 Determine the support reactions at A for the beam with built-in ends.

7.28 The properties of the cantilever beam AB are $E = 200$ MPa and $I = 60 \times 10^6$ mm^4. The stiffness of the spring supporting end A is $k = 660$ kN/m. If the spring is initially undeformed, determine the force in the spring when the 75-kN load is applied to the beam.

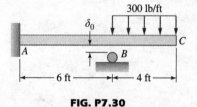

FIG. P7.30

7.29 The midpoint B of the steel cantilever beam ABC is supported by a vertical aluminum rod. Find the maximum allowable value of the applied force P if the stress in the rod is not to exceed 120 MPa. Use $E_{st} = 200$ GPa, $I = 50 \times 10^6$ mm^4 for the beam, $E_{al} = 70$ GPa, and $A = 40$ mm^2 for the rod.

7.30 When the cantilever beam ABC is unloaded, there is a gap $\delta_0 = 0.2$ in. between the beam and the support at B. Determine the support reaction at B when the 300-lb/ft distributed load is applied. Use $E = 29 \times 10^6$ psi and $I = 32$ in.^{4}.

7.5 Method of Superposition

We have used the method of superposition to solve problems involving statically indeterminate bars and shafts. The application of this method to statically indeterminate beams requires the following steps:

- Determine the degree of static indeterminacy and choose the redundant reactions. This choice is not unique—any support reaction can be deemed to be redundant provided the remaining reactions can support the loading.
- Release the constraints associated with the redundant reactions so that the beam becomes statically determinate. Consider the redundant reactions as applied loads.
- Using superposition, compute the deflections or rotations associated with the released constraints.
- Write the compatibility equations by equating the deflections or rotations found in the previous step to those imposed by the supports on the original beam.
- Solve the compatibility equations for the redundant reactions.

After the redundant reactions have been found, the remaining reactions can be computed from the equilibrium equations, as demonstrated in the sample problems.

Sample Problem 7.8

The propped cantilever beam AB in Fig. (a) carries a uniformly distributed load over its entire length L. Use the method of superposition to determine all the support reactions. (*Note*: This beam was analyzed by the method of double integration in Sample Problem 7.1 on page 246.)

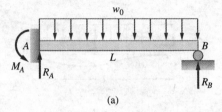

(a)

Solution

Compatibility Because there are three support reactions (R_A, R_B, and M_A) but only two independent equilibrium equations, the beam is statically indeterminate of degree one. Therefore, any one of the support reactions shown in Fig. (a) can be viewed as being redundant (you can verify that any two of the reactions can support the load). Choosing R_B as the redundant reaction, we remove the support at B and treat the reaction R_B as an applied force. The result is a cantilever beam loaded as shown in Fig. (b). The problem is now to find R_B for which the deflection at B is zero.

From Fig. (c) and Table 6.2, the displacement at B due to the load w_0 acting alone is

$$\delta_1 = \frac{w_0 L^4}{8EI} \downarrow$$

and the displacement at B due to R_B alone is

$$\delta_2 = \frac{R_B L^3}{3EI} \uparrow$$

The displacement at B of the original beam is obtained by superimposing δ_1 and δ_2. Because the result must be zero, the compatibility equation is

$$\delta_1 - \delta_2 = \frac{w_0 L^4}{8EI} - \frac{R_B L^3}{3EI} = 0$$

yielding

$$R_B = \frac{3w_0 L}{8} \qquad \text{Answer}$$

Equilibrium The reactions at A can now be obtained from the equilibrium equations. The results are (see the solution of Sample Problem 7.1 for details)

$$R_A = \frac{5w_0 L}{8} \qquad M_A = \frac{w_0 L^2}{8} \qquad \text{Answer}$$

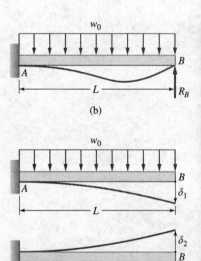

(b)

(c)

Alternative Solution

As we have mentioned, any one of the reactions can be chosen as being redundant. As an illustration, let us treat M_A as the redundant reaction. We must now release the rotational constraint at A and treat M_A as an applied couple, resulting in the simply supported beam shown in Fig. (d). The value of M_A is determined from the constraint $\theta_A = 0$, where θ_A is the slope of the elastic curve at A.

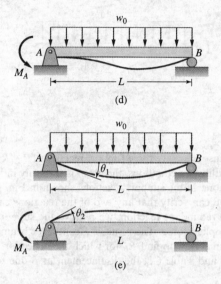

By superposition, $\theta_A = \theta_1 - \theta_2$, where θ_1 and θ_2 are the slopes caused by the two loads acting separately, as illustrated in Fig. (e). Using Table 6.3, we obtain

$$\theta_1 = \frac{w_0 L^3}{24EI} \circlearrowright \qquad \theta_2 = \frac{M_A L}{3EI} \circlearrowright$$

Hence, the compatibility equation is

$$\theta_1 - \theta_2 = \frac{w_0 L^3}{24EI} - \frac{M_A L}{3EI} = 0$$

which gives

$$M_A = \frac{w_0 L^2}{8} \qquad \text{Answer}$$

The other reactions could now be computed from the equilibrium equations.

Sample Problem 7.9

The beam in Fig. (a) has built-in supports at both ends and carries a uniformly distributed load over part of its length. Using the method of superposition, compute all of the support reactions acting on the beam. (*Note:* This problem was solved by other methods in Sample Problems 7.4 and 7.6.)

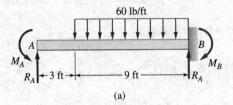

(a)

Solution

Compatibility The beam is statically indeterminate of degree two—there are four support reactions (R_A, R_B, M_A, and M_B) shown in Fig. (a) but only two independent equilibrium equations. Therefore, two of the reactions are redundant. Choosing R_A and M_A as the redundant reactions, we release the deflection and slope constraints at A and consider R_A and M_A to be applied loads, resulting in the cantilever beam shown in Fig. (b). Our task is now to determine R_A and M_A so that the deflection and the slope at A are zero.

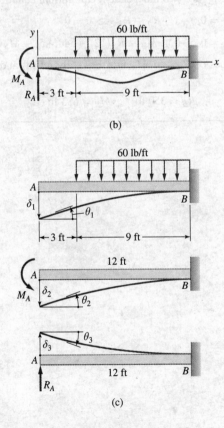

(b)

(c)

Figure (c) shows the slopes and deflections at A due to each of the loads. Using Table 6.2, we have

$$EI\theta_1 = \frac{w_0 L^3}{6} = \frac{60(9)^3}{6} = 7290 \text{ lb} \cdot \text{ft}^2$$

$$EI\theta_2 = M_A L = M_A(12) = 12 M_A \text{ lb} \cdot \text{ft}^2$$

$$EI\theta_3 = \frac{R_A L^2}{2} = \frac{R_A(12)^2}{2} = 72 R_A \text{ lb} \cdot \text{ft}^2$$

$$EI\delta_1 = \frac{w_0 a^3}{24}(4L - a) = \frac{60(9)^3}{24}[4(12) - 9] = 71\,078 \text{ lb} \cdot \text{ft}^3$$

$$EI\delta_2 = \frac{M_A L^2}{2} = \frac{M_A(12)^2}{2} = 72 M_A \text{ lb} \cdot \text{ft}^3$$

$$EI\delta_3 = \frac{R_A L^3}{3} = \frac{R_A(12)^3}{3} = 576 R_A \text{ lb} \cdot \text{ft}^3$$

where the forces and couples are measured in pounds and pound-feet, respectively.

From Fig. (c), the conditions of zero slope and zero deflection at A become (the common factor EI cancels out)

$$\theta_1 + \theta_2 - \theta_3 = 0 \qquad 7290 + 12 M_A - 72 R_A = 0 \qquad \text{(a)}$$

$$\delta_1 + \delta_2 - \delta_3 = 0 \qquad 71\,078 + 72 M_A - 576 R_A = 0 \qquad \text{(b)}$$

Solving Eqs. (a) and (b) for the reactions, we obtain

$$R_A = 189.8 \text{ lb} \qquad M_A = 532 \text{ lb} \cdot \text{ft} \qquad \text{Answer}$$

Equilibrium From Fig. (a), two independent equilibrium equations are

$$\Sigma F_y = 0 \quad +\uparrow \quad R_A + R_B - 540 = 0 \qquad \text{(c)}$$

$$\Sigma M_B = 0 \quad +\circlearrowleft \quad M_A - R_A(12) + 540(4.5) - M_B = 0 \qquad \text{(d)}$$

Substituting the values of R_A and M_A into Eqs. (a) and (b) and solving for the reactions at B yield

$$R_B = 350 \text{ lb} \qquad M_B = 683 \text{ lb} \cdot \text{ft} \qquad \text{Answer}$$

Problems

7.31 Solve Sample Problem 7.9 by choosing M_A and M_B as the redundant reactions.

7.32 Solve Prob. 7.15 by superposition.

7.33 Solve Prob. 7.16 by superposition.

7.34 Solve Prob. 7.18 by superposition.

7.35 Solve Prob. 7.20 by superposition.

7.36 Solve Prob. 7.22 by superposition.

7.37 Solve Prob. 7.23 by superposition.

7.38 Solve Prob. 7.28 by superposition.

7.39 Before the 30-kN·m couple is applied to the beam, there is a gap $\delta_0 = 3.75$ mm between the beam and the support at B. Determine the support reaction at B after the couple is applied. The beam is a W200 × 22 shape with $E = 200$ GPa.

7.40 When unloaded, the two identical cantilever beams just make contact at B. Determine the reactive couples acting on the beams at A and C when the uniformly distributed load is applied to BC.

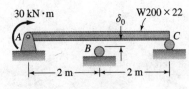

FIG. P7.39

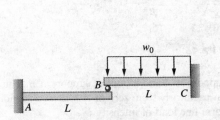

FIG. P7.40

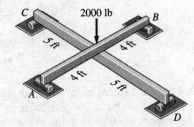

FIG. P7.41

7.41 The two simply supported timber beams are mounted so that they just make contact at their midpoints when unloaded. Beam AB is 2 in. wide and 4 in. deep; beam CD is 3 in. wide and 8 in. deep. Determine the contact force between the beams when the 2000-lb load is applied at the crossover point.

7.42 When the steel cantilever beams AB and CD are mounted, there is a 2-mm gap between their free ends A and C. Determine the contact force between A and C when the 3-N load is applied. Use $E = 200$ GPa for steel.

7.43 The beam AB has built-in supports at both ends. Find the bending moments at A and B.

7.44 The beam $ABCD$ has four equally spaced supports. Find all the support reactions.

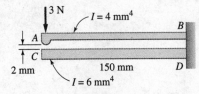

FIG. P7.42

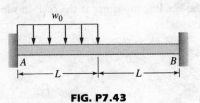

FIG. P7.43

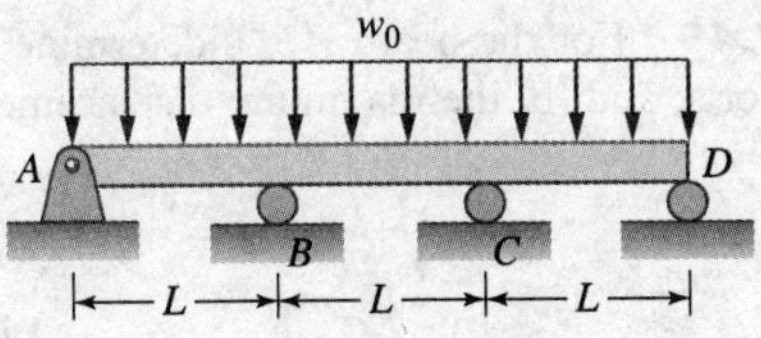

FIG. P7.44

7.45 The overhanging beam rests on three supports. Determine the length b of the overhangs so that the bending moments over all three supports have the same magnitude.

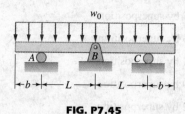

FIG. P7.45

Review Problems

7.46 The beam AB has a sliding support at A that prevents rotation but allows vertical displacement. The support at B is built in. Determine the reactive couple acting on the beam at A.

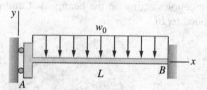

FIG. P7.46

7.47 The tapered beam has a simple support at A and a built-in support at B. The moment of inertia of the cross sections varies linearly from zero at A to I_0 at B. Find the support reaction at A due to the uniform line load of intensity w_0.

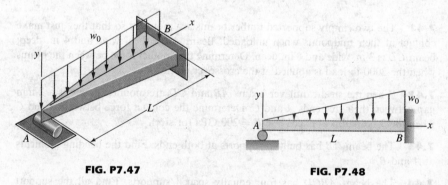

FIG. P7.47   **FIG. P7.48**

7.48 The propped cantilever beam carries a distributed load of triangular shape. Find the support reaction at A.

7.49 For the beam ABC, determine (a) the bending moments at the built-in supports; and (b) the maximum displacement.

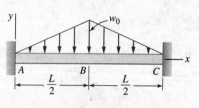

FIG. P7.49

7.50 Determine all support reactions for the propped cantilever beam *ABC*.

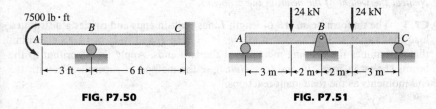

FIG. P7.50   FIG. P7.51

7.51 The beam *ABC* rests on three supports. Find the bending moment over the support at *B*.

7.52 The overhanging beam has three supports. Determine all the support reactions.

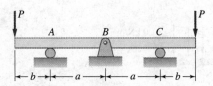

FIG. P7.52

7.53 The end of the cantilever beam *BD* rests on the simply supported beam *ABC*. The two beams have identical cross sections and are made of the same material. Find the maximum bending moment in each beam when the 1400-lb load is applied.

7.54 When the beam *ABC* is unloaded, there is a gap of length δ_0 between the beam and the support at *B*. Determine δ_0 for which all three support reactions are equal when the uniformly distributed load of intensity w_0 is applied.

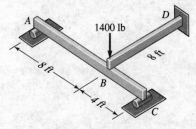

FIG. P7.53

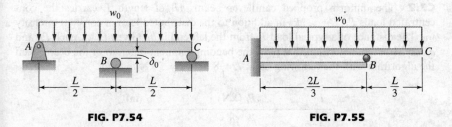

FIG. P7.54   FIG. P7.55

7.55 The two cantilever beams have the same flexural rigidity *EI*. When unloaded, the beams just make contact at *B*. Find the contact force between the beams at *B* when the uniformly distributed load is applied.

7.56 The overhanging beam *ABC* has a flexural rigidity *EI* and length *L*. End *C* is attached to a spring of stiffness *k*. Show that the force in the spring due to the applied couple M_0 is

$$P = \frac{M_0}{2L}\left(1 + \frac{12EI}{kL^3}\right)^{-1}$$

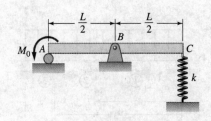

FIG. P7.56

Computer Problems

Neglect the weight of the beam in the problems.

C7.1 The uniform beam AB of length L has built-in ends and carries a distributed load that varies with x, as shown in Fig. (a). Given L and $w(x)$, write an algorithm that determines the bending moments at the two ends. Apply the algorithm to the beams shown in Figs. (b) and (c). (*Hint*: Use the method of superposition with the end moments as the redundant reactions.)

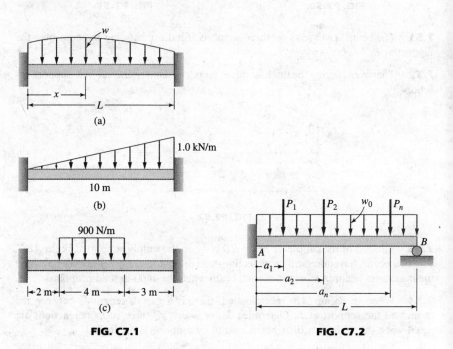

FIG. C7.1

FIG. C7.2

C7.2 The uniform propped cantilever beam AB of length L carries the concentrated loads $P_1, P_2, \ldots, P_n$ in addition to the distributed load of constant intensity w_0. The distance of a typical load P_i from the left end is a_i. Given L, w_0, each P_i, and a_i, construct an algorithm that plots the bending moment diagram of the beam. Run the algorithm with the following data: $L = 8$ m, $w_0 = 5$ kN/m, and

i	P_i (kN)	a_i (m)
1	10	2
2	12	4
3	-8	5
4	15	6

(*Hint*: Use superposition with the reaction at B as the redundant reaction.)

C7.3 The uniform beam of length L rests on five supports. The three middle supports, denoted by ①, ②, and ③, are located at distances a_1, a_2, and a_3 from the left end. A uniformly distributed load of intensity w_0 acts on the beam. Given L, a_1, a_2, a_3, and w_0, write an algorithm that computes the reactions at the three middle supports. (a) Run the algorithm with the following data: $L = 16$ m, $a_1 = 5$ m, $a_2 = 9$ m, $a_3 = 12$ m, and $w_0 = 100$ kN/m. (b) Determine by trial-and-error the approximate locations of the middle supports for which their reactions are equal. (*Hint*: Use the method of superposition.)

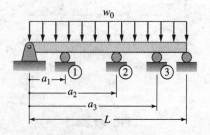

FIG. C7.3

C7.4 The laminated timber beam AOB of length $2L$ has built-in ends. It carries a uniformly distributed load of intensity w_0. The cross section of the beam is rectangular with constant width b, but the height h varies as

$$h = h_1 + (h_2 - h_1)\left(\frac{x}{L}\right)^2$$

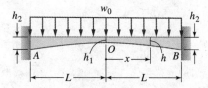

FIG. C7.4

Given L, b, h_1, and h_2, write an algorithm to plot the maximum bending stress acting on a cross section as a function of x from $x = 0$ to L. Run the algorithm with the following data: $L = 18$ ft, $w_0 = 360$ lb/ft, $b = 8$ in., and (a) $h_1 = 15$ in., $h_2 = 36$ in.; and (b) $h_1 = h_2 = 22$ in. (These two beams have the same volume.) (*Hint*: First find the moment at O using the moment-area method and utilizing symmetry.)

C7.5 The uniform beam ABC of length L carries a linearly distributed load of maximum intensity w_0. The distance between A and the simple support at B is a. Given L, w_0, and a, write an algorithm to plot the bending moment of the beam. (a) Run the algorithm with the following data: $L = 4$ m, $w_0 = 6$ kN/m, and $a = 2$ m. (b) By trial-and error, find the approximate value of a that minimizes the maximum bending moment in the beam. (*Hint*: Use the method of superposition.)

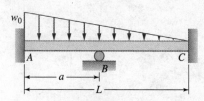

FIG. C7.5

C7.6 The loads P_1 and P_2, a fixed distance b apart, travel across the uniform beam ABC of length L. Given P_1, P_2, b, and L, construct an algorithm that plots the reaction at B versus the distance x from $x = -b$ (when P_2 enters the span) to $x = L$ (when P_1 leaves the span). Run the algorithm with the following data: $L = 80$ ft, $P_1 = 40$ kips, $P_2 = 60$ kips, and (a) $b = 20$ ft; and (b) $b = 40$ ft. (*Hint*: Use the method of superposition.)

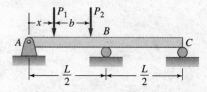

FIG. C7.6

8 Stresses Due to Combined Loads

Natural gas storage tanks. The pressure of the gas in the tank causes biaxial tension (tension in two directions) in the wall of the vessel.

8.1 Introduction

In preceding chapters, we studied stress analysis of various structural members carrying fundamental loads: bars with axial loading, torsion of circular and thin-walled shafts, and bending of beams. This chapter begins by considering two additional topics. The first deals with stresses in thin-walled pressure vessels (Art. 8.2), which introduces us to biaxial loading. The second topic is combined axial and lateral loading of bars (Art. 8.3), which is a straightforward application of superimposing stresses caused by an axial force and a bending moment.

To design a load-carrying member, we must be able to compute the stress components not only at any point in the member but also on *any plane* passing through a point. Being able to determine the stresses acting on an arbitrary plane at a given point is referred to as knowing the *state of stress* at a point. Our discussion of the state of stress at a point begins in Art. 8.4 and continues for the next three articles. Article 8.8 brings together all of the knowledge you have acquired about stress analysis—we analyze the state of stress at various points in members that carry different combinations of the fundamental loads.

This chapter concludes with the study of the *state of strain* at a point. Strain is important in experimental studies because it can be measured, whereas direct determination of stress is not possible. However, as you will see, the state of stress at a point can be calculated from the state of strain and the mechanical properties of the material.

8.2 Thin-Walled Pressure Vessels

A pressure vessel is a pressurized container, often cylindrical or spherical. The pressure acting on the inner surface is resisted by tensile stresses in the walls of the vessel. If the wall thickness t is sufficiently small compared to the radius r of the vessel, these stresses are almost uniform throughout the wall thickness. It can be shown that if $r/t \geq 10$, the stresses between the inner and outer surfaces of the wall vary by less than 5%. In this article, we consider only vessels for which this inequality applies.

a. Cylindrical vessels

Consider the cylindrical tank of inner radius r and wall thickness t shown in Fig. 8.1(a). The tank contains a fluid (or gas) under pressure p. In this simplified analysis, we assume that the weights of the fluid and the vessel can be neglected compared to the other forces that act on the vessel. The tensile stresses in the wall that resist the internal pressure are the *longitudinal stress* σ_ℓ and the *circumferential stress* σ_c (also known as the *hoop stress*), as shown in Fig. 8.1(a).

The circumferential stress can be obtained from the free-body diagram in Fig. 8.1(b). This free body is obtained by taking the slice of infinitesimal length dx shown in Fig. 8.1(a) and cutting it in half along a diametral plane.

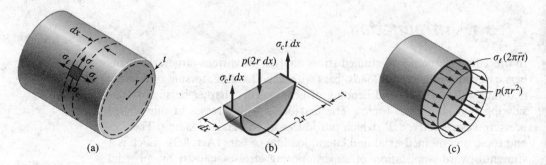

FIG. 8.1 (a) Cylindrical pressure vessel; (b) free-body diagram for computing the circumferential stress σ_c; (c) free-body diagram for computing the longitudinal stress σ_l.

The fluid isolated by the cuts is considered to be part of the free-body diagram. The resultant force due to the pressure acting on the diametral plane is $p(2r\,dx)$, where $2r\,dx$ is the area of the plane. If we assume the circumferential stress σ_c in the wall of the cylinder is constant throughout the thickness, then its resultant force is $2(\sigma_c t\,dx)$. Neglecting the weight of the fluid, we find that the equilibrium of vertical forces becomes

$$\Sigma F = 0 \quad +\uparrow \quad 2(\sigma_c t\,dx) - p(2r\,dx) = 0$$

which yields for the circumferential stress

$$\boxed{\sigma_c = \frac{pr}{t}} \tag{8.1}$$

To obtain the longitudinal stress σ_ℓ, we cut the cylinder into two parts along a cross-sectional plane. Isolating the cylinder and the fluid to the left of the cut gives the free-body diagram in Fig. 8.1(c). For thin-walled cylinders, the cross-sectional area of the wall can be approximated by (mean circumference) × (thickness) = $(2\pi\bar{r})t$, where $\bar{r} = r + t/2$ is the mean radius of the vessel. Therefore, the resultant of the longitudinal stress is $\sigma_\ell(2\pi\bar{r}t)$. The resultant of the pressure acting on the cross section is $p(\pi r^2)$. From the equilibrium of axial forces, we get

$$\Sigma F = 0 \quad +\searrow \quad \sigma_\ell(2\pi\bar{r}t) - p(\pi r^2) = 0$$

Therefore, the longitudinal stress is $\sigma_\ell = pr^2/(2\bar{r}t)$. For thin-walled vessels, we can use the approximation $\bar{r} \approx r$, which results in

$$\boxed{\sigma_\ell = \frac{pr}{2t}} \tag{8.2}$$

Comparing Eqs. (8.1) and (8.2), we see that the circumferential stress is twice as large as the longitudinal stress. It follows that if the pressure in a cylinder is raised to the bursting point, the vessel will split along a longitudinal line. When a cylindrical tank is manufactured from curved sheets that are riveted together, as in Fig. 8.2, the strength of longitudinal joints should be twice the strength of girth joints.

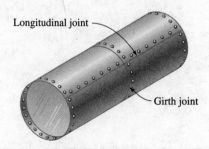

FIG. 8.2 Cylindrical pressure vessel made of curved sheets.

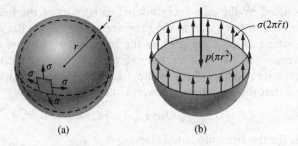

FIG. 8.3 (a) Spherical pressure vessel; (b) free-body diagram for computing the stress σ.

b. Spherical vessels

Using an analysis similar to that used for cylinders, we can derive the expression for the tensile stress σ in the wall of the thin-walled, spherical pressure vessel in Fig. 8.3(a). Because of symmetry, different directions on the surface of the sphere are indistinguishable. Therefore, the stress is constant throughout the vessel. As shown in Fig. 8.3(b), we use half of the vessel as the free-body diagram. The fluid is included in the free-body diagram, but its weight is neglected together with the weight of the vessel. The resultant force due to the pressure acting on the circular surface of the fluid is $p(\pi r^2)$, where r is the inner radius of the vessel. We use again the approximation $2\pi \bar{r} t$ for the cross-sectional area of the wall, where $\bar{r}$ denotes the mean radius of the vessel and t is the wall thickness. Therefore, the resultant force due to σ is $\sigma(2\pi \bar{r} t)$. The equilibrium equation

$$\Sigma F = 0 \quad +\uparrow \quad \sigma(2\pi \bar{r} t) = p(\pi r^2)$$

yields $\sigma = pr^2/(2\bar{r}t)$. If we again neglect the small difference between $\bar{r}$ and r, the stress becomes

$$\boxed{\sigma = \frac{pr}{2t}} \quad (8.3)$$

Note on the Choice of Radius in Eqs. (8.1)–(8.3) As pointed out before, the difference between the inner radius r and the mean radius $\bar{r}$ of a thin-walled vessel ($r/t \geq 10$) is insignificant, so that either radius may be substituted for r in Eqs. (8.1)–(8.3). The stresses computed using $\bar{r}$ rather than r would be different, of course, but the discrepancy is at most a few percent.[1]

---

[1] Some engineers prefer to use the inner radius because it yields stresses that are marginally closer to the exact theoretical values.

Sample Problem 8.1

A cylindrical steel pressure vessel has hemispherical end-caps. The inner radius of the vessel is 24 in. and the wall thickness is constant at 0.25 in. When the vessel is pressurized to 125 psi, determine the stresses and the change in the radius of (1) the cylinder; and (2) the end-caps. Use $E = 29 \times 10^6$ psi and $v = 0.28$ for steel.

Solution

Part 1

The circumferential and longitudinal stresses in the cylinder are

$$\sigma_c = \frac{pr}{t} = \frac{(125)(24)}{0.25} = 12\,000 \text{ psi} \qquad \text{Answer}$$

$$\sigma_\ell = \frac{\sigma_c}{2} = 6000 \text{ psi} \qquad \text{Answer}$$

The circumferential strain is obtained from biaxial Hooke's law—see Eq. (2.10):

$$\varepsilon_c = \frac{1}{E}(\sigma_c - v\sigma_\ell) = \frac{12\,000 - 0.28(6000)}{29 \times 10^6} = 355.9 \times 10^{-6}$$

Because the radius is proportional to the circumference, ε_c is also the strain of the radius (change of radius per unit length); that is, $\varepsilon_c = \Delta r/r$. Therefore, the change in the radius of the cylinder is

$$\Delta r = \varepsilon_c r = (355.9 \times 10^{-6})(24) = 8.45 \times 10^{-3} \text{ in.} \qquad \text{Answer}$$

Part 2

The stress in the spherical end-caps is

$$\sigma = \frac{pr}{2t} = \frac{(125)(24)}{2(0.25)} = 6000 \text{ psi} \qquad \text{Answer}$$

Because σ acts biaxially, the strain must again be computed from biaxial Hooke's law, which yields

$$\varepsilon = \frac{1}{E}(\sigma - v\sigma) = \frac{(1-v)\sigma}{E} = \frac{(1-0.28)(6000)}{29 \times 10^6} = 148.97 \times 10^{-6}$$

Therefore, the change in the radius of an end-cap is

$$\Delta r = \varepsilon r = (148.97 \times 10^{-6})(24) = 3.58 \times 10^{-3} \text{ in.} \qquad \text{Answer}$$

Note on Incompatibility at the Joints

According to our analysis, the radii of the cylinder and the end-caps change by different amounts. Because this discrepancy is a violation of compatibility (displacements and slopes of the walls must be continuous), we conclude that our solution is not valid near the joints between the cylinder and the end-caps. Continuity of displacements and slopes requires the presence of bending stresses in the vicinity of the joints. The analysis of these bending stresses, which are localized in the sense that they decay rapidly with distance from the joints (Saint Venant's principle), is beyond the scope of this text. It can be shown that in this vessel the bending stresses become insignificant approximately 5 in. from each joint.

Problems

8.1 A spherical shell with 72-in. outer diameter and 69-in. inner diameter contains helium at a pressure of 1500 psi. Compute the stress in the shell.

8.2 A spherical pressure vessel has a 2-ft inner radius and 5/16-in. wall thickness. If the working tensile stress of the material is 8000 psi, determine the maximum allowable internal pressure.

8.3 A cylindrical container of 2-ft inner diameter is pressurized internally to 1400 psi. Given that the working stress is 12 ksi, find the smallest allowable thickness of the cylinder wall.

8.4 To determine the strength of the riveted joints in a cylindrical vessel, tensile tests were performed on the 6-in.-wide specimens, as shown in the figure. The tensile force P at failure was found to be 32 kips for the longitudinal joint specimen and 16 kips for the circumferential joint specimen. Determine the largest allowable inner diameter of the cylinder that can support a pressure of 150 psi with a factor of safety of 2.0.

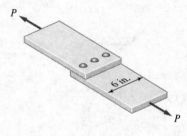

FIG. P8.4

8.5 A cylindrical pressure vessel is fabricated from steel plating that has a thickness of 20 mm. The inner diameter of the vessel is 450 mm, and its length is 2 m. Determine the maximum internal pressure that can be applied if the longitudinal stress is limited to 140 MPa and the circumferential stress is limited to 60 MPa.

8.6 A spherical weather balloon is made of 0.2-mm-thick fabric that has a tensile strength of 10 MPa. The balloon is designed to reach an altitude where the interior pressure is 1500 Pa above the atmospheric pressure. Find the largest allowable diameter of the balloon, using 1.2 as the factor of safety.

8.7 The hydraulic cylinder has a 12-in. inner diameter and a 0.25-in. wall thickness. The left end of the cylinder is attached to a rigid wall, whereas the right end is open. Determine the longitudinal and circumferential stresses in the wall of the cylinder when the 10-kip force is applied to the piston. Neglect local bending of the cylinder at the walls.

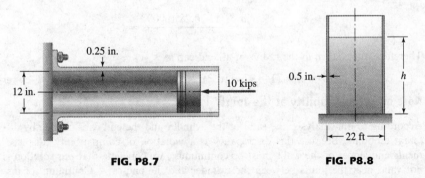

FIG. P8.7   **FIG. P8.8**

8.8 The cylindrical water tank has a 22-ft mean diameter and 0.5-in. wall thickness. If the working tensile stress is 6000 psi, find the maximum height h to which the tank may be filled. Use 62.4 lb/ft³ for the weight of water, and neglect local bending stresses at the base of the tank.

8.9 A cylindrical steel pressure vessel has a 600-mm inner radius and 10-mm-thick walls. Find the change in the inner radius when the vessel is pressurized to 1.5 MPa. Use $E = 200$ GPa and $\nu = 0.3$ for steel.

8.10 The pipe carrying steam at 3.5 MPa has an outer diameter of 450 mm and a wall thickness of 10 mm. A gasket is inserted between the flange at one end of the pipe, and a flat plate is used to cap the end. (a) How many 40-mm-diameter bolts must be used to hold the cap on if the allowable stress in the bolts is 80 MPa, of which 55 MPa is the initial stress? (b) What circumferential stress is developed in the pipe?

8.11 The ends of the 3-in. inner diameter bronze tube are attached to rigid walls. Determine the longitudinal and circumferential stresses when the tube is pressurized to 600 psi. Use $E = 12 \times 10^6$ psi and $v = 1/3$ for bronze. Neglect localized bending at the ends of the tube.

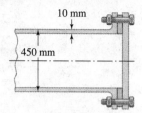

FIG. P8.10

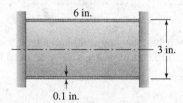

FIG. P8.11

8.12 The cylindrical pressure vessel with hemispherical end-caps is made of steel. The vessel has a uniform thickness of 20 mm and an outer diameter of 400 mm. When the vessel is pressurized to 4.5 MPa, determine the change in the overall length of the vessel. Use $E = 200$ GPa and $v = 0.3$ for steel. Neglect localized bending.

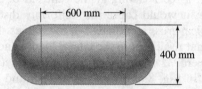

FIG. P8.12

8.13 The thin-walled pressure vessel has an elliptical cross section with the dimensions shown in the figure. Assuming that $a > b$ and that the wall thickness t is constant, derive the expressions for the maximum and minimum circumferential stresses in the vessel caused by an internal pressure p.

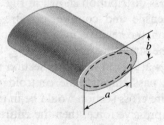

FIG. P8.13

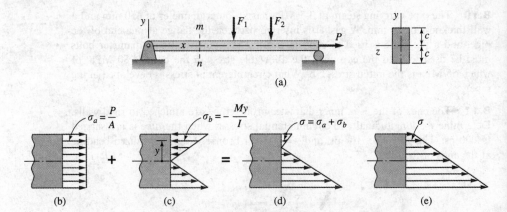

FIG. 8.4 (a) Rectangular bar carrying axial and lateral loads; (b)–(d) stress distribution obtained by superimposing stresses due to axial load and bending; (e) stress distribution if $P/A > |M|c/I$.

8.3 Combined Axial and Lateral Loads

Figure 8.4(a) shows a bar of rectangular cross section that carries lateral loading and an axial force P acting at the centroid of the cross section. If P were acting alone, it would cause the uniformly distributed axial stress $\sigma_a = P/A$ on the typical cross section m-n of the bar, as shown in Fig. 8.4(b). The bending stress that results from the lateral loading acting by itself would be $\sigma_b = -My/I$, where M is the bending moment acting at section m-n. This stress is shown in Fig. 8.4(c). When the axial and lateral loads act simultaneously, the stress σ at any point on section m-n is obtained by superimposing the two separate effects:

$$\sigma = \sigma_a + \sigma_b = \frac{P}{A} - \frac{My}{I} \qquad (8.4)$$

which results in the stress distribution shown in Fig. 8.4(d).

The maximum tensile and compressive stresses on a cross section depend, of course, upon the relative magnitudes of the two terms in Eq. (8.4). When drawing the stress distribution in Fig. 8.4(d), we assumed that $P/A < |M|c/I$. There is a line on the cross section where the stress is zero, but this line does not pass through the centroid of the cross section. If $P/A > |M|c/I$, the entire cross section would be in tension, as shown in Fig. 8.4(e) (if P were a compressive force, then the entire cross section would be in compression). If the cross section were not symmetric about the neutral axis, the distances to the top and bottom fibers would, of course, have to be considered when sketching stress distributions such as those in Figs. 8.4(c) and (d).

The superposition implied in Eq. (8.4) is valid only when the deformation of the bar is sufficiently small so that displacements can be neglected

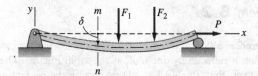

FIG. 8.5 When beam deflections are large, the contribution of the axial load to the bending moment cannot be neglected.

in the computation of M. Referring to Fig. 8.5, we see that if deformation is not neglected, P contributes to the bending moment at section m-n by the amount $-P\delta$, where δ is the lateral displacement of the bar at that section. Consequently, we have

$$\sigma = \frac{P}{A} - \frac{(M - P\delta)y}{I}$$

where, as before, M is to be interpreted as the bending moment due to lateral loading acting alone. We now see that Eq. (8.4) is valid only if $P\delta$ is small compared to M. Note that if P is tensile (positive), its moment reduces the bending stress. The opposite effect occurs when P is compressive (negative), when its moment increases the bending stress. These effects are negligible for most structural members, which are usually so stiff that the additional bending stresses caused by P can be ignored. However, in slender compression members (columns), the effects can be very significant, requiring more exact methods of analysis.

Before Eq. (8.4) can be applied, equilibrium analysis must be used to determine the axial load P and the bending moment M at the cross section of interest. When the normal stresses at a particular cross section are needed, a free-body diagram exposing the force system at that section will suffice. However, if the maximum normal stress is to be found, axial force and bending moment diagrams will usually be required to locate the critical section.

Sample Problem 8.2

To reduce interference, a link in a machine is designed so that its cross-sectional area in the center section is reduced by one-half, as shown in Fig. (a). The thickness of the link is 50 mm. Given that $P = 40$ kN, (1) determine the maximum and minimum values of the normal stress acting on section m-n; and (2) sketch the stress distribution on section m-n.

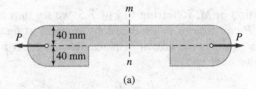

(a)

Solution

Part 1

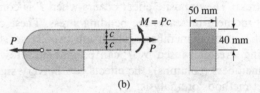

(b)

The free-body diagram in Fig. (b) shows that the internal force system at section m-n can be represented as the normal force P acting at the centroid of the section and the bending moment $M = Pc$. Therefore, the extremum values of the normal stress are

$$\left.\begin{array}{c}\sigma_{\max}\\ \sigma_{\min}\end{array}\right\} = \frac{P}{A} \pm \frac{Mc}{I} = \frac{P}{A} \pm \frac{Pc^2}{I}$$

Substituting $P = 40$ kN and

$$A = bh = 50(40) = 2000 \text{ mm}^2 = 2.0 \times 10^{-3} \text{ m}^2$$

$$I = \frac{bh^3}{12} = \frac{50(40)^3}{12} = 266.7 \times 10^3 \text{ mm}^4 = 266.7 \times 10^{-9} \text{ m}^4$$

$$c = \frac{h}{2} = \frac{40}{2} = 20 \text{ mm} = 0.020 \text{ m}$$

gives

$$\frac{P}{A} = \frac{40 \times 10^3}{2.0 \times 10^{-3}} = 20 \times 10^6 \text{ Pa} = 20 \text{ MPa}$$

$$\frac{Pc^2}{I} = \frac{(40 \times 10^3)(0.020)^2}{266.7 \times 10^{-9}} = 60 \times 10^6 \text{ Pa} = 60 \text{ MPa}$$

Therefore, the maximum and minimum normal stresses acting on section m-n are

$$\sigma_{\max} = 20 + 60 = 80 \text{ MPa} \qquad \text{Answer}$$

$$\sigma_{\min} = 20 - 60 = -40 \text{ MPa} \qquad \text{Answer}$$

where the positive value indicates tension and the negative value indicates compression.

Part 2

The stress distribution on section *m-n* is shown in Fig. (c). The 20-MPa tensile stress due to P is uniformly distributed over the entire cross section. The bending moment M causes a linear stress distribution that reaches a magnitude of 60 MPa in the extreme fibers (compression at the top and tension at the bottom). Superimposing the stresses due to P and M results in stress that varies linearly between $\sigma_{max} = 80$ MPa (at the bottom) and $\sigma_{min} = -40$ MPa (at the top). The distance between the line of zero stress and the top of the section can be located from similar triangles: $d/40 = (40 - d)/80$ yields $d = 13.33$ mm.

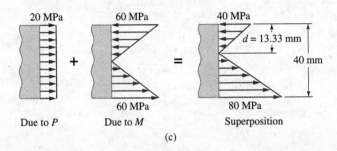

Due to *P*   Due to *M*   Superposition

(c)

Sample Problem 8.3

The timber beam *ABCD* in Fig. (a) carries two vertical loads. The beam is supported by a pin at *A* and the horizontal cable *CE*. Determine the magnitude of the largest stress (tensile or compressive) in the beam and its location. Neglect the weight of the beam.

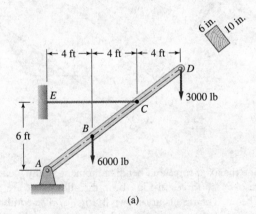

(a)

Solution

Preliminary Calculations The cross-sectional dimensions of the beam are $b = 6$ in. and $h = 10$ in., which yield the following cross-sectional properties:

$$A = bh = 6(10) = 60 \text{ in.}^2$$

$$I = \frac{bh^3}{12} = \frac{6(10)^3}{12} = 500 \text{ in.}^4$$

Equilibrium Analysis The free-body diagram of the beam is shown in Fig. (b). From the equilibrium equation

$$\Sigma M_A = 0 \quad +\circlearrowleft \quad 6T - 6000(4) - 3000(12) = 0$$

we obtain $T = 10\,000$ lb for the tension in the cable. The reactions at A can now be computed from $\Sigma \mathbf{F} = \mathbf{0}$, which yields $A_h = 10\,000$ lb and $A_v = 9000$ lb.

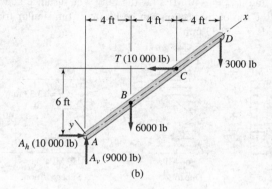

(b)

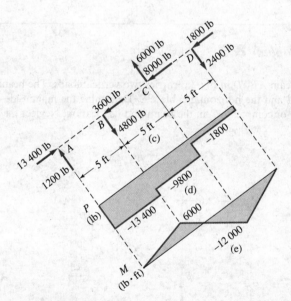

To determine the axial force and bending moment at any location in the beam, it is convenient resolve the forces in Fig. (b) in directions that are parallel and perpendicular to the beam. The results are shown in Fig. (c). The equilibrium analysis of the beam is now completed by constructing the axial force and bending moment diagrams in Figs. (d) and (e), respectively.

Computation of the Largest Stress Because the axial force is negative (compressive) everywhere in the beam, the maximum compressive stress in the beam has a larger magnitude than the maximum tensile stress. Inspection of the axial force and bending moment diagrams leads us to conclude that the largest compressive stress occurs either on the cross section immediately below point B or on the cross section immediately below point C. Which stress is larger can be determined by computing the stresses at both sections.

At the section immediately below B, we have $P = -13\,400$ lb and $M = +6000$ lb·ft. The maximum compressive stress σ_B occurs at the top of the section, where the compressive stress caused by P adds to the maximum compressive stress caused by the *positive* bending moment. If we use $y = h/2 = 5$ in. in Eq. (8.4), this stress is

$$\sigma_B = \frac{P}{A} - \frac{My}{I} = -\frac{13\,400}{60} - \frac{(6000 \times 12)(5)}{500} = -943 \text{ psi}$$

The axial force and the bending moment that act at the section immediately below C are $P = -9800$ lb and $M = -12\,000$ lb·ft. For this case, the maximum compressive stress σ_C occurs at the bottom of the section, where the compressive stress caused by P adds to the maximum compressive stress caused by the *negative* bending moment. Using $y = -h/2 = -5$ in. in Eq. (8.4) yields

$$\sigma_C = \frac{P}{A} - \frac{My}{I} = -\frac{9800}{60} - \frac{(-12\,000 \times 12)(-5)}{500} = -1603 \text{ psi}$$

Comparing the two values, we see that the largest stress in the beam has the magnitude

$$|\sigma|_{\max} = 1603 \text{ psi} \qquad \text{Answer}$$

and it acts at the bottom of the section just below C.

Problems

8.14 The bent rod has a 1/2-in.-square cross section. Compute the maximum normal stress in the rod, and compare the result with the normal stress if the rod were straight.

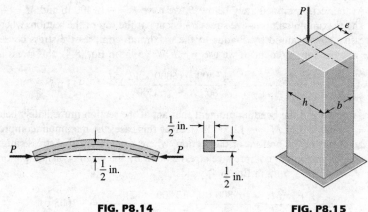

FIG. P8.14 **FIG. P8.15**

8.15 The force P acting on the concrete column has an eccentricity e. Because concrete is weak in tension, it is desirable to have all parts of the column in compression. Determine the largest value of e for which there is no tensile stress anywhere in the column. (The area that is the locus of points through which P can act without causing tensile stress is called the *kern of the cross section*.)

8.16 Find the largest clamping force that can be applied by the cast iron C-clamp if the allowable normal stresses on section *m-n* are 20 MPa in tension and 40 MPa in compression.

8.17 The frame of the bow saw is a bent tube of 3/4-in. outer diameter and 1/32-in. wall thickness. If the sawblade is pre-tensioned to 30 lb, determine the normal stresses at points A and B.

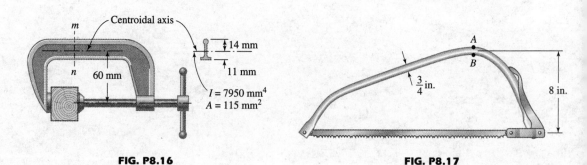

FIG. P8.16 **FIG. P8.17**

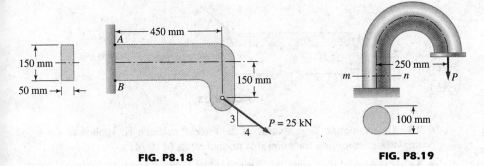

FIG. P8.18 **FIG. P8.19**

8.18 Calculate the normal stresses at points *A* and *B* of the bracket caused by the 25-kN force.

8.19 Determine the largest load *P* that can be applied to the bent circular bar if the magnitude of the normal stress on section *m-n* is limited to 80 MPa.

8.20 Find the maximum normal stress acting on section *m-n* of the bent bar.

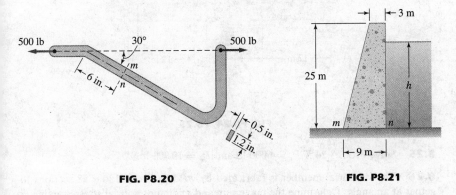

FIG. P8.20 **FIG. P8.21**

8.21 A concrete dam has the cross section shown in the figure. Determine the maximum compressive stress on section *m-n* if the depth of water behind the dam is $h = 15$ m. The density of concrete is 2400 kg/m^3 and that of water is 1000 kg/m^3.

8.22 The steel column is fabricated by welding a 9-in. by 1/2-in. plate to a W12 × 50 section. The axial load *P* acts at the centroid *C* of the W-section. If the normal working stress is 18 ksi, find the maximum allowable value of *P*. If the plate were removed, would the allowable *P* be larger or smaller?

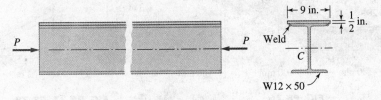

FIG. P8.22

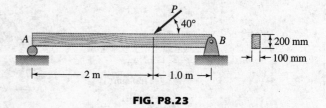

FIG. P8.23

8.23 Determine the largest value of the force P that can be applied to the wood beam without exceeding the allowable normal stress of 10 MPa.

8.24 The forces $P_1 = 9000$ lb and $P_2 = 3000$ lb are applied to the W6 × 12 section. Compute the maximum and minimum normal stresses acting on section m-n.

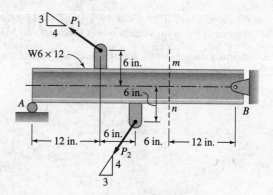

FIG. P8.24, P8.25

8.25 Solve Prob. 8.24 if $P_1 = 6000$ lb and $P_2 = 10\,000$ lb.

8.26 The structural member is fabricated by welding two W130 × 28 sections together at an angle. Determine the maximum and minimum normal stresses acting on section m-n.

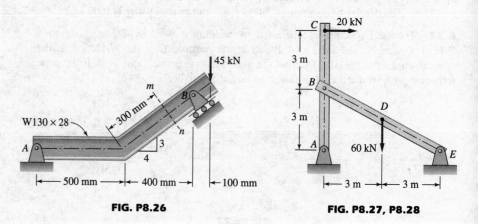

FIG. P8.26   **FIG. P8.27, P8.28**

8.27 The cross sections of the members of the pin-jointed structure are 200-mm square. Find the maximum compressive stress in member BDE.

8.28 The cross sections of the members of the pin-jointed structure are 200-mm square. Calculate the maximum compressive stress in member ABC.

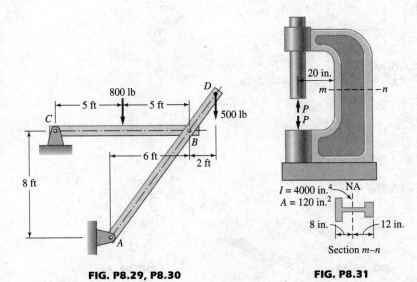

FIG. P8.29, P8.30   **FIG. P8.31**

8.29 Each member of the pin-jointed structure has a 4-in. by 4-in. cross section. Determine the maximum compressive stress in member ABD.

8.30 Each member of the pin-jointed structure has a 4-in. by 4-in. cross section. Compute the maximum compressive and tensile stresses developed in member CB.

8.31 Determine the largest force P that can be exerted at the jaws of the punch without exceeding a stress of 18 ksi on section m-n of the frame.

8.32 The force $P = 100$ kN is applied to the bracket as shown in the figure. Compute the normal stresses developed at points A and B.

8.33 Determine the largest force P, directed as shown in the figure, that can be applied to the bracket if the allowable normal stresses on section A-B are 8 MPa in tension and 12 MPa in compression.

8.34 The cross section of the bent steel bar is a 200-mm by 200-mm square. Compute the normal stress acting at points A and B.

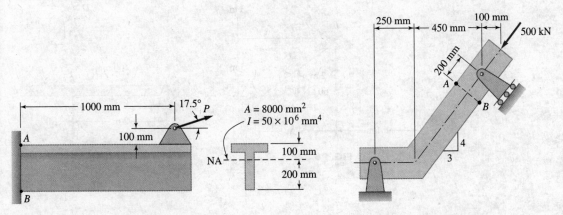

FIG. P8.32, P8.33   **FIG. P8.34**

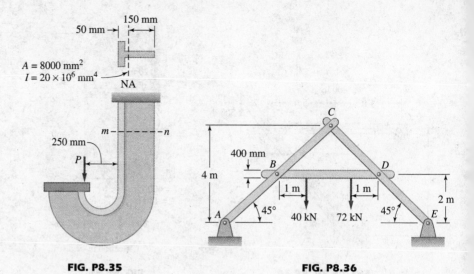

FIG. P8.35   **FIG. P8.36**

8.35 Find the largest force P that can be supported by the cast iron bracket if the allowable normal stresses on section m-n are 30 MPa in tension and 70 MPa in compression.

8.36 Member BD of the pin-connected frame has a rectangular cross section 100 mm wide by 400 mm deep. Determine the maximum normal stress in this member.

8.37 The structure consists of two bars that are joined with a pin at B. Compute the maximum compressive normal stress in bar BDE if its cross section is a 200-mm by 200-mm square.

8.38 The rectangular beam ABC, 100 mm wide by 400 mm deep, is supported by a pin at A and the cable CD. Determine the largest vertical force P that can be applied at B if the normal stress in the beam is limited to 120 MPa.

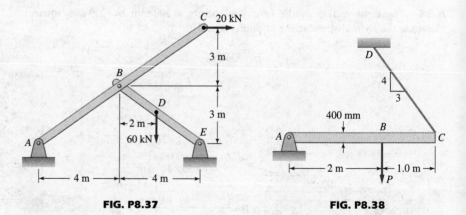

FIG. P8.37   **FIG. P8.38**

8.39 Find the maximum compressive normal stress in the inclined beam *ABC*.

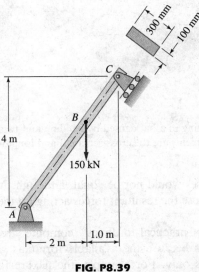

FIG. P8.39

8.4 State of Stress at a Point (Plane Stress)

In this article, we formalize the concept of stress at a point, which requires the introduction of a sign convention and a subscript notation for stress components.

a. Reference planes

In Art. 1.3, we saw that the stresses acting at a point in a body depend on the orientation of the reference plane. As a review of that discussion, consider the body in Fig. 8.6(a) that is acted upon by a system of forces in equilibrium. Assume that we first introduce the reference plane *a-a* and compute the stresses σ and τ acting on that plane at point *O*, as illustrated in Fig. 8.6(b). We then pass the reference plane *b-b* through *O* and repeat the computations, obtaining the stresses σ' and τ' shown in Fig. 8.6(c). In general,

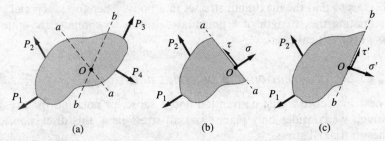

FIG. 8.6 (a) Body in equilibrium; (b) stresses acting on plane *a-a* at point *O*; (c) stresses acting on plane *b-b* at point *O*.

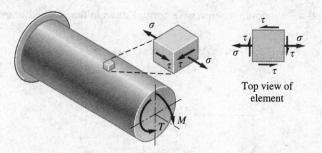

FIG. 8.7 Stresses acting in a bar caused by bending and twisting. The reference planes (faces of the element) are the cross-sectional and longitudinal planes.

the two sets of stresses would not be equal, although they are computed at the same point, because the resultant forces acting on the two planes are not equal.

It is usually not practical to directly compute stresses acting on arbitrarily chosen planes because the available formulas give stresses on certain reference planes only. For example, the flexure formula $\sigma = -My/I$ is restricted to the normal stress on the cross-sectional plane of the beam. Similarly, the shear stress formulas, $\tau = VQ/(Ib)$ for beams and $\tau = T\rho/J$ for shafts, apply to only cross-sectional and longitudinal (complementary) planes. Therefore, if a bar is subjected to simultaneous bending and twisting, as in Fig. 8.7, we can readily calculate the stresses on the sides of the element shown because the sides coincide with the reference planes used in the formulas.

The stresses σ and τ shown in Fig. 8.7 are not the *maximum* normal and *maximum* shear stress that act at the location of the element. As we shall see, the maximum values of these stresses occur on planes that are inclined to the sides of the element. Because maximum stresses are important in design, we must examine how the stress components at a point vary with the orientation of the reference planes, which is the subject of the next article.

b. State of stress at a point

A basic concept of stress analysis is the state of stress at a point:

The state of stress at a point is defined by the stress components acting on the sides of a differential volume element that encloses the point.

Knowing the state of stress at a point enables us to calculate the stress components that act on any plane passing through that point. This, in turn, enables us to find the maximum stresses in a body. Therefore, a crucial step in evaluating the strength of a potential design is to compute the state of stress at the critical points.

c. Sign convention and subscript notation

We next introduce a sign convention and a subscript notation for stresses. Although we consider only plane (biaxial) stress here, this discussion also applies to triaxial stress.

Consider the differential element in Fig. 8.8, where the faces of the element coincide with the coordinate planes. A face takes its name from the

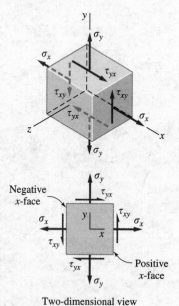

FIG. 8.8 Biaxial (plane) state of stress. Stress components are shown acting in their positive directions.

coordinate axis that is normal to it. For example, the *x*-face of the element is perpendicular to the *x*-axis. A face is also considered to be positive or negative, depending on the direction of its *outward normal* (directed away from the element). If the outward normal points in the positive coordinate direction, the face is positive. Conversely, if the outward normal points in the negative coordinate direction, the face is negative. The sign convention for stresses is as follows:

- Positive stresses act in positive coordinate directions on positive faces of the element, as shown in Fig. 8.8.

The stresses acting on the negative faces are, of course, equal and opposite to their counterparts on the positive faces. Note that this sign convention considers a tensile normal stress as positive and a compressive stress as negative. Here are the rules for the subscripts on the stresses:

- The single subscript on the normal stress indicates the face on which it acts.
- The first subscript on the shear stress indicates the face on which it acts; the second subscript shows the direction of the stress.

Thus, σ_x denotes the normal stress acting on an *x*-face; τ_{xy} is the shear stress on the *x*-face acting in the *y*-direction; and so on. Because the magnitudes of the shear stresses on complementary planes are equal (this was proven in Art. 5.4), we have $\tau_{xy} = \tau_{yx}$.

In a triaxial state of stress, each face of an element is generally subject to three stress components—a normal stress and two shear stress components, as shown in Fig. 8.9. The stress components acting on the *x*-face, for example, are the normal stress σ_x and the shear stresses τ_{xy} and τ_{xz}. These shear stresses are accompanied by the numerically equal shear stresses acting on complementary planes (τ_{yx} on the *y*-face and τ_{zx} on the *z*-face).

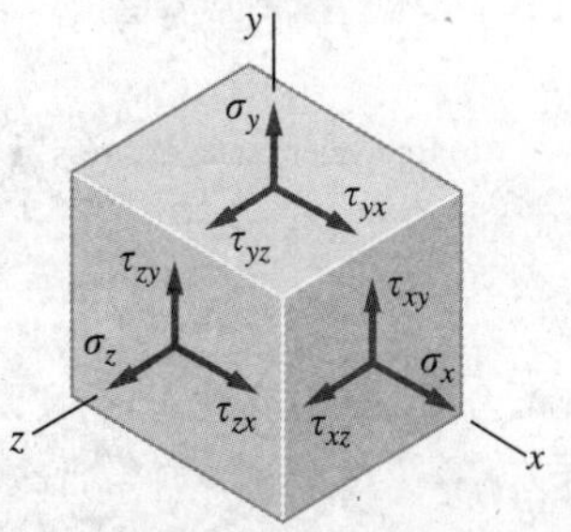

FIG. 8.9 Triaxial state of stress. Stress components are shown acting in their positive directions.

8.5 Transformation of Plane Stress

As discussed in the previous article, the state of stress at a point is represented by the stresses that act on the mutually perpendicular faces of a volume element enclosing the point. Because the stresses at a point depend on the inclinations of the planes on which they act, the stresses on the faces of the element vary as the orientation of the element is changed. The mathematical relationships that describe this variation are called the *transformation equations for stress*. In this article, we derive the transformation equations for plane stress and then use them to find the maximum and minimum stresses at a point.

a. Transformation equations

Figure 8.10(a) shows the state of plane stress at a point, where the reference planes (faces of the element) are perpendicular to the *x*- and *y*-axes. Figure 8.10(b) represents the state of stress at the same point, but now the faces of the element are perpendicular to the *x'*- and *y'*-axes, where the orientation of the two sets of axes differs by the angle θ. Note that positive θ is measured in a *counterclockwise* direction from the *x*-axis to the *x'*-axis. The stress components in both figures are drawn in their positive directions in accordance

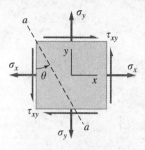

(a) Original state of stress

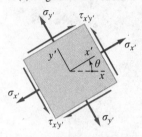

(b) Equivalent state of stress

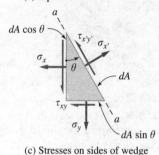

(c) Stresses on sides of wedge

(d) FBD showing forces

FIG. 8.10 Transforming the state of stress from the xy-coordinate planes to the $x'y'$-coordinate planes.

with the established sign convention. We also use the equality of shear stresses on complementary planes ($\tau_{yx} = \tau_{xy}$ and $\tau_{x'y'} = \tau_{y'x'}$) in labeling the figures. The stress states in Figs. 8.10(a) and (b) are said to be *equivalent* because they represent the same state of stress referred to two different sets of coordinate axes.

Let us now cut the element in Fig. 8.10(a) into two parts along the plane a-a and isolate the left portion as shown in Fig. 8.10(c). The inclined plane a-a coincides with the x'-face of the element in Fig. 8.10(b), so that the stresses acting on that plane are $\sigma_{x'}$ and $\tau_{x'y'}$. We can now apply equilibrium equations to the wedge-shaped element and derive $\sigma_{x'}$ and $\tau_{x'y'}$ in terms of θ and the stresses acting on the x- and y-faces. Letting dA be the area of the inclined face, we find that the areas of the x- and y-faces are $dA \cos \theta$ and $dA \sin \theta$, as shown in Fig. 8.10(c). By multiplying the stresses by the areas on which they act, we obtain the forces shown on the free-body diagram of the wedge in Fig. 8.10(d). The equilibrium equations are

$$\Sigma F_x = 0 \quad \sigma_{x'} dA \cos \theta - \tau_{x'y'} dA \sin \theta - \sigma_x dA \cos \theta - \tau_{xy} dA \sin \theta = 0$$

$$\Sigma F_y = 0 \quad \sigma_{x'} dA \sin \theta + \tau_{x'y'} dA \cos \theta - \sigma_y dA \sin \theta - \tau_{xy} dA \cos \theta = 0$$

Solving for the stresses on the inclined plane, we get

$$\sigma_{x'} = \sigma_x \cos^2 \theta + \sigma_y \sin^2 \theta + 2\tau_{xy} \sin \theta \cos \theta \quad \text{(a)}$$

$$\tau_{x'y'} = -(\sigma_x - \sigma_y) \sin \theta \cos \theta + \tau_{xy}(\cos^2 \theta - \sin^2 \theta) \quad \text{(b)}$$

The normal stress acting on the y'-face of the element in Fig. 8.10(b) can be obtained by replacing θ with $\theta + 90°$ in Eq. (a). Noting that $\cos(\theta + 90°) = -\sin \theta$ and $\sin(\theta + 90°) = \cos \theta$, we get

$$\sigma_{y'} = \sigma_x \sin^2 \theta + \sigma_y \cos^2 \theta - 2\tau_{xy} \sin \theta \cos \theta \quad \text{(c)}$$

Equations (a)–(c) are the stress transformation equations. Another form of these equations is obtained by substituting the trigonometric relationships

$$\cos^2 \theta = \frac{1 + \cos 2\theta}{2} \quad \sin^2 \theta = \frac{1 - \cos 2\theta}{2} \quad \sin \theta \cos \theta = \frac{1}{2} \sin 2\theta$$

into Eqs. (a)–(c), which yields the more commonly used form of the stress transformation equations:

$$\sigma_{x'} = \frac{\sigma_x + \sigma_y}{2} + \frac{\sigma_x - \sigma_y}{2} \cos 2\theta + \tau_{xy} \sin 2\theta \quad \text{(8.5a)}$$

$$\sigma_{y'} = \frac{\sigma_x + \sigma_y}{2} - \frac{\sigma_x - \sigma_y}{2} \cos 2\theta - \tau_{xy} \sin 2\theta \quad \text{(8.5b)}$$

$$\tau_{x'y'} = -\frac{\sigma_x - \sigma_y}{2} \sin 2\theta + \tau_{xy} \cos 2\theta \quad \text{(8.5c)}$$

The stress transformation equations (Eqs. (a)–(c) or equivalently Eqs. (8.5)) show that if the state of stress (σ_x, σ_y, and τ_{xy}) at a point is known, we can calculate the stresses that act on any plane passing through that point. It follows that if the original state of stress in Fig. 8.10(a) is known, then the stress transformation equations enable us to obtain the equivalent state of stress in Fig. 8.10(b).

Inspection of Eqs. (8.5a) and (8.5b) reveals that

$$\sigma_{x'} + \sigma_{y'} = \sigma_x + \sigma_y \tag{8.6}$$

In other words, the *sum of the normal stresses is an invariant*; that is, it does not depend on the orientation of the element.

b. Principal stresses and principal planes

The maximum and minimum normal stresses at a point are called the *principal stresses* at that point. The planes on which the principal stresses act are referred to as the *principal planes*. The directions that are perpendicular to the principal planes are called the *principal directions*. The values of the angle θ that define the principal directions are found from the condition $d\sigma_{x'}/d\theta = 0$. If we use the expression for $\sigma_{x'}$ from Eq. (8.5a), this condition becomes

$$\frac{d\sigma_{x'}}{d\theta} = -(\sigma_x - \sigma_y)\sin 2\theta + 2\tau_{xy}\cos 2\theta = 0$$

which yields

$$\tan 2\theta = \frac{2\tau_{xy}}{\sigma_x - \sigma_y} \tag{8.7}$$

Equation (8.7) yields two solutions for 2θ that differ by 180°. If we denote one solution by $2\theta_1$, the second solution is $2\theta_2 = 2\theta_1 + 180°$. Hence, the *two principal directions differ by 90°*.

The sines and cosines of $2\theta_1$ and $2\theta_2$ can be obtained from the right triangle in Fig. 8.11:

$$\left.\begin{array}{l}\sin 2\theta_1 \\ \sin 2\theta_2\end{array}\right\} = \pm\frac{\tau_{xy}}{R} \tag{8.8a}$$

$$\left.\begin{array}{l}\cos 2\theta_1 \\ \cos 2\theta_2\end{array}\right\} = \pm\frac{\sigma_x - \sigma_y}{2R} \tag{8.8b}$$

where

$$R = \sqrt{\left(\frac{\sigma_x - \sigma_y}{2}\right)^2 + \tau_{xy}^2} \tag{8.9}$$

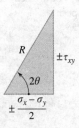

FIG. 8.11 Geometric method for determining the angles θ_1 and θ_2 that define the principal directions.

Substituting Eqs. (8.8) and (8.9) into Eq. (8.5a) and simplifying, we obtain for the principal stresses $\sigma_1 = [(\sigma_x + \sigma_y)/2] + R$ and $\sigma_2 = [(\sigma_x + \sigma_y)/2] - R$, or

$$\left.\begin{array}{l}\sigma_1 \\ \sigma_2\end{array}\right\} = \frac{\sigma_x + \sigma_y}{2} \pm \sqrt{\left(\frac{\sigma_x - \sigma_y}{2}\right)^2 + \tau_{xy}^2} \tag{8.10}$$

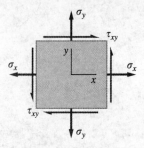

(a) Original state of stress

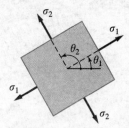

(b) Principal stresses

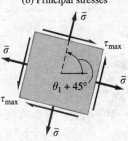

(c) Stresses on planes of maximum shear

FIG. 8.12 Equivalent states of stress at a point.

where σ_1 and σ_2 act on the planes defined by θ_1 and θ_2, respectively. Note that Eq. (8.10) defines σ_1 to be the *larger* of the principal stresses; that is, $\sigma_1 \geq \sigma_2$.

The shear stresses acting on the principal planes are obtained by substituting Eqs. (8.8) into Eq. (8.5c). The result is

$$\left.\begin{array}{c}\tau_{12}\\ \tau_{21}\end{array}\right\} = -\frac{\sigma_x - \sigma_y}{2}\left(\pm\frac{\tau_{xy}}{R}\right) + \tau_{xy}\left(\pm\frac{\sigma_x - \sigma_y}{2R}\right) = 0$$

which shows that there are *no shear stresses on the principal planes*.

Figure 8.12(a) shows an original state of stress relative to an arbitrary xy-coordinate system. The equivalent state of stress referred to the principal directions is shown in Fig. 8.12(b). Note that the two principal planes differ by 90° and are free of shear stress. We mention, once again, that Figs. 8.12(a) and (b) display *two different representations of the same state of stress*. By transforming the original state of stress to the principal directions, we obtain a representation that is more meaningful in design. When illustrating a state of stress, be sure to show a *complete* sketch of the element; that is, include the stresses that act on each of the faces, and indicate the angle that the element makes with a reference axis.

c. Maximum in-plane shear stress

The largest magnitude of $\tau_{x'y'}$ at a point, denoted by τ_{max}, is called the *maximum in-plane shear stress*. The values of θ that define the planes of maximum in-plane shear are found from the equation $d\tau_{x'y'}/d\theta = 0$, where $\tau_{x'y'}$ is given in Eq. (8.5c). Setting the derivative equal to zero and solving for the angle θ give

$$\tan 2\theta = -\frac{\sigma_x - \sigma_y}{2\tau_{xy}} \quad (8.11)$$

Equation (8.11) has two solutions for 2θ that differ by 180°. Hence, there are two values of θ that differ by 90°. We also note that Eq. (8.11) is the negative reciprocal of Eq. (8.7), meaning that the angles 2θ defined by these two equations differ by 90°. We thus conclude that *the planes of maximum in-plane shear stress are inclined at 45° to the principal planes*.

Using a triangle similar to that in Fig. 8.11, we could find $\sin 2\theta$ and $\cos 2\theta$, and substitute the results into Eqs. (8.5). Omitting the details, we obtain for the maximum in-plane shear stress

$$\tau_{max} = R = \sqrt{\left(\frac{\sigma_x - \sigma_y}{2}\right)^2 + \tau_{xy}^2} \quad (8.12)$$

The normal stresses acting on the planes of maximum shear are found to be

$$\bar{\sigma} = \frac{\sigma_x + \sigma_y}{2} \quad (8.13)$$

These results are shown in Fig. 8.12(c). If we compare Eqs. (8.10) and (8.12), we see that the maximum in-plane shear stress can also be expressed as

$$\boxed{\tau_{max} = \frac{|\sigma_1 - \sigma_2|}{2}} \quad (8.14)$$

d. Summary of stress transformation procedures

Identifying Given Stress Components Before applying any of the equations derived in this article, you must identify the given stress components σ_x, σ_y, and τ_{xy} using the sign convention introduced in Art. 8.4: *Positive stress components act in positive coordinate directions on positive faces of the element.*

Computing Stresses on Inclined Planes The transformation equations

$$\left.\begin{matrix}\sigma_{x'}\\ \sigma_{y'}\end{matrix}\right\} = \frac{\sigma_x + \sigma_y}{2} \pm \frac{\sigma_x - \sigma_y}{2} \cos 2\theta \pm \tau_{xy} \sin 2\theta \quad \text{(8.5a, b, repeated)}$$

$$\tau_{x'y'} = -\frac{\sigma_x - \sigma_y}{2} \sin 2\theta + \tau_{xy} \cos 2\theta \quad \text{(8.5c, repeated)}$$

can be used to compute the stress components $\sigma_{x'}$, $\sigma_{y'}$, and $\tau_{x'y'}$ acting on the sides of an inclined element. Remember that the angle θ that defines the inclination is measured from the x-axis to the x'-axis in the *counterclockwise* direction.

Computing Principal Stresses The principal stresses σ_1 and σ_2 are given by

$$\left.\begin{matrix}\sigma_1\\ \sigma_2\end{matrix}\right\} = \frac{\sigma_x + \sigma_y}{2} \pm R \quad \text{(8.10, repeated)}$$

where

$$R = \sqrt{\left(\frac{\sigma_x - \sigma_y}{2}\right)^2 + \tau_{xy}^2} \quad \text{(8.9, repeated)}$$

Recall that the shear stress vanishes on the principal planes (the planes on which σ_1 and σ_2 act).

Computing Principal Directions The principal directions can be found from

$$\tan 2\theta = \frac{2\tau_{xy}}{\sigma_x - \sigma_y} \quad \text{(8.7, repeated)}$$

The angle θ is measured *counterclockwise* from the x-axis to a principal axis. Equation (8.7) yields two values of θ that differ by 90°. Substituting one of these angles into Eq. (8.5a) yields the value of either σ_1 or σ_2, thereby identifying the principal stress associated with that angle.

Computing Maximum In-plane Shear Stress The magnitude of the maximum in-plane shear stress is

$$\tau_{max} = \frac{|\sigma_1 - \sigma_2|}{2} \qquad (8.14, \text{ repeated})$$

and the planes of maximum shear are inclined at 45° to the principal planes. If the principal stresses are not known, the maximum in-plane shear stress and the orientation of the shear planes can be obtained from

$$\tau_{max} = R \qquad (8.12, \text{ repeated})$$

$$\tan 2\theta = -\frac{\sigma_x - \sigma_y}{2\tau_{xy}} \qquad (8.11, \text{ repeated})$$

Equation (8.11) has two solutions for θ that differ by 90°. Each solution represents the angle measured from the x-axis to an axis of maximum shear (the normal to the plane of maximum shear) in the *counterclockwise* direction. The sense of the maximum shear stress can be obtained by substituting one of these angles into Eq. (8.5c). The result is either $+\tau_{max}$ or $-\tau_{max}$, with the sign determining the direction of the shear stress on the plane defined by the angle.

The normal stresses acting on the planes of maximum shear are given by

$$\bar{\sigma} = \frac{\sigma_x + \sigma_y}{2} \qquad (8.13, \text{ repeated})$$

Because the sum of the normal stresses does not change with transformation —see Eq. (8.6), we also have $\bar{\sigma} = (\sigma_1 + \sigma_2)/2$.

Sample Problem 8.4

The state of plane stress at a point with respect to the xy-axes is shown in Fig. (a). Determine the equivalent state of stress with respect to the $x'y'$-axes. Show the results on a sketch of an element aligned with the x'- and y'-axes.

Solution

According to our sign convention (positive stresses act in the positive coordinate directions on positive faces of the element), all the stress components in Fig. (a) are positive: $\sigma_x = 30$ MPa, $\sigma_y = 60$ MPa, and $\tau_{xy} = 40$ MPa. To transform these stresses to the $x'y'$-coordinate system, we use Eqs. (8.5). The angle 2θ used in transformation equations is twice the angle measured counterclockwise from the x-axis to the x'-axis. Substituting $2\theta = 2(30°) = 60°$ into Eqs. (8.5), we obtain

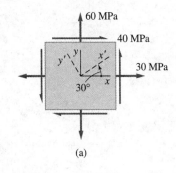

(a)

$$\sigma_{x'} = \frac{\sigma_x + \sigma_y}{2} + \frac{\sigma_x - \sigma_y}{2} \cos 2\theta + \tau_{xy} \sin 2\theta$$

$$= \frac{30 + 60}{2} + \frac{30 - 60}{2} \cos 60° + 40 \sin 60°$$

$$= 72.1 \text{ MPa} \qquad \text{Answer}$$

$$\sigma_{y'} = \frac{\sigma_x + \sigma_y}{2} - \frac{\sigma_x - \sigma_y}{2} \cos 2\theta - \tau_{xy} \sin 2\theta$$

$$= \frac{30 + 60}{2} - \frac{30 - 60}{2} \cos 60° - 40 \sin 60°$$

$$= 17.9 \text{ MPa} \qquad \text{Answer}$$

$$\tau_{x'y'} = -\frac{\sigma_x - \sigma_y}{2} \sin 2\theta + \tau_{xy} \cos 2\theta$$

$$= -\frac{30 - 60}{2} \sin 60° + 40 \cos 60°$$

$$= 33.0 \text{ MPa} \qquad \text{Answer}$$

The results are shown in Fig. (b). Because all the calculated stress components are positive, they act in the positive coordinate directions on the positive x'- and y'-faces.

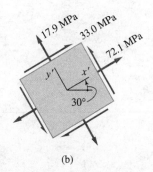

(b)

Sample Problem 8.5

Determine the principal stresses and the principal directions for the state of plane stress given in Fig. (a). Show the results on a sketch of an element aligned with the principal directions.

Solution

If we use the established sign convention (positive stresses act in the positive coordinate directions on positive faces of the element), the stress components shown in Fig. (a) are $\sigma_x = 8000$ psi, $\sigma_y = 4000$ psi, and $\tau_{xy} = 3000$ psi. Substituting these values into Eq. (8.9), we get

$$R = \sqrt{\left(\frac{\sigma_x - \sigma_y}{2}\right)^2 + \tau_{xy}^2} = \sqrt{\left(\frac{8000 - 4000}{2}\right)^2 + (3000)^2} = 3606 \text{ psi}$$

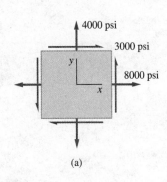

(a)

297

The principal stresses are obtained from Eq. (8.10):

$$\left.\begin{array}{c}\sigma_1 \\ \sigma_2\end{array}\right\} = \frac{\sigma_x + \sigma_y}{2} \pm R = \frac{8000 + 4000}{2} \pm 3606$$

which yields

$$\sigma_1 = 9610 \text{ psi} \qquad \sigma_2 = 2390 \text{ psi} \qquad \textbf{Answer}$$

The principal directions are given by Eq. (8.7):

$$\tan 2\theta = \frac{2\tau_{xy}}{\sigma_x - \sigma_y} = \frac{2(3000)}{8000 - 4000} = 1.500$$

The two solutions are

$$2\theta = 56.31° \quad \text{and} \quad 56.31° + 180° = 236.31°$$
$$\theta = 28.16° \quad \text{and} \quad 118.16°$$

To determine which of the two angles is θ_1 (associated with σ_1) and which is θ_2 (associated with σ_2), we use Eq. (8.5a) to compute the normal stress $\sigma_{x'}$ that corresponds to one of the angles. The result, which will be equal to either σ_1 or σ_2, identifies the principal stress associated with that angle. With $\theta = 28.16°$, Eq. (8.5a) yields

$$\sigma_{x'} = \frac{\sigma_x + \sigma_y}{2} + \frac{\sigma_x - \sigma_y}{2} \cos 2\theta + \tau_{xy} \sin 2\theta$$

$$= \frac{8000 + 4000}{2} + \frac{8000 - 4000}{2} \cos[2(28.16°)] + 3000 \sin[2(28.16°)]$$

$$= 9610 \text{ psi}$$

which is equal to σ_1. Therefore, we conclude that

$$\theta_1 = 28.2° \qquad \theta_2 = 118.2° \qquad \textbf{Answer}$$

The sketch of the differential element in Fig. (b) shows the principal stresses and the principal planes. Note that there is no shear stress on the principal planes, which may be verified by substituting the values for θ_1 and θ_2 into Eq. (8.5c).

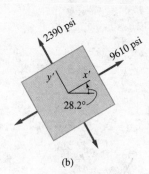

(b)

Sample Problem 8.6

For the state of plane stress shown in Fig. (a), determine the maximum in-plane shear stress and the planes on which it acts. Show the results on a sketch of an element aligned with the planes of maximum shear.

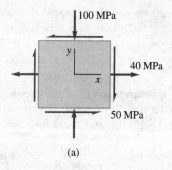

(a)

Solution

With the established sign convention (positive stresses act in the positive coordinate directions on positive faces of the element), the stress components shown in Fig. (a) are $\sigma_x = 40$ MPa, $\sigma_y = -100$ MPa, and $\tau_{xy} = -50$ MPa. Substituting these stresses into Eq. (8.12), we get for the maximum in-plane shear stress

$$\tau_{max} = \sqrt{\left(\frac{\sigma_x - \sigma_y}{2}\right)^2 + \tau_{xy}^2} = \sqrt{\left(\frac{40 - (-100)}{2}\right)^2 + (-50)^2}$$

$$= 86.0 \text{ MPa} \qquad\qquad\qquad\qquad\qquad\qquad\qquad\qquad \textit{Answer}$$

The orientation of planes that carry the maximum in-plane shear stress are found from Eq. (8.11):

$$\tan 2\theta = -\frac{\sigma_x - \sigma_y}{2\tau_{xy}} = -\frac{40 - (-100)}{2(-50)} = 1.400$$

which has the solutions

$$2\theta = 54.46° \quad \text{and} \quad 54.46° + 180° = 234.46°$$

$$\theta = 27.23° \quad \text{and} \quad 117.23° \qquad\qquad\qquad \textit{Answer}$$

To determine the directions of the maximum in-plane shear stresses on the sides of the element, we must find the sign of the shear stress on one of the planes—say, on the plane defined by $\theta = 27.23°$. Substituting the given stress components and $\theta = 27.23°$ into Eq. (8.5c), we obtain

$$\tau_{x'y'} = -\frac{\sigma_x - \sigma_y}{2} \sin 2\theta + \tau_{xy} \cos 2\theta$$

$$= -\frac{40 - (-100)}{2} \sin[2(27.23°)] + (-50) \cos[2(27.23°)]$$

$$= -86.0 \text{ MPa}$$

The negative sign indicates that the shear stress on the positive x'-face acts in the negative y'-direction, as shown in Fig. (b). Once this result has been obtained, the directions of the remaining shear stresses can be determined by inspection. The normal stresses acting on the element are computed from Eq. (8.13), which yields

$$\bar{\sigma} = \frac{\sigma_x + \sigma_y}{2} = \frac{40 + (-100)}{2} = -30 \text{ MPa}$$

As shown in Fig. (b), the normal stresses are equal on all faces of the element.

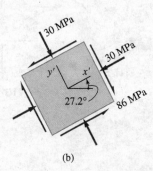

(b)

Problems

8.40–8.42 Given the state of stress shown, determine the stress components acting on the inclined plane *a-a*. Solve by drawing the free-body diagram of the shaded wedge and applying the equilibrium equations.

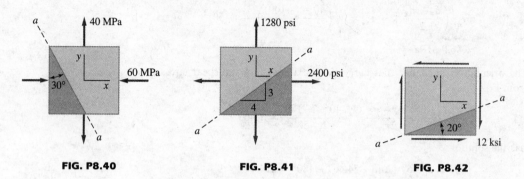

FIG. P8.40 **FIG. P8.41** **FIG. P8.42**

8.43–8.47 The state of stress at a point is shown with respect to the *xy*-axes. Determine the equivalent state of stress with respect to the $x'y'$-axes. Show the results on a sketch of an element aligned with the $x'y'$-axes.

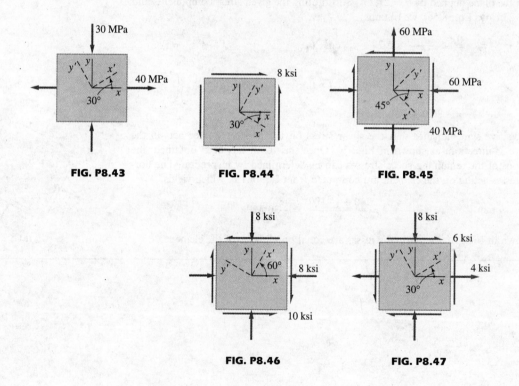

FIG. P8.43 **FIG. P8.44** **FIG. P8.45**

FIG. P8.46 **FIG. P8.47**

8.6 Mohr's Circle for Plane Stress

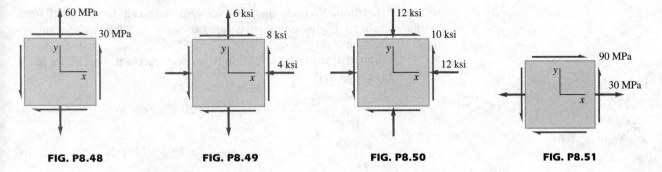

FIG. P8.48 **FIG. P8.49** **FIG. P8.50** **FIG. P8.51**

8.48–8.51 For the state of stress shown, determine the principal stresses and the principal directions. Show the results on a sketch of an element aligned with the principal directions.

8.52–8.55 For the state of stress shown, determine the maximum in-plane shear stress. Show the results on a sketch of an element aligned with the planes of maximum in-plane shear stress.

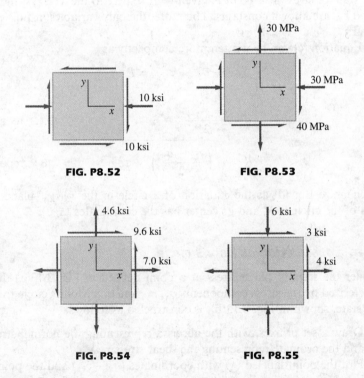

FIG. P8.52 **FIG. P8.53**

FIG. P8.54 **FIG. P8.55**

8.6 Mohr's Circle for Plane Stress

Mohr's circle, a graphical representation of the transformation equations,[2] is a popular method of stress transformation for two reasons. First, it allows us to visualize how the normal and shear stress components vary with the angle

---

[2] This graphical method was developed in 1882 by Otto Mohr, a German engineer. The method is also applicable to the transformation of strain and moment of inertia of area.

of transformation. Second, since all relevant data can be obtained from Mohr's circle by trigonometry, it is not necessary to refer to the transformation equations.

To show that the transformation equations represent a circle, we recall Eqs. (8.5a) and (8.5c):

$$\sigma_{x'} - \frac{\sigma_x + \sigma_y}{2} = \frac{\sigma_x - \sigma_y}{2} \cos 2\theta + \tau_{xy} \sin 2\theta \quad \text{(a)}$$

$$\tau_{x'y'} = -\frac{\sigma_x - \sigma_y}{2} \sin 2\theta + \tau_{xy} \cos 2\theta \quad \text{(b)}$$

where the term $(\sigma_x + \sigma_y)/2$ in Eq. (a) was moved from the right to the left side of the equation. We eliminate 2θ by squaring both sides of Eqs. (a) and (b) and then adding the equations, which yields

$$\left(\sigma_{x'} - \frac{\sigma_x + \sigma_y}{2}\right)^2 + \tau_{x'y'}^2 = \left(\frac{\sigma_x - \sigma_y}{2}\right)^2 + \tau_{xy}^2 \quad \text{(c)}$$

If the state of stress at a point is given with respect to the xy-axes, then σ_x, σ_y, and τ_{xy} are known constants. Therefore, the only variables in Eq. (c) are $\sigma_{x'}$ and $\tau_{x'y'}$.

Equation (c) can be written more compactly as

$$(\sigma_{x'} - \bar{\sigma})^2 + \tau_{x'y'}^2 = R^2 \quad \text{(d)}$$

where

$$\bar{\sigma} = \frac{\sigma_x + \sigma_y}{2} \quad \text{(8.13, repeated)}$$

$$R = \sqrt{\left(\frac{\sigma_x - \sigma_y}{2}\right)^2 + \tau_{xy}^2} \quad \text{(8.9, repeated)}$$

We recognize Eq. (d) as the equation of a circle in the $\sigma_{x'}\tau_{x'y'}$-plane. The radius of the circle is R, and its center has the coordinates $(\bar{\sigma}, 0)$.

a. Construction of Mohr's circle

Consider the state of plane stress at a point defined in Fig. 8.13(a) that is characterized by the stress components σ_x, σ_y, and τ_{xy}. Mohr's circle for this stress state, shown in Fig. 8.13(b), is constructed as follows:

1. Draw a set of axes, with the abscissa representing the normal stress σ and the ordinate representing the shear stress τ.
2. Plot the point labeled $\widehat{x}$ with coordinates $(\sigma_x, -\tau_{xy})$ and the point labeled $\widehat{y}$ with coordinates (σ_y, τ_{xy}). The coordinates of these points are the stresses acting on the x- and y-faces on the element, respectively. It is important to label these points to avoid confusion later on.
3. Join the points $\widehat{x}$ and $\widehat{y}$ with a straight line, and draw a circle with this line as its diameter.

Mohr's circle is now complete. Note that the radius of the circle is

$$R = \sqrt{\left(\frac{\sigma_x - \sigma_y}{2}\right)^2 + \tau_{xy}^2}$$

and its center is located at $(\bar{\sigma}, 0)$, where $\bar{\sigma} = (\sigma_x + \sigma_y)/2$.

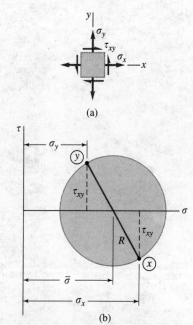

FIG. 8.13 Construction of Mohr's circle from given stress components.

b. Properties of Mohr's circle

The properties of Mohr's circle follow:

- The coordinates of each point on the circle represent the normal and shear stresses that act on a specific plane that passes through the selected point in the body. (For this reason, each time you plot a point on the circle, you should immediately label it to identify the plane that it represents.)
- Here is a convenient method for keeping track of the sense of shear stress: Shear stress that has a *clockwise* moment about the center of the element, as shown in Fig. 8.14(a), is plotted *up* (above the σ-axis). If the moment of the shear stress is *counterclockwise*, as in Fig. 8.14(b), the point is plotted *down* (below the σ-axis).[3]
- The angle 2θ between two diameters on the circle is *twice* the transformation angle θ, with both angles measured in the *same direction* (clockwise or counterclockwise).

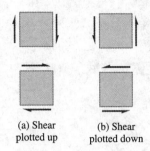

(a) Shear plotted up (b) Shear plotted down

FIG. 8.14 Convention for plotting shear stress on Mohr's circle.

The procedure for using Mohr's circle to transform the stress components from the xy-axes in Fig. 8.15(a) to the $x'y'$-axes in Fig. 8.15(b) is described below. We use the Mohr's circle shown in Fig. 8.15(d), which was drawn by using the points $\circled{x}$ and $\circled{y}$ as the diameter.

- Note the sense and magnitude of the angle θ between the xy- and the $x'y'$-axes in Fig. 8.15(b). (The sense of θ is the direction in which the xy-axes must be rotated to coincide with the $x'y'$-axes.)
- Rotate the diameter $\circled{x}$-$\circled{y}$ of Mohr's circle through the angle 2θ in the same sense as θ. Label the endpoints of this new diameter as $\circled{x'}$ and $\circled{y'}$ as shown in Fig. 8.15(d). The coordinates of $\circled{y'}$ are $(\sigma_{x'}, -\tau_{x'y'})$, and the coordinates of $\circled{y'}$ are $(\sigma_{y'}, \tau_{x'y'})$.

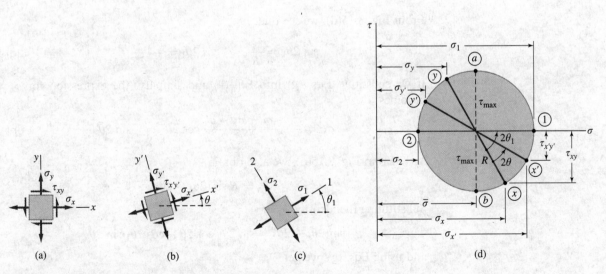

FIG. 8.15 Using Mohr's circle to transform stress components from the xy-axes to the $x'y'$-axes.

---

[3] This convention is different from the standard sign convention for shear stress and applies only to Mohr's circle.

Mohr's circle can also be used to find the principal stresses and principal directions. Referring to Fig. 8.15(d), we see that the principal planes (planes of maximum and minimum normal stress) are labeled ① and ②, respectively. The normal stress coordinates of these points are $\sigma_1 = \bar{\sigma} + R$ and $\sigma_2 = \bar{\sigma} - R$, and their shear stress coordinates are zero, as expected. The principal directions differ from the xy-coordinate directions by the angle θ_1 shown in Fig. 8.15(c). The magnitude and sense of θ_1 are determined by the angle $2\theta_1$ on the Mohr's circle.

In addition, we see by inspection of the Mohr's circle that the maximum in-plane shear stress τ_{max} equals the radius R of the circle. The planes on which τ_{max} acts are represented by points ⓐ and ⓑ on the circle. Observe that the σ-coordinates of both ⓐ and ⓑ are $\bar{\sigma}$. The points on the circle that correspond to the planes of maximum shear stress differ by 90° from the points that represent the principal planes. Therefore, the difference between the planes of maximum shear and the principal planes is 45°, as expected.

c. Verification of Mohr's circle

To prove that Mohr's circle is a valid representation of the transformation equations, we must show that the coordinates of the point ⓧ agree with Eqs. (8.5a) and (8.5c). From Fig. 8.15(d), we see that

$$\sigma_{x'} = \bar{\sigma} + R\cos(2\theta_1 - 2\theta) \tag{e}$$

Using the identity

$$\cos(2\theta_1 - 2\theta) = \cos 2\theta_1 \cos 2\theta + \sin 2\theta_1 \sin 2\theta$$

and substituting $\bar{\sigma} = (\sigma_x + \sigma_y)/2$, we obtain from Eq. (e)

$$\sigma_{x'} = \frac{(\sigma_x + \sigma_y)}{2} + R(\cos 2\theta_1 \cos 2\theta + \sin 2\theta_1 \sin 2\theta) \tag{f}$$

From Fig. 8.15(d), we see that

$$\sin 2\theta_1 = \frac{\tau_{xy}}{R} \qquad \cos 2\theta_1 = \frac{(\sigma_x - \sigma_y)}{2R} \tag{g}$$

If we substitute Eqs. (g) into Eq. (f) and simplify, the expression for $\sigma_{x'}$ becomes

$$\sigma_{x'} = \frac{\sigma_x + \sigma_y}{2} + \frac{\sigma_x - \sigma_y}{2}\cos 2\theta + \tau_{xy}\sin 2\theta \tag{h}$$

From Fig. 8.15(d), we also obtain

$$\tau_{x'y'} = R\sin(2\theta_1 - 2\theta)$$

Substituting the identity

$$\sin(2\theta_1 - 2\theta) = \sin 2\theta_1 \cos 2\theta - \cos 2\theta_1 \sin 2\theta$$

and using Eqs. (g), we get

$$\tau_{x'y'} = -\frac{\sigma_x - \sigma_y}{2}\sin 2\theta + \tau_{xy}\cos 2\theta \tag{i}$$

Because Eqs. (h) and (i) are identical to the transformation equations, Eqs. (8.5a) and (8.5c), we conclude that Mohr's circle is a valid representation of the transformation equations.

Sample Problem 8.7

The state of plane stress at a point with respect to the xy-axes is shown in Fig. (a). Using Mohr's circle, determine (1) the principal stresses and principal planes; (2) the maximum in-plane shear stress; and (3) the equivalent state of stress with respect to the $x'y'$-axes. Show all results on sketches of properly oriented elements.

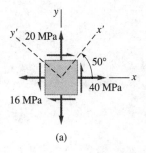

(a)

Solution

Construction of Mohr's Circle

From the established sign convention (positive stresses act in the positive coordinate directions on positive faces of the element), the stress components in Fig. (a) are $\sigma_x = 40$ MPa, $\sigma_y = 20$ MPa, and $\tau_{xy} = 16$ MPa. Using these stresses and the procedure explained in Art. 8.6, we obtain the Mohr's circle shown in Fig. (b). The coordinates of point $\widehat{x}$ on the circle are the stress components acting on the x-face of the element. Because the shear stress on this face has a counterclockwise moment about the center of the element, it is plotted below the σ-axis. The coordinates of point $\widehat{y}$ are the stress components acting on the y-face of the element. The shear stress on this face has a clockwise moment about the center of the element; thus, it is plotted above the σ-axis.

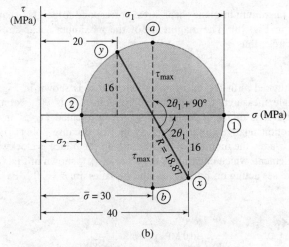

(b)

The characteristics of the circle, computed from Fig. (b), are

$$\bar{\sigma} = \frac{40 + 20}{2} = 30 \text{ MPa}$$

$$R = \sqrt{(10)^2 + (16)^2} = 18.87 \text{ MPa}$$

If the circle were drawn to scale, all the requested quantities could be determined by direct measurements. However, we choose to compute them here using trigonometry.

Part 1

By inspection of Fig. (b), we see that points $\widehat{1}$ and $\widehat{2}$ represent the principal planes. The principal stresses are

$$\sigma_1 = \bar{\sigma} + R = 30 + 18.87 = 48.9 \text{ MPa} \qquad \text{Answer}$$

$$\sigma_2 = \bar{\sigma} - R = 30 - 18.87 = 11.13 \text{ MPa} \qquad \text{Answer}$$

305

The principal directions are determined by the angle θ_1 where θ_1 is the angle measured counterclockwise from the x-face of the element to the plane on which σ_1 acts (recall that the angles on Mohr's circle are twice the angles between the physical planes, measured in the same direction). From Fig. (b), we see that

$$\tan 2\theta_1 = \frac{16}{10} \qquad 2\theta_1 = 58.0°$$

$$\theta_1 = 29.0° \qquad \qquad \text{Answer}$$

Figure (c) shows the principal stresses on an element aligned with the principal directions, labeled 1 and 2, respectively.

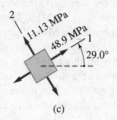

(c)

Part 2

The planes of maximum in-plane shear stress correspond to points ⓐ and ⓑ on the Mohr's circle in Fig. (b). The magnitude of the maximum shear stress equals the radius of the circle; thus,

$$\tau_{max} = 18.87 \text{ MPa} \qquad \text{Answer}$$

The element aligned with the maximum shear planes is shown in Fig. (d). On the circle, the angle measured from point ① to point ⓐ is 90°, counterclockwise. Therefore, the a-axis of the element in Fig. (d) is oriented at 45° in the counterclockwise direction relative to the 1-axis, as shown. Because point ⓐ on the circle lies above the σ-axis, the moment of τ_{max} acting on the a-face is clockwise about the center of the element, which determines the sense of τ_{max} shown in Fig. (d). Note that the normal stresses acting on both the a- and b-planes are $\bar{\sigma} = 30$ MPa.

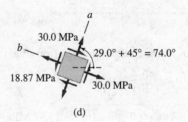

(d)

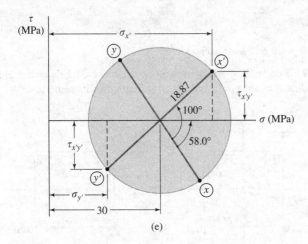

(e)

Part 3

Figure (a) shows that if we were to rotate the xy-coordinate axes through $50°$ in the counterclockwise direction, they would coincide with the $x'y'$-axes. This corresponds to a $100°$ counterclockwise rotation of the diameter ⓧ-ⓨ of the Mohr's circle to the position ⓧ'-ⓨ', as shown in Fig. (e). The coordinates of ⓧ' and ⓨ' are the stress components acting on the faces of the element that is aligned with the $x'y'$-axes.

From the geometry of the Mohr's circle in Fig. (e), we obtain

$$\sigma_{x'} = 30 + 18.87 \cos(100° - 58.0°) = 44.0 \text{ MPa} \qquad \text{Answer}$$

$$\sigma_{y'} = 30 - 18.87 \cos(100° - 58.0°) = 15.98 \text{ MPa} \qquad \text{Answer}$$

$$\tau_{x'y'} = 18.87 \sin(100° - 58.0°) = 12.63 \text{ MPa} \qquad \text{Answer}$$

The element showing these stresses is illustrated in Fig. (f). Again, the sense of the shear stress was found from the convention: If the shear stress is plotted above the σ-axis, its moment about the center of the element is clockwise, and vice versa. Because point ⓧ' lies above the σ-axis, the shear stress on the x'-face of the element applies a clockwise moment about the center of the element.

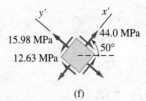

(f)

Problems

Solve the following problems using Mohr's circle.

8.56–8.59 For state of stress shown, (a) draw the Mohr's circle; and (b) determine the radius R of the circle and the coordinate $\bar{\sigma}$ of its center.

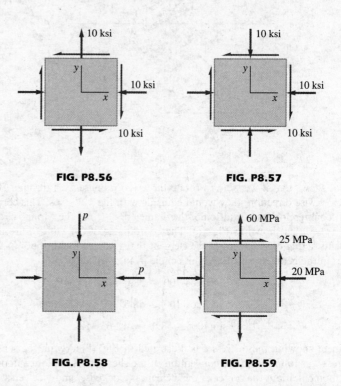

FIG. P8.56

FIG. P8.57

FIG. P8.58

FIG. P8.59

8.60–8.66 The state of stress at a point is shown with respect to the xy-axes. Determine the equivalent state of stress with respect to the $x'y'$-axes. Display the results on a sketch of an element aligned with the $x'y'$-axes.

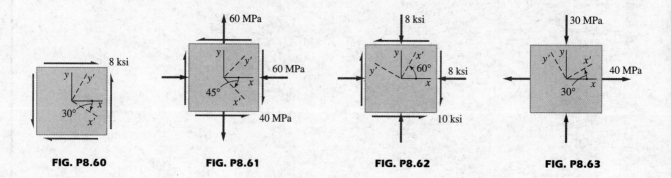

FIG. P8.60

FIG. P8.61

FIG. P8.62

FIG. P8.63

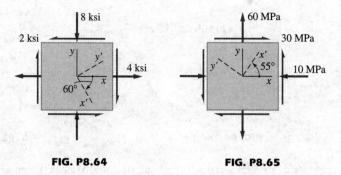
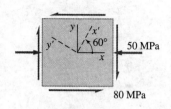

FIG. P8.64   **FIG. P8.65**   **FIG. P8.66**

8.67–8.71 For the state of stress shown, determine (a) the principal stresses; and (b) the maximum in-plane shear stress. Show the results on properly oriented elements.

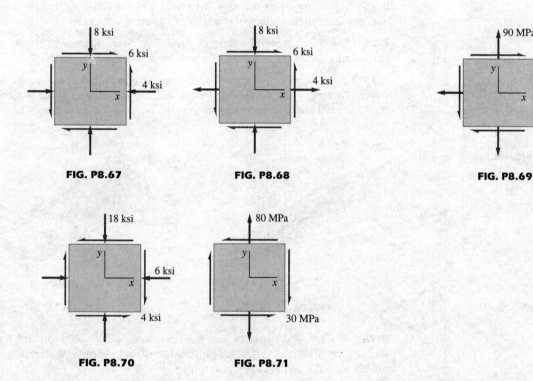

FIG. P8.67   **FIG. P8.68**   **FIG. P8.69**

FIG. P8.70   **FIG. P8.71**

8.72 The state of stress at a point is the result of two loadings. When acting alone, the first loading produces the 3-ksi pure shear with respect to the xy-axes. The second loading alone results in the 4-ksi pure shear with respect to the $x'y'$-axes. The angle between the two sets of axes is $\theta = 30°$ as shown. If the two loadings act simultaneously, determine (a) the state of stress at this point with respect to the xy-axis; and (b) the principal stresses and the principal planes. Show the results on properly oriented elements.

FIG. P8.72, P8.73

8.73 Solve Prob. 8.72 if $\theta = 45°$.

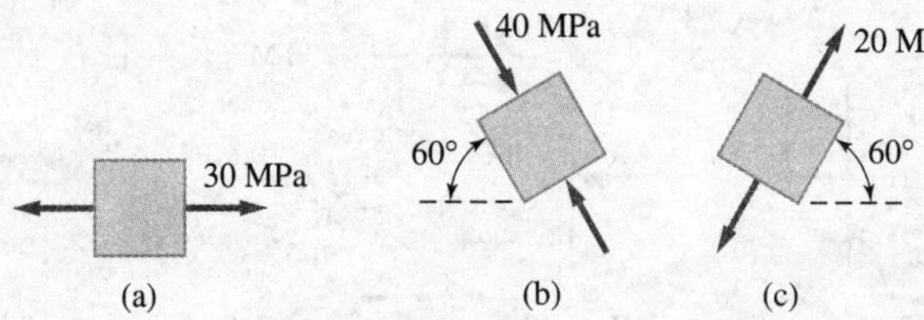

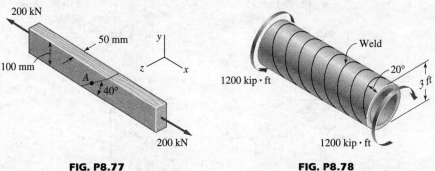

FIG. P8.74

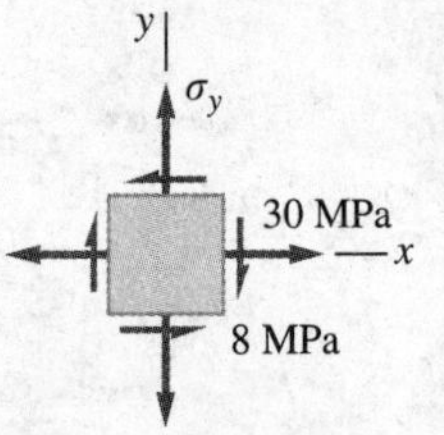

FIG. P8.75

8.74 The state of stress at a point is the result of three loadings. When acting separately, the loadings produce the three states of stress shown in the figure. If the three loadings are applied simultaneously, find the principal stresses and the principal planes. Show the results on properly oriented elements.

8.75 The figure shows the state of stress at a point. Knowing that the maximum in-plane shear stress at this point is 10 MPa, determine the value of σ_y.

8.76 The figure shows the state of stress at a point. Knowing that the normal stress acting on the plane *a-a* is 15 ksi tension, determine the value of σ_x.

8.77 Two 50-mm by 100-mm wooden joists are glued together along a 40° joint as shown. Determine (a) the state of stress at point A with respect to the *xyz*-axis; and (b) the normal and shear stresses acting on the plane of the glued joint.

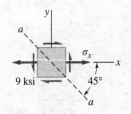

FIG. P8.76

FIG. P8.77

FIG. P8.78

8.78 A 1/4-in.-thick strip of steel is welded along a 20° spiral to make the thin-walled tube. Find the magnitudes of the normal and shear stresses acting in the welded seam due to the 1200-kip·ft torque.

8.7 Absolute Maximum Shear Stress

Up to this point, our discussion has been limited to in-plane transformation of stress (transformation in the *xy*-plane). The largest shear stress encountered in this transformation is called the *maximum in-plane shear stress* and is denoted by τ_{max}. However, τ_{max} is not necessarily the largest shear stress at a point. To find the largest shear stress, called the *absolute maximum shear stress*, we must also consider transformations in the other two coordinate planes.

a. Plane state of stress

Consider the state of plane stress shown in Fig. 8.16, where σ_1 and σ_2 are the principal stresses and the xy-axes coincide with the principal directions. Mohr's circle for transformation in the xy-plane is shown in Fig. 8.17(a). The radius of this circle is the maximum in-plane shear stress $\tau_{max} = |\sigma_1 - \sigma_2|/2$. Figures 8.17(b) and (c) show Mohr's circles representing stress transformation in the zx- and yz-planes.[4] The absolute maximum shear stress τ_{abs} is the radius of the largest circle; that is,

$$\tau_{abs} = \max\left(\frac{|\sigma_1 - \sigma_2|}{2}, \frac{|\sigma_1|}{2}, \frac{|\sigma_2|}{2}\right) \quad (8.15)$$

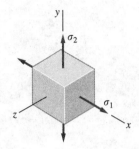

FIG. 8.16 State of plane stress referred to the principal planes.

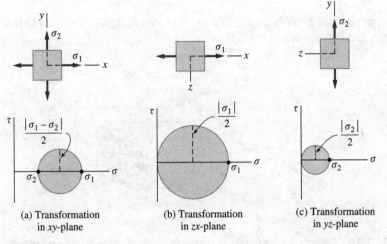

FIG. 8.17 Mohr's circles for stress transformation in three coordinate planes (plane stress).

It is standard practice to draw the three Mohr's circles on a single set of axes, as shown in Fig. 8.18.

From Eq. (8.15), we can draw the following conclusions:

1. If σ_1 and σ_2 have the same sign (both tension or both compression), the absolute maximum shear stress is $|\sigma_1|/2$ or $|\sigma_2|/2$, whichever is larger.
2. If σ_1 and σ_2 have opposite signs (one tension and the other compression), the absolute maximum shear stress is $|\sigma_1 - \sigma_2|/2$, which equals the maximum in-plane shear stress τ_{max}.

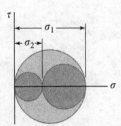

FIG. 8.18 Mohr's circles for stress transformation in three coordinate planes drawn on a single set of axes. The radius of the largest circle is the absolute maximum shear stress.

---

[4] Although the transformation equations in Eqs. (8.5) were derived for plane stress, they remain valid even if the out-of-plane normal stress is not zero. Because this normal stress does not have an in-plane component, it does not affect the equilibrium equations from which the transformation equations were derived.

b. General state of stress

A complete discussion of a general (three-dimensional) state of stress at a point, as opposed to plane stress, is beyond the scope of this text. It can be shown that any state of stress can be represented by three principal stresses (σ_1, σ_2, and σ_3) that act on mutually perpendicular planes, as shown in Fig. 8.19(a). The corresponding Mohr's circles are shown in Fig. 8.19(b). The absolute maximum shear stress is again equal to the radius of the largest circle:

$$\tau_{\text{abs}} = \max\left(\frac{|\sigma_1 - \sigma_2|}{2}, \frac{|\sigma_2 - \sigma_3|}{2}, \frac{|\sigma_3 - \sigma_1|}{2}\right) \quad (8.16)$$

Comparing the Mohr's circles in Figs. 8.18 and 8.19, we see that plane stress is the special case where $\sigma_3 = 0$.

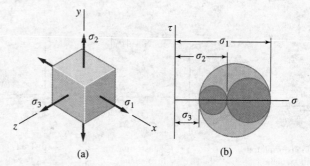

FIG. 8.19 Mohr's circles for stress transformation in three coordinate planes (triaxial stress).

Sample Problem 8.8

For the state of plane stress shown in Fig. (a), determine the maximum in-plane shear stress and the absolute maximum shear stress.

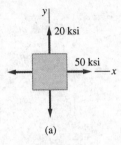

(a)

Solution

The given stresses are $\sigma_1 = \sigma_x = 50$ ksi and $\sigma_2 = \sigma_y = 20$ ksi. The Mohr's circles representing stress transformation in the three coordinate planes are shown in Fig. (b). The maximum in-plane shear stress τ_{max} is equal to the radius of the circle that represents transformation in the xy-plane. Thus,

$$\tau_{max} = 15 \text{ ksi} \qquad \text{Answer}$$

The absolute maximum shear stress equals the radius of the largest circle, which represents the transformation in the zx-plane:

$$\tau_{abs} = 25 \text{ ksi} \qquad \text{Answer}$$

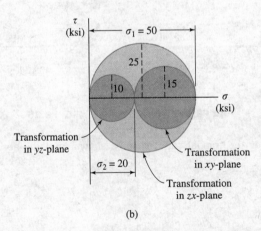

(b)

Problems

8.79–8.85 For the state of plane stress shown, determine (a) the maximum in-plane shear stress and (b) the absolute maximum shear stress.

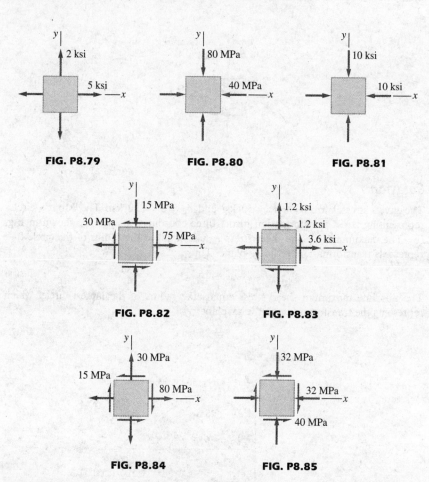

FIG. P8.79

FIG. P8.80

FIG. P8.81

FIG. P8.82

FIG. P8.83

FIG. P8.84

FIG. P8.85

8.86–8.88 For the triaxial state of stress shown, find the absolute maximum shear stress.

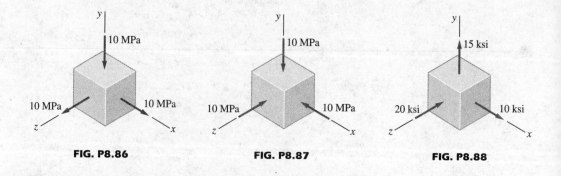

FIG. P8.86

FIG. P8.87

FIG. P8.88

8.8 Applications of Stress Transformation to Combined Loads

The most important use of stress transformation is in the analysis of members that are subjected to combined loads (axial load, torsion, and bending). The general procedure for analysis is as follows:

- Compute the state of stress at the critical point (the most highly stressed point). Sometimes the location of the critical point is uncertain, in which case stresses at several points in the member must be compared to determine which is *the* critical point.
- Draw a Mohr's circle for the state of stress at the critical point.
- Use the Mohr's circle to calculate the relevant stresses, such as the principal stresses or maximum shear stress.

The transformation formulas developed in Art. 8.5 could be used to analyze the state of stress. However, because Mohr's circle has the advantage of visualization, its use is preferred by many engineers.

Sample Problem 8.9

Find the smallest diameter d of the steel shaft $ABCD$ that is capable of carrying the loads shown in Fig. (a) if the working stresses are $\sigma_w = 120$ MPa and $\tau_w = 70$ MPa. Neglect the weights of the pulleys and the shaft. The pulleys are rigidly attached to the shaft. Assume that the stress caused by direct shear force is negligible.

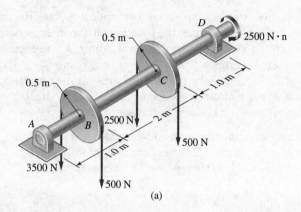

(a)

Solution

The first step is to construct the bending moment (M) and torque (T) diagrams produced by the loading, as shown in Fig. (b). These diagrams indicate that the largest bending moment occurs at B, but the largest torque occurs in segment CD. Therefore, the critical point lies on either section a-a or section b-b, as shown in Fig. (b).

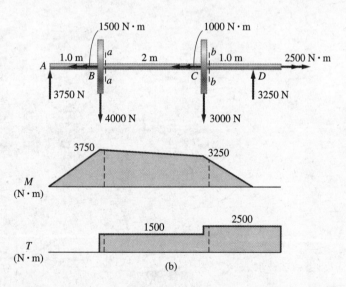

(b)

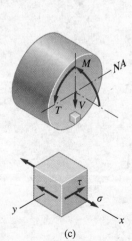

(c)

Referring to Fig. (c), we deduce that the most highly stressed points on each section are located at the top and bottom of the cross section where the magnitude of the bending stress is largest (compression at the top and tension at the bottom). Because the given design criterion for normal stress ($\sigma_w \leq 120$ MPa) does not distinguish between tensile and compressive stresses, we may choose either the bottom or top points. Choosing the bottom points, we will now analyze the state of stress at each section.

Section a-a At this section, we have $M = 3750$ N·m and $T = 1500$ N·m. Therefore, the stresses at the bottom of the section are

$$\sigma = \frac{32M}{\pi d^3} = \frac{32(3750)}{\pi d^3} = \frac{120 \times 10^3}{\pi d^3} \text{ Pa} = \frac{120}{\pi d^3} \text{ kPa}$$

$$\tau = \frac{16T}{\pi d^3} = \frac{16(1500)}{\pi d^3} = \frac{24 \times 10^3}{\pi d^3} \text{ Pa} = \frac{24}{\pi d^3} \text{ kPa}$$

where the diameter d is in meters. The corresponding Mohr's circle is shown in Fig. (d). From the circle, we see that the maximum normal and in-plane shear stresses are

$$\sigma_{max} = \frac{60}{\pi d^3} + \frac{64.62}{\pi d^3} = \frac{124.62}{\pi d^3} \text{ kPa}$$

$$\tau_{max} = \frac{64.62}{\pi d^3} \text{ kPa}$$

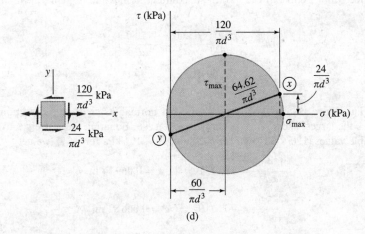

(d)

Section b-b At this section, $M = 3250$ N·m and $T = 2500$ N·m, which result in the following stresses at the bottom of the section:

$$\sigma = \frac{32M}{\pi d^3} = \frac{32(3250)}{\pi d^3} = \frac{104 \times 10^3}{\pi d^3} \text{ Pa} = \frac{104}{\pi d^3} \text{ kPa}$$

$$\tau = \frac{16T}{\pi d^3} = \frac{16(2500)}{\pi d^3} = \frac{40 \times 10^3}{\pi d^3} \text{ Pa} = \frac{40}{\pi d^3} \text{ kPa}$$

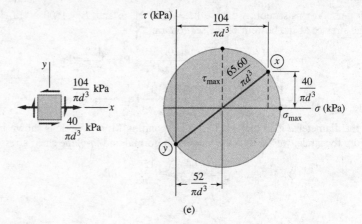

(e)

From the Mohr's circle in Fig. (e), the maximum normal and in-plane shear stresses are

$$\sigma_{max} = \frac{52}{\pi d^3} + \frac{65.60}{\pi d^3} = \frac{117.60}{\pi d^3} \text{ kPa}$$

$$\tau_{max} = \frac{65.60}{\pi d^3} \text{ kPa}$$

The above results show that the largest normal stress occurs at section a-a, whereas the largest shear stress occurs at section b-b. Equating these stresses to their allowable values (120×10^3 kPa for σ_{max} and 70×10^3 kPa for τ_{max}), we get

$$\frac{124.62}{\pi d^3} = 120 \times 10^3 \qquad d = 0.069\,14 \text{ m}$$

$$\frac{65.60}{\pi d^3} = 70 \times 10^3 \qquad d = 0.066\,82 \text{ m}$$

The smallest safe diameter is the larger of these two values:

$$d = 0.0691 \text{ m} = 69.1 \text{ mm} \qquad \textbf{Answer}$$

Note The shear stress design criterion is $\tau_{abs} \leq \tau_w$, where τ_{abs} is the absolute maximum shear stress. In this problem, the principal stresses have opposite signs at both points investigated, so that τ_{abs} is equal to the maximum in-plane shear stress τ_{max}.

Sample Problem 8.10

The thin-walled cylindrical pressure vessel with closed ends has a mean radius of 450 mm and a wall thickness of 10 mm. In addition to an internal pressure $p = 2$ MPa, the vessel carries the torque T as shown in Fig. (a). Determine the largest allowable value of T if the working shear stress is $\tau_w = 50$ MPa.

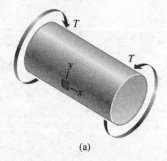

(a)

Solution

The state of stress in the wall of the vessel is shown on the element in Fig. (b). The normal stresses, which are caused by the pressure, were obtained from Eqs. (8.1) and (8.2):

$$\sigma_x = \sigma_\ell = \frac{pr}{2t} = \frac{(2 \times 10^6)(0.45)}{2(0.01)} = 45.0 \text{ MPa}$$

$$\sigma_y = \sigma_c = \frac{pr}{t} = 2\sigma_\ell = 90.0 \text{ MPa}$$

where we used the mean radius for r. The shear stress τ_{xy} due to torsion is unknown at this stage.

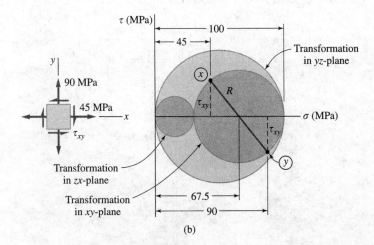

(b)

In this case, the maximum in-plane shear stress is not the largest shear stress in the wall of the cylinder. This can be verified by drawing the three Mohr's circles that represent stress transformation in the three coordinate planes, as was explained in Art. 8.7. The result, shown in Fig. (b), reveals that the largest circle is the one that represents transformation in the yz-plane.

When drawing the Mohr's circles in Fig. (b), we set the radius of the largest circle equal to 50 MPa, which is the prescribed limit of shear stress. The value of τ_{xy} that causes this limit to be reached can now be calculated from the circles in Fig. (b). We have

$$R = 100 - 67.5 = 32.5 \text{ MPa}$$

$$\tau_{xy} = \sqrt{R^2 - (90 - 67.5)^2} = \sqrt{32.5^2 - 22.5^2} = 23.45 \text{ MPa}$$

We now compute the torque that produces this shear stress. Because the vessel is thin-walled, all parts of its cross-sectional area A are located approximately at a distance r from the centroid. Therefore, a good approximation of the polar moment of inertia is $J = Ar^2$. Using $A \approx 2\pi rt$, we thus obtain

$$J = 2\pi r^3 t = 2\pi (450)^3 (10) = 5.726 \times 10^9 \text{ mm}^4 = 5.726 \times 10^{-3} \text{ m}^4$$

The largest allowable torque is obtained from $\tau_{xy} = Tr/J$, which yields

$$T = \frac{\tau_{xy} J}{r} = \frac{(23.45 \times 10^6)(5.726 \times 10^{-3})}{0.45}$$

$$= 298 \times 10^3 \text{ N} \cdot \text{m} = 298 \text{ kN} \cdot \text{m} \qquad \textit{Answer}$$

Sample Problem 8.11

The radius of the 15-in.-long bar in Fig. (a) is 3/8 in. Determine the maximum normal stress in the bar at (1) point A; and (2) point B.

Solution

Preliminary Calculations

The internal force system acting on the cross section at the base of the rod is shown in Fig. (b). It consists of the torque $T = 540$ lb·in., the bending moment $M = 15P = 15(30) = 450$ lb·in. (acting about the x-axis), and the transverse shear force $V = P = 30$ lb.

The cross-sectional properties of the bar are

$$I = \frac{\pi r^4}{4} = \frac{\pi(3/8)^4}{4} = 15.532 \times 10^{-3} \text{ in.}^4$$

$$J = 2I = 2(15.532 \times 10^{-3}) = 31.06 \times 10^{-3} \text{ in.}^4$$

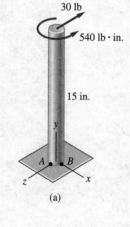

(a)

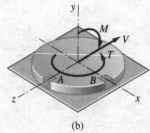

(b)

Part 1

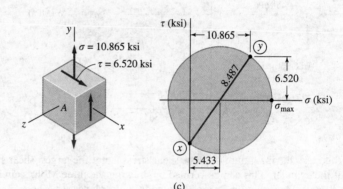

(c)

Figure (c) shows the state of stress at point A together with the corresponding Mohr's circle. The bending stress is

$$\sigma = \frac{Mr}{I} = \frac{(450)(3/8)}{15.532 \times 10^{-3}} = 10\,865 \text{ psi} = 10.865 \text{ ksi}$$

and the torque causes the shear stress

$$\tau_T = \frac{Tr}{J} = \frac{540(3/8)}{31.06 \times 10^{-3}} = 6520 \text{ psi} = 6.520 \text{ ksi}$$

The shear stress due to transverse shear force V is zero at A.

From the Mohr's circle for the state of stress at point A in Fig. (c), we see that the maximum normal stress at point A is

$$\sigma_{max} = 5.433 + 8.487 = 13.92 \text{ ksi} \qquad \text{Answer}$$

Part 2

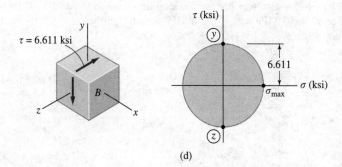

(d)

The state of stress at point B is shown in Fig. (d). The shear stress due to torque is $\tau_T = 6.520$ ksi, as before. But because the point lies on the neutral axis, the bending stress is zero. There is, however, an additional shear stress caused by the transverse shear force V. The magnitude of this shear stress is $\tau_V = VQ/(Ib)$, where $b = 2r = 3/4$ in. and Q is the first moment of half the cross-sectional area about the neutral axis. Referring to Fig. (e), we get

$$Q = A'\bar{z}' = \left(\frac{\pi r^2}{2}\right)\left(\frac{4r}{3\pi}\right) = \frac{2r^3}{3} = \frac{2(3/8)^3}{3} = 35.16 \times 10^{-3} \text{ in.}^3$$

Therefore,

$$\tau_V = \frac{VQ}{Ib} = \frac{30(35.16 \times 10^{-3})}{(15.532 \times 10^{-3})(3/4)} = 90.5 \text{ psi} = 0.091 \text{ ksi}$$

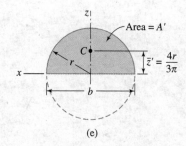

(e)

Because τ_T and τ_V act on the same planes and have the same sense, they can be added. Hence, the total shear stress is

$$\tau = \tau_T + \tau_V = 6.520 + 0.091 = 6.611 \text{ ksi}$$

The Mohr's circle for this state of pure shear, shown in Fig. (d), yields for the maximum normal stress at B

$$\sigma_{max} = 6.61 \text{ ksi} \qquad \textit{Answer}$$

Note that the shear stress due to V is small compared to the stresses caused by M and T. As pointed out before, the effect of V on the state of stress in a slender bar is seldom significant, unless the bar is thin-walled.

∎

Problems

8.89 The 60-mm-diameter bar carries a compressive axial force P. If the working stresses are $\sigma_w = 80$ MPa and $\tau_w = 30$ MPa, determine the largest P that can be applied safely.

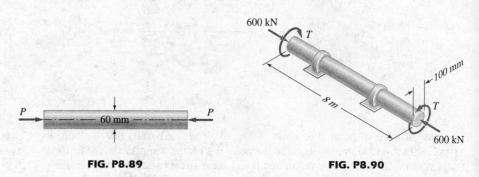

FIG. P8.89

FIG. P8.90

8.90 The solid steel shaft 100 mm in diameter and 8 m long is subjected simultaneously to an axial compressive force of 600 kN and a torque T that twists the shaft through 1.5°. If the shear modulus of steel is 80 GPa, find the maximum normal and shear stresses in the shaft.

8.91 The solid shaft of a small turbine is 4 in. in diameter and supports an axial compressive load of 100 kips. Determine the horsepower that the shaft can transmit at 250 rev/min without exceeding the working stresses $\sigma_w = 13$ ksi and $\tau_w = 10$ ksi.

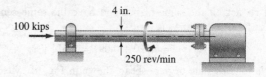

FIG. P8.91

8.92 A hollow propeller shaft has a 12-in. outer diameter and a 11-in. inner diameter. When transmitting maximum power, the thrust of the propeller exerts a compressive force of 200 kips on the shaft, which is also twisted through 1.15° in a length of 20 ft. Using $G = 12 \times 10^6$ psi, compute the maximum normal and shear stresses in the shaft.

8.93 A solid shaft of a turbine, 100 mm in diameter, carries an axial compressive force of 440 kN. Determine the maximum power that can be developed at 4 Hz without exceeding a shear stress of 70 MPa or a normal stress of 90 MPa.

8.94 A solid shaft, 100 mm in diameter, is subjected to bending loads that cause a maximum bending moment of 8.0 kN·m. What torque can act simultaneously on the shaft without exceeding a shear stress of 80 MPa or a normal stress of 100 MPa?

8.95 The 3-in.-diameter rod carries a bending moment of 2210 lb·ft and the torque T. Determine the largest T that can be applied if the working stresses are $\sigma_w = 17$ ksi and $\tau_w = 13$ ksi.

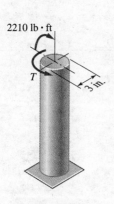

FIG. P8.95

8.96 The solid 80-mm-diameter bar carries a torque T and a tensile force of 125 kN acting 20 mm from the centerline of the bar. Find the largest safe value of T if the working stresses are $\sigma_w = 100$ MPa and $\tau_w = 80$ MPa.

8.97 The shaft 50 mm in diameter carries a 1.0-kN lateral force and a torque T. Determine the largest T that can be applied if the working stresses are $\sigma_w = 120$ MPa and $\tau_w = 60$ MPa. Neglect the stress due to the transverse shear force.

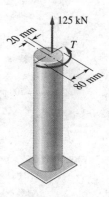

FIG. P8.96

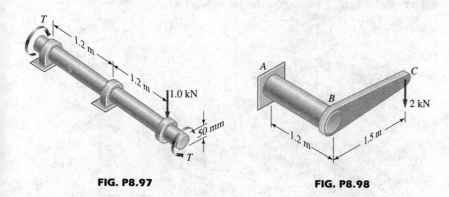

FIG. P8.97   **FIG. P8.98**

8.98 The working stresses for the circular bar AB are $\sigma_w = 120$ MPa and $\tau_w = 60$ MPa. Determine the smallest radius of the bar that can carry the 2-kN load. Neglect the stress due to the transverse shear force.

8.99 A solid 100-mm-diameter shaft carries simultaneously an axial tensile force of 160 kN, a maximum bending moment of 6 kN·m, and a torque of 9 kN·m. Compute the maximum tensile, compressive, and shear stresses in the shaft.

8.100 Solve Prob. 8.99 if the axial tensile force is replaced by a compressive force of 125 kN.

8.101 The cylindrical pressure vessel with closed ends has an external diameter of 16 in. and a wall thickness of 3/4 in. The vessel carries simultaneously an internal pressure of 600 psi, a torque of 48 kip·ft, and a bending moment of 12 kip·ft. Calculate the maximum normal and shear stresses in the wall of the vessel.

8.102 The closed cylindrical tank fabricated from 1/2-in. plate is subjected to an internal pressure of 240 psi. Determine the largest permissible diameter of the tank if the working stresses are $\sigma_w = 12$ ksi and $\tau_w = 4.8$ ksi.

8.103 The cylindrical tank has an external diameter of 20 in. and a wall thickness of 1/2 in. The tank is pressurized to 200 psi and carries an axial tensile force of 24 kips. Determine the maximum normal and shear stresses in the wall of the tank.

8.104 Solve Prob. 8.103 if the pressure is 120 psi and the axial force is 40 kips compression.

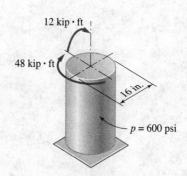

FIG. P8.101

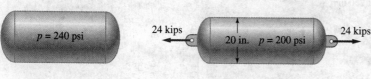

FIG. P8.102   **FIG. P8.103, P8.104**

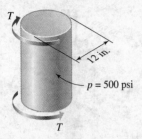

FIG. P8.105

8.105 The closed cylindrical tank of 12-in. mean diameter is fabricated from 1/4-in. plate. The tank is subjected to an internal pressure of 500 psi and a torque T. Find the largest safe value of T if the working stresses are $\sigma_w = 18$ ksi and $\tau_w = 10$ ksi.

8.106 The 2-in.-diameter bracket carries the horizontal and vertical forces shown in the figure. (a) Find the principal stresses and the maximum shear stress at point A. (b) If the direction of the 800-lb load is reversed, do the results change?

8.107 Solve Prob. 8.106 for the stresses at point B.

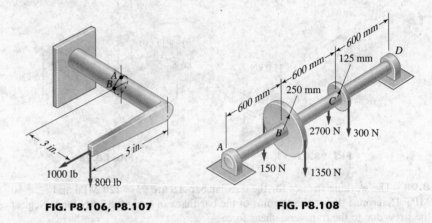

FIG. P8.106, P8.107   **FIG. P8.108**

8.108 A shaft carries the loads shown in the figure. If the working shear stress is $\tau_w = 80$ MPa, determine the smallest allowable diameter of the shaft. Neglect the weights of the pulleys and the shaft as well as the stress due to the transverse shear force.

8.109 The 50-mm-diameter shaft is subject to the loads shown in the figure. The belt tensions are horizontal on pulley B and vertical on pulley C. Calculate the maximum normal and shear stresses in the shaft. Neglect the weights of the pulleys and the shaft as well as the stress due to the transverse shear force.

8.110 Determine the smallest allowable diameter of the shaft that carries the three pulleys. The working stresses are $\sigma_w = 18$ ksi and $\tau_w = 8$ ksi. Neglect the weights of the pulleys and the shaft as well as the stress due to the transverse shear force.

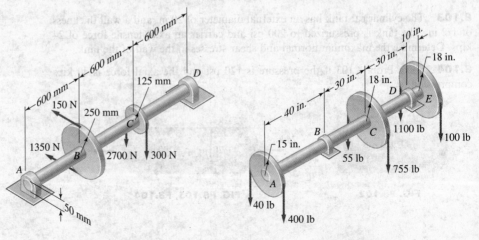

FIG. P8.109   **FIG. P8.110**

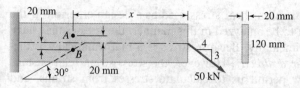
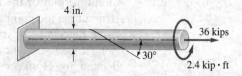

FIG. P8.111, P8.112   **FIG. P8.113**

8.111 Determine the principal stresses and the maximum in-plane shear stress at point A of the cantilever beam. The point is located at $x = 250$ mm. Show the answers on sketches of properly oriented elements.

8.112 Compute the stress components on the 30° plane at point B of the cantilever beam. Assume that $x = 300$ mm. Show the results on a properly oriented element.

8.113 The 4-in.-diameter shaft carries a 36-kip axial load and 2.4-kip·ft torque. Determine the normal and shear stresses acting on the spiral weld that makes a 30° angle with the axis of the shaft.

8.114 For the cantilever beam in the figure, determine the principal stresses at point A located just below the flange. Show the results on a properly oriented element.

8.115 Determine the maximum in-plane shear stress at point A located just below the flange of the cantilever beam. Show the results on a properly oriented element.

8.116 The plastic cylinder, which has a 10-in. inner diameter and a 10.5-in. outer diameter, is filled with oil. The cylinder is sealed at both ends by gaskets that are fastened in place by two rigid end-plates held together by four bolts. After the oil has been pressurized to 200 psi, the bolts are tightened until each carries a tensile force of 7950 lb. Determine the resulting maximum shear stress in the cylinder.

8.117 The plastic cylinder has an inner diameter of 10 in. and an outer diameter 10.5 in. The cylinder is sealed at both ends by gaskets fastened in place by two rigid end-plates that are held together by four bolts. The allowable stresses for the plastic are 8000 psi in tension, 5000 psi in compression, and 3500 psi in shear. Find the maximum internal pressure that can be applied if each of the four bolts carries an initial tensile force of 10 kips.

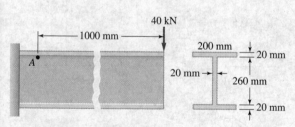
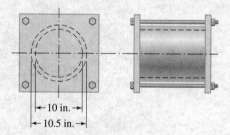

FIG. P8.114, P8.115   **FIG. P8.116, P8.117**

8.9 Transformation of Strain; Mohr's Circle for Strain

Many problems encountered in engineering design involve a combination of axial, torsional, and bending loads applied to elastic bars. In such cases, the stresses may be computed as described in the preceding articles and the maximum normal or the maximum shear stress used as a design criterion. If the structure is too complex to be analyzed in this manner, the stresses may have to be determined experimentally. Because stress is a mathematical abstraction, it *cannot be measured directly*. However, the stress-strain relationships defined by Hooke's law permit us to calculate stresses from strains, and strains *can be measured*. In this article, we derive the transformation equations for plane strain. If we know the strain components at a point associated with a given set of axes, these equations enable us to calculate the strain components with respect to any set of axes at that point. With this information, we are able to find the principal strains at the point. In this article, we also consider the conversion of strain measurements into stresses using Hooke's law.

a. Review of strain

Figure 8.20(a) shows an infinitesimal element of dimensions dx by dy undergoing the *normal strains* ϵ_x and ϵ_y. Recalling that positive normal strain is the elongation per unit length, we know the element elongates in the x-direction by $\epsilon_x \, dx$. Similarly, the elongation of the element in the y-direction is $\epsilon_y \, dy$. Compressive normal strains are considered to be negative.

The *shear strain* γ_{xy}, shown in Fig. 8.20(b), measures the change (in radians) in the original right angle between the edges of the element. The subscripts xy indicate that γ_{xy} is the angular change between the edges that coincide with, or are parallel to, the x- and y-axes. Shear strains are considered to be positive when the angle decreases, and negative when the angle increases. With this sign convention for strain, positive shear stresses cause positive shear strains.

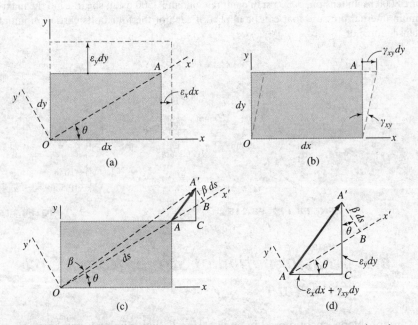

FIG. 8.20 (a) Element undergoing normal strains; (b) element undergoing shear strain; (c) change in the diagonal of the element due to normal and shear strains; (d) components of displacement vector of corner A.

b. Transformation equations for plane strain

The state of plane strain at a point O is defined by the strain components ϵ_x, ϵ_y, and γ_{xy} at that point. If the state of strain at a point is known, the transformation equations for strain can be used to calculate the normal and shear strains associated with respect to any set of axes at that point.

We now derive the equations that transform the strains from the xy-axes to the $x'y'$-axes shown in Fig. 8.20(a). To facilitate the derivation, we choose the dimensions of the element so that the diagonal OA coincides with the x'-axis. It is also convenient to assume that corner O is fixed and the edge formed by the x-axis does not rotate. These assumptions eliminate arbitrary rigid-body motions without impeding the deformation of the element.

Transformation Equations for Normal Strain When all three strains components occur simultaneously, corner A of the element is displaced to position A' as shown in Fig. 8.20(c). This displacement is obtained by superimposing the displacements shown in Figs. 8.20(a) and (b). Thus, the horizontal and vertical components of $\overline{AA'}$ are

$$\overline{AC} = \epsilon_x\, dx + \gamma_{xy}\, dy \quad \text{and} \quad \overline{CA'} = \epsilon_y\, dy \tag{a}$$

We now resolve $\overline{AA'}$ into components that are parallel and perpendicular to the diagonal OA. The parallel component $\overline{AB}$ represents the change in length of OA, and the perpendicular component $\overline{BA'}$ is caused by a rotation of OA through the small angle β. From Fig. 8.20(c), we see that $\overline{BA'} = \beta\, ds$.

The displacements in Fig. 8.20(c) are shown enlarged in Fig. 8.20(d). The increase in length of the diagonal OA can be found by adding the projections of $\overline{AC}$ and $\overline{CA'}$ onto the x'-direction. This yields $\overline{AB} = \overline{AC}\cos\theta + \overline{CA'}\sin\theta$, where θ is the angle between OA and the x-axis. Upon substitution from Eqs. (a), we get

$$\overline{AB} = (\epsilon_x\, dx + \gamma_{xy}\, dy)\cos\theta + \epsilon_y\, dy\, \sin\theta \tag{b}$$

Dividing $\overline{AB}$ by the original length ds of the diagonal OA, we obtain for the strain in the x'-direction

$$\epsilon_{x'} = \frac{\epsilon_x\, dx\, \cos\theta}{ds} + \frac{\gamma_{xy}\, dy\, \cos\theta}{ds} + \frac{\epsilon_y\, dy\, \sin\theta}{ds} \tag{c}$$

We see from Fig. 8.20(a) that $dx/ds = \cos\theta$ and $dy/ds = \sin\theta$, so that Eq. (c) becomes

$$\epsilon_{x'} = \epsilon_x \cos^2\theta + \gamma_{xy}\sin\theta\cos\theta + \epsilon_y \sin^2\theta \tag{d}$$

The standard form of this transformation equation is obtained by substituting the trigonometric identities

$$\cos^2\theta = \frac{1+\cos 2\theta}{2} \qquad \sin^2\theta = \frac{1-\cos 2\theta}{2} \qquad \sin\theta\cos\theta = \frac{1}{2}\sin 2\theta$$

into Eq. (d), which yields

$$\boxed{\epsilon_{x'} = \frac{\epsilon_x + \epsilon_y}{2} + \frac{\epsilon_x - \epsilon_y}{2}\cos 2\theta + \frac{1}{2}\gamma_{xy}\sin 2\theta} \tag{8.17a}$$

The expression for $\epsilon_{y'}$ can be found by replacing θ in this equation by $\theta + 90°$, which results in

$$\epsilon_{y'} = \frac{\epsilon_x + \epsilon_y}{2} - \frac{\epsilon_x - \epsilon_y}{2}\sin 2\theta - \frac{1}{2}\gamma_{xy}\cos 2\theta \qquad (8.17b)$$

Transformation Equation for Shear Strain Referring again to Fig. 8.20(d), we see that the displacement component $\overline{BA'}$ can be found by projecting $\overline{AC}$ and $\overline{CA'}$ onto the y'-direction: $\overline{BA'} = \overline{CA'}\cos\theta - \overline{AC}\sin\theta$, which yields, after substituting from Eq. (a),

$$\overline{BA'} = \epsilon_y\,dy\cos\theta - (\epsilon_x\,dx + \gamma_{xy}\,dy)\sin\theta \qquad (e)$$

Therefore, the angle of rotation of OA is

$$\beta = \frac{\overline{BA'}}{ds} = \frac{\epsilon_y\,dy\cos\theta}{ds} - \frac{\epsilon_x\,dx\sin\theta}{ds} - \frac{\gamma_{xy}\,dy\sin\theta}{ds}$$

$$= \epsilon_y\sin\theta\cos\theta - \epsilon_x\sin\theta\cos\theta - \gamma_{xy}\sin^2\theta \qquad (f)$$

The rotation angle β' of the line element at right angles to OA (coincident with the y'-axis) may be found by substituting $\theta + 90°$ for θ in Eq. (f), yielding

$$\beta' = -\epsilon_y\sin\theta\cos\theta + \epsilon_x\sin\theta\cos\theta - \gamma_{xy}\cos^2\theta \qquad (g)$$

Because the positive direction for both β and β' is counterclockwise, the shear strain $\gamma_{x'y'}$, which is the decrease in the right angle formed by the x'- and y'-axes, is the difference in the two angles. Thus, we have

$$\gamma_{x'y'} = \beta - \beta' = \epsilon_y(2\sin\theta\cos\theta) - \epsilon_x(2\sin\theta\cos\theta)$$
$$+ \gamma_{xy}(\cos^2\theta - \sin^2\theta) \qquad (h)$$

Substituting $2\sin\theta\cos\theta = \sin 2\theta$ and $\cos^2\theta - \sin^2\theta = \cos 2\theta$, we obtain the following standard form of the transformation equations for shear strain:

$$\frac{1}{2}\gamma_{x'y'} = -\frac{\epsilon_x - \epsilon_y}{2}\sin 2\theta + \frac{1}{2}\gamma_{xy}\cos 2\theta \qquad (8.17c)$$

c. Mohr's circle for strain

Comparing Eqs. (8.17) with the stress transformation equations in Eqs. (8.5) shows that they are identical in form, the association between the stresses and strains being

$$(\sigma_x, \sigma_y, \tau_{xy}) \Leftrightarrow (\epsilon_x, \epsilon_y, \gamma_{xy}/2)$$

We conclude that the transformation equations for strain can also be represented by a Mohr's circle, constructed in the same manner as the Mohr's circle for stress. The notable difference is that *half of the shear strain* $(\gamma/2)$ *is plotted on the ordinate* instead of the shear strain. Thus, for Mohr's circle for strain, the endpoints of the diameter have the coordinates $(\epsilon_x, -\gamma_{xy}/2)$ and

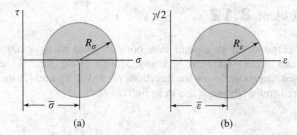

FIG. 8.21 Mohr's circles for (a) stress and (b) strain.

(ϵ_y, $\gamma_{xy}/2$). This is consistent with the convention used for Mohr's circle for stress, where the corresponding coordinates are (σ_x, $-\tau_{xy}$) and (σ_y, τ_{xy}).

The similarities between stress and strain are further exemplified by the fact that Mohr's circle for strain can be transformed into Mohr's circle for stress using the scale transformations

$$\boxed{R_\sigma = R_\epsilon \frac{E}{1+v} \qquad \bar{\sigma} = \bar{\epsilon}\frac{E}{1-v}} \qquad (8.18)$$

where E is the modulus of elasticity and v represents Poisson's ratio. As shown in Fig. 8.21, R_σ and R_ϵ are the radii of the stress and strain circles, respectively, and $\bar{\sigma} = (\sigma_x + \sigma_y)/2$ and $\bar{\epsilon} = (\epsilon_x + \epsilon_y)/2$ locate the centers of the circles. (The proof of these relationships is requested in Prob. 8.118.)

Sample Problem 8.12

The state of plane strain at a point in a body is given by $\epsilon_x = 800 \times 10^{-6}$, $\epsilon_y = 200 \times 10^{-6}$, and $\gamma_{xy} = -600 \times 10^{-6}$. Using Mohr's circle, find (1) the principal strains and their directions (show the directions on a sketch); and (2) the strain components referred to the $x'y'$-axes shown in Fig. (a).

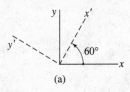

(a)

Solution

Using the given strain components, we plot the Mohr's circle shown in Fig. (b). We start by plotting the points $\textcircled{x}$ with the coordinates $(\epsilon_x, -\gamma_{xy}/2) = (800, 300) \times 10^{-6}$ and $\textcircled{y}$ with the coordinates $(\epsilon_y, \gamma_{xy}/2) = (200, -300) \times 10^{-6}$. Joining these two points with a straight line gives us the diameter of the circle. If the circle were drawn to scale, all unknown values could be determined from it by direct measurements. However, we choose to solve the problem using trigonometry.

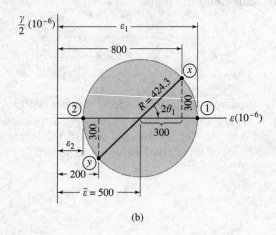

(b)

Referring to Fig. (b), we compute the following parameters of the circle:

$$\bar{\epsilon} = \left(\frac{800 + 200}{2}\right) \times 10^{-6} = 500 \times 10^{-6}$$

$$R = \left(\sqrt{(300)^2 + (300)^2}\right) \times 10^{-6} = 424.3 \times 10^{-6}$$

Part 1

Indicating the principal axes as $\textcircled{1}$ and $\textcircled{2}$ on the Mohr's circle in Fig. (b), we see that the principal strains are

$$\epsilon_1 = \bar{\epsilon} + R = (500 + 424.3) \times 10^{-6} = 924.3 \times 10^{-6} \qquad \textit{Answer}$$

$$\epsilon_2 = \bar{\epsilon} - R = (500 - 424.3) \times 10^{-6} = 75.7 \times 10^{-6} \qquad \textit{Answer}$$

We note in Fig. (b) that the angle measured from $\textcircled{x}$ to $\textcircled{1}$ on the Mohr's circle is $2\theta_1 = 45°$ in the clockwise direction. Therefore, the angle between the x-axis and the 1-axis is $\theta_1 = 22.5°$, also measured clockwise from the x-axis, as shown in Fig. (c). The principal direction corresponding to ϵ_2 is, of course, perpendicular to the direction of ϵ_1.

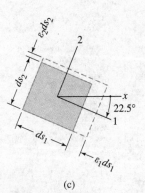

(c)

Part 2

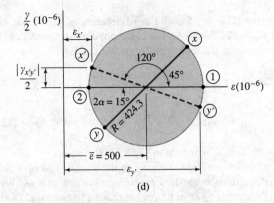

(d)

The Mohr's circle for the given state of strain is redrawn in Fig. (d). To determine strains relative to the $x'y'$-axes, we must identify the corresponding points $\widehat{x'}$ and $\widehat{y'}$ on the circle. Because the angle from the x-axis to the x'-axis is 60° in the counterclockwise direction, point $\widehat{x'}$ on the circle is 120° counterclockwise from $\widehat{x}$, as shown in Fig. (d). Of course, $\widehat{y'}$ is located at the opposite end of the diameter from $\widehat{x'}$. To facilitate our computations, we have introduced the central angle

$$2\alpha = 180° - 45° - 120° = 15°$$

between the points $\widehat{x'}$ and ②. Referring to the circle, we find that the normal strains in the x'- and y'-directions are

$$\epsilon_{x'} = \bar{\epsilon} - R\cos 2\alpha = (500 - 424.3\cos 15°) \times 10^{-6}$$
$$= 90.2 \times 10^{-6} \qquad \text{Answer}$$

$$\epsilon_{y'} = \bar{\epsilon} + R\cos 2\alpha = (500 + 424.3\cos 15°) \times 10^{-6}$$
$$= 910 \times 10^{-6} \qquad \text{Answer}$$

The magnitude of the shear strain is obtained from

$$\frac{|\gamma_{x'y'}|}{2} = R\sin 2\alpha = 424.3\sin 15° = 109.82 \times 10^{-6}$$

Noting that point $\widehat{y'}$ lies below the ϵ-axis, we conclude that $\gamma_{x'y'}$ is negative. Therefore,

$$\gamma_{x'y'} = -2(109.82) \times 10^{-6} = -220 \times 10^{-6} \qquad \text{Answer}$$

The positive values for the normal strains indicate elongations, whereas the negative sign for the shear strain means that the angle between the $x'y'$-axes is increased, as indicated in Fig. (e).

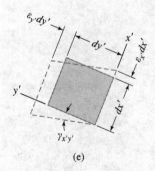

(e)

■

Sample Problem 8.13

The state of plane strain at a point is $\epsilon_x = 800 \times 10^{-6}$, $\epsilon_y = 200 \times 10^{-6}$, and $\gamma_{xy} = -600 \times 10^{-6}$. If the material properties are $E = 200$ GPa and $\nu = 0.30$, (1) use the Mohr's circle for strain obtained in the solution of Sample Problem 8.12 to construct Mohr's circle for stress; and (2) from the stress circle, determine the principal stresses and principal directions.

Solution

Part 1

In Sample Problem 8.12, we found the parameters of the strain circle to be $\bar{\epsilon} = 500 \times 10^{-6}$ and $R_\epsilon = 424.3 \times 10^{-6}$. The corresponding parameters of the stress circle are obtained by applying the scale transformations in Eqs. (8.18):

$$R_\sigma = R_\epsilon \frac{E}{1+\nu} = (424.3 \times 10^{-6}) \frac{200 \times 10^9}{1+0.3} = 65.27 \times 10^6 \text{ Pa} = 65.27 \text{ MPa}$$

$$\bar{\sigma} = \bar{\epsilon} \frac{E}{1-\nu} = (500 \times 10^{-6}) \frac{200 \times 10^9}{1-0.3} = 142.86 \times 10^6 \text{ Pa} = 142.86 \text{ MPa}$$

The resulting Mohr's circle is shown in Fig. (a). The points (x) and (y) on the circle, the coordinates of which are stresses acting on the x- and y-faces of the element, have the same angular positions as on the strain circle. We can now use the stress circle to find the stress components relative to any set of axes in the usual manner.

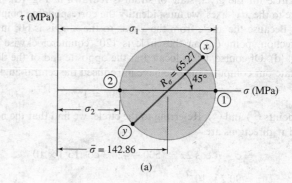

(a)

Part 2

From Fig. (a), we see that the principal stresses are

$$\sigma_1 = \bar{\sigma} + R_\sigma = 142.86 + 65.27 = 208 \text{ MPa} \qquad \text{Answer}$$

$$\sigma_2 = \bar{\sigma} + R_\sigma = 142.86 - 65.27 = 77.6 \text{ MPa} \qquad \text{Answer}$$

The principal directions, also found from the stress circle, are shown in Fig. (b).

Alternative Solution for Stresses

Instead of using the transformed circle for stress, we can calculate the stress components from the strains using generalized Hooke's law for a biaxial state of stress (see Art. 2.4):

$$\sigma_x = \frac{(\epsilon_x + \nu\epsilon_y)E}{1-\nu^2} \qquad \sigma_y = \frac{(\epsilon_y + \nu\epsilon_x)E}{1-\nu^2} \qquad \tau_{xy} = G\gamma_{xy} = \frac{E}{2(1+\nu)}\gamma_{xy}$$

When we substitute the principal strains $\epsilon_1 = 924.3 \times 10^{-6}$ and $\epsilon_2 = 75.7 \times 10^{-6}$ found in the solution of Sample Problem 8.12, the principal stresses become

$$\sigma_1 = \frac{[924.3 + 0.3(75.7)](10^{-6})(200 \times 10^9)}{1 - (0.3)^2} = 208 \text{ MPa}$$

$$\sigma_2 = \frac{[75.7 + 0.3(924.3)](10^{-6})(200 \times 10^9)}{1 - (0.3)^2} = 77.6 \text{ MPa}$$

which agree with the results in Part 2 above.

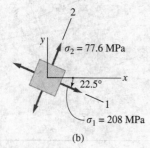

(b)

Problems

8.118 Prove that Eqs. (8.18) transform Mohr's circle for strain into Mohr's circle for stress.

8.119 Given that the state of strain at the point O is $\epsilon_x = \epsilon_2$, $\epsilon_y = \epsilon_1$, and $\gamma_{xy} = 0$, show that the angle of rotation of the line OA given by Eq. (f) of Art. 8.9 becomes $\beta = \gamma_{x'y'}/2$.

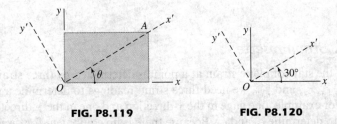

FIG. P8.119   FIG. P8.120

8.120 The state of strain at the point O is $\epsilon_x = -400 \times 10^{-6}$, $\epsilon_y = 200 \times 10^{-6}$, and $\gamma_{xy} = 800 \times 10^{-6}$. If $E = 200$ GPa and $\nu = 0.3$, determine (a) the principal stresses; and (b) the stress components acting on the x'-plane. Show your results on properly oriented sketches of an element.

8.121 The state of strain at a point is $\epsilon_x = 600 \times 10^{-6}$, $\epsilon_y = -300 \times 10^{-6}$, and $\gamma_{xy} = -400 \times 10^{-6}$. If $E = 30 \times 10^6$ psi and $\nu = 0.3$, determine the principal stresses at that point.

8.122 The state of strain at the point O is $\epsilon_x = -533 \times 10^{-6}$, $\epsilon_y = 67 \times 10^{-6}$, and $\gamma_{xy} = -626 \times 10^{-6}$. Find the stress components acting on the x'-plane, where the x'-axis is inclined at $\theta = 45°$ to the x-axis. Use $E = 30 \times 10^6$ psi and $\nu = 0.3$. Show your results on a properly oriented sketch of an element.

8.123 The state of strain at the point O is $\epsilon_x = -800 \times 10^{-6}$, $\epsilon_y = 200 \times 10^{-6}$, and $\gamma_{xy} = -800 \times 10^{-6}$. Determine the stress components acting on the x'-plane, where the x'-axis is inclined at $\theta = 20°$ to the x-axis. Use $E = 200$ GPa and $\nu = 0.3$. Show your results on a properly oriented sketch of an element.

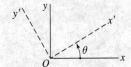

FIG. P8.122, P8.123

8.10 The Strain Rosette

a. Strain gages

The electrical-resistance strain gage is a device for measuring normal strain in a specific direction. Gages of this type operate on the principle that the change in electrical resistance of wires or foil strips is directly related to a change in their lengths. The gage is cemented to the object, so that the gage and the object undergo the same normal strain. The resulting change in the electrical resistance of the gage element is measured and converted into strain. Figure 8.22 shows a typical foil strain gage. Commercially available gages have gage lengths that vary from 0.008 in. to 4 in. A wide variety of other strain gages are available that depend upon electrical properties other than resistance, such as capacitance and inductance. However, the electrical-resistance gages are by far the most widely used because they are relatively inexpensive while at the same time very accurate and durable. Electrical-resistance strain gages are useful for measuring both static and dynamic strains.

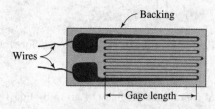

FIG. 8.22 Electrical-resistance strain gage (shown greatly enlarged).

b. Strain rosette

Because the state of plane strain at a point is determined by three strain components, ϵ_x, ϵ_y, and γ_{xy}, we need three strain readings to determine a state of strain. For example, one gage in the x-direction and one in the y-direction can be used to determine ϵ_x and ϵ_y. Because there is no equipment for direct measurement of shear strain, we must determine γ_{xy} indirectly. This can be done by using a third gage to measure the normal strain in a direction different from the x- or y-axis.

We now show how a state of plane strain can be determined from three normal strain measurements. The *strain rosette*, shown in Fig. 8.23, contains three strain gages oriented at angles θ_a, θ_b, and θ_c with respect to a reference line, such as the x-axis. We denote their strain readings by ϵ_a, ϵ_b, and ϵ_c. Substituting these strains and angles into Eq. (8.17a), we obtain the following set of simultaneous equations:

$$\epsilon_a = \frac{\epsilon_x + \epsilon_y}{2} + \frac{\epsilon_x - \epsilon_y}{2} \cos 2\theta_a + \frac{\gamma_{xy}}{2} \sin 2\theta_a \qquad \text{(a)}$$

$$\epsilon_b = \frac{\epsilon_x + \epsilon_y}{2} + \frac{\epsilon_x - \epsilon_y}{2} \cos 2\theta_b + \frac{\gamma_{xy}}{2} \sin 2\theta_b \qquad \text{(b)}$$

$$\epsilon_c = \frac{\epsilon_x + \epsilon_y}{2} + \frac{\epsilon_x - \epsilon_y}{2} \cos 2\theta_c + \frac{\gamma_{xy}}{2} \sin 2\theta_c \qquad \text{(c)}$$

Assuming that ϵ_a, ϵ_b, ϵ_c, θ_a, θ_b, and θ_c are known, Eqs. (a)–(c) represent three linear algebraic equations that can be solved for the three unknowns ϵ_x, ϵ_y, and γ_{xy}. After these unknowns have been found, we can construct Mohr's circles for strain and stress as explained in the preceding article.

As a matter of practical convenience, the normal strains are usually obtained by using one of the two strain rosettes described next.

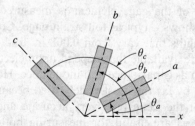

FIG. 8.23 Strain rosette with arbitrary orientation of gages.

c. The 45° strain rosette

The 45° strain rosette is shown in Fig. 8.24. The orientation of the strain gages are $\theta_a = 0$, $\theta_b = 45°$, and $\theta_c = 90°$. Substituting these angles into Eqs. (a)–(c) and solving, we obtain

$$\epsilon_x = \epsilon_a \qquad \epsilon_y = \epsilon_c \qquad \frac{\gamma_{xy}}{2} = \epsilon_b - \frac{\epsilon_a + \epsilon_c}{2} \qquad (8.19)$$

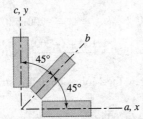

FIG. 8.24 45° strain rosette.

d. The 60° strain rosette

For the 60° rosette in Fig. 8.25, the strain gages are oriented at $\theta_a = 0$, $\theta_b = 60°$, and $\theta_c = 120°$. Using these values in Eqs. (a)–(c), we get for the solution

$$\epsilon_x = \epsilon_a \qquad \epsilon_y = \frac{2\epsilon_b + 2\epsilon_c - \epsilon_a}{3} \qquad \frac{\gamma_{xy}}{2} = \frac{\epsilon_b - \epsilon_c}{\sqrt{3}} \qquad (8.20)$$

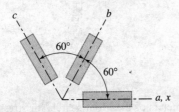

FIG. 8.25 60° strain rosette.

Sample Problem 8.14

The readings from a 45° strain rosette are $\epsilon_a = 100 \times 10^{-6}$, $\epsilon_b = 300 \times 10^{-6}$, and $\epsilon_c = -200 \times 10^{-6}$. If the material properties are $E = 180$ GPa and $\nu = 0.28$, determine the principal stresses and their directions. Show the results on a sketch of an element.

Solution

From Eqs. (8.19), we get

$$\epsilon_x = \epsilon_a = 100 \times 10^{-6} \qquad \epsilon_y = \epsilon_c = -200 \times 10^{-6}$$

$$\frac{\gamma_{xy}}{2} = \epsilon_b - \frac{\epsilon_a + \epsilon_c}{2} = \left[300 - \frac{100 + (-200)}{2}\right] \times 10^{-6} = 350 \times 10^{-6}$$

These results enable us to draw the Mohr's circle for strain shown in Fig. (a). The parameters of the circle are

$$\bar{\epsilon} = \frac{100 - 200}{2} \times 10^{-6} = -50 \times 10^{-6}$$

$$R_\epsilon = \sqrt{350^2 + 150^2} \times 10^{-6} = 380.8 \times 10^{-6}$$

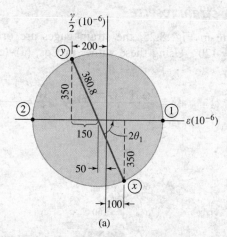

(a)

The corresponding parameters of the Mohr's circle for stress are obtained from Eqs. (8.18):

$$\bar{\sigma} = \frac{E}{1-\nu}\bar{\epsilon} = \frac{180 \times 10^9}{1 - 0.28}(-50 \times 10^{-6}) = -12.50 \times 10^6 \text{ Pa} = -12.50 \text{ MPa}$$

$$R_\sigma = \frac{E}{1+\nu}R_\epsilon = \frac{180 \times 10^9}{1 + 0.28}(380.8 \times 10^{-6}) = 53.55 \times 10^6 \text{ Pa} = 53.55 \text{ MPa}$$

Therefore, the principal stresses are

$$\sigma_1 = \bar{\sigma} + R_\sigma = -12.50 + 53.55 = 41.1 \text{ MPa} \qquad \text{Answer}$$

$$\sigma_2 = \bar{\sigma} - R_\sigma = -12.50 - 53.55 = -66.1 \text{ MPa} \qquad \text{Answer}$$

Because the principal directions for stress and strain coincide, we can find the former from the Mohr's circle for strain. The circle in Fig. (a) yields

$$2\theta_1 = \tan^{-1}\frac{350}{150} = 66.80° \qquad \theta_1 = 33.4° \qquad \text{Answer}$$

Figure (b) shows the principal stresses and the principal directions.

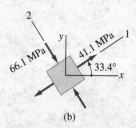

(b)

Problems

8.124 Prove Eqs. (8.19) for the 45° strain rosette.

8.125 Prove Eqs. (8.20) for the 60° strain rosette.

8.126 If gages a and c of a 45° strain rosette are aligned with the principal strain directions, what is the relationship among ϵ_a, ϵ_b, and ϵ_c?

8.127 A 60° strain rosette measures the following strains at a point on the aluminum skin of an airplane: $\epsilon_a = 100 \times 10^{-6}$, $\epsilon_b = -200 \times 10^{-6}$, and $\epsilon_c = 400 \times 10^{-6}$. Using $E = 10 \times 10^6$ psi and $\nu = 1/3$, determine the principal stresses and the maximum in-plane shear stress.

8.128 Solve Prob. 8.127 if the strain readings are $\epsilon_a = 300 \times 10^{-6}$, $\epsilon_b = -600 \times 10^{-6}$, and $\epsilon_c = -300 \times 10^{-6}$.

8.129 The strain readings from a 45° strain rosette are $\epsilon_a = 400 \times 10^{-6}$, $\epsilon_b = -200 \times 10^{-6}$, and $\epsilon_c = -100 \times 10^{-6}$. If $E = 200$ GPa and $\nu = 0.3$, find the principal stresses and their directions. Show the results on a sketch of a properly oriented element.

8.130 Solve Prob. 8.129 if the strain readings are $\epsilon_a = 300 \times 10^{-6}$, $\epsilon_b = 600 \times 10^{-6}$, and $\epsilon_c = 100 \times 10^{-6}$.

8.131 The strains measured with a 60° strain rosette are $\epsilon_a = 300 \times 10^{-6}$, $\epsilon_b = -400 \times 10^{-6}$, and $\epsilon_c = 100 \times 10^{-6}$. Using $E = 200$ GPa and $\nu = 0.3$, find the principal stresses and their directions. Show the results on a sketch of a properly oriented element.

8.11 Relationship Between Shear Modulus and Modulus of Elasticity

In Art. 2.4, we stated that the shear modulus G of a material is related to its modulus of elasticity E and Poisson's ratio ν by

$$G = \frac{E}{2(1+\nu)} \qquad \text{(2.14, repeated)}$$

We can now prove this relationship.

Consider the state of pure shear illustrated in Fig. 8.26(a). The Mohr's circle for this stress state in Fig. 8.26(b) shows that the principal stresses are

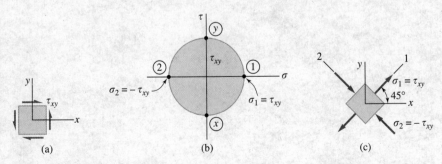

FIG. 8.26 (a) State of pure shear; (b) Mohr's circle for pure shear; (c) principal stresses associated with pure shear.

$\sigma_1 = \tau_{xy}$ and $\sigma_2 = -\tau_{xy}$. Furthermore, we note that the principal directions are inclined at 45° to the xy-axes, as indicated in Fig. 8.26(c). Our proof of Eq. (2.14) starts by deriving the principal strain ϵ_1 in terms of τ_{xy} by two different methods. Equating the two results will complete the proof.

Method 1 If we refer to Fig. 8.26(c), Hooke's law for biaxial stress, Eq. (2.10), yields

$$\epsilon_1 = \frac{1}{E}(\sigma_1 - v\sigma_2) = \frac{1}{E}[\tau_{xy} - v(-\tau_{xy})] = \frac{(1+v)\tau_{xy}}{E} \qquad \text{(a)}$$

Method 2 Because the principal directions for stress and strain are the same, the direction of ϵ_1 is inclined at the angle $\theta = 45°$ to the x-axis. We can now relate ϵ_1 to γ_{xy} by using the transformation equation for normal strain:

$$\epsilon_{x'} = \frac{\epsilon_x + \epsilon_y}{2} + \frac{\epsilon_x - \epsilon_y}{2}\cos 2\theta + \frac{\gamma_{xy}}{2}\sin 2\theta \qquad \text{(8.17a, repeated)}$$

Substituting $\epsilon_x = \epsilon_y = 0$ (this is a consequence of $\sigma_x = \sigma_y = 0$) and $2\theta = 90°$ into Eq. (8.17a), we get

$$\epsilon_1 = \frac{\gamma_{xy}}{2} = \frac{\tau_{xy}}{2G} \qquad \text{(b)}$$

where in the last step we used Hooke's law for shear: $\gamma_{xy} = \tau_{xy}/G$.

Proof Comparing Eqs. (a) and (b), we conclude that $(1+v)/E = 1/(2G)$, or

$$G = \frac{E}{2(1+v)}$$

which completes the proof. Although it is often convenient to analyze the deformation of an elastic body using three elastic constants (E, G, and v), we should realize that only two of the constants are independent.

Review Problems

8.132 The concrete post of radius r carries a concentrated load P that has an eccentricity e relative to the axis of the post. Find the largest e for which there is no tensile stress in the post.

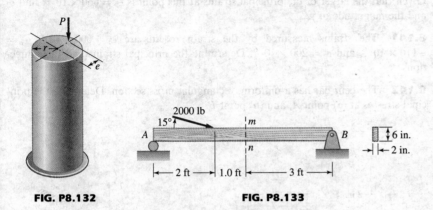

FIG. P8.132   **FIG. P8.133**

8.133 For the wooden beam shown in the figure, determine the maximum tensile and compressive normal stresses acting at section m-n.

8.134 The state of plane stress at a point is $\sigma_x = -15$ ksi, $\sigma_y = -5$ ksi, and τ_{xy}. If no tensile stress is permitted at this point, find the largest allowable magnitude of τ_{xy}.

8.135 The state of plane stress at a point is $\sigma_x = -180$ MPa, $\sigma_y = 120$ MPa, and $\tau_{xy} = -80$ MPa. Determine the principal stresses and the principal directions. Show the results on a sketch of an element oriented in the principal directions.

8.136 A state of plane stress is the result of two loadings. When acting separately, the loadings produce the stresses shown. Determine the state of stress with respect to the xy-axes when the loads act together. Show the results on a properly oriented element.

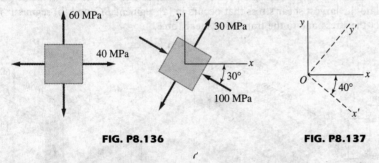

FIG. P8.136   **FIG. P8.137**

8.137 The state of plane stress at the point O is $\sigma_x = 0$, $\sigma_y = 12$ ksi, and $\tau_{xy} = -8$ ksi. Calculate the equivalent state of stress with respect to the $x'y'$-axes. Show your results on a sketch of a properly oriented element.

8.138 The state of plane stress at a point is σ_x, σ_y, and $\tau_{xy} = 30$ MPa. Knowing that the principal stresses are $\sigma_1 = 40$ MPa and $\sigma_2 = -80$ MPa, determine (a) the values of σ_x and σ_y; and (b) the principal directions relative to the xy-axes. Show both results on sketches of properly oriented elements.

8.139 The state of plane strain at a point is $\epsilon_x = 1600 \times 10^{-6}$, $\epsilon_y = -800 \times 10^{-6}$, and $\gamma_{xy} = 1000 \times 10^{-6}$. Determine the principal stresses and the principal directions at this point. Use $E = 10 \times 10^6$ psi and $\nu = 0.28$. Show the results on a sketch of a properly oriented element.

8.140 The normal strains at a point are $\epsilon_x = 540 \times 10^{-6}$ and $\epsilon_y = 180 \times 10^{-6}$. Given that the larger of the principal strains at this point is $\epsilon_1 = 660 \times 10^{-6}$, find ϵ_2 and the magnitude of γ_{xy}.

8.141 The strains measured by the strain rosette are $\epsilon_a = 600 \times 10^{-6}$, $\epsilon_b = -110 \times 10^{-6}$, and $\epsilon_c = 200 \times 10^{-6}$. Determine the principal strains and their directions.

8.142 The bent bar has a uniform, rectangular cross section. Determine the principal stresses at (a) point A; and (b) point B.

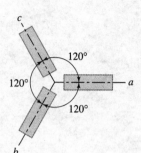

FIG. P8.141

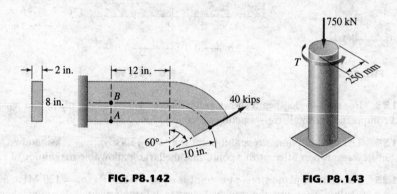

FIG. P8.142   FIG. P8.143

8.143 The 250-mm-diameter concrete pile is pushed into the soil by the 750-kN force while being rotated about its axis by the torque T. Find the largest allowable value of T for which the tensile stress in the pile does not exceed 1.5 MPa.

8.144 Determine the maximum shear stress in the thin-walled, square tube at (a) point A; and (b) point B.

8.145 The bent pipe has a 1.25-in. outer diameter and a 0.125-in. wall thickness. Calculate the largest shear stress that occurs in (a) segment BC; and (b) segment AB. Neglect the stress due to the transverse shear force.

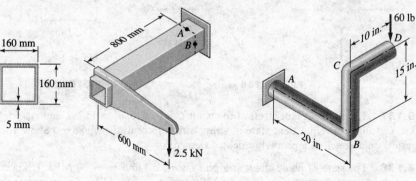

FIG. P8.144   FIG. P8.145

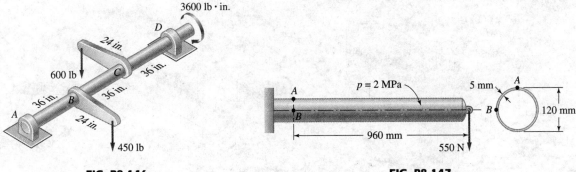

FIG. P8.146 **FIG. P8.147**

8.146 The shaft, supported by bearings at A and D, is loaded as shown in the figure. Determine the smallest allowable radius of the shaft if the normal stress is not to exceed 18 ksi. Neglect the stress due to the transverse shear force.

8.147 The plastic tube carries the 550-N load in addition to an internal pressure of 2 MPa. Find the principal stresses in the tube at (a) point A; and (b) point B.

Computer Problems

C8.1 The link shown in the figure carries a tensile force P with an offset d. Given P, d, and the dimensions of the cross section, write an algorithm that computes the maximum tensile and compressive stresses on section m-n. (a) Run the algorithm with the following data: $P = 4000$ lb, $d = 2$ in., $h = 3$ in., $t_w = 0.375$ in., $t = 0.5$ in., $a = 1.5$ in., and $b = 1.0$ in. (b) If the working normal stress is 10 000 psi, find by experimentation the optimal values of a and b (the values that minimize $a+b$), the other data being as in part (a).

C8.2 Given the stresses σ_x, σ_y, and τ_{xy}, construct an algorithm that plots the normal and shear stresses acting on the inclined plane m-n from $\theta = 0$ to $180°$, and compute the principal stresses and the corresponding values of θ. Run the algorithm with (a) $\sigma_x = 60$ MPa, $\sigma_y = -30$ MPa, and $\tau_{xy} = 80$ MPa; and (b) $\sigma_x = 80$ MPa, $\sigma_y = 80$ MPa, and $\tau_{xy} = -30$ MPa.

C8.3 Three strain gages are arranged as shown. Given the three angles (θ_a, θ_b, θ_c), the strain readings (ϵ_a, ϵ_b, ϵ_c), and the material constants (E, v), write an algorithm that computes the principal stresses and the principal directions (angles θ). Run the algorithm with $E = 200$ GPa, $v = 0.3$, and (a) $\theta_a = 0$, $\theta_b = 60°$, $\theta_c = 120°$, $\epsilon_a = 300 \times 10^{-6}$, $\epsilon_b = -400 \times 10^{-6}$, $\epsilon_c = 100 \times 10^{-6}$; and (b) $\theta_a = 30°$, $\theta_b = 75°$, $\theta_c = 120°$, $\epsilon_a = 100 \times 10^{-6}$, $\epsilon_b = 300 \times 10^{-6}$, $\epsilon_c = -200 \times 10^{-6}$.

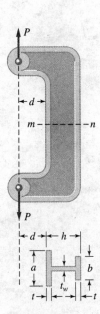

FIG. C8.1

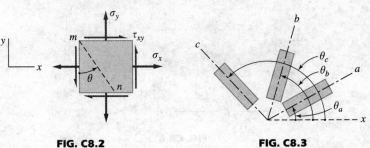

FIG. C8.2 **FIG. C8.3**

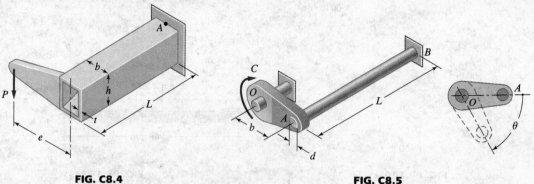

FIG. C8.4 **FIG. C8.5**

C8.4 The thin-walled rectangular tube of length L is loaded by the force P that has an eccentricity e. The circumference of the tube is $S = 2(h+b)$, where h and b are the mean dimensions of the cross section. The wall thickness is t. Given P, L, e, t, and S, devise an algorithm that plots the maximum normal stress in the tube against h in the range $h = 0.1S$ to $0.4S$. Experiment with the algorithm to determine the smallest possible S and the corresponding values of h and b if $P = 2000$ lb, $t = 0.125$ in., $\sigma_w = 10\,000$ psi, and (a) $L = e = 1.5$ ft; and (b) $L = 2$ ft, $e = 1.0$ ft. (*Hint*: The maximum normal stress occurs at point A.)

C8.5 The monel alloy bar AB of diameter d and length L is built into a fixed support at B and rigidly attached to the arm OA. When the couple C is applied to OA, the arm rotates about O though the angle θ as shown, causing bending and torsion of bar AB. Write an algorithm that plots the absolute maximum shear stress in bar AB as a function of θ from $\theta = 0$ to $180°$. Assume that the bar remains elastic and neglect deformation of the arm. Also determine the maximum allowable values of θ and C if the maximum shear stress is not to exceed the yield stress τ_{yp} of the alloy. Use the following data: $L = 60$ in., $b = 6$ in., $d = 0.25$ in., $E = 26 \times 10^6$ psi, $\nu = 0.28$, and $\tau_{yp} = 50 \times 10^3$ psi. Neglect the stress due to the transverse shear force.

C8.6 The steel, thin-walled pressure vessel of length L, mean diameter D, and wall thickness t is simply supported at each end. The vessel is filled with water under pressure p. There is additional loading due to the weights of the vessel and the water. Write an algorithm that computes the maximum in-plane shear stress τ_{max} and the absolute maximum shear stress τ_{abs} at any point in the vessel defined by the coordinates x and θ. Plot τ_{max} and τ_{abs} as functions of θ ($0 \leq \theta \leq 360°$) at $x = 0$ and $x = L$. Use the following data: $L = 10$ m, $D = 2$ m, $t = 10$ mm, and $p = 250 \times 10^6$ Pa. The mass densities are 7850 kg/m^3 for steel and 1000 kg/m^3 for water. (*Hint*: Model the vessel as a simply supported beam carrying a constant internal pressure p and a uniformly distributed load due to the weights of the vessel and the water.)

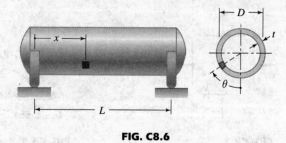

FIG. C8.6

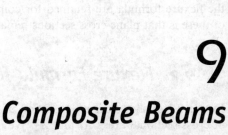

9
Composite Beams

Steel reinforcing rods for concrete columns protrude from a construction site. The reinforcement considerably increases the strength of beams and columns.

9.1 Introduction

The terms *composite beam* and *reinforced beam* are applied to beams that are made of two or more different materials. Usually, the material that forms the bulk of a composite beam is inexpensive but not sufficiently strong in bending to carry the loading by itself. The function of the reinforcement is to increase the flexural strength of the beam.

The flexure formula $\sigma = -My/I$ derived in Art. 5.2 does not apply to composite beams because it is based on the assumption that the beam is homogeneous. However, we can modify the formula by transforming the composite cross section into an equivalent homogeneous section that has the same bending stiffness. The other assumptions made in the derivation of

9.2 Flexure Formula for Composite Beams

Figure 9.1(a) shows the cross section of a beam composed of two materials of cross-sectional areas A_1 and A_2. We denote the corresponding moduli of elasticity by E_1 and E_2. The materials are assumed to be bonded together so that no slip occurs between them during bending. If we retain the assumption that plane cross sections remain plane, the normal strain is, as for homogeneous beams,

$$\epsilon = -\frac{y}{\rho}$$

where y is the distance above the neutral axis and ρ represents the radius of curvature of the beam, as illustrated in Fig. 9.1(b). It follows that the bending stresses in the two materials are

$$\sigma_1 = -\frac{E_1}{\rho} y \qquad \sigma_2 = -\frac{E_2}{\rho} y \qquad \text{(a)}$$

which result in the stress distribution shown in Fig. 9.1(c) (in drawing the figure, we assumed that $E_2 > E_1$).

To locate the neutral axis, we apply the equilibrium condition that the axial force acting on the cross section is zero:

$$\int_{A_1} \sigma_1 \, dA + \int_{A_2} \sigma_2 \, dA = 0$$

Substituting for the stresses from Eq. (a), we get

$$\int_{A_1} \frac{E_1}{\rho} y \, dA + \int_{A_2} \frac{E_2}{\rho} y \, dA = 0 \qquad \text{(b)}$$

When we cancel ρ and let

$$\boxed{\frac{E_2}{E_1} = n} \qquad (9.1)$$

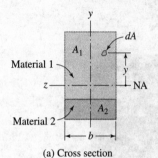

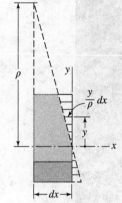

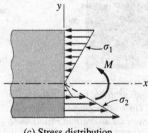

FIG. 9.1 Bending of a composite beam made of two materials.

(a) Cross section

(b) Strain distribution

(c) Stress distribution

Eq. (b) can be written as

$$\int_{A_1} y \, dA + \int_{A_2} y (n \, dA) = 0 \qquad \text{(c)}$$

Equation (c) shows that the area of material 2 is weighted by the factor n when we determine the location of the neutral axis. A convenient way to account for this weighting is to introduce the *equivalent*, or *transformed*, cross section shown in Fig. 9.2, which is made entirely of material 1. The weighting factor is taken into account by multiplying the width of area A_2 by n. Thus, *the neutral axis of the original cross section passes through the centroid C of the transformed cross-sectional area.*

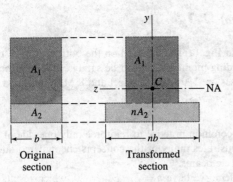

FIG. 9.2 Transforming the cross section to an equivalent section made entirely of material 1.

The equilibrium condition that the resultant moment of the stress distribution equals the bending moment M acting on the cross section is

$$\int_{A_1} \sigma_1 y \, dA + \int_{A_2} \sigma_2 y \, dA = -M$$

(the minus sign is due to the sign convention for M—positive M causes compression in the region $y > 0$). Substituting the stresses from Eq. (a), we obtain

$$\int_{A_1} \frac{E_1}{\rho} y^2 \, dA + \int_{A_2} \frac{E_2}{\rho} y^2 \, dA = M$$

which can be written in the form

$$\frac{1}{\rho} = \frac{M}{E_1 I} \qquad (9.2a)$$

where

$$I = \int_{A_1} y^2 \, dA + \int_{A_2} y^2 (n \, dA) \qquad (9.2b)$$

is the moment of inertia of the *transformed cross-sectional area* about the neutral axis.

Substituting $1/\rho$ from Eq. (9.2a) into Eqs. (a) yields the modified flexure formulas

$$\boxed{\sigma_1 = -\frac{My}{I} \qquad \sigma_2 = -n\frac{My}{I}} \qquad (9.3)$$

Sample Problem 9.1

The timber beam in Fig. (a) is reinforced on the bottom with a steel strip. Determine the maximum bending moment that can be safely carried by the beam if the allowable bending stresses are 120 MPa for steel and 8 MPa for wood. Use $E_{st}/E_{wd} = 20$.

Solution

Although it is not common practice to reinforce only one side of a timber beam, this sample problem illustrates many of the concepts encountered later in our discussion of reinforced concrete beams.

In this problem, we have $n = E_{st}/E_{wd} = 20$, which results in the transformed section shown in Fig. (b), which is made of wood. The transformed section consists of two rectangles with the areas $A_{wd} = 150(300) = 45 \times 10^3$ mm² and $nA_{st} = 20(75 \times 10) = 15 \times 10^3$ mm². The centroids of these areas are located at $\bar{y}_{wd} = 150 + 10 = 160$ mm and $\bar{y}_{st} = 10/2 = 5$ mm from the bottom of the section. The centroidal coordinate $\bar{y}$ of the entire transformed section is given by (the common factor 10^3 has been canceled)

$$\bar{y} = \frac{A_{wd} \bar{y}_{wd} + nA_{st} \bar{y}_{st}}{A_{wd} + nA_{st}} = \frac{45(160) + 15(5)}{45 + 15} = 121.25 \text{ mm}$$

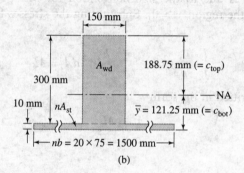

(b)

Recalling that the moment of inertia of a rectangle is $I = bh^3/12 + Ad^2$, we find that the moment of inertia of the transformed cross section about the neutral axis becomes

$$I = \left[\frac{150(300)^3}{12} + (45 \times 10^3)(160 - 121.25)^2\right]$$

$$+ \left[\frac{1500(10)^3}{12} + (15 \times 10^3)(121.25 - 5)^2\right]$$

$$= 607.9 \times 10^6 \text{ mm}^4 = 607.9 \times 10^{-6} \text{ m}^4$$

From Fig. (b), the distances from the neutral axis to the top and bottom of the cross section are

$$c_{top} = (300 + 10) - 121.25 = 188.75 \text{ mm}$$

$$c_{bot} = 121.25 \text{ mm}$$

The largest bending stress in each material is obtained from the modified flexure formula

$$\sigma_{wd} = \frac{Mc_{top}}{I} \quad \text{(a)}$$

$$\sigma_{st} = n\frac{Mc_{bot}}{I} \quad \text{(b)}$$

Replacing σ_{wd} in Eq. (a) with the working stress for wood and solving for the bending moment, we get

$$M = \frac{\sigma_{wd} I}{c_{top}} = \frac{(8 \times 10^6)(607.9 \times 10^{-6})}{188.75 \times 10^{-3}} = 25.8 \times 10^3 \text{ N} \cdot \text{m}$$

Similarly, replacing σ_{st} in Eq. (b) with the working stress for steel gives

$$M = \frac{\sigma_{st} I}{nc_{bot}} = \frac{(120 \times 10^6)(607.9 \times 10^{-6})}{20(121.25 \times 10^{-3})} = 30.1 \times 10^3 \text{ N} \cdot \text{m}$$

The smaller of the two values is the maximum bending moment that can be safely carried by the beam. Thus,

$$M_{max} = 25.8 \text{ kN} \cdot \text{m} \quad \textit{Answer}$$

as determined by the stress in wood. In this case, the beam is be said to be over-reinforced because there is an excess of steel.

■

Problems

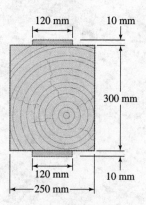

FIG. P9.1–P9.3

9.1 The timber beam is reinforced with steel plates rigidly attached at the top and bottom. The allowable stresses are 8 MPa for wood and 120 MPa for steel, and the ratio of the elastic moduli is $E_{st}/E_{wd} = 15$. Determine the increase in the allowable bending moment due to the reinforcement.

9.2 Solve Prob. 9.1 assuming that the reinforcing plates are made of aluminum for which the working stress is 80 MPa. Use $E_{al}/E_{wd} = 5$.

9.3 A simply supported wood beam, reinforced with steel plates, has the cross section shown. The beam carries a uniformly distributed load of 20 kN/m over the middle half of its 4-m span. If $E_{st}/E_{wd} = 15$, determine the largest bending stresses in the wood and the steel.

9.4 The timber beam is reinforced at the bottom by a steel plate of width $b = 4$ in. If $E_{st}/E_{wd} = 20$, determine the largest vertical concentrated load that can be applied at the center of a 18-ft simply supported span. The working stresses are 1.2 ksi for wood and 18 ksi for steel.

9.5 Determine the width b of the steel plate fastened to the bottom of the timber beam so that the working stresses 1.2 ksi for wood and 18 ksi for steel are reached simultaneously. Use $E_{st}/E_{wd} = 20$.

9.6 A simply supported wood beam is reinforced with a steel plate of width b as shown in the figure. The beam carries a uniformly distributed load (including the weight of the beam) of intensity 300 lb/ft over its 20-ft span. Using $b = 4$ in. and $E_{st}/E_{wd} = 20$, determine the maximum bending stresses in the wood and the steel.

9.7 The timber beam is reinforced by steel plates of width b at the top and bottom. Determine the smallest value of b necessary to resist a 40-kN·m bending moment. Assume that $E_{st}/E_{wd} = 15$ and that the allowable bending stresses are 10 MPa for wood and 120 MPa for steel.

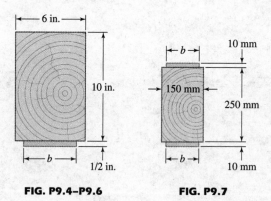

FIG. P9.4–P9.6 **FIG. P9.7**

9.8 A timber beam, 150 mm wide and 200 mm deep, is reinforced at the top and bottom with 6-mm-thick aluminum plates of width b. The maximum bending moment in the beam is 14 kN·m. If the working stresses in bending are 10 MPa for wood and 80 MPa for aluminum, determine the smallest allowable value of b. Use $E_{al}/E_{wd} = 5$.

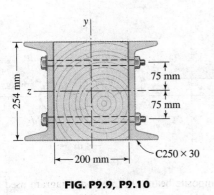
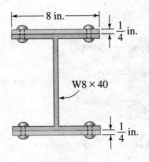

FIG. P9.9, P9.10 **FIG. P9.11**

9.9 A pair of C250 × 30 steel channels are secured to the wood beam (the depth of the wood is the same as the 254-mm depth of a channel) by two rows of bolts. If bending occurs about the z-axis, determine the largest allowable bending moment. Use $E_{st}/E_{wd} = 20$ and the working bending stresses of 8 MPa for wood and 120 MPa for steel.

9.10 For the beam in Prob. 9.9, find the largest allowable bending moment if the bending occurs about the y-axis.

9.11 The aluminum beam with the dimensions of a W8 × 40 section is reinforced by bolting steel plates to its flanges. The allowable bending stresses are 18 ksi for steel and 15 ksi for aluminum. Using $E_{st}/E_{al} = 3$, determine (a) the percentage increase in the allowable bending moment due to the reinforcement; and (b) the percentage increase in the flexural rigidity EI.

9.12 A rectangular beam, 150 mm wide and 250 mm deep, carries a bending moment of 140 kN·m. If the compressive modulus of elasticity is 1.5 times the tensile modulus, determine the maximum tensile and compressive stresses in the beam.

9.13 The experimental beam is composed of three materials firmly fastened together. Using the elastic moduli and working bending stresses given below, find the largest allowable bending moment that the beam can carry.

Material	E (ksi)	σ_w (ksi)
Wood	1.5×10^3	1.5
Aluminum	10×10^3	12
Steel	30×10^3	18

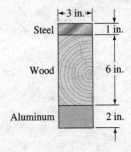

FIG. P9.13

9.3 Shear Stress and Deflection in Composite Beams

Article 9.2 discussed the bending stresses in composite beams. Here we consider the shear stresses and deflections in beams of this type.

a. Shear stress

The equation for shear stress

$$\tau = \frac{VQ}{Ib} \qquad \text{(5.8, repeated)}$$

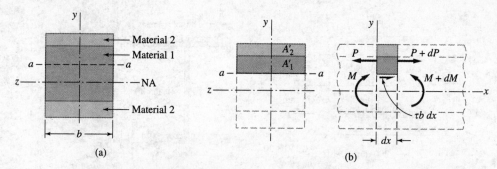

FIG. 9.3 (a) Cross section of a composite beam; (b) free-body diagram used to determine the longitudinal shear stress.

derived in Art. 5.4 also applies to composite beams, provided that Q and I are calculated using the *transformed* section and b is the width of the *original* cross section.

To arrive at this equation, consider a composite beam with the cross section shown in Fig. 9.3(a). The shear stress τ acting on the horizontal section a-a can be obtained from equilibrium by using the free-body diagram in Fig. 9.3(b). The free body is bounded by two cross sections a distance dx apart and the section a-a. The equilibrium equation $\Sigma F_x = 0$ yields $\tau b\, dx + dP = 0$, or

$$\tau = -\frac{1}{b}\frac{dP}{dx} \qquad\text{(a)}$$

where P is the normal force acting on the cross-sectional area above a-a (the area below a-a could also be used). From the expressions for the bending stresses in Eqs. (9.3), this force is

$$P = \int_{A'_1} \sigma_1\, dA + \int_{A'_2} \sigma_2\, dA = -\int_{A'_1} \frac{My}{I}\, dA - \int_{A'_2} n\frac{My}{I}\, dA$$

where the areas A'_1 and A'_2 are identified in Fig. 9.3(b), I is the moment of inertia of the *transformed* cross section, and $n = E_2/E_1$. When we rewrite this equation as

$$P = -\frac{M}{I}\left[\int_{A'_1} y\, dA + \int_{A'_2} y(n\, dA)\right] = -\frac{M}{I}Q \qquad\text{(b)}$$

we see that Q is the first moment of the *transformed* cross-sectional area above a-a, taken about the neutral axis. Therefore, $dP/dx = -(dM/dx)(Q/I) = -VQ/I$, which, upon substitution in Eq. (a), yields Eq. (5.8).

b. Deflection

According to Eq. (9.2a), the moment-curvature relationship for a composite beam is $1/\rho = M/(E_1 I)$. Because $1/\rho = d^2v/dx^2$, the differential equation for the deflection v becomes

$$\frac{d^2v}{dx^2} = \frac{M}{E_1 I} \qquad (9.4)$$

Therefore, the deflections of composite beams can be computed by the methods used for homogeneous beams, provided we use the flexural rigidity $E_1 I$ of the *transformed* cross section.

Sample Problem 9.2

The cross section of a simply supported beam in Fig. (a) has a wood core and aluminum face plates. The beam is 72 in. long and carries a 6000-lb concentrated load 24 in. from one of the supports. Determine (1) the maximum vertical shear stress in the beam and (2) the displacement at midspan. Use $E_{wd} = 1.5 \times 10^6$ psi and $E_{al} = 10 \times 10^6$ psi.

Solution

The transformed cross section, consisting of wood, is shown in Fig. (b). Note that the widths of the rectangles originally occupied by the aluminum plates are increased by the factor $E_{al}/E_{wd} = 20/3$.[1] The neutral axis is located by symmetry. The moment of inertia of the transformed section about the neutral axis is

$$I = \sum \left(\frac{bh^3}{12} + Ad^2 \right) = \frac{2(8)^3}{12} + 2\left[\frac{(40/3)(0.4)^3}{12} + \left(\frac{40}{3} \times 0.4 \right)(4.2)^2 \right] = 273.6 \text{ in.}^4$$

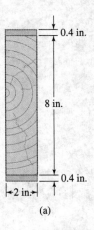

(a)

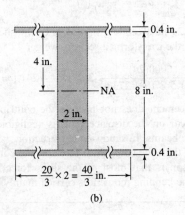

(b)

Part 1

Because the maximum vertical shear stress occurs at the neutral axis, we need the first moment Q of the upper (or lower) half of the transformed section about the neutral axis. Referring to Fig. (b), we have

$$Q = \sum A'\bar{y}' = (2 \times 4)(2) + \left(\frac{40}{3} \times 0.4 \right)(4.2) = 38.40 \text{ in.}^3$$

From the shear force diagram in Fig. (c), we see that the largest magnitude of the shear force is $V_{max} = 4000$ lb. Therefore, the maximum vertical shear stress in the beam becomes

$$\tau_{max} = \frac{V_{max} Q}{Ib} = \frac{4000(38.40)}{273.6(2)} = 281 \text{ psi} \qquad \text{Answer}$$

---

[1] We could also use a transformed cross section consisting of aluminum. In that case, the face plates would retain their original widths, but the original width of the core would be multiplied by the factor $E_{wd}/E_{al} = 3/20$.

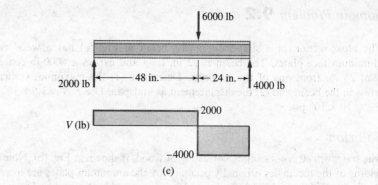

Part 2

The formula given in Table 6.3 for the midspan displacement of a simply supported beam carrying a concentrated load is

$$\delta_{\text{mid}} = \frac{Pb}{48EI}(3L^2 - 4b^2)$$

where, in our case, $b = 24$ in. Recalling that for a composite beam we must use the flexural rigidity $E_{\text{wd}}I$ of the transformed section, we get

$$\delta_{\text{mid}} = \frac{6000(24)}{48(1.5 \times 10^6)(273.6)}[3(72)^2 - 4(24)^2] = 0.0968 \text{ in.} \qquad \textit{Answer}$$

The above displacement does not include the contribution of vertical shear. Usually the effect of shear on the displacements is negligible, but this is not necessarily true for composite beams. In sandwich construction, where the faces are made of much stiffer material than the core, shear stress in the core may cause significant displacements. The reason is that most of the shear force is resisted by the core, which is soft, whereas the bending moment is carried mainly by the stiffer faces.

Problems

9.14 Compute the largest allowable shear force in a beam with a cross section as described in Prob. 9.4. Use $b = 4$ in., $E_{st}/E_{wd} = 20$, and a working shear stress of 120 psi for wood.

9.15 For the beam in Prob. 9.9, assume that the channels are fastened to the wood by 20-mm bolts spaced 300 mm apart along the beam. Assuming $E_{st}/E_{wd} = 20$, determine the average shear stress in the bolts caused by a 40-kN shear force acting along the y-axis.

9.16 Solve Prob. 9.15 if the 40-kN shear force acts along the z-axis.

9.17 The beam in Prob. 9.1 carries a uniformly distributed load of 40 kN/m on its entire simply supported span of length 4 m. If $E_{st} = 200$ GPa and $E_{wd} = 10$ GPa, compute the midspan deflection.

9.18 For the beam described in Prob. 9.13, determine the shear stress developed between wood and steel and between wood and aluminum. Express the results in terms of the shear force V.

9.19 A cantilever beam with the cross section described in Prob. 9.13 is 10 ft long and carries a 25-kip concentrated load at its free end. Calculate the displacement of the beam under the load.

9.20 The beam described in Prob. 9.4 carries a 2500-lb concentrated force at the middle of its 18-ft simply supported span. Determine the midspan deflection of the beam using $E = 29 \times 10^6$ psi for steel.

9.4 Reinforced Concrete Beams

Concrete is a popular building material because it is relatively inexpensive. Although concrete has approximately the same compressive strength as soft wood, its tensile strength is practically zero. For this reason, concrete beams are reinforced with longitudinal steel bars embedded in the tensile side of the beam. Fortunately, there is a natural bond between concrete and steel, so that no slipping occurs between them during bending. This allows us to apply the principles developed in the preceding article.[2]

It is usually assumed that concrete carries no tensile stress whatsoever. The tensile side of the concrete thus serves merely to position the steel that carries the entire tensile load. If there is only one row of steel rods, as shown in Fig. 9.4(a), the steel can be assumed to be uniformly stressed (the diameters of the rods are small compared to the depth of the cross section). Consequently, the transformed cross section of the beam is as shown in Fig. 9.4(b). The shaded portions indicate areas that are effective in resisting bending. The ratio $n = E_{st}/E_{co}$ is usually between 6 and 10, depending upon the quality of the concrete.

---

[2] Sufficient bond is developed in long beams to permit the steel bars to be laid straight. However, in short beams, the ends are usually bent over to anchor the steel more securely in the concrete.

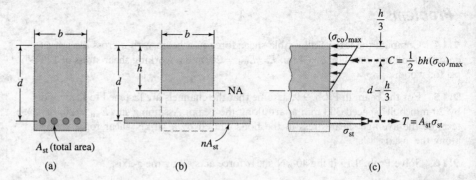

FIG. 9.4 (a) Cross section of a reinforced concrete beam; (b) equivalent concrete cross section; (c) normal stresses acting on the cross section.

a. Analysis

As shown in Fig. 9.4, we denote the distance between the reinforcement and the top of the beam by d and the depth of the concrete in the compressive zone by h. Because h also locates the neutral axis of the cross section, it can be found from the condition that the first moment of the transformed cross section about the neutral axis is zero. This yields

$$(bh)\frac{h}{2} - nA_{st}(d - h) = 0$$

which can be written as

$$\left(\frac{h}{d}\right)^2 + \frac{2nA_{st}}{bd}\frac{h}{d} - \frac{2nA_{st}}{bd} = 0 \qquad (9.5)$$

where b is the width of the original beam and A_{st} denotes the total area of the reinforcement.

After solving this quadratic equation for h/d, we could compute the moment of inertia I of the transformed section and then find the maximum compressive stress in concrete and the tensile stress in steel from

$$(\sigma_{co})_{max} = \frac{Mh}{I} \qquad \sigma_{st} = n\frac{M(d - h)}{I} \qquad (9.6)$$

However, for the rectangular cross section in Fig. 9.4(a), it is often easier to obtain the stresses from the formulas derived below.

From the stress distribution on the cross section in Fig. 9.4(c), we see that the resultant force of the compressive stress in the concrete is

$$C = \frac{1}{2}bh(\sigma_{co})_{max}$$

which acts at the centroid of the stress diagram—that is, at the distance $h/3$ from the top of the section. The tensile force

$$T = A_{st}\sigma_{st}$$

carried by the steel has the same magnitude as C, so that the two forces form a couple with the moment arm $d - h/3$. Because this couple is equal to the bending moment M acting on the cross section, we have the following useful relationships between the bending moment and the stresses:

$$M = \frac{1}{2}bh\left(d - \frac{h}{3}\right)(\sigma_{co})_{max} = \left(d - \frac{h}{3}\right)A_{st}\sigma_{st} \qquad (9.7)$$

Either Eqs. (9.6) or (9.7) can be used for the computation of stresses.

b. Balanced-stress reinforcement

For maximum economy, the stresses in concrete and steel should reach their allowable limits simultaneously, a condition known as *balanced-stress reinforcement*. From Eqs. (9.6), we obtain

$$\frac{\sigma_{st}}{(\sigma_{co})_{max}} = \frac{n(d-h)}{h} = n\frac{1 - h/d}{h/d} \qquad (a)$$

The ratio h/d for balanced-stress reinforcement can be found from Eq. (a) by replacing the stresses by their working stresses $(\sigma_{st})_w$ and $(\sigma_{co})_w$. The result is

$$\frac{h}{d} = \frac{n}{n + [(\sigma_{st})_w/(\sigma_{co})_w]} \qquad (9.8)$$

The common practice is to use values of h/d slightly larger than that given by Eq. (9.8), so that the steel will reach its working stress before the concrete. This so-called under-reinforcement is desirable because failure of steel, caused by its ductility, is not as catastrophic as the brittle fracture of concrete.

Sample Problem 9.3

A reinforced concrete beam with the cross section shown in Fig. 9.4(a) has the properties $b = 300$ mm, $d = 500$ mm, $A_{st} = 1500$ mm², and $n = E_{st}/E_{co} = 8$. Determine the maximum stress in the concrete and the stress in the steel produced by a bending moment of 70 kN·m.

Solution

We must first determine the distance h of the neutral axis from the top of the beam using Eq. (9.5):

$$\left(\frac{h}{d}\right)^2 + \frac{2nA_{st}}{bd}\frac{h}{d} - \frac{2nA_{st}}{bd} = 0 \qquad (a)$$

For the given beam, we have

$$\frac{2nA_{st}}{bd} = \frac{2(8)(1500)}{300(500)} = 0.160$$

so that Eq. (a) becomes

$$\left(\frac{h}{500}\right)^2 + 0.160\left(\frac{h}{500}\right) - 0.160 = 0$$

The positive root of this equation is $h = 163.96$ mm $= 0.163\,96$ m.

Using the first expression in Eq. (9.7), we have

$$M = \frac{1}{2}bh\left(d - \frac{h}{3}\right)(\sigma_{co})_{max}$$

$$70 \times 10^3 = \frac{1}{2}(0.3)(0.163\,96)\left(0.5 - \frac{0.163\,96}{3}\right)(\sigma_{co})_{max}$$

which yields for the maximum compressive stress in the concrete

$$(\sigma_{co})_{max} = 6.39 \times 10^6 \text{ Pa} = 6.39 \text{ MPa} \qquad \text{Answer}$$

The second expression in Eq. (9.7) gives

$$M = \left(d - \frac{h}{3}\right)A_{st}\sigma_{st}$$

$$70 \times 10^3 = \left(0.5 - \frac{0.163\,96}{3}\right)(1500 \times 10^{-6})\sigma_{st}$$

Solving, we get for the tensile stress in the steel reinforcement

$$\sigma_{st} = 104.8 \times 10^6 \text{ Pa} = 104.8 \text{ MPa} \qquad \text{Answer}$$

Sample Problem 9.4

Design a rectangular concrete beam with balanced-stress reinforcement to carry a bending moment of 90 kN·m. Use $(\sigma_{co})_w = 12$ MPa, $(\sigma_{st})_w = 140$ MPa, $n = E_{st}/E_{co} = 8$, and $d = 1.5b$.

Solution

In balanced-stress reinforcement, the steel and concrete reach their working stresses simultaneously. The location of the neutral axis for this design is specified in Eq. (9.8):

$$\frac{h}{d} = \frac{n}{n + [(\sigma_{st})_w/(\sigma_{co})_w]} = \frac{8}{8 + (140/12)} = 0.4068$$

or

$$h = 0.4068 d$$

According to Eq. (9.7), the maximum stress in concrete equals its working value when the bending moment is

$$M = \frac{1}{2} bh(\sigma_{co})_w \left(d - \frac{h}{3}\right)$$

$$90 \times 10^3 = \frac{1}{2} b(0.4068d)(12 \times 10^6)\left(d - \frac{0.4068d}{3}\right)$$

from which

$$bd^2 = 0.04266 \text{ m}^3$$

With $d = 1.5b$, we have

$$d = 0.400 \text{ m} = 400 \text{ mm} \quad \textit{Answer}$$

$$b = 0.267 \text{ m} = 267 \text{ mm} \quad \textit{Answer}$$

Therefore,

$$h = 0.4068 d = 0.4068(0.400) = 0.16272 \text{ m}$$

Applying Eq. (9.7) to the steel reinforcement, we get

$$M = \left(d - \frac{h}{3}\right) A_{st}(\sigma_{st})_w$$

$$90 \times 10^3 = \left(0.400 - \frac{0.16272}{3}\right) A_{st}(140 \times 10^6)$$

which gives for the required area of steel

$$A_{st} = 1.859 \times 10^{-3} \text{ m}^2 = 1859 \text{ mm}^2 \quad \textit{Answer}$$

Sample Problem **9.5**

Figure (a) shows a reinforced concrete T-beam, where the cross-sectional area of the steel reinforcement is 2400 mm^2. Using $n = E_{st}/E_{co} = 8$ and the working stresses of 12 MPa for concrete and 140 MPa for steel, determine the largest bending moment that the beam can carry.

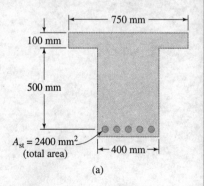

(a)

Solution

We first locate the neutral axis of the transformed cross section shown in Fig. (b). Because the cross section is not rectangular, we cannot use Eq. (9.5). Therefore, we find the neutral axis from the condition that the first moment of the transformed cross section about the neutral axis is zero. Letting h be the distance between the neutral axis and the top of the beam, we see that the transformed section consists of three rectangles. The areas and the centroidal coordinates of these rectangles are

$$A_1 = 750 \times 100 = 75 \times 10^3 \text{ mm}^2 \qquad \bar{y}_1 = (h - 50) \text{ mm}$$

$$A_2 = 400(h - 100) \text{ mm}^2 \qquad \bar{y}_2 = (h - 100)/2 \text{ mm}$$

$$A_3 = 19.20 \times 10^3 \text{ mm}^2 \qquad \bar{y}_3 = -(600 - h) \text{ mm}$$

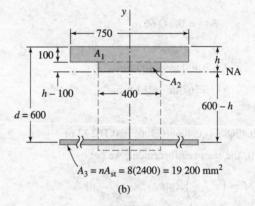

(b)

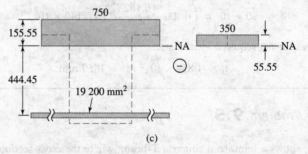

(c)

Setting the first moment $\sum A_i \bar{y}_i$ of the transformed section about the neutral axis to zero yields

$$(75 \times 10^3)(h - 50) + 400\frac{(h-100)^2}{2} - (19.20 \times 10^3)(600 - h) = 0$$

which has the positive root

$$h = 155.55 \text{ mm}$$

It is convenient to compute the moment of inertia by considering the area above the neutral axis to be the difference of the two rectangles shown in Fig. (c). This yields

$$I = \frac{750(155.55)^3}{3} - \frac{350(55.55)^3}{3} + (19.20 \times 10^3)(444.45)^2$$
$$= 4714 \times 10^6 \text{ mm}^4 = 4714 \times 10^{-6} \text{ m}^4$$

From Eq. (9.6), the largest bending moment that can be supported without exceeding the working stress in concrete is

$$M = \frac{(\sigma_{co})_w I}{h} = \frac{(12 \times 10^6)(4714 \times 10^{-6})}{155.55 \times 10^{-3}} = 364 \times 10^3 \text{ N} \cdot \text{m}$$

The largest safe bending moment determined by the working stress in steel is

$$M = \frac{(\sigma_{st})_w I}{n(d - h)} = \frac{(140 \times 10^6)(4714 \times 10^{-6})}{8(444.45 \times 10^{-3})} = 185.6 \times 10^3 \text{ N} \cdot \text{m}$$

The maximum allowable bending moment is the smaller of the preceding two values; namely,

$$M_{\max} = 185.6 \text{ kN} \cdot \text{m} \qquad \text{Answer}$$

Because the stress in the steel is the limiting condition, the beam is under-reinforced.

∎

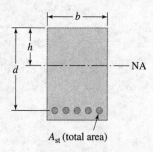

A_{st} (total area)

FIG. P9.21–P9.32

Problems

9.21 The properties of the reinforced concrete beam are $b = 250$ mm, $d = 450$ mm, and $A_{st} = 1500$ mm^2. Compute h using (a) $E_{st}/E_{co} = 6$; and (b) $E_{st}/E_{co} = 10$.

9.22 The reinforced concrete beam has the properties $b = 10$ in., $d = 16$ in., and $E_{st}/E_{co} = 8$. Determine h and A_{st} for balanced-stress reinforcement. The working stresses are 1.6 ksi for concrete and 24 ksi for steel.

9.23 Solve Prob. 9.22 if $d = 18$ in., with all other data remaining unchanged.

9.24 For the reinforced concrete beam, determine the maximum stresses in the steel and the concrete caused by a 70-kN·m bending moment. Use $b = 300$ mm, $d = 500$ mm, $A_{st} = 1200$ mm^2, and $E_{st}/E_{co} = 8$.

9.25 For the reinforced concrete beam, $b = 500$ mm, $d = 750$ mm, $A_{st} = 6000$ mm^2, and $E_{st}/E_{co} = 10$. Find the maximum stresses in the steel and the concrete caused by a 270-kN·m bending moment.

9.26 The properties of the reinforced concrete beam are $b = 300$ mm, $d = 450$ mm, $A_{st} = 1400$ mm^2, and $E_{st}/E_{co} = 8$. If the working stresses are 12 MPa for concrete and 140 MPa for steel, determine the largest bending moment that may be safely applied.

9.27 For the reinforced concrete beam, $b = 10$ in., $d = 18$ in., $A_{st} = 2$ in.^{2}, and $E_{st}/E_{co} = 10$. There is 2 in. of concrete below the reinforcing rods. In addition to its own weight, the beam carries a uniformly distributed load of intensity w_0 on a simply supported span 12 ft long. Determine the largest allowable value of w_0 if the working stresses are 1.8 ksi for concrete and 20 ksi for steel. The concrete weighs 150 lb/ft^3 (the weight of steel may be neglected).

9.28 The reinforced concrete beam has the properties $b = 12$ in., $d = 18$ in., and $E_{st}/E_{co} = 8$. When a 80-kip·ft bending moment is applied, the maximum compressive stress in the concrete is 1400 psi. Determine the stress in the steel and the cross-sectional area of the steel reinforcement.

9.29 For the reinforced concrete beam, $d = 18$ in. and $E_{st}/E_{co} = 10$. The allowable stresses are 1.6 ksi for concrete and 20 ksi for steel. Find b and A_{st} so that the beam can carry a maximum bending moment of 60 kip·ft with balanced-stress reinforcement.

9.30 For the reinforced concrete beam, $b = 200$ mm, $d = 400$ mm, and $E_{st}/E_{co} = 10$. The allowable stresses are 10 MPa for concrete and 140 MPa for steel. Determine A_{st} for balanced-stress design and the largest safe bending moment that the beam can carry.

9.31 Design the beam for balanced-stress reinforcement to carry a bending moment of 140 kN·m. Use $d = 1.5b$, $E_{st}/E_{co} = 8$, and the allowable stresses of 12 MPa for concrete and 160 MPa for steel.

9.32 Solve Prob. 9.31 if $d = 4b/3$.

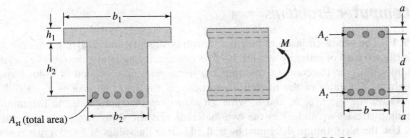

FIG. P9.33–P9.35   **FIG. P9.36–P9.39**

9.33 The properties of the reinforced concrete T-beam are $b_1 = 750$ mm, $h_1 = 100$ mm, $b_2 = 300$ mm, $h_2 = 450$ mm, $A_{st} = 3000$ mm^2, and $E_{st}/E_{co} = 10$. Find the maximum stresses in the concrete and the steel produced by a 120-kN·m bending moment.

9.34 For the reinforced concrete T-beam, $b_1 = 500$ mm, $h_1 = 150$ mm, $b_2 = 250$ mm, $h_2 = 500$ mm, $A_{st} = 3300$ mm^2, and $E_{st}/E_{co} = 8$. Determine the maximum bending moment that can be applied so that the stresses do not exceed 12 MPa in concrete and 140 MPa in steel.

9.35 The properties of the reinforced concrete T-beam are $b_1 = 36$ in., $h_1 = 3$ in., $b_2 = 12$ in., $h_2 = 21$ in., and $E_{st}/E_{co} = 10$. If the working stresses are 1.6 ksi for concrete and 24 ksi for steel, find A_{st} and the maximum allowable bending moment for balanced-stress reinforcement.

9.36 The dimensions of the reinforced concrete beam are $b = 300$ mm, $d = 500$ mm, and $a = 75$ mm. The total cross-sectional areas of the steel reinforcing rods are $A_t = 1200$ mm^2 on the tension side and $A_c = 400$ mm^2 on the compression side. If the allowable stresses are 12 MPa for concrete and 140 MPa for steel, determine the maximum bending moment M that the beam can carry. Use $E_{st}/E_{co} = 8$.

9.37 Solve Prob. 9.36 if the sense of the bending moment is reversed (the direction of M is opposite to that shown in the figure).

9.38 The reinforced concrete beam with dimensions $b = 12$ in., $d = 18$ in., and $a = 3$ in. carries a bending moment $M = 80$ kip·ft as shown in the figure. The cross-sectional areas of the steel reinforcement are $A_t = 4$ in.^{2} on the tension side and $A_c = 2$ in.^{2} on the compression side. Using $E_{st}/E_{co} = 8$, calculate the maximum compressive stress in the concrete and the stresses in the reinforcing rods.

9.39 Solve Prob. 9.38 if $A_c = 4$ in.^{2}, with all other data remaining unchanged.

Computer Problems

C9.1 The sandwich beam of length L is simply supported and carries a uniformly distributed load of intensity w_0. The beam is constructed by gluing the faces ②, each of thickness t, to the core ①, resulting in a rectangular cross section of width b and height h. The ratio of the moduli of elasticity of the two materials is $n = E_2/E_1$. (a) Given L, w_0, t, b, h, and n, write an algorithm that computes the maximum bending stresses σ_1 and σ_2 in the two materials and the shear stress τ in the glue. (b) Use the algorithm to determine by trial-and-error the values of b and h that result in the minimum cost design, assuming that material ① costs eight times more than material ②. Use $L = 18$ ft, $w_0 = 600$ lb/ft, $t = 0.5$ in., and $n = 5$, and the working stresses $(\sigma_1)_w = 2000$ psi, $(\sigma_2)_w = 12\,000$ psi, and $\tau_w = 150$ psi.

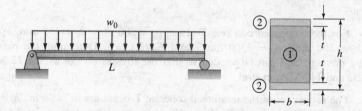

FIG. C9.1

C9.2 The concrete beam shown in Fig. P9.36–P9.39 contains both tensile and compressive steel reinforcement of cross-sectional areas A_t and A_c, respectively. The width of the cross section is b, the distance between the reinforcing bars is d, and the thickness of the concrete covering the bars is a. (a) Given A_t, A_c, b, d, a, $n = E_{st}/E_{co}$, and the bending moment M, write an algorithm that computes the maximum compressive stress in concrete and the stresses in the steel bars. (b) Run the program with the following data: $A_t = 1859$ mm^2, $A_c = 0$, $b = 267$ mm, $d = 325$ mm, $a = 75$ mm, $n = 8$, and $M = 90$ kN·m. (c) If $M = 100$ kN·m, determine by trial-and-error the smallest possible values of A_t and A_c for which the stresses do not exceed 12 MPa in concrete and 140 MPa in steel. Use the same n and the cross-sectional dimensions as in part (b).

10
Columns

Columns supporting the decks of a stadium. Although a column is a compression member, it must have sufficient bending rigidity to prevent failure by lateral instability known as buckling.

10.1 Introduction

The term *column* is applied to a member that carries a compressive axial load. Columns are generally subdivided into the following three types according to how they fail:

- *Short columns* fail by crushing (e.g., yielding). Even if loaded eccentrically, a short column undergoes negligible lateral deflection, so that it can be analyzed as a member subjected to combined axial loading and bending, as described in Art. 8.3.
- *Long columns* fail by buckling. If the axial load is increased to a critical value, the initially straight shape of a slender column becomes unsta-

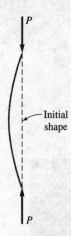

FIG. 10.1 Buckling of a slender column under axial loading.

ble, causing the column to deflect laterally, as shown in Fig. 10.1, and eventually collapse. This phenomenon, which is known as *buckling*, can occur at stresses that are smaller (often much smaller) than the yield stress or the proportional limit of the material.

- *Intermediate columns* fail by a combination of crushing and buckling. Because this mechanism of failure is difficult to analyze, intermediate columns are designed using empirical formulas derived from experiments.

This chapter discusses the analysis and design of long and intermediate columns. In analysis, these columns are treated as beams subject to axial load and bending, but with one major difference: The effect of lateral deflections on equilibrium is no longer ignored. In other words, the free-body diagrams are drawn using the deformed rather than the undeformed column.

10.2 Critical Load

a. Definition of critical load

Figure 10.2 shows an *idealized model* of a simply supported column. In this model, the column is initially straight with the axial load perfectly aligned with the centroidal axis of the column. We also assume that the stresses remain below the proportional limit. When the end moments M_0 are applied, as shown in Fig. 10.2(a), the column deflects laterally, with the maximum displacement δ_{max} being proportional to M_0. Now suppose that we gradually apply the axial load P while at the same time decreasing the end moments so that the maximum displacement δ_{max} does not change, as illustrated in Fig. 10.2(b). When the end moments become zero, as in Fig. 10.2(c), δ_{max} is maintained by the axial load alone. The axial load required to hold the column in its deflected position without any lateral loading (such as the end moments) is called the *critical load*, or *buckling load*, and is denoted by P_{cr}.

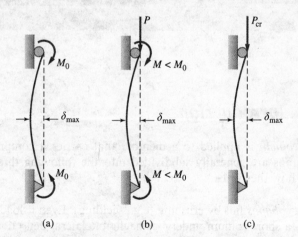

FIG. 10.2 (a) Slender column bent by couples $M = M_0$; (b) if $M < M_0$, the lateral displacement can be maintained by introducing an axial load P of appropriate magnitude; (c) when P reaches its critical value P_{cr}, the lateral displacement is maintained with $M = 0$.

Any increase in the axial load beyond P_{cr} increases the deflection δ_{max} catastrophically, causing the column to fail. On the other hand, if the axial load is decreased slightly below the critical value, the opposite effect occurs—the column becomes straight. The critical load can thus be defined as the maximum axial load that a column can carry and still remain straight. However, at the critical load, the straight position of the column is unstable because the smallest sideways force would cause the column to deflect laterally. In other words, the lateral stiffness of the column is zero when $P = P_{cr}$.

b. Euler's formula

The formula for the critical load of a column was derived in 1757 by Leonhard Euler, the great Swiss mathematician. Euler's analysis was based on the differential equation of the elastic curve

$$\frac{d^2v}{dx^2} = \frac{M}{EI} \tag{a}$$

which we used in the analysis of beam deflections in Chapter 6.

Figure 10.3(a) shows an ideal *simply supported column AB* subjected to the axial load P. We assume that this load is capable of keeping the column in a laterally displaced position. As in the analysis of beams, we let x be the distance measured along the column and denote the lateral deflection by v. The bending moment M acting at an arbitrary section can be obtained from the free-body diagram in Fig. 10.3(b). (M and v shown on the diagram are positive according to the sign conventions introduced in Chapter 4.) The equilibrium equation $\Sigma M_A = 0$ gives $M = -Pv$, which upon substitution into Eq. (a) yields

$$\frac{d^2v}{dx^2} + \frac{P}{EI}v = 0 \tag{b}$$

Equation (b) is a homogeneous, linear differential equation with constant coefficients. The solution, which may be verified by direct substitution, is

$$v = C_1 \sin\left(\sqrt{\frac{P}{EI}}x\right) + C_2 \cos\left(\sqrt{\frac{P}{EI}}x\right) \tag{c}$$

The constants of integration, C_1 and C_2, are determined by the constraints imposed by the supports:

1. $v|_{x=0} = 0$, which yields $C_2 = 0$.
2. $v|_{x=L} = 0$, resulting in

$$0 = C_1 \sin\sqrt{\frac{PL^2}{EI}} \tag{d}$$

Equation (d) can be satisfied with $C_1 = 0$, but this solution is of no interest because it represents the trivial case $P = v = 0$. Other solutions are $\sqrt{PL^2/(EI)} = 0, \pi, 2\pi, 3\pi, \ldots$, or

$$P = n^2 \frac{\pi^2 EI}{L^2} \quad (n = 0, 1, 2, 3, \ldots) \tag{e}$$

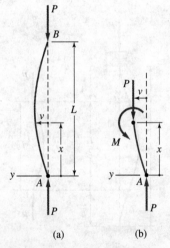

FIG. 10.3 (a) Buckling of a simply supported column; (b) free-body diagram for determining the bending moment M.

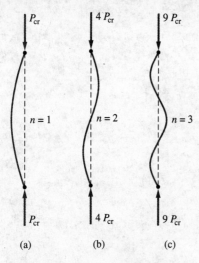

FIG. 10.4 First three buckling mode shapes of a simply supported column.

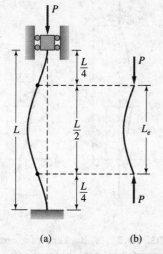

FIG. 10.5 Buckling of a column with built-in ends.

The case $n = 0$ can be discarded because it again yields the trivial case $P = v = 0$. The critical load is obtained by setting $n = 1$, yielding *Euler's formula*:

$$P_{cr} = \frac{\pi^2 EI}{L^2} \tag{10.1}$$

This is the smallest value of P that is capable of maintaining the lateral displacement. The corresponding equation of the elastic curve, called the *mode shape*, is

$$v = C_1 \sin \frac{\pi x}{L}$$

as shown in Fig. 10.4(a). The constant C_1 is indeterminate, implying that the magnitude of the displacement is arbitrary.

The elastic curves corresponding to $n = 2$ and $n = 3$ are shown in Figs. 10.4(b) and (c). Because these mode shapes require axial loads larger than P_{cr}, they can be realized only if the column is braced at its midpoint (for $n = 2$) or at its third points (for $n = 3$).

The critical loads of columns with other end supports can be expressed in terms of the critical load for a simply supported column. Consider, for example, the column with *built-in ends* in Fig. 10.5(a). Its mode shape has inflection points at the distance $L/4$ from each support. Because the bending moment is zero at a point of inflection (due to zero curvature), the free-body diagram in Fig. 10.5(b) shows that the middle half of the column is equivalent to a simply supported column with an effective length $L_e = L/2$. Therefore, the critical load for this column is

$$P_{cr} = \frac{\pi^2 EI}{L_e^2} = \frac{\pi^2 EI}{(L/2)^2} = 4\frac{\pi^2 EI}{L^2} \tag{10.2}$$

This is four times the critical load for a simply supported column.

The critical load for a *cantilever column* of length L can be determined by the same method. This column behaves as the bottom (or top) half of a simply supported column, as shown in Fig. 10.6. Therefore, its effective length is $L_e = 2L$, which results in the critical load

$$P_{cr} = \frac{\pi^2 EI}{L_e^2} = \frac{1}{4} \frac{\pi^2 EI}{L^2} \tag{10.3}$$

This is one-quarter of the critical load for a simply supported column of the same length.

The effective length of the *propped cantilever column* in Fig. 10.7 can be shown to be approximately $L_e = 0.7L$, which is the distance between the point of inflection and the simple support. This value yields for the critical load

$$P_{cr} = \frac{\pi^2 EI}{L_e^2} = \frac{\pi^2 EI}{(0.7L)^2} \approx 2\frac{\pi^2 EI}{L^2} \tag{10.4}$$

which is twice the critical load for a simply supported column.

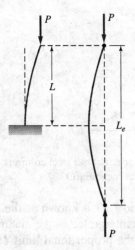
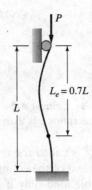

FIG. 10.6 Buckling of a cantilever column.

FIG. 10.7 Buckling of a propped cantilever column.

10.3 Discussion of Critical Loads

In the previous article, we discovered that the critical, or buckling, load of a column is

$$P_{cr} = \frac{\pi^2 EI}{L_e^2} \qquad (10.5)$$

where the effective length L_e of the column is determined by the types of end supports. For a simply supported column, we have $L_e = L$.

Equation (10.5) shows that P_{cr} does not depend on the strength of the material but only on the modulus of elasticity and the dimensions of the column. Thus, two dimensionally identical slender columns, one of high-strength steel and the other of ordinary steel, will buckle under the same critical load because they have the same modulus of elasticity.

The critical load obtained from Eq. (10.5) is physically meaningful only if the stress at buckling does not exceed the proportional limit. The stress in the column just before it buckles may be found by substituting $I = Ar^2$ into Eq. (10.5), where A is the cross-sectional area and r is the *least radius of gyration* of the cross section.[1] We get

$$\sigma_{cr} = \frac{P_{cr}}{A} = \frac{\pi^2 E}{(L_e/r)^2} \qquad (10.6)$$

---

[1] Here, we are using r to denote the radius of gyration to conform to AISC notation. Do not confuse this r with the radius of a circle.

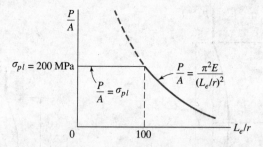

FIG. 10.8 Critical stress versus slenderness ratio for structural steel columns. For slenderness ratios less than 100 the critical stress is not meaningful.

where σ_{cr} is called the *critical stress* and the ratio L_e/r is known as the *slenderness ratio* of the column. Thus, P_{cr} should be interpreted as the maximum sustainable load only if $\sigma_{cr} < \sigma_{pl}$, where σ_{pl} is the proportional limit of the material.

Long columns are defined as columns for which σ_{cr} is less than σ_{pl}. Therefore, the dividing line between long and intermediate columns is the slenderness ratio that satisfies Eq. (10.6) when $\sigma_{cr} = \sigma_{pl}$. This limiting slenderness ratio varies with different materials and even with different grades of the same material. For example, for steel that has a proportional limit of 200 MPa and a modulus of elasticity $E = 200$ GPa, the limiting slenderness ratio is

$$\left(\frac{L_e}{r}\right)^2 = \frac{\pi^2 E}{\sigma_{pl}} = \frac{\pi^2 (200 \times 10^9)}{200 \times 10^6} \approx 10\,000 \qquad \frac{L_e}{r} \approx 100$$

For slenderness ratios below this value, the critical stress given by Eq. (10.6) exceeds the proportional limit of the material. Hence, the load-carrying capacity of a steel column is determined by the critical stress only if $L_e/r > 100$, as illustrated by the plot in Fig. 10.8. The plot also shows that the critical stress rapidly decreases as the slenderness ratio increases. It must be pointed out that Fig. 10.8 shows the stress at failure, not the working stress. Therefore, it is necessary to divide the critical stress by a suitable factor of safety to obtain the allowable stress. The factor of safety should allow for unavoidable imperfections always present in a real column, such as manufacturing flaws and eccentricity of loading.

A column always tends to buckle in the direction that offers the least resistance to bending. For this reason, buckling occurs about the axis that yields the largest slenderness ratio L_e/r, which is usually the axis of least moment of inertia of the cross section.

Sample Problem 10.1

Select the lightest W-shape that can be used as a steel column 7 m long to support an axial load of 450 kN with a factor of safety of 3. Use $\sigma_{pl} = 200$ MPa and $E = 200$ GPa. Assume that the column is (1) simply supported; and (2) a propped cantilever.

Solution

Multiplying the given design load by the factor of safety, we get $P_{cr} = 450(3) = 1350$ kN for the minimum allowable critical load. The selected section must be able to carry this load without buckling or crushing. The crushing criterion is $P_{cr}/A < \sigma_{pl}$, which yields for the minimum required cross-sectional area

$$A = \frac{P_{cr}}{\sigma_{pl}} = \frac{1350 \times 10^3}{200 \times 10^6} = 6.75 \times 10^{-3} \text{ m}^2 = 6750 \text{ mm}^2$$

Due to the different support conditions, the buckling criteria for the two columns must be treated separately.

Part 1

The effective length of the simply supported column is $L_e = L = 7$ m. Solving Eq. (10.5) for I, we obtain for the smallest allowable moment of inertia

$$I = \frac{P_{cr} L_e^2}{\pi^2 E} = \frac{(1350 \times 10^3)(7)^2}{\pi^2 (200 \times 10^9)} = 33.5 \times 10^{-6} \text{ m}^4 = 33.5 \times 10^6 \text{ mm}^4$$

Searching the W-shapes in Table B-2, we find that the lightest section that has the required moment of inertia is a W250 × 73 section. Its properties are $I_{min} = 38.8 \times 10^6$ mm^4 (moment of inertia about the weakest axis) and $A = 9280$ mm^2. Because the cross-sectional area exceeds the value required to prevent crushing (6750 mm^2), the lightest acceptable choice is the

W250 × 73 section   Answer

Part 2

Noting that the effective length of the propped cantilever column is $L_e = 0.7L = 0.7(7) = 4.90$ m, we find the smallest moment of inertia that would prevent buckling is

$$I = \frac{P_{cr} L_e^2}{\pi^2 E} = \frac{(1350 \times 10^3)(4.90)^2}{\pi^2 (200 \times 10^9)} = 16.421 \times 10^{-6} \text{ m}^4 = 16.42 \times 10^6 \text{ mm}^4$$

The lightest W-shape that meets this requirement is a W200 × 52 section, which has the cross-sectional properties $I_{min} = 17.8 \times 10^6$ mm^4 and $A = 6660$ mm^2. However, this section is not acceptable because its cross-sectional area does not satisfy the crushing criterion $A > 6750$ mm^2.

The lightest section that meets both the buckling and the crushing criteria is the

W250 × 58 section   Answer

which has $I_{min} = 18.8 \times 10^6$ mm^4 and $A = 7240$ mm^2.

Problems

10.1 A simply supported steel column is 10 ft long and has a square cross section of side length b. If the column is to support a 20-kip axial load, determine the smallest value of b that would prevent buckling. Use $E = 29 \times 10^6$ psi for steel.

10.2 Solve Prob. 10.1 if the column is made of wood, for which $E = 1.6 \times 10^6$ psi.

10.3 A 50-mm by 100-mm timber, 2.5 m long, is used as a column with built-in ends. If $E = 10$ GPa and $\sigma_{pl} = 30$ MPa, determine the largest axial load that can be carried with a factor of safety of 2.

10.4 An aluminum tube of length 8 m is used as a simply supported column carrying a 1.2-kN axial load. If the outer diameter of the tube is 50 mm, compute the inner diameter that would provide a factor of safety of 2 against buckling. Use $E = 70$ GPa for aluminum.

10.5 An aluminum column 6 ft long has a solid rectangular cross section 3/4 in. by 2 in. The column is secured at each end with a bolt parallel to the 3/4-in. direction as shown in the figure. Thus, the ends can rotate about the z-axis but not about the y-axis. Find the largest allowable axial load using $E = 10.3 \times 10^6$ psi, $\sigma_{pl} = 6000$ psi, and a factor of safety of 2.

10.6 Two C310 × 45 channels are laced together as shown to form a section with equal moments of inertia about the x- and y-axes (the lacing does not contribute to the bending stiffness). If this section is used as a simply supported column, determine (a) the shortest length for which the column would fail by buckling; and (b) the largest allowable axial load that can be supported by a 12-m-long column with a factor of safety of 2.5. Use $E = 200$ GPa and $\sigma_{pl} = 240$ MPa.

FIG. P10.5

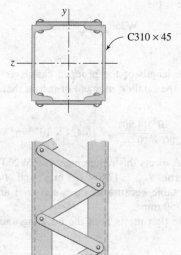

FIG. P10.6, P10.7

10.7 Solve Prob. 10.6 assuming that the column is a propped cantilever.

10.8 A W-shape is used as a simply supported column 8 m long. Select the lightest shape than can carry an axial load of 270 kN with a factor of safety of 2.5. Use $E = 200$ GPa and $\sigma_{pl} = 200$ MPa.

10.9 Select the lightest W-shape for a 40-ft-long column with built-in ends that can carry an axial load of 150 kips with a factor of safety of 2. Assume that $E = 29 \times 10^6$ psi and $\sigma_{pl} = 30 \times 10^3$ psi.

10.10 The two members of the pin-jointed wood structure ABC have identical square cross sections of dimensions $b \times b$. A 800-lb vertical load acts at B. Determine the smallest value of b that would provide a factor of safety of 3 against buckling. Use $E = 1.5 \times 10^6$ psi for wood and assume that joint B is braced so that it can move only in the plane of the structure.

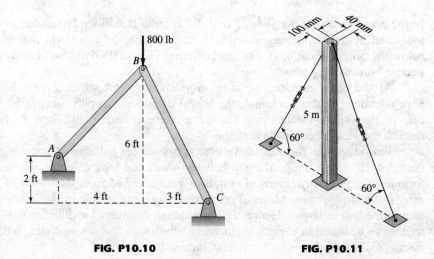

FIG. P10.10

FIG. P10.11

10.11 The 5-m-long timber column is built in at its base and stayed by two cables at the top. The turnbuckles in the cables are turned until the tensile force in each cable is T. Determine the value of T that would cause the column to buckle. Use $E = 10$ GPa for timber.

10.12 The L75 × 75 × 13 angle section is used as a cantilever column of length L. Find the maximum allowable value of L if the column is not to buckle when the 12-kN axial load is applied. Use $E = 200$ MPa.

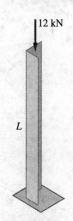

FIG. P10.12

10.13 The 24-ft-long steel column is an S8 × 23 section that is built in at both ends. The midpoint of the column is braced by two cables that prevent displacement in the x-direction. Determine the critical value of the axial load P. Use $E = 29 \times 10^6$ psi for steel.

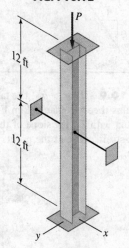

FIG. P10.13

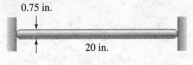

FIG. P10.14

10.14 The solid aluminum bar of circular cross section is fitted snugly between two immovable walls. Determine the temperature increase that would cause the bar to buckle. Use $E = 10.3 \times 10^6$ psi and $\alpha = 12.8 \times 10^{-6}/°F$ for aluminum.

10.4 Design Formulas for Intermediate Columns

In the previous article, we showed that if a column is sufficiently slender, as measured by the slenderness ratio L_e/r, buckling occurs at a stress that is below the proportional limit. For steel columns ($E = 200$ GPa), "sufficiently slender" means $L_e/r > 100$.

At the other extreme, we have short columns, for which $L_e/r < 30$ approximately. In short columns, the lateral displacements play a negligible role in the failure mechanism. Therefore, these columns fail when P/A reaches the yield stress of the material.

Various design formulas have been proposed for columns of intermediate length, which bridge the gap between short and long columns. These formulas are primarily empirical in nature, being derived from the results of extensive test programs. Material properties play a major role in the failure of intermediate columns. Hence, different design formulas for different materials can be found in various engineering handbooks and design codes. We discuss mainly the formulas for steel columns.

a. Tangent modulus theory

Suppose that the compressive stress σ_{cr} just prior to buckling exceeds the proportional limit σ_{pl}, as indicated in the compressive stress-strain diagram in Fig. 10.9. Any additional increment of strain $d\epsilon$ would result in the stress increment $d\sigma = E_t \, d\epsilon$, where E_t is the slope of the diagram at the point $\sigma = \sigma_{cr}$. Thus, the bending stiffness of the column at buckling is determined by E_t, which is smaller than the elastic modulus E. To account for the reduced stiffness, the tangent modulus theory replaces E by E_t in Eq. (10.6), resulting in the following expression for the critical stress:

$$\sigma_{cr} = \frac{P_{cr}}{A} = \frac{\pi^2 E_t}{(L_e/r)^2} \qquad (10.7)$$

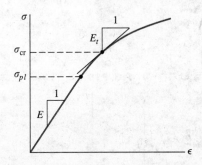

FIG. 10.9 If $\sigma_{cr} > \sigma_{pl}$, the tangent modulus theory replaces E in Euler's formula with E_t, the slope of the stress-strain curve at the point $\sigma = \sigma_{cr}$.

Although Eq. (10.7) accounts for the nonlinearity of the stress-strain diagram beyond the proportional limit, its theoretical basis is somewhat weak. Therefore, this equation should be viewed as an empirical formula. However, the results obtained from Eq. (10.7) are in satisfactory agreement with experimental results.

Because the slope of the stress-strain diagram, called the *tangent modulus*, is not constant beyond the proportional limit, the evaluation of the critical load from Eq. (10.7) is not straightforward. The difficulty is that the slope E_t at the critical stress is not known beforehand (after all, its value

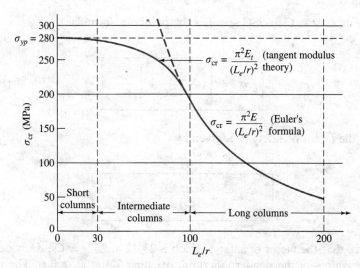

FIG. 10.10 Critical stress versus slenderness ratio for structural steel columns. Tangent modulus theory was used for slenderness ratios less than 100.

depends on the critical stress). Therefore, Eq. (10.7) must be solved for σ_{cr} and E_t simultaneously by trial-and-error.

Figure 10.10 shows the plot of the critical stress obtained from tangent modulus theory against the slenderness ratio for structural-grade steel columns. As can be seen, the tangent modulus theory smoothly connects the curves for short and long columns.

b. AISC column specifications

The American Institute of Steel Construction (AISC) defines the dividing line between intermediate and long columns to be the slenderness ratio $L_e/r = C_c$, where

$$C_c = \sqrt{\frac{2\pi^2 E}{\sigma_{yp}}} \quad (10.8)$$

in which $E = 200$ GPa $= 29 \times 10^6$ psi for most grades of steel and σ_{yp} is the yield stress, which depends on the grade of steel used.

For columns for which $L_e/r > C_c$, the working stress σ_w is given by

$$\sigma_w = \frac{12\pi^2 E}{23(L_e/r)^2} \quad (10.9)$$

This is the buckling formula for long columns in Eq. (10.6) with a factor of safety of $23/12 = 1.92$.

For $L_e/r < C_c$, AISC specifies the parabolic formula

$$\sigma_w = \left[1 - \frac{(L_e/r)^2}{2C_c^2}\right] \frac{\sigma_{yp}}{N} \tag{10.10}$$

where the factor of safety N is given by

$$N = \frac{5}{3} + \frac{3(L_e/r)}{8C_c} - \frac{(L_e/r)^3}{8C_c^3} \tag{10.11}$$

Observe that the factor of safety, which is 23/12 at $L_e/r = C_c$, decreases for smaller values of the slenderness ratio, reaching 5/3 at $L_e/r = 0$. The variation of σ_w with L_e/r for several grades of steel is shown in Fig. 10.11.

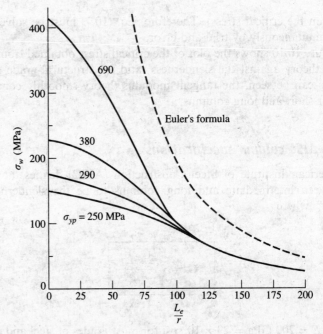

FIG. 10.11 Working stress versus slenderness ratio using AISC column specifications.

Sample Problem 10.2

Using AISC column specifications, determine the allowable axial loads on a W360 × 122 section used as a column under the following conditions: (1) simply supported ends with a length of 9 m; (2) built-in ends with a length of 10 m; and (3) built-in ends with a length of 10 m and braced at the midpoint. Use $\sigma_{yp} = 380$ MPa and $E = 200$ GPa.

Solution

According to Table B-2 in Appendix B, a W360 × 122 section has the properties $A = 15.5 \times 10^3$ mm² and $r = 63.0$ mm (minimum radius of gyration). For the specified material properties, the slenderness ratio that separates long and intermediate columns is, from Eq. (10.8),

$$C_c = \sqrt{\frac{2\pi^2 E}{\sigma_{yp}}} = \sqrt{\frac{2\pi^2 (200 \times 10^9)}{380 \times 10^6}} = 101.9$$

Part 1

The slenderness ratio of the simply supported column is $L_e/r = L/r = 9000/63.0 = 142.9$. Because $L_e/r > C_c$, the column is considered to be "long," in which case the working stress is obtained from Eq. (10.9):

$$\sigma_w = \frac{12\pi^2 E}{23(L_e/r)^2} = \frac{12\pi^2 (200 \times 10^9)}{23(142.9)^2} = 50.43 \times 10^6 \text{ Pa}$$

Therefore, the largest safe axial load is (converting A from mm² to m²)

$$P = \sigma_w A = (50.43 \times 10^6)(15.5 \times 10^{-3}) = 782 \times 10^3 \text{ N} = 782 \text{ kN} \quad \textit{Answer}$$

Part 2

For the column with built-in ends, the effective length is $L_e = 0.5L = 0.5(10) = 5.0$ m. Hence, the slenderness ratio $L_e/r = 5000/63.0 = 79.37$ is less than C_c, so that the column is of intermediate length, for which Eqs. (10.10) and (10.11) apply. These equations yield the factor of safety

$$N = \frac{5}{3} + \frac{3(L_e/r)}{8C_c} - \frac{(L_e/r)^3}{8C_c^3} = \frac{5}{3} + \frac{3(79.37)}{8(101.9)} - \frac{(79.37)^3}{8(101.9)^3} = 1.900$$

and the working stress

$$\sigma_w = \left[1 - \frac{(L_e/r)^2}{2C_c^2}\right]\frac{\sigma_{yp}}{N} = \left[1 - \frac{(79.37)^2}{2(101.9)^2}\right]\frac{(380 \times 10^6)}{1.900} = 139.3 \times 10^6 \text{ Pa}$$

The largest allowable load thus becomes

$$P = \sigma_w A = (139.3 \times 10^6)(15.5 \times 10^{-3}) = 2160 \times 10^3 \text{ N} = 2160 \text{ kN} \quad \textit{Answer}$$

Note The above answers illustrate the greatly increased strength of a column when its ends are rigidly built in. Because perfect rigidity is not realized in practice, it is advisable to determine the allowable load using an effective length larger than the ideal value of $0.5L$. Choosing, for example, $L_e = 0.75L$ would yield a more realistic result.

Part 3

A column of length L with built-in ends and midpoint bracing behaves as a propped cantilever column of length $L/2$. Thus, the effective length of our column is $L_e = 0.7(L/2) = 0.7(10/2) = 3.5$ m, giving the slenderness ratio $L_e/r = 3500/63.0 = 55.56$. Because $L_e < C_c$, the column falls in the intermediate length category. Proceeding as in Part 2, we obtain

$$N = \frac{5}{3} + \frac{3(L_e/r)}{8C_c} - \frac{(L_e/r)^3}{8C_c^3} = \frac{5}{3} + \frac{3(55.56)}{8(101.9)} - \frac{(55.56)^3}{8(101.9)^3} = 1.851$$

$$\sigma_w = \left[1 - \frac{(L_e/r)^2}{2C_c^2}\right]\frac{\sigma_{yp}}{N} = \left[1 - \frac{(55.56)^2}{2(101.9)^2}\right]\frac{(380 \times 10^6)}{1.851} = 174.8 \times 10^6 \text{ Pa}$$

which yields for the largest safe axial load

$$P = \sigma_w A = (174.8 \times 10^6)(15.5 \times 10^{-3}) = 2710 \times 10^3 \text{ N} = 2710 \text{ kN} \quad \textit{Answer}$$

Problems

10.15 Determine the slenderness ratio of a 5-ft-long column with built-in ends if the cross section is (a) circular with a radius of 40 mm; and (b) 50 mm square.

10.16 Find the slenderness ratio of a 12-ft propped cantilever column if the cross section is (a) circular with a radius of 2 in.; and (b) a 2-in. × 3-in. rectangle.

10.17 A W250 × 167 section is to be used as a simply supported column to support a 1600-kN axial load. Use AISC specifications with $E = 200$ GPa and $\sigma_{yp} = 380$ MPa to determine the maximum allowable length of the column.

10.18 Solve Prob. 10.17 if the axial load is 3200 kN.

10.19 A solid circular steel rod is used as a column with an effective length of 18 ft. Use AISC specifications to find the smallest allowable diameter of the rod if the axial load is 40 kips. The properties of steel are $E = 29 \times 10^6$ psi and $\sigma_{yp} = 30 \times 10^3$ psi.

10.20 Solve Prob. 10.19 if the axial load is 600 kips.

10.21 A W14 × 82 section is used as a column with an effective length of 30 ft. Compute the maximum axial load that the column can carry using AISC specifications with $E = 29 \times 10^6$ psi and $\sigma_{yp} = 50 \times 10^3$ psi.

10.22 Determine the largest safe load that can be applied to a 30-ft simply supported column made of a W310 × 52 section if its length is (a) 10 m; and (b) 14 m. Use AISC specifications with $E = 200$ GPa and $\sigma_{yp} = 250$ MPa.

10.23 Four L4 × 4 × 1/2 angle sections latticed together form a simply supported column with the cross section shown in the figure. Use AISC specifications with $E = 29 \times 10^6$ psi and $\sigma_{yp} = 60 \times 10^3$ psi to determine the maximum allowable length of the column that can carry a 200-kip axial load.

10.24 The truss carries three loads of magnitude P each. The upper chord BCD is composed of two C9 × 20 channels that are laced together so that the cross section has equal moments of inertia about both axes of symmetry (see Fig. P10.6, P10.7). Considering the buckling of BCD only, find the maximum safe value of P using AISC specifications with $E = 29 \times 10^6$ psi and $\sigma_{yp} = 36 \times 10^3$ psi. Assume that the truss has cross-bracing that allows only displacements in the plane of the truss.

10.25 Solve Prob. 10.24 using two C10 × 30 channels and $\sigma_{yp} = 50$ ksi.

10.26 For the truss shown in the figure, find the lightest W-shape that can be used for member AB if $P = 60$ kips. Use AISC specifications with $E = 29 \times 10^6$ psi and $\sigma_{yp} = 50 \times 10^3$ psi. Assume that cross-bracing prevents the joints from deflecting perpendicular to the plane of the truss.

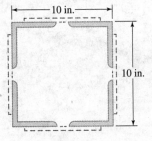

FIG. P10.23

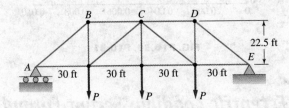

FIG. P10.24–P10.26

10.27 A steel tube with a 100-mm outer diameter and an 85-mm inner diameter is used as a simply supported column of length 1.2 m. Determine the maximum allowable axial load using AISC specifications with $E = 200$ GPa and $\sigma_{yp} = 250$ MPa.

10.28 The Aluminum Association specifies the following working stress for columns made of 2014-T6 aluminum alloy that fall in the slenderness range $12 < L_e/r < 55$:

$$\sigma_w = 212 - 1.59 \frac{L_e}{r} \text{ MPa}$$

If the column described in Prob. 10.27 were made of 2014-T6 alloy, determine the largest allowable axial load using (a) the Aluminum Association specification; and (b) AISC specifications with $E = 73$ GPa and $\sigma_{yp} = 410$ MPa (use of this specification for aluminum is not recommended in practice). Which specification gives the more conservative result?

10.29 A simply supported column 30 ft long is fabricated from a W8 × 31 section and two C12 × 30 channels arranged as shown in the figure. Determine the maximum safe axial load using AISC specifications with $E = 29 \times 10^6$ psi and $\sigma_{yp} = 36 \times 10^3$ psi.

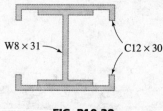

FIG. P10.29

10.30 A square 2024-T3 aluminum alloy tube with 2-in. by 2-in. exterior dimensions and 1/4-in. wall thickness is used as a simply supported column. The alloy has the compressive stress-strain diagram shown. Use the tangent modulus theory to determine the length of the column for which the critical stress is (a) 35 ksi; and (b) 25 ksi.

10.31 The figure shows the compressive stress-strain diagram for 2024-T3 aluminum alloy. Use the tangent modulus theory to estimate the critical stress for a column of slenderness ratio $L_e/r = 40$ made of this alloy.

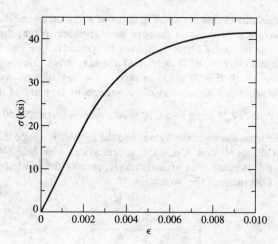

FIG. P10.30, P10.31

10.5 Eccentric Loading: Secant Formula

Most columns are designed to support purely axial loads. Designers usually take great care in arranging the structural details so that the loads act along the centroidal axes of columns. Small eccentricities of loading are of course unavoidable, but they are accidental. We now consider columns with definite

10.5 Eccentric Loading: Secant Formula

and deliberate load eccentricities. For example, in building frames where beams are connected to the flanges of a column, the point of loading is on the flange, not the centroid of the column. The results of this study are also useful for accidental eccentricities with magnitudes that can be estimated.

a. Derivation of the secant formula

Consider the simply supported column of length L shown in Fig. 10.12(a), where the load P has an eccentricity e with respect to the centroid of the cross section. The eccentricity gives rise to end moments of magnitude Pe, which cause the column to bend. If the column is slender, the displacements accompanying bending can be quite large so that their effect on equilibrium cannot be ignored. Therefore, the bending moment in the column must be computed from a free-body diagram of the *deformed* column, as shown in Fig. 10.12(b). This figure yields for the bending moment

$$M = -P(v + e)$$

where v is the lateral deflection. Substituting M into the differential equation of the elastic curve, $d^2v/dx^2 = M/(EI)$, and rearranging terms, we get

$$\frac{d^2v}{dx^2} + \frac{P}{EI}v = -\frac{Pe}{EI} \qquad \text{(a)}$$

Because this is a nonhomogeneous differential equation, its solution is the sum of the complementary and particular solutions. The complementary solution is identical to Eq. (c) of Art. 10.2. It can be readily verified by substitution that a particular solution is $v = -e$. Therefore, the general solution of Eq. (a) is

$$v = C_1 \sin\left(\sqrt{\frac{P}{EI}}x\right) + C_2 \cos\left(\sqrt{\frac{P}{EI}}x\right) - e \qquad \text{(b)}$$

The constants of integration, C_1 and C_2, are determined from the zero displacement requirements at the supports:

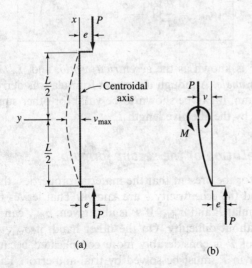

FIG. 10.12 (a) Deformation of a column due to eccentric loading; (b) free-body diagram for determining the bending moment M.

$$v|_{x=-L/2} = 0 \qquad -C_1 \sin\sqrt{\frac{PL^2}{4EI}} + C_2 \cos\sqrt{\frac{PL^2}{4EI}} - e = 0 \qquad \text{(c)}$$

$$v|_{x=L/2} = 0 \qquad C_1 \sin\sqrt{\frac{PL^2}{4EI}} + C_2 \cos\sqrt{\frac{PL^2}{4EI}} - e = 0 \qquad \text{(d)}$$

The solution of Eqs. (c) and (d) is

$$C_1 = 0 \qquad C_2 = e \sec\sqrt{\frac{PL^2}{4EI}}$$

Substituting these constants into Eq. (b), we get the equation of the elastic curve

$$v = e\left[\sec\sqrt{\frac{PL^2}{4EI}} \cos\left(\sqrt{\frac{P}{EI}}\,x\right) - 1\right] \qquad \text{(e)}$$

The maximum deflection is

$$v_{\max} = v|_{x=0} = e\left(\sec\sqrt{\frac{PL^2}{4EI}} - 1\right) = e\left[\sec\left(\frac{L}{2r}\sqrt{\frac{P}{EA}}\right) - 1\right] \qquad (10.12)$$

where $r = \sqrt{I/A}$ is the smallest radius of gyration of the cross section. Observe that when $P \to P_{cr} = \pi^2 EI/L^2$, we have $v_{\max} \to e[\sec(\pi/2) - 1] = \infty$.

The maximum bending moment in the column occurs at midspan. Its magnitude is $M_{\max} = P(v_{\max} + e)$. Therefore, the highest *compressive stress* in the column is given by

$$\sigma_{\max} = \frac{P}{A} + \frac{M_{\max} c}{I} = \frac{P}{A} + \frac{P(v_{\max} + e)c}{Ar^2}$$

where c is the distance from the centroidal axis to the outermost compression fiber. Substituting $v_{\max}$ from Eq. (10.12), we obtain the *secant formula*

$$\boxed{\sigma_{\max} = \frac{P}{A}\left[1 + \frac{ec}{r^2}\sec\left(\frac{L}{2r}\sqrt{\frac{P}{EA}}\right)\right]} \qquad (10.13)$$

The term ec/r^2 is known as the *eccentricity ratio*, and $[L/(2r)]\sqrt{P/(EA)}$ is called the *Euler angle*. Although the secant formula was derived for a simply supported column, it can be shown to be valid for other support conditions if we replace L by the effective length L_e.

b. Application of the secant formula

Let us assume for the present that the material properties, the dimensions of the column, and the eccentricity e are known. That leaves two variables in the secant formula: P and $\sigma_{\max}$. If P is also given, $\sigma_{\max}$ can computed from the formula without difficulty. On the other hand, if $\sigma_{\max}$ is specified, the determination of P is considerably more complicated because Eq. (10.13), being nonlinear in P, must be solved by trial-and-error. The selection of a structural section that can safely carry a given load P is similarly a trial-and-error procedure.

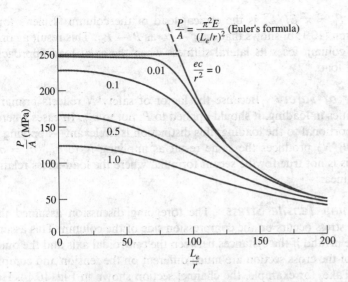

FIG. 10.13 Average stress at yielding versus slenderness ratio for eccentrically loaded structural steel columns (obtained be equating the maximum stress given by the secant formula to the yield stress).

Determining the Working Load
A common problem in column design is to determine the working load (the largest allowable load) P_w that a given column can carry with a factor of safety N against yielding. Substituting $\sigma_{max} = \sigma_{yp}$ into Eq. (10.13), we get

$$\sigma_{yp} = \frac{P}{A}\left[1 + \frac{ec}{r^2}\sec\left(\frac{L_e}{2r}\sqrt{\frac{P}{EA}}\right)\right] \quad (10.14)$$

which can be solved by trial-and error for P/A. Because P in the solution represents the load that initiates yielding, the working load that provides a factor of safety N against yielding is $P_w = P/N$.

Figure 10.13 shows plots of the solutions P/A in Eq. (10.14) versus L_e/r for standard-grade steel columns ($E = 200$ GPa and $\sigma_{yp} = 250$ MPa) for various values of the eccentricity ratios. These plots provide a rough estimate of P, which can be used as the starting value in the trial-and-error solution of Eq. (10.14).

Selecting a Standard Structural Section
Equation (10.13) can also be used to find a standard section that can safely carry a given working load P_w. The procedure is to choose a trial section and substitute its cross-sectional properties together with $P = NP_w$ into Eq. (10.13). If the resulting σ_{max} exceeds σ_{yp}, a larger section must be chosen and the procedure repeated. On the other hand, if σ_{max} is much less than σ_{yp}, a lighter section should be tried. When we work with standard sections, it is more convenient to perform the calculations by rewriting Eq. (10.13) in the form

$$\sigma_{max} = \frac{P}{A} + \frac{Pe}{S}\sec\left(\frac{L_e}{2}\sqrt{\frac{P}{EI}}\right) \quad (10.15a)$$

where S is the section modulus. Another useful form of this equation is

$$\sigma_{max} = \frac{P}{A} + \frac{Pe}{S}\sec\left(\frac{\pi}{2}\sqrt{\frac{P}{P_{cr}}}\right) \quad (10.15b)$$

where $P_{cr} = \pi^2 EI/L_e^2$ is the critical load of the column (Euler's formula). Equation (10.15b) shows that $\sigma_{max} \to \infty$ as $P \to P_{cr}$. This result again shows that a column loses its lateral stiffness when the axial load approaches the critical load.

Factor of Safety Because the factor of safety N reflects primarily uncertainties in loading, it should applied to P, not to σ_{yp}. In cases where stress is proportional to the loading, this distinction is irrelevant—applying N to P ($P_w = P/N$) produces the same result as applying N to σ_{yp} ($\sigma_w = \sigma_{yp}/N$). But this is not true for the secant formula, where the load-stress relationship is nonlinear.

Maximum Tensile Stress The foregoing discussion assumed that the largest stress occurs on the compression side of the column. This assumption may be invalid if the distances between the centroidal axis and the outermost fibers of the cross section are much different on the tension and compression sides. Take, for example, the channel section shown in Fig. 10.14. Here the maximum distance c_t from the centroid C to the outermost tension fibers is much larger that its counterpart c_c on the compression side. Therefore, the maximum tensile stress Mc_t/I due to bending is considerably larger than the maximum compressive bending stress Mc_c/I. Consequently, the net maximum stress in tension $(\sigma_t)_{max} = Mc_t/I - P/A$ may exceed the maximum compressive stress $(\sigma_c)_{max} = Mc_c/I + P/A$.

Equations (10.13)–(10.15) can also be used to compute the maximum tensile stress, provided we *subtract* the direct stress P/A from the bending stress represented by the secant term; that is, we must reverse the sign of the first term on the right-hand side of each equation. Of course, we must also use c_t for the maximum fiber distance.

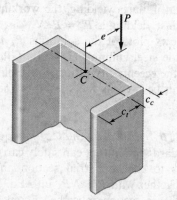

FIG. 10.14 If c_t is much larger than c_c, the maximum tensile stress may exceed the maximum compressive stress.

Sample Problem 10.3

A W14 × 61 section is used as a simply supported column 25 ft long. When the 150-kip load is applied with the 4-in. eccentricity shown, determine (1) the maximum compressive stress in the column; (2) the factor of safety against yielding; and (3) the maximum lateral deflection. Assume that the column does not buckle about the y-axis. Use $E = 29 \times 10^6$ psi and $\sigma_{yp} = 36 \times 10^3$ psi.

Solution

From Table B-7 in Appendix B, we find that the properties of the section are $A = 17.9$ in.^{2}, $I_z = 640$ in.^{4}, and $S_z = 92.2$ in.^{3}. Since the column is simply supported, its effective length is $L_e = L = 25$ ft.

Part 1

Using Eq. (10.15a), we obtain for the maximum compressive stress

$$\sigma_{max} = \frac{P}{A} + \frac{Pe}{S_z} \sec\left(\frac{L_e}{2}\sqrt{\frac{P}{EI_z}}\right)$$

$$= \frac{150 \times 10^3}{17.9} + \frac{(150 \times 10^3)(4)}{92.2} \sec\left(\frac{25 \times 12}{2}\sqrt{\frac{150 \times 10^3}{(29 \times 10^6)(640)}}\right)$$

$$= 15.528 \times 10^3 \text{ psi} \qquad \text{Answer}$$

Part 2

It is tempting to compute the factor of safety against yielding as $N = \sigma_{yp}/\sigma_{max} = 36/15.1528 = 2.32$, but this is incorrect. The factor of safety must be applied to the loading, not to the stress. Thus, N is determined by the solution of the equation

$$\sigma_{yp} = \frac{NP}{A} + \frac{NPe}{S_z} \sec\left(\frac{L}{2}\sqrt{\frac{NP}{EI_z}}\right)$$

When we substitute the known data, this equation becomes

$$36 \times 10^3 = \frac{N(150 \times 10^3)}{17.9} + \frac{N(150 \times 10^3)(4)}{92.2} \sec\left[\frac{25 \times 12}{2}\sqrt{\frac{N(150 \times 10^3)}{(29 \times 10^6)(640)}}\right]$$

$$36 \times 10^3 = N[8380 + 6508 \sec(0.4264\sqrt{N})]$$

By trial-and-error, the solution is

$$N = 2.19 \qquad \text{Answer}$$

which is smaller than the "factor of safety" of 2.32 based on the maximum stress.

Part 3

According to Eq. (10.12), the maximum lateral deflection is

$$v_{max} = e\left(\sec\sqrt{\frac{PL_e^2}{4EI}} - 1\right)$$

$$= 4\left[\sec\sqrt{\frac{(150 \times 10^3)(25 \times 12)^2}{4(29 \times 10^6)(640)}} - 1\right] = 0.393 \text{ in.} \qquad \text{Answer}$$

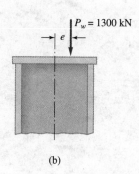

Sample Problem 10.4

The lower column in Fig. (a), which has an effective length $L_e = 7$ m, is to carry the two loads with a factor of safety of 2 against yielding. Due to space limitations, the depth h of the section must be kept under 400 mm. (1) Considering bending about the z-axis of the section, find the lightest suitable W-shape for the lower column. (2) What factor of safety does the chosen section have against buckling about the y-axis? Use $E = 200$ GPa and $\sigma_{yp} = 250$ MPa.

Solution

The two loads are statically equivalent to the force $P_w = 1300$ kN shown in Fig. (b). The eccentricity e is obtained by equating the moments of the forces in Figs. (a) and (b) about the centroid C of the section. This yields

$$125(900) - 75(400) = 1300e \quad e = 63.46 \text{ mm}$$

When we apply the factor of safety, the load to be used in Eq. (10.15a) is $P = NP_w = 2(1300) = 2600$ kN.

Part 1

The lightest W-shape that satisfies the space limitation must have a depth slightly less than 400 mm. Inspection of Table B-2 in Appendix B reveals that the lightest sections with depths just below the 400-mm limit are the W360 shapes. Therefore, we try a succession of W360 shapes by computing σ_{max} from Eq. (10.15a) and comparing the result with σ_{yp}. The results of the trials are summarized in the table.

Section	A (mm^2)	I (mm^4)	S (mm^3)	σ_{max} (MPa)
W360 × 110	14.0 × 10^3	331 × 10^6	1840 × 10^3	302
W360 × 122	15.5 × 10^3	365 × 10^6	2010 × 10^3	272
W360 × 134	17.1 × 10^3	415 × 10^6	2330 × 10^3	239

We see that the lightest acceptable section—that is, the section for which $\sigma_{max} \leq \sigma_{yp} = 250$ MPa—is

$$\text{W360} \times 134 \qquad \textbf{Answer}$$

Its depth is 356 mm.

Part 2

The moment of inertia of a W360 × 134 section about the y-axis is $I = 415 \times 10^6$ mm^4, giving for the critical load (note that the load has no eccentricity about the y-axis)

$$P_{cr} = \frac{\pi^2 EI}{L_e^2} = \frac{\pi^2 (200 \times 10^9)(415 \times 10^{-6})}{(7)^2} = 16.72 \times 10^6 \text{ N} = 16\,720 \text{ kN}$$

Thus, the factor of safety against buckling about the y-axis is

$$N = \frac{P_{cr}}{P_w} = \frac{16\,720}{1300} = 12.9 \qquad \textbf{Answer}$$

Problems

10.32 A W14 × 82 section acts as a simply supported column of length 24 ft. If two loads are applied as shown in the figure, determine (a) the maximum compressive stress in the column; and (b) the factor of safety against yielding. Use $E = 29 \times 10^3$ ksi and $\sigma_{yp} = 40$ ksi.

10.33 For the column shown in the figure, find the lightest W-shape, no more than 15 in. deep, that can carry the loading shown with a factor of safety of 2 against yielding. Use $E = 29 \times 10^3$ ksi and $\sigma_{yp} = 40$ ksi.

10.34 The aluminum T-section is used as a simply supported column 6 m long. Compute the maximum compressive and tensile stresses in the column caused by the load $P = 40$ kN. Use $E = 70$ GPa for aluminum.

10.35 Determine the maximum allowable load P for the column described in Prob. 10.34 if the working stress is $\sigma_w = 130$ MPa.

10.36 The 8-ft cantilever column is made of a steel tube with a 4.5-in. outer diameter and the cross-sectional properties $A = 3.17$ in.^{2} and $I = 7.23$ in.^{4}. Determine the eccentricity e of the 6500-lb load that provides a factor of safety of 2 against yielding. Use $E = 29 \times 10^6$ psi and $\sigma_{yp} = 50 \times 10^3$ psi.

10.37 Solve Prob. 10.36 if the tube is made of 2014-T6 aluminum alloy with the properties $E = 10.6 \times 10^6$ psi and $\sigma_{yp} = 60 \times 10^3$ psi.

10.38 The simply supported column is a W360 × 122 section 10 m long. If $P = 500$ kN, find (a) the maximum compressive stress in the column; and (b) the factor of safety against yielding. Use $E = 200$ GPa and $\sigma_{yp} = 250$ MPa.

10.39 Find the largest allowable P that the column in Prob. 10.38 can carry if the bottom is built in and the top is simply supported. Use a factor of safety of 2 against yielding and neglect the possibility of buckling about the y-axis.

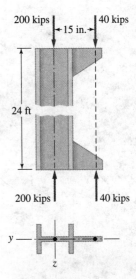

FIG. P10.32, P10.33

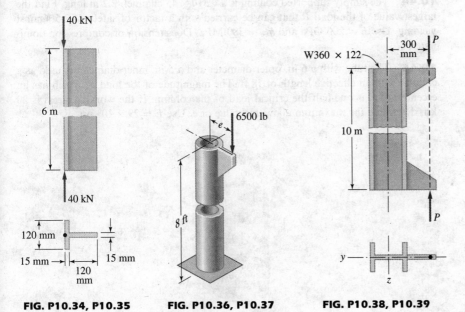

FIG. P10.34, P10.35   **FIG. P10.36, P10.37**   **FIG. P10.38, P10.39**

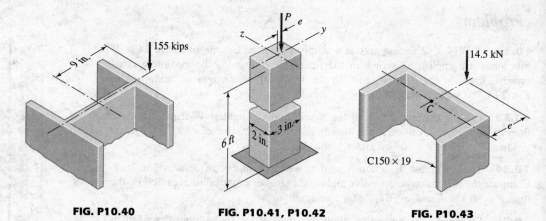

FIG. P10.40 **FIG. P10.41, P10.42** **FIG. P10.43**

10.40 A W shape is used as a column with an effective length of 30 ft. The 155-kip load has an eccentricity of 9 in. as shown in the figure. Find the lightest section that can carry the load with a factor of safety of 2.5 against yielding. Due to space limitations, the depth of the section must not exceed 20 in. Use $E = 29 \times 10^3$ ksi and $\sigma_{yp} = 50$ ksi.

10.41 The load P acting on the steel cantilever column has an eccentricity $e = 0.5$ in. with respect to the z-axis. Determine the maximum allowable value of P using a factor of safety of 2.5 against yielding and buckling. Which failure mode governs the solution? The material properties are $E = 29 \times 10^6$ psi and $\sigma_{yp} = 36 \times 10^3$ psi.

10.42 Solve Prob. 10.41 if $e = 5$ in.

10.43 The C150 × 19 channel is used as a column with an effective length of 3 m. Determine the largest eccentricity e of the 14.5-kN load for which the maximum tensile stress does not exceed the maximum compressive stress. Use $E = 200$ GPa.

10.44 The simply supported column is a C310 × 45 channel, 2.2 m long. Find the largest value of the load P that can be carried with a factor of safety of 2.5 against yielding. Use $E = 200$ GPa and $\sigma_{yp} = 380$ MPa. Does tension or compression dominate?

10.45 The tube with a 6-in. outer diameter and a 5-in. inner diameter is used as a column with an effective length of 18 ft. The magnitude of the load P, which has an eccentricity e, is one-half the critical load of the column. If the working stress is 20 ksi, determine the maximum allowable value of e. Use $E = 29 \times 10^6$ psi.

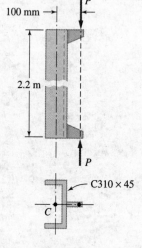

FIG. P10.44

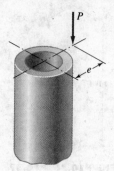

FIG. P10.45

Computer Problems

C10.1 A circular steel tube with outer diameter D and inner diameter d is used as a column with an effective length L_e. The modulus of elasticity of steel is E and the yield stress is σ_{yp}. Given $\beta = d/D$, L_e, E, and σ_{yp}, write an algorithm that uses the AISC specifications to plot the maximum allowable axial force P versus D from $D = 0$ to D_{max}. Run the algorithm with the following data: $L_e = 18$ ft, $E = 29 \times 10^6$ psi, $\sigma_{yp} = 36 \times 10^3$ psi, and (a) $\beta = 0$ (solid rod), $D_{max} = 9$ in.; and (b) $\beta = 0.9$, $D_{max} = 16$ in. If $P = 600$ kips, use the plots to estimate the smallest allowable D in each case.

C10.2 The column consisting of a W-section has an effective length L_e and carries an axial load P with an eccentricity e. Given L_e, e, E, σ_{yp}, and the cross-sectional properties A, S, and I, write an algorithm that uses the secant formula to (1) calculate P_{yp}, the value of P that initiates yielding; and (2) plot the maximum compressive normal stress σ_{max} versus P from $P = 0$ to P_{yp}. Run the algorithm for a W360 × 122 section with $e = 300$ mm, $E = 200$ GPa, $\sigma_{yp} = 250$ MPa; and (a) $L_e = 10$ m; and (b) $L_e = 20$ m. In each case, use the plots to estimate the value of P that results in $\sigma_{max} = 150$ MPa.

C10.3 The outer and inner diameters of a hollow tube are D and d, respectively. The tube is used as a column with an effective length L_e to carry an axial load P. The relationship between the stress σ and the strain ϵ of the material may be approximated by

$$\sigma = 68.8\epsilon - (2.36 \times 10^3)\epsilon^2 - (2.06 \times 10^6)\epsilon^3 \text{ GPa}$$

Given D, d, and L_e, write an algorithm that computes the critical value of P using the tangent modulus theory of buckling. Run the algorithm with the following data: $D = 80$ mm, $d = 60$ mm, and (a) $L_e = 1.2$ m; and (b) $L_e = 2$ m.

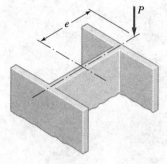

FIG. C10.2

11
Additional Beam Topics

Fuselage of an aircraft under construction. The rib sections are curved, thin-walled beams. Highly curved beams and beams with thin walls require special methods of analysis.

11.1 Introduction

The theory of bending in Chapter 5 was based on several assumptions that placed restrictions on its use. Here are two of these assumptions:

- The beam has a plane of symmetry and the applied loads act in this plane.
- The beam is initially straight, or its radius of curvature is large compared to the depth of the beam.

If the loads do not act in a plane of symmetry of the beam, the shear stresses may cause the beam to twist, unless the plane of loading passes through a certain point on the cross section known as the *shear center*. This problem is analyzed in Arts. 11.2 and 11.3, where we also learn how to locate the shear center of a given cross section.

The topic of Art 11.4 is *unsymmetrical bending*, where the bending moment acts about an axis that is not a principal axis of the cross section. As a consequence, the neutral axis of the cross section does not generally coincide with the axis of bending, so that the flexure formula $\sigma = -My/I$ is not directly applicable. Article 11.4 shows how to modify the analysis so that the flexure formula can be used for unsymmetrical bending.

Article 11.5 discusses stresses in beams with significant initial curvature. In straight beams, the assumption that plane sections remain plane gave rise to linear stress distribution over a cross section, as seen in the flexure formula. If the beam is curved, the same assumption results in a nonlinear distribution of stress, which is described by the curved beam formula derived in the article.

11.2 Shear Flow in Thin-Walled Beams

In Art 5.4, we derived the equation $\tau = VQ/(Ib)$ for calculating the vertical shear stresses induced by the transverse shear force in symmetric beams. A similar formula can be obtained for the shear stress in the flanges of structural shapes, such as wide-flange and channel sections. Deriving this shear stress is essentially an extension of the arguments that were used in Art. 5.4.

Figures 11.1(a) and (b) show an infinitesimal segment of a wide-flange beam. The segment is bounded by the sections ① and ②, a distance dx apart. The bending moments acting on the two sections are denoted by M and $M + dM$. In Fig. 11.1(c), we have isolated the shaded portion of the flange by a longitudinal (vertical) cutting plane. The normal forces P and $P + dP$ are the resultants of the bending stresses acting over the area A' of the flange. In Art. 5.4, we showed that

$$P = -\frac{MQ}{I} \qquad \text{(5.6, repeated)}$$

where Q is the first moment of area A' about the neutral axis, and I represents the moment of inertia of the cross-sectional area of the beam about the neutral axis. Therefore,

$$dP = -\frac{dM}{dx}\frac{Q}{I}dx = -\frac{VQ}{I}dx$$

where we have substituted $V = dM/dx$.

Equilibrium of the free-body diagram in Fig. 11.1(c) (only forces acting in the x-direction are shown) requires the presence of a longitudinal shear force dF on the longitudinal plane. Assuming that the shear stress τ is uniformly distributed over the thickness t of the flange, we have $dF = \tau t\, dx$. Therefore, the equilibrium equation

$$\Sigma F_x = 0: \quad (P + dP) - P + dF = 0$$

becomes

$$-\frac{VQ}{I}dx + \tau t\, dx = 0$$

FIG. 11.1 (a)–(b) Infinitesimal segment of a W-section; (c) free-body diagram used in determining the shear stress in the flange.

yielding

$$\tau = \frac{VQ}{It} \qquad (11.1)$$

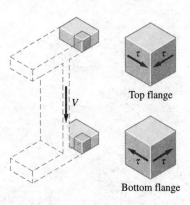

FIG. 11.2 Shear stresses in the flanges of a W-section.

Longitudinal shear stress in a flange is accompanied by shear stress of equal magnitude on the cross-sectional plane, as illustrated in Fig. 11.2. Because Q in Eq. (11.1) becomes negative when A' is below the neutral axis, the shear stresses in the top and bottom flanges have opposite directions as shown.

For thin-walled members, it is convenient to introduce the concept of *shear flow*. Whereas shear stress represents the force per unit area, shear flow q refers to the *force per unit length*. In terms of the shear stress τ and wall thickness t, the shear flow is

$$q = \tau t = \frac{VQ}{I} \qquad (11.2)$$

Using Eq. (11.2), we can find the shear flow distribution in the flanges of wide-flange and channel sections. Referring to Fig. 11.3(a), we have

$$q = \frac{VQ}{I} = \frac{V(tz)\bar{y}}{I} = \frac{Vht}{2I} z \qquad \text{(a)}$$

This equation shows that the shear flow q varies linearly with the distance from the free edge of the flange. The variation and direction of the shear flow on the cross sections are illustrated by the *shear flow diagrams* in Fig. 11.3(b). (The shear force V acting on each cross section is assumed to act downward.)

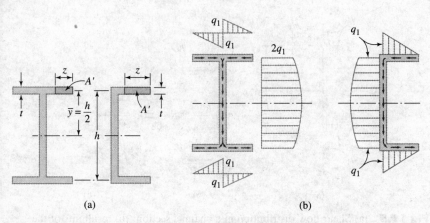

FIG. 11.3 (a) Computation of shear flow in wide-flange and channel sections; (b) distribution of shear flow.

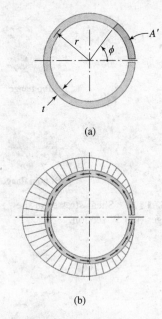

FIG. 11.4 (a) Computation of shear flow in a split tube; (b) distribution of shear flow.

Similarly, for the split tube in Fig. 11.4(a), the shear flow at the location defined by the angle ϕ is

$$q = \frac{VQ}{I} = \frac{V}{I}\int_0^\phi (r \sin \phi)(tr\,d\phi) = \frac{Vtr^2}{I}(1 - \cos \phi) \tag{b}$$

which gives the shear flow diagram in Fig. 11.4(b). (The shear force V is again assumed to act downward on the cross section.)

11.3 Shear Center

We now consider the bending of *thin-walled* sections that have only one axis of symmetry. In previous chapters, we always assumed that this axis of symmetry was in the plane of loading. In this article, we examine the conditions under which bending theory can be applied to sections for which the axis of symmetry is the neutral axis.

An example of a section with a single axis of symmetry is the channel section in Fig. 11.5. We assume the loading is vertical, so that the axis of symmetry is the neutral axis of the cross section. Figure 11.5(a) shows the shear flow induced by a vertical shear force V acting on the cross section. The maximum shear flow $q_1 = Vhtb/(2I)$ in the flanges is obtained from Eq. (a) of Art. 11.2 by substituting $z = b$, where the dimensions h, t, and b are defined in the figure. The resultant force H of the shear flow in a flange, shown in Fig. 11.5(b), is the area of the shear flow diagram:

$$H = \frac{1}{2}q_1 b = \frac{Vhtb^2}{4I} \tag{c}$$

The shear flow resultant in the web is equal to the vertical shear force V acting on the section. Noting that the forces H in the flanges form a couple of magnitude Hh, we can replace the force system in Fig. 11.5(b) by a statically equivalent force V acting through the point O, as shown in Fig. 11.5(c). The location of O is determined from the requirement that the moments of the two force systems about any point must be equal. Choosing B as the moment center, we get $Hh = Ve$, which yields

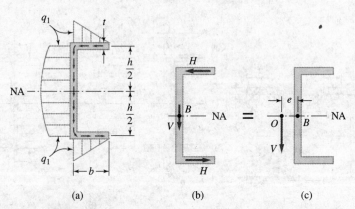

Fig 11.5 (a) Shear flow distribution in a channel section; (b) resultants of the shear flows; (c) statically equivalent force system consisting of the shear force V acting at the shear center O.

$$e = \frac{Hh}{V} = \frac{h^2 b^2 t}{4I} \tag{11.3}$$

The point O is known as the *shear center*, or the *flexural center*, of the cross section.

The foregoing results show that for the bending theory developed in Chapters 5 and 6 to be applicable, the shear force must act through the shear center. To satisfy this requirement, the *plane of external loading must pass through the shear center*. If the loading does not comply with this condition, bending will be accompanied by twisting. The cantilever beam in Fig. 11.6(a), for example, will not twist because the load P acts through the shear center of the end section. On the other hand, if P is placed at any other point, such as the centroid C of the section, the deformation of the beam will consist of twisting as well as bending, as illustrated in Fig. 11.6(b).

The shear centers of other thin-walled sections can be determined by similar analyses. In some cases, the location of the shear center can be determined by inspection. The T-section in Fig. 11.7(a) carries practically all the vertical shear in its flange (because the web lies along the neutral axis, its contribution is negligible). Therefore, the shear center of the section is located on the centerline of the flange. The shear center of the equal angle in Fig. 11.7(b) is clearly at the corner of the section, where the shear flows in the legs intersect.

If the axis of symmetry is vertical (in the plane of loading), the shear center always lies on that axis. As illustrated in Fig. 11.7(c), the horizontal shear flows are symmetric about the axis of symmetry. Therefore, they have no resultant. The vertical shear flow, on the other hand, has the resultant V that lies on the axis of symmetry.

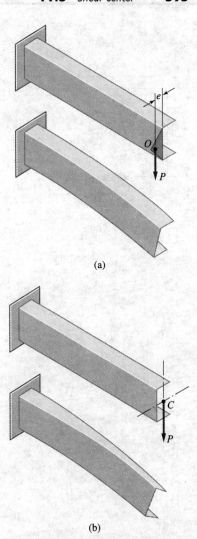

FIG. 11.6 (a) A channel section bends without twisting when the load acts at the shear center O; (b) if the load is not at the shear center, bending is accompanied by twisting.

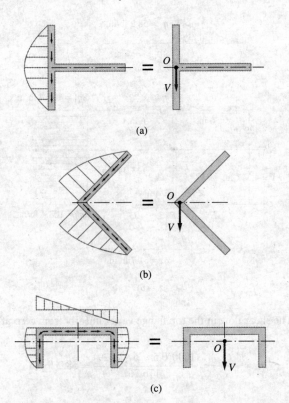

FIG. 11.7 Sections for which the shear center O can be located by inspection.

Sample Problem 11.1

The thin-walled section in Fig. (a) has a constant wall thickness of 0.5 in. If the shear force acting on the section is $V = 1000$ lb directed in the negative y-direction, draw the shear flow diagram for the cross section.

Solution

Referring to Fig. (a), we see that the location of the neutral axis is given by

$$\bar{y} = \frac{\sum A_i \bar{y}_i}{\sum A_i} = \frac{4(0) + 4(4) + 6(8)}{4 + 4 + 6} = 4.571 \text{ in.}$$

as shown in Fig. (b).

The moment of inertia of the cross-sectional area about the neutral axis can be calculated from

$$I = \sum \left[\frac{b_i h_i^3}{12} + A_i(\bar{y}_i - \bar{y})^2 \right]$$

For thin-walled sections, it is customary to neglect the first term if the longer side of the component area is parallel to the neutral axis—that is, if $b_i \gg h_i$. Applying this simplification to the flanges and referring to Fig. (b), we get

$$I = 4(4.571)^2 + \left[\frac{0.5(8)^3}{12} + 4(4 - 4.571)^2 \right] + 6(8 - 4.571)^2$$

$$= 176.76 \text{ in.}^4$$

The shear flow diagram is shown in Fig. (b). The computational details are explained below.

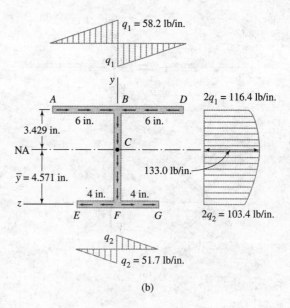

(b)

Top Flange The shear flow in the top flange varies linearly from zero at A to q_1 at B, where

$$q_1 = \frac{V Q_{AB}}{I} = \frac{1000(0.5 \times 6)(3.429)}{176.76} = 58.2 \text{ lb/in.}$$

The shear flow in the other half of the flange has the same distribution but it acts in the opposite direction.

Bottom Flange In the bottom flange, the shear is zero at E and increases linearly to

$$q_2 = \frac{VQ_{EF}}{I} = \frac{1000(0.5 \times 4)(4.571)}{176.76} = 51.7 \text{ lb/in.}$$

at point F. Again, the shear flow is symmetric in the two halves of the flange.

Web The shear flow at B in the web is

$$q_B = \frac{VQ_{AD}}{I} = \frac{V(2Q_{AB})}{I} = 2q_1 = 2(58.2) = 116.4 \text{ lb/in.}$$

Similarly, we get for the shear flow at F

$$q_F = \frac{VQ_{EG}}{I} = \frac{V(2Q_{EF})}{I} = 2q_2 = 2(51.7) = 103.4 \text{ lb/in.}$$

The maximum shear flow, which occurs at the centroid C of the cross section, is

$$q_{max} = \frac{V(Q_{AD} + Q_{BC})}{I} = 116.4 + \frac{1000(0.5 \times 3.429)(3.429/2)}{176.76} = 133.0 \text{ lb/in.}$$

■

Sample Problem 11.2

The section in Sample Problem 11.1 is rotated 90°, with the shear force $V = 1000$ lb still acting vertically downward. Determine (1) the shear force carried by each flange; and (2) the distance between the shear center and the 12-in.-wide flange.

Solution

From Fig. (a), the moment of inertia of the cross section about the neutral axis is (we neglect the small contribution of the web)

$$I = \frac{0.5(12)^3}{12} + \frac{0.5(8)^3}{12} = 93.33 \text{ in.}^4$$

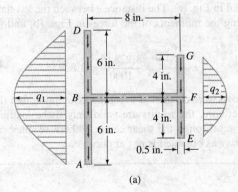

(a)

Part 1

The entire vertical shear force V is carried by the flanges, where the shear flow distribution is parabolic, as illustrated in Fig. (a). The maximum values of the shear flow are obtained from

$$q_1 = \frac{VQ_{BD}}{I} = \frac{1000(0.5 \times 6)(3)}{93.33} = 96.43 \text{ lb/in.}$$

$$q_2 = \frac{VQ_{FG}}{I} = \frac{1000(0.5 \times 4)(2)}{93.33} = 42.86 \text{ lb/in.}$$

The shear forces in the flanges, shown in Fig. (b), are equal to areas of the respective shear flow diagrams. Thus,

$$V_1 = \frac{2}{3}q_1\overline{AD} = \frac{2}{3}(96.43)(12) = 771.4 \text{ lb} \qquad \text{Answer}$$

$$V_2 = \frac{2}{3}q_2\overline{EG} = \frac{2}{3}(42.86)(8) = 228.6 \text{ lb} \qquad \text{Answer}$$

As a check, we note that $V_1 + V_2 = 771.4 + 228.6 = 1000.0$ lb, which is equal to the vertical shear force V acting on the cross section.

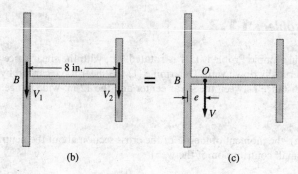

Part 2

The forces V_1 and V_2 are statically equivalent to the shear force V acting at the shear center O as indicated in Fig. (c). The distance e between the left flange and point O is obtained by equating the moments of the forces in Figs. (b) and (c) about B, which yields $8V_2 = eV$ or

$$e = \frac{8V_2}{V} = \frac{8(228.6)}{1000} = 1.829 \text{ in.} \qquad \text{Answer}$$

Note The shear flows were calculated using the assumption that the beam bends but does not twist. Therefore, the results are valid only if the loading is such that the shear force V does act through the shear center. Otherwise, the bending theory from which the shear flow was obtained is not applicable.

Problems

11.1–11.3 The vertical shear force V acts on the thin-walled section shown. Draw the shear flow diagram for the cross section. Assume that the thickness of the section is constant.

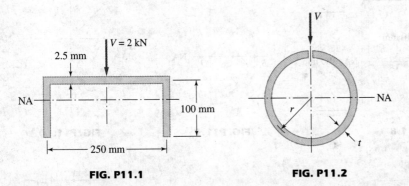
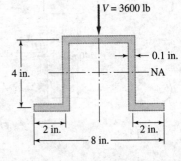

FIG. P11.1   FIG. P11.2   FIG. P11.3

11.4–11.7 For the cross section shown, (a) draw the shear flow diagram due to the vertical shear force V; and (b) calculate the distance e locating the shear center O. Assume that the thickness of the section is constant.

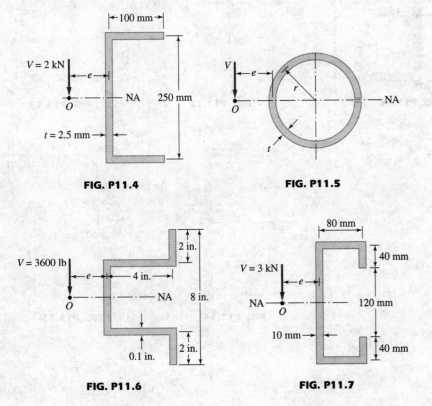

FIG. P11.4   FIG. P11.5

FIG. P11.6   FIG. P11.7

11.8–11.17 The cross section of the beam shown has a uniform wall thickness. Determine the location of the shear center relative to point B.

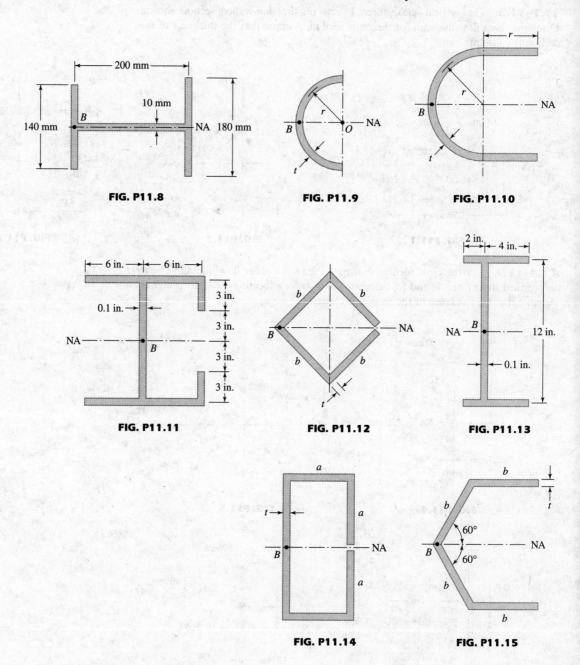

FIG. P11.8

FIG. P11.9

FIG. P11.10

FIG. P11.11

FIG. P11.12

FIG. P11.13

FIG. P11.14

FIG. P11.15

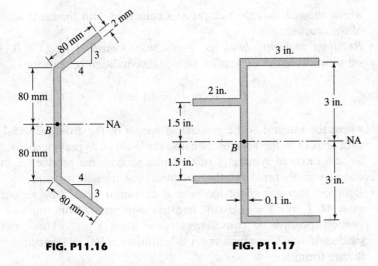

FIG. P11.16   **FIG. P11.17**

11.4 Unsymmetrical Bending

a. Review of symmetrical bending

The theory of bending developed in Chapter 5 was restricted to loads lying in a plane that contains an axis of symmetry of the cross section. The derivation of the equations that govern symmetrical bending was based on Fig. 5.3, which is repeated here as Fig. 11.8. The assumption that plane cross sections remain plane, combined with Hooke's law, led to the normal stress distribution

$$\sigma = -\frac{E}{\rho} y \qquad \text{(5.1, repeated)}$$

where ρ is the radius of curvature of the beam. The following equilibrium conditions had to be satisfied:

- *Resultant axial force must vanish*: $\int_A \sigma \, dA = 0$. Substituting σ from Eq. (5.1), we obtained

$$\int_A y \, dA = 0 \qquad \text{(a)}$$

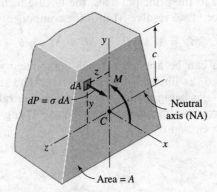

FIG. 11.8 Symmetrical bending, where the plane of loading (the *xy*-plane) is also a plane of symmetry of the beam.

which showed that the neutral axis coincided with the centroidal axis of the cross section.

- *Resultant moment about the y-axis must vanish*: $\int_A \sigma z \, dA = 0$. After substitution of the expression for σ, this condition became

$$I_{yz} = \int_A yz \, dA = 0 \tag{b}$$

where the integral is the *product of inertia* of the cross-sectional area with respect to the yz-axes. Because the y-axis was previously assumed to be an axis of symmetry of the cross section, the product of inertia was identically zero, and this condition was trivially satisfied.

- *Resultant moment about the neutral axis must equal the bending moment M*: $\int_A \sigma y \, dA = -M$ (the negative sign results from our sign convention: positive M causes negative σ when $y > 0$). This condition yielded $M = EI/\rho$, which upon substitution in Eq. (5.1) resulted in the flexure formula

$$\sigma = -\frac{My}{I} \tag{c}$$

We can now see that the requirement that the y-axis must be an axis of symmetry of the cross section is overly restrictive. According to Eq. (b), the flexure formula is valid as long as $I_{yz} = 0$, which is the case when the yz-axes are the *principal axes of inertia* of the cross section.[1] Therefore, the flexure formula is applicable if M acts about one of the principal axes of the cross section. The planes that are parallel to the principal axes and pass through the shear center are called the *principal planes of bending*. For the flexure formula to be valid, the external loads must lie in the principal planes of bending.

b. Symmetrical sections

We are now ready to discuss *unsymmetrical bending*. Unsymmetrical bending is caused by loads that pass through the shear center but do not lie in a principal plane of bending. Consider the symmetric cross section that carries the bending moment **M** as shown in Fig. 11.9 (we use a double-headed arrow to represent the bending moment as a vector). The yz-axes are the principal axes of inertia, and θ is the angle between **M** and the z-axis. Because **M** is inclined to the principal axes, the flexure formula is not directly applicable. However, if we resolve **M** into the components

$$M_y = M \sin \theta \qquad M_z = M \cos \theta \tag{d}$$

the flexure formula can be applied to each component separately and the results superimposed, as illustrated in Fig. 11.10. The stress due to M_y is $M_y z / I_y$, with the y-axis being the neutral axis. The neutral axis for M_z is the z-axis, and the corresponding stress is $-M_z y / I_z$. Using superposition, we obtain for the stress caused by **M**

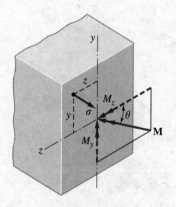

FIG. 11.9 Unsymmetrical bending of a symmetrical section.

---

[1] See Appendix A for a discussion of the inertial properties of areas, including product of inertia and principal moments of inertia.

$$\sigma = \frac{M_y z}{I_y} - \frac{M_z y}{I_z} \qquad (11.4)$$

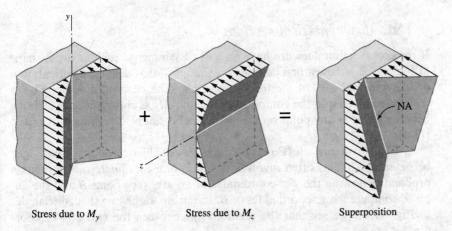

Stress due to M_y Stress due to M_z Superposition

FIG. 11.10 Normal stress distribution caused by unsymmetrical bending.

c. Inclination of the neutral axis

In general, the neutral axis for unsymmetrical bending is not parallel to the bending moment **M**. Because the neutral axis is the line where the bending stress is zero, its equation can be determined by setting $\sigma = 0$ in Eq. (11.4), which yields

$$\frac{M_y z}{I_y} - \frac{M_z y}{I_z} = 0$$

Substituting the components of the bending moment from Eq. (d) gives

$$M\left(\frac{z \sin \theta}{I_y} - \frac{y \cos \theta}{I_z}\right) = 0$$

which, after we cancel M, can be rearranged in the form

$$\frac{y}{z} = \frac{I_z}{I_y} \tan \theta \qquad (e)$$

Referring to Fig. 11.11, we see that y/z is the slope of the neutral axis. Therefore, Eq. (e) can be written as

$$\boxed{\tan \alpha = \frac{I_z}{I_y} \tan \theta} \qquad (11.5)$$

where α is the slope angle of the neutral axis (the angle between the neutral axis and the z-axis). Equation (11.5) shows that unless we have symmetrical bending ($\theta = 0$ or $90°$), the neutral axis will be parallel to the moment vector **M** only if $I_y = I_z$.

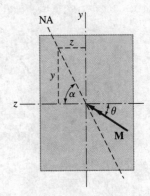

FIG. 11.11 In unsymmetrical bending the neutral axis is not necessarily parallel to the bending moment vector.

Because the bending stress is proportional to the distance from the neutral axis, the maximum stress occurs at the point that is farthest from the neutral axis. Therefore, locating the neutral axis can be useful in determining the location of the maximum bending stress on a cross section.

d. Unsymmetrical sections

If the cross section does not have an axis of symmetry, such as the Z-shape in Fig. 11.12, we must first determine the angle θ that defines the orientation of the principal axes of inertia (the yz-axes). The bending moment **M** can then be resolved into the components M_y and M_z as shown, after which the bending stress at any point on the cross section can be determined from Eq. (11.4).

Before we can use Eq. (11.4), the yz-coordinates of the point must be determined. This often involves the following *coordinate transformation* problem: Knowing the $y'z'$-coordinates of an arbitrary point B and the angle θ, compute the yz-coordinates of B. Referring to the two shaded triangles in Fig. 11.12, we see that the relationships between the two sets of coordinates are

$$y_B = y'_B \cos \theta + z'_B \sin \theta \qquad z_B = z'_B \cos \theta - y'_B \sin \theta \qquad (11.6)$$

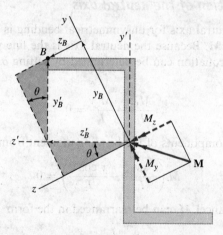

FIG. 11.12 Bending of an unsymmetrical section. The yz-axes are the principal axes of inertia of the cross section. The shaded triangles are used to derive coordinate transformations for points on the cross section.

Sample Problem 11.3

The W250 × 33 section carries a 32-kN · m bending moment inclined at 16.2° to the z-axis, as shown in Fig. (a). Determine (1) the angle between the neutral axis and the z-axis; and (2) the largest bending stress acting on the section.

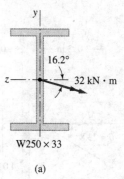

(a)

Solution

From Table B-2 in Appendix B, we obtain the following properties of the section:

$$I_y = 4.73 \times 10^6 \text{ mm}^4 \qquad I_z = 48.9 \times 10^6 \text{ mm}^4$$
$$S_y = 64.7 \times 10^3 \text{ mm}^3 \qquad S_z = 379 \times 10^3 \text{ mm}^3$$

Part 1

The angle α between the neutral axis and the z-axis can be computed from Eq. (11.5):

$$\tan \alpha = \frac{I_z}{I_y} \tan \theta = \frac{48.9}{4.73} \tan 16.2° = 3.004$$

$$\alpha = 71.6° \qquad \textit{Answer}$$

The neutral axis is shown in Fig. (b).

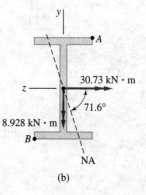

(b)

Part 2

Resolving the bending moment in Fig. (a) into components parallel to the principal axes, we get

$$M_y = -32 \sin 16.2° = -8.928 \text{ kN} \cdot \text{m}$$
$$M_z = -32 \cos 16.2° = -30.73 \text{ kN} \cdot \text{m}$$

By inspection of Fig. (b), we see that the largest bending stress occurs at A and B because these points are farthest from the neutral axis. We also note that both components of the bending moment cause tension at A and compression at B. Therefore, the largest bending stress acting on the section is

$$\sigma_{max} = \sigma_A = |\sigma_B| = \frac{|M_y|}{S_y} + \frac{|M_z|}{S_z} = \frac{8.928 \times 10^3}{64.7 \times 10^{-6}} + \frac{30.73 \times 10^3}{379 \times 10^{-6}}$$

$$= 219 \times 10^6 \text{ Pa} = 219 \text{ MPa} \qquad \textit{Answer}$$

Sample Problem 11.4

An L6 × 4 × 1/2 angle is used as a simply supported beam, 48 in. long. The beam carries a 1.0-kip load at its midspan as shown in Fig. (a). Determine (1) the angle between the neutral axis and the horizontal; and (2) the maximum bending stress in the beam.

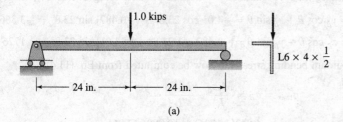

(a)

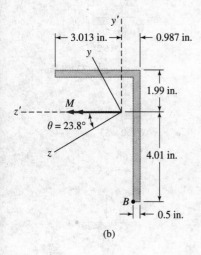

(b)

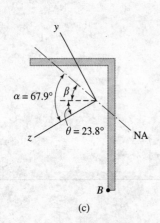

(c)

Solution

Preliminary Computations

The cross section is shown in Fig. (b), where the yz-axes are the principal axes of inertia. The relevant dimensions shown in the figure were obtained from the data in Table B-10 in Appendix B. The table also lists the following geometric properties of the cross section: $A = 4.75$ in.^{2}, $I_{y'} = 6.27$ in.^{4}, $I_{z'} = 17.4$ in.^{4}, $r_y = 0.870$ in. (radius of gyration about the y-axis), and $\tan \theta = 0.440$.

From the given data, we obtain

$$\theta = \tan^{-1} 0.440 = 23.8°$$

and

$$I_y = A r_y^2 = 4.75(0.870)^2 = 3.595 \text{ in.}^4$$

The moment of inertia about the z-axis can be calculated from the property $I_y + I_z = I_{y'} + I_{z'}$ (the sum of the moments of inertia does not vary with coordinate transformation). Thus,

$$I_z = I_{y'} + I_{z'} - I_y = 6.27 + 17.4 - 3.595 = 20.08 \text{ in.}^4$$

The bending moment **M** acts about the z'-axis as shown in Fig. (b).

Part 1

The angle of inclination of the neutral axis with respect to the z-axis is given by Eq. (11.5):

$$\tan \alpha = \frac{I_z}{I_y} \tan \theta = \frac{20.08}{3.595}(0.440) = 2.458 \qquad \alpha = 67.9°$$

Thus, the angle between the neutral axis and the horizontal is—see Fig. (c):

$$\beta = \alpha - \theta = 67.9° - 23.8° = 44.1° \qquad \textbf{Answer}$$

Part 2

The maximum bending moment is located at the midspan of the beam. Its magnitude is

$$M = \frac{PL}{4} = \frac{1.0(48)}{4} = 12 \text{ kip} \cdot \text{in.}$$

From Fig. (b), the components of this moment along the principal axes of inertia are

$$M_y = M \sin \theta = 12 \sin 23.8° = 4.843 \text{ kip} \cdot \text{in.}$$

$$M_z = M \cos \theta = 12 \cos 23.8° = 10.980 \text{ kip} \cdot \text{in.}$$

From Fig. (c), we see that the maximum bending stress occurs at point B because B is the most distant point from the neutral axis. From Fig. (b), the coordinates of B are

$$y'_B = -4.01 \text{ in.} \qquad z'_B = -(0.987 - 0.5) = -0.487 \text{ in.}$$

Transforming these coordinates to the principal axes using Eqs. (11.6), we get

$$y_B = y'_B \cos \theta + z'_B \sin \theta = -4.01 \cos 23.8° + (-0.487) \sin 23.8° = -3.866 \text{ in.}$$

$$z_B = z'_B \cos \theta - y'_B \sin \theta = -0.487 \cos 23.8° - (-4.01) \sin 23.8° = 1.1726 \text{ in.}$$

The maximum bending stress can now be computed from Eq. (11.4):

$$\sigma_{max} = \sigma_B = \frac{M_y z_B}{I_y} - \frac{M_z y_B}{I_z}$$

$$= \frac{4.843(1.1726)}{3.595} - \frac{10.980(-3.866)}{20.08} = 3.69 \text{ ksi} \qquad \textbf{Answer}$$

Problems

11.18 The cross section of a wood beam carries a bending moment **M** of magnitude 18 kip·in. acting at 10° to the horizontal. Determine (a) the angle between the neutral axis and the horizontal; and (b) the maximum bending stress acting on the cross section.

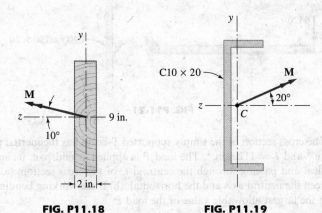

FIG. P11.18   FIG. P11.19

11.19 The magnitude of the bending moment **M** acting on the C10 × 20 section is 60 kip·in. Calculate (a) the angle between the neutral axis and the horizontal; and (b) the largest bending stress acting on the section.

11.20 The simply supported beam is loaded by a force P that is inclined at 40° to the vertical and passes through the centroid C of the cross section. If the working bending stress is 20 ksi, determine the largest allowable value of P.

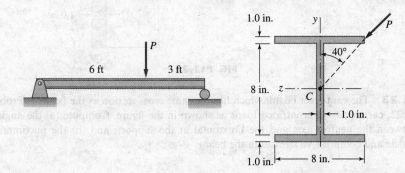

FIG. P11.20

11.21 The L200 × 100 × 20 section is used as a cantilever beam supporting the 6-kN load. Determine (a) the angle between the neutral axis and the vertical; and (b) the maximum bending stress in the beam.

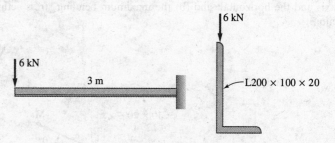

FIG. P11.21

11.22 The cross section of the simply supported T-beam has the inertial properties $I_y = 18.7$ in.^{4} and $I_z = 112.6$ in.^{4}. The load P is applied at midspan, inclined at 30° to the vertical and passing through the centroid C of the cross section. (a) Find the angle between the neutral axis and the horizontal. (b) If the working bending stress is 12 ksi, find the largest allowable value of the load P.

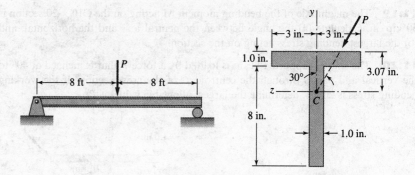

FIG. P11.22

11.23 The cantilever beam, which has the same cross section as the beam in Prob. 11.22, carries two concentrated loads as shown in the figure. Compute (a) the angle between the neutral axis and the horizontal at the support; and (b) the maximum tensile and compressive stresses in the beam.

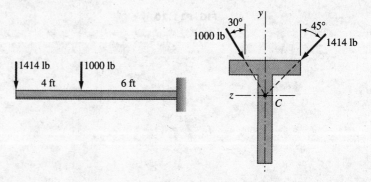

FIG. P11.23

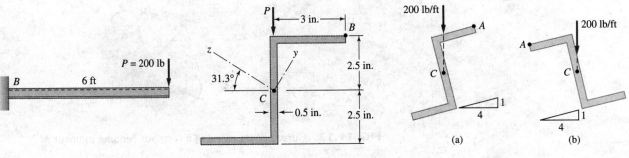

FIG. P11.24   **FIG. P11.25**

11.24 The Z-section is used as a cantilever beam that carries the force $P = 200$ lb at its free end. The principal moments of inertia of the cross section are $I_y = 2.95$ in.^{4} and $I_z = 25.25$ in.^{4}. Compute the bending stress at point B.

11.25 The Z-section described in Prob. 11.24 is used as a simply supported roof purlin, 12 ft long, carrying a distributed vertical load of 200 lb/ft. The slope of the roof is 1:4, as indicated in the figure. Determine the maximum bending stress at corner A for the orientations (a); and (b) of the purlin.

11.26 The masonry column carries an eccentric load $P = 15$ kN as shown in the figure. (a) Locate the points on the cross section where the neutral axis crosses the y- and the z-axes. (b) Determine the maximum tensile and compressive normal stresses.

11.27 The short column is made of a thin-walled equal-angle section. The principal moments of inertia of the section are $I_y = tb^3/3$ and $I_z = tb^3/24$. If the load P acts at the tip of a leg as shown, determine the compressive stress at point B on a cross section.

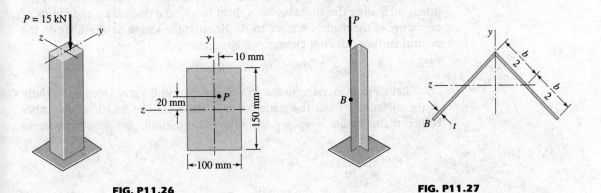

FIG. P11.26   **FIG. P11.27**

11.5 Curved Beams

a. Background

When deriving the flexure formula $\sigma = -My/I$, we assumed that the beam was initially straight. If a beam is curved, as shown in Fig. 11.13, the bending stress distribution is no longer linear. Therefore, we must derive another

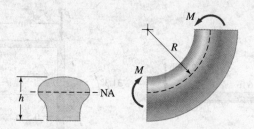

FIG. 11.13 Curved beam carrying a constant bending moment M.

formula for computing the bending stress. The difference between the two formulas will be negligible if the curvature of the beam is small. As a general rule, the flexure formula gives acceptable results for curved beams if $R/h > 5$, where R is the radius of curvature of the neutral surface and h is the depth of the beam. For more sharply curved beams, the flexure formula should not be used.

b. Compatibility

Consider the deformation of the infinitesimal segment of a curved beam shown in Fig. 11.14. We denote the radius of curvature of the neutral surface by R and the angle between the two cross sections by $d\theta$ before deformation. We assume, as we did for straight beams, that plane cross sections remain plane after bending. Although this assumption is not strictly accurate, it gives results that agree closely with strain measurements. When the bending moment M is applied, the cross sections rotate relative to each other, increasing the angle between them to $d\theta'$ and decreasing the radius of curvature of the neutral surface to R'. Because the length ds of a fiber on the neutral surface does not change, we have

$$R\, d\theta = R'\, d\theta' \tag{a}$$

Let us now investigate the strain of a typical fiber ab, located initially at the distance r from the center of curvature O. The length of this fiber before deformation is $\overline{ab} = r\, d\theta$. After deformation, the length becomes

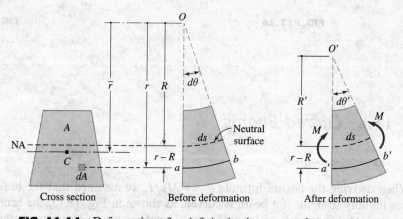

FIG. 11.14 Deformation of an infinitesimal segment of a curved beam.

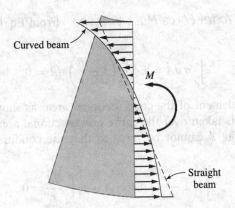

FIG. 11.15 Normal stress distribution in a curved beam.

$\overline{a'b'} = [R' + (r - R)] d\theta' = R \, d\theta + (r - R) \, d\theta'$, where in the last step we used Eq. (a). Thus, the normal strain of the fiber is

$$\epsilon = \frac{\overline{a'b'} - \overline{ab}}{\overline{ab}} = \frac{[R \, d\theta + (r - R) \, d\theta'] - r \, d\theta}{r \, d\theta}$$

$$= \frac{r(d\theta' - d\theta) - R(d\theta' - d\theta)}{r \, d\theta} = \frac{d\theta' - d\theta}{d\theta}\left(1 - \frac{R}{r}\right)$$

If we let $\phi = (d\theta' - d\theta)/d\theta$, the strain of the fiber ab becomes

$$\epsilon = \phi\left(1 - \frac{R}{r}\right)$$

The normal stress in the fiber is

$$\sigma = E\epsilon = E\phi\left(1 - \frac{R}{r}\right) \quad \text{(b)}$$

The resulting stress distribution is shown in Fig. 11.15. Because the stress distribution is nonlinear, the tensile and compressive forces over a cross section cannot be balanced if the neutral surface passes through the centroid of the cross section. Therefore, the neutral surface must shift from the centroid of the section toward the center of curvature O. Comparing the stress in Eq. (b) with the linear stress distribution obtained from the flexure formula (dashed line) shows not only this shift but also the increased stress at the inner fibers and the decreased stress in the outer fibers.

c. Equilibrium

We next locate the neutral axis and derive the relationship between the stress and the applied bending moment. Because we consider only the effects of bending (no axial force), the normal stress acting on a cross section must satisfy the two equilibrium conditions stated next.

The Resultant Axial Force Must Vanish.
From Eq. (b), this condition becomes

$$\int_A \sigma \, dA = E\phi \int_A \left(1 - \frac{R}{r}\right) dA = 0 \qquad \text{(c)}$$

where dA is an element of the cross-sectional area, as shown in Fig. 11.14, and the integral is taken over the entire cross-sectional area A. Unless there is no deformation, ϕ cannot be zero, so that the condition for zero axial force becomes

$$\int_A \left(1 - \frac{R}{r}\right) dA = A - R \int_A \frac{dA}{r} = 0$$

Thus, the distance from the center of curvature O to the neutral axis is

$$R = \frac{A}{\int_A (1/r) \, dA} \qquad (11.7)$$

Evaluation of the integral $\int_A (1/r) \, dA$ by analytical means is possible only if the cross section has a simple shape. For complex cross sections, the integral must be computed by numerical methods.

The Resultant Moment Must Equal M.
Because the resultant of the stress distribution is a couple (there is no axial force), we can use any convenient moment center for computing its magnitude. Choosing the center of curvature O in Fig. 11.14 as the moment center, we get

$$\int_A r\sigma \, dA = M$$

which becomes, after substituting the stress from Eq. (b),

$$E\phi \int_A (r - R) \, dA = M$$

We recognize that $\int_A R \, dA = RA$ and $\int_A r \, dA = A\bar{r}$, where $\bar{r}$ is the distance from O to the centroid C of the cross section, as shown in Fig. 11.14. Therefore, the relationship between ϕ and M is

$$EA\phi(\bar{r} - R) = M \qquad (11.8)$$

d. Curved beam formula

Solving Eq. (11.8) for $E\phi$ and substituting the result into Eq. (b) yields the *curved beam formula*:

$$\sigma = \frac{M}{A(\bar{r} - R)}\left(1 - \frac{R}{r}\right) \qquad (11.9)$$

Equations (11.7) and (11.9) are sufficient for determining bending stresses in curved beams. As mentioned before, a potential difficulty is the computation

Cross section	$\int_A \dfrac{dA}{r}$
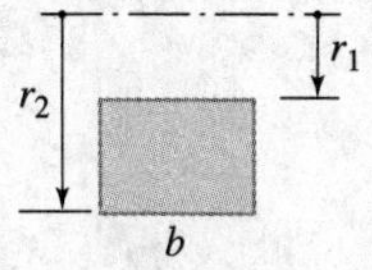 rectangle, width b, inner radius r_1, outer radius r_2	$b \ln \dfrac{r_2}{r_1}$
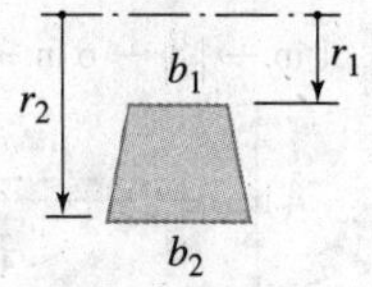 trapezoid with widths b_1 (top) and b_2 (bottom), radii r_1, r_2	$b_2 - b_1 + \dfrac{r_2 b_1 - r_1 b_2}{r_2 - r_1} \ln \dfrac{r_2}{r_1}$
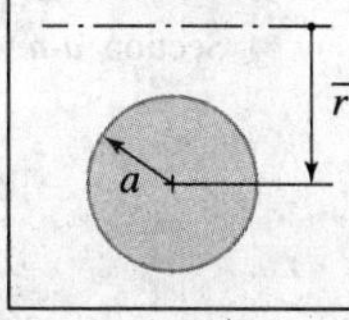 circle of radius a, centroidal radius $\bar{r}$	$2\pi(\bar{r} - \sqrt{\bar{r}^2 - a^2})$

FIG. 11.16 Values of $\int_A (1/r)\, dA$ for several cross sections of curved beams.

of the integral $\int_A (1/r)\, dA$. However, the integrals for some cross-sectional shapes are known; several of these are listed in Fig. 11.16.

Finally, we must point out that the difference between $\bar{r}$ and R is often very small, so that the subtraction $\bar{r} - R$ in the denominator of Eq. (11.9) can introduce a significant roundoff error. For this reason, $\bar{r}$ and R should be computed with two or three additional significant figures.

Sample Problem 11.5

The T-section in Fig. (a) is formed into a ring of 7.8-in. inner radius. Determine the largest force P that may be applied to the ring if the working normal stress is $\sigma_w = 18$ ksi.

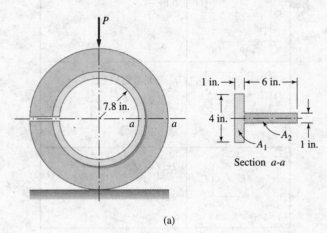

Section a-a

(a)

Solution

The critical section is a-a, where both the bending moment and the axial force reach their maximum values. The free-body diagram of the upper half of the ring in Fig. (b) shows the internal force system acting on section a-a. The compressive force P is placed at the centroid C of the cross section so that its contribution to normal stress is simply $-P/A$. The bending moment is $M = P\bar{r}$, where $\bar{r}$ is the radius of the centroidal axis of the ring. The bending stress due to M must be computed from the curved beam formula. The normal stress acting on a-a is obtained by superimposing the contributions of P and M. Because we do not know beforehand whether the highest stress occurs at the inner or outer radius of the ring, we must investigate the stresses at both locations.

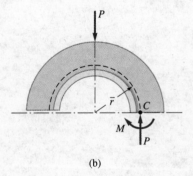

(b)

Cross-sectional Properties We consider the cross-sectional area as a composite of two rectangles of areas $A_1 = 4$ in.^{2} and $A_2 = 6$ in.^{2}, as shown in Fig. (a). Thus, the area of the section is

$$A = A_1 + A_2 = 4 + 6 = 10 \text{ in.}^2$$

From Fig. (a), the inner and outer radii of the ring are

$$r_1 = 7.8 \text{ in.} \qquad r_2 = 7.8 + 7 = 14.8 \text{ in.}$$

The radius of the centroidal axis is given by

$$\bar{r} = \frac{A_1\bar{r}_1 + A_2\bar{r}_2}{A_1 + A_2} = \frac{4(7.8 + 0.5) + 6(7.8 + 4)}{10} = 10.40 \text{ in.}$$

To calculate the radius R of the neutral surface for bending from Eq. (11.7), we need the integral $\int_A (1/r)\, dA$. Applying the formula in Fig. 11.16 for each rectangle, we get

$$\int_A \frac{dA}{r} = \int_{A_1} \frac{dA}{r} + \int_{A_2} \frac{dA}{r} = 4 \ln \frac{7.8 + 1}{7.8} + 1.0 \ln \frac{7.8 + 7}{7.8 + 1} = 1.002\,39 \text{ in.}$$

Equation (11.7) now yields

$$R = \frac{A}{\int_A (1/r)\, dA} = \frac{10}{1.002\,39} = 9.9762 \text{ in.}$$

Note that we used an extra significant figure in the computation of R.

Maximum Stress The bending stresses at the inner and outer radii are computed from Eq. (11.9). This equation contains the constant

$$\frac{M}{A(\bar{r} - R)} = \frac{P\bar{r}}{A(\bar{r} - R)} = \frac{P(10.40)}{10(10.40 - 9.9762)} = 2.454P$$

If we superimpose the uniform stress $-P/A$ and the bending stress, the normal stresses at points on the inner radius r_1 and the outer radius r_2 are

$$\sigma_{\text{inner}} = -\frac{P}{A} + \frac{M}{A(\bar{r} - R)}\left(1 - \frac{R}{r_1}\right) = -\frac{P}{10} + 2.454P\left(1 - \frac{9.9762}{7.8}\right)$$

$$= -0.7847P \text{ ksi}$$

$$\sigma_{\text{outer}} = -\frac{P}{A} + \frac{M}{A(\bar{r} - R)}\left(1 - \frac{R}{r_2}\right) = -\frac{P}{10} + 2.454P\left(1 - \frac{9.9762}{14.8}\right)$$

$$= 0.6998P \text{ ksi}$$

The stress distribution on section a-a is shown in Fig. (c). We see that the largest stress occurs at the inner radius.

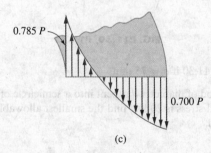

(c)

The maximum allowable load is given by

$$|\sigma_{\text{inner}}| = \sigma_w \qquad 0.7847P = 18 \text{ ksi}$$

which yields

$$P = 22.9 \text{ kips} \qquad\qquad \textit{Answer}$$

Problems

11.28 Derive the expression for $\int_A (1/r)\, dA$ in Fig. 11.16 for a rectangle.

11.29 The bending moment acting on the curved beam with a rectangular cross section is $M = 800$ N · m. Calculate the bending stress at point B.

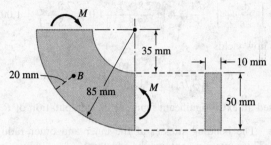

FIG. P11.29

11.30 The hook has a circular cross section of diameter $d = 100$ mm. Determine the maximum allowable value of load P if the working normal stress is 120 MPa.

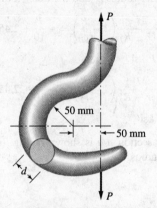

FIG. P11.30, P11.31

11.31 Solve Prob. 11.30 if $d = 75$ mm.

11.32 A circular rod of diameter d is bent into a semicircle of mean radius $\bar{r} = 2d$. If the working normal stress is 20 ksi, find the smallest allowable value of d that can carry the 1800-lb load.

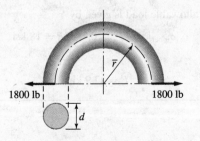

FIG. P11.32

11.33–11.37 Determine the maximum tensile and compressive stresses acting on section *a-a* of the curved beam shown.

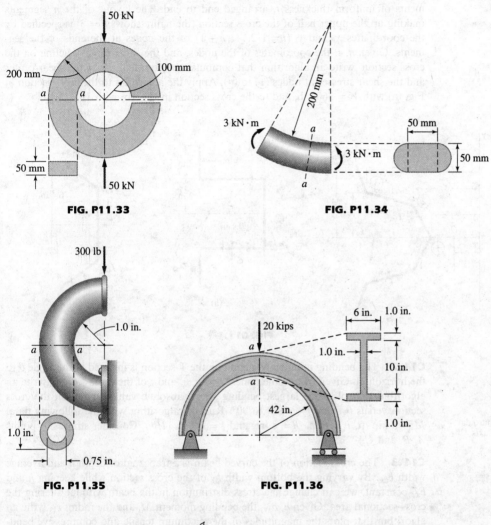

FIG. P11.33

FIG. P11.34

FIG. P11.35

FIG. P11.36

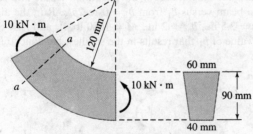

FIG. P11.37

Computer Problems

C11.1 The symmetric cross section of a thin-walled beam consists of straight segments of uniform thickness t, arranged end to end. The layout of the n segments making up the upper half of the cross section (the figure shows $n = 3$) is specified by the coordinates x_i and y_i ($i = 1, 2, \ldots, n+1$) of the nodes at the ends of the segments. Given n, t, the coordinates of the nodes, and the shear force V acting on the cross section, write an algorithm that computes the coordinate e of the shear center and the shear stresses at nodes ① to ⓝ. Apply the algorithm to the cross section in Fig. (a) with $V = 3600$ lb; and to the cross section in Fig. (b) with $V = 5000$ lb.

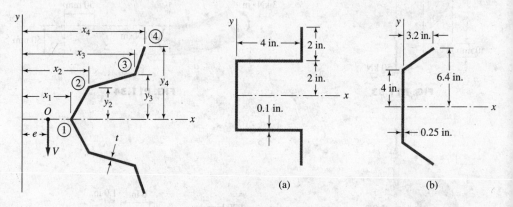

FIG. C11.1

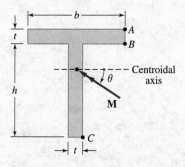

FIG. C11.2

C11.2 The bending moment **M** acting on the T-section is inclined at the angle θ to the horizontal. Given **M** and the dimensions b, h, and t of the cross section, write an algorithm that plots the largest bending stress (absolute value) acting on the cross section versus θ from $\theta = -90°$ to $90°$. Run the algorithm with the following data: $M = 8$ kip·ft, $b = 6$ in., $h = 8$ in., and $t = 1.0$ in. (*Hint*: Consider stresses at points A, B, and C.)

C11.3 The cross section of the curved beam is a trapezoid of depth h and average width b_{av}. By varying the bottom width b_1 of the cross section while keeping h and b_{av} constant, we can change the stress distribution in the beam without altering the cross-sectional area. Given h, b_{av}, the bending moment M, and the radius r_1, write an algorithm that plots the magnitudes of the maximum tensile and compressive bending stresses in the beam versus b_1 from $b_1 = 0$ to $2b_{av}$. Run the algorithm with the following data: $h = 2.5$ in., $b_{av} = 2$ in., $M = 1000$ lb·ft, and $r_1 = 6$ in. Use the plot to determine the value of b_1 that results in the smallest bending stress.

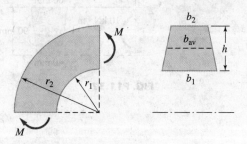

FIG. C11.3

12
Special Topics

Failure of a bridge caused by an earthquake. The source of the loading was ground movement, which is a form of dynamic loading.

12.1 Introduction

The preceding chapters covered the material that forms the core of a typical undergraduate course in mechanics of materials. Here we discuss briefly several topics that serve as a basis for more advanced studies. Because each topic constitutes an extensive field of study, the articles in this chapter should be considered as brief introductions to complex subjects. The articles are independent of one another, so that any one of them may be studied without reference to the others.

12.2 Energy Methods

a. Work and strain energy

When a force F is gradually applied to an elastic body that is adequately supported (no rigid-body displacements permitted), the force does work as the body deforms. This work can be calculated from $U = \int_0^\delta F \, d\delta$, where δ is the *work-absorbing displacement* of the point of application of F—that is, the displacement component in the direction of F. If the stress is below the proportional limit, then F is proportional to δ, as shown in Fig. 12.1, and the work becomes

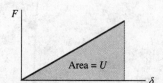

FIG. 12.1 Force-displacement diagram for an elastic body. The area U is the work done by the force.

$$U = \frac{1}{2} F \delta \qquad (12.1)$$

Note that U is the area under the force-displacement diagram. The work of a couple C has the same form as Eq. (12.1): $U = C\theta/2$, where θ is the angle of rotation (in radians) in the plane of the couple. Therefore, we view Eq. (12.1) as a generalized expression for work, where F can represent a force or a couple and δ is the work-absorbing displacement or rotation.

The work of several loads (forces and couples) $F_1, F_2, F_3, \ldots$ acting on an elastic body is independent of the order in which the loads are applied. It is often convenient to compute the work by assuming that all the loads are applied simultaneously, which results in

$$U = \frac{1}{2} \sum F_i \delta_i \qquad (12.2)$$

where δ_i is the work-absorbing displacement of F_i caused by all the loads.

If the body is elastic, the work of external loads is stored in the body as mechanical energy, called *strain energy*. If the loads are removed, the strain energy is released as the body returns to its original shape. Using the concepts of work and strain energy, we can develop powerful procedures for computing the displacements of elastic bodies. We begin by deriving expressions for the strain energy stored in bars and beams under various loadings.

b. Strain energy of bars and beams

Axial Loading Consider the bar of constant cross-sectional area A, length L, and modulus of elasticity E in Fig. 12.2. If the axial load P is applied gradually, the work-absorbing displacement of its point of application is the elongation $\delta = PL/(EA)$ of the bar. Therefore, the strain energy of the bar is

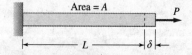

FIG. 12.2 Elongation of a bar due to an axial load P.

$$U = \frac{1}{2} P \delta = \frac{1}{2} \frac{P^2 L}{EA} \qquad (12.3)$$

If $P^2/(EA)$ is not constant along the length of the bar, Eq. (12.3) can be applied only to a segment of length dx. The strain energy of the bar can then be obtained by adding the strain energies of the segments—that is, by integration. The result is

$$U = \frac{1}{2}\int_0^L \frac{P^2\,dx}{EA} \qquad (12.4)$$

Torsion Figure 12.3 shows a circular bar of length L and constant cross section. If the torque T is applied gradually, the free end of the bar rotates through the angle $\theta = TL/(GJ)$, where J is the polar moment of inertia of the cross-sectional area. Because the work-absorbing displacement of the couple T is the rotation θ, the strain energy of the bar is

$$U = \frac{1}{2}T\theta = \frac{1}{2}\frac{T^2 L}{GJ} \qquad (12.5)$$

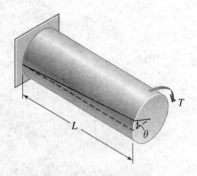

FIG. 12.3 Twisting of a bar due to a torque T.

When $T^2/(GJ)$ varies along the bar, Eq. (12.5) can be applied to a segment of length dx and the result integrated, yielding

$$U = \frac{1}{2}\int_0^L \frac{T^2\,dx}{GJ} \qquad (12.6)$$

Bending When the couple M is applied to the beam of length L in Fig. 12.4, the beam is deformed into an arc of radius ρ. As a result, the free end of the beam rotates through the angle $\theta = L/\rho$, which is the work-absorbing displacement of M. Thus, the strain energy of the beam is

$$U = \frac{1}{2}M\theta = \frac{1}{2}\frac{ML}{\rho}$$

Substituting $1/\rho = M/(EI)$ from Eq. (5.2b), where I is the moment of inertia of the cross-sectional area about the neutral axis, we get

$$U = \frac{1}{2}\frac{M^2 L}{EI} \qquad (12.7)$$

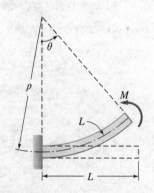

FIG. 12.4 Bending of a bar due to a couple M.

In general, M is not constant along the beam, so the strain energy of the beam must be obtained by applying Eq. (12.7) to a segment of length dx and integrating. Therefore,

$$U = \frac{1}{2}\int_0^L \frac{M^2\,dx}{EI} \qquad (12.8)$$

c. Deflections by Castigliano's theorem

Castigliano's Theorem Castigliano's theorem states that if an elastic body is in equilibrium under the external loads $F_1, F_2, F_3, \ldots$, then

$$\delta_i = \frac{\partial U}{\partial F_i} \qquad (12.9)$$

where δ_i is the work-absorbing displacement associated with F_i and U is the strain energy of the body (expressed in terms of the external loads).

In Eq. (12.9), F_i can be a force or a couple. If F_i is a force acting at a point A, then δ_i is the displacement of A in the direction of the force. If F_i represents a couple applied at a point A, then δ_i is the rotation of the body at A in the direction of the couple.

Proof Let the body in Fig. 12.5 be elastic and sufficiently supported so that it can maintain equilibrium after the loads $F_1, F_2, F_3, \ldots$ are applied (only three of the loads are shown). The strain energy U of the body is equal to the work done by these loads during the deformation of the body. Thus, the strain energy can be expressed as a function of the applied loads: $U = U(F_1, F_2, F_3, \ldots)$. Note that the reactions do no work because there are no corresponding work-absorbing displacements (the displacement of the roller support is perpendicular to the reaction). Assume now that after all the loads are applied, one of the loads, say F_i, is increased by an infinitesimal amount dF_i. The corresponding change in U is

$$dU = \frac{\partial U}{\partial F_i} dF_i \qquad (a)$$

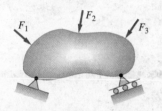

FIG. 12.5 Elastic body carrying several loads.

Consider next the case where the order of the loading is reversed. We apply dF_i first, followed by $F_1, F_2, F_3, \ldots$. The contribution of dF_i to the strain energy is now

$$dU = dF_i \delta_i \qquad (b)$$

where δ_i is the work-absorbing displacement of dF_i caused by all the loads. Note that the factor $1/2$ is absent because dF_i remains constant as the displacement δ_i occurs. Since the order of loading does not affect the strain energy, Eqs. (a) and (b) must give the same result; that is,

$$dU = \frac{\partial U}{\partial F_i} dF_i = dF_i \delta_i$$

which yields $\delta_i = \partial U / \partial F_i$, thereby completing the proof of Castigliano's theorem.

Application of Castigliano's Theorem In general, the strain energy of a bar subjected to combined loading is obtained by superimposing the contributions of axial loading, torsion, and bending:

$$U = \int_0^L \frac{P^2}{2EA} dx + \int_0^L \frac{T^2}{2GJ} dx + \int_0^L \frac{M^2}{2EI} dx$$

The deflection $\delta_i = \partial U/\partial F_i$ is best evaluated by differentiating inside the integral signs before integrating. This procedure is permissible because F_i is not a function of x. With this simplification, we obtain

$$\delta_i = \frac{\partial U}{\partial F_i} = \int_0^L \frac{P}{EA}\frac{\partial P}{\partial F_i}\,dx + \int_0^L \frac{T}{GJ}\frac{\partial T}{\partial F_i}\,dx + \int_0^L \frac{M}{EI}\frac{\partial M}{\partial F_i}\,dx \quad (12.10\text{a})$$

If no load acts at the point where the deflection is desired, a *dummy load* in the direction of the desired deflection must be added at that point. Then, *after differentiating* but before integrating, we set the dummy load equal to zero (this avoids integration of terms that will eventually be set equal to zero). If we denote the dummy load by Q, the displacement in the direction of Q thus is

$$\begin{aligned}\delta_Q &= \left.\frac{\partial U}{\partial Q}\right|_{Q=0} \\ &= \int_0^L \left[\frac{P}{EA}\frac{\partial P}{\partial Q}\right]_{Q=0} dx + \int_0^L \left[\frac{T}{GJ}\frac{\partial T}{\partial Q}\right]_{Q=0} dx \\ &\quad + \int_0^L \left[\frac{M}{EI}\frac{\partial M}{\partial Q}\right]_{Q=0} dx\end{aligned} \quad (12.10\text{b})$$

Castigliano's theorem can also be used to find redundant reactions in statically indeterminate problems. If we let Q be a redundant reaction that imposes the displacement constraint $\delta_Q = \Delta$ in the direction of Q, the equation for computing Q is $\Delta = \partial U/\partial Q$, or

$$\Delta = \int_0^L \frac{P}{EA}\frac{\partial P}{\partial Q}\,dx + \int_0^L \frac{T}{GJ}\frac{\partial T}{\partial Q}\,dx + \int_0^L \frac{M}{EI}\frac{\partial M}{\partial Q}\,dx \quad (12.10\text{c})$$

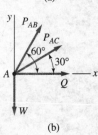

Sample Problem 12.1

For the steel truss in Fig. (a), find the horizontal displacement of point A due to the applied load $W = 24$ kips. Use $E = 29 \times 10^6$ psi and the cross-sectional areas shown in the figure.

Solution

The lengths of the members are

$$L_{AB} = \frac{72}{\sin 30°} = 144 \text{ in.} \qquad L_{AC} = \frac{72}{\sin 60°} = 83.14 \text{ in.}$$

The free-body diagram of joint A of the truss is shown in Fig. (b). In addition to the applied load W, the diagram contains the horizontal dummy load Q, which is required for the computation of the horizontal displacement. Using the free-body diagram, we obtain the equilibrium equations

$$\Sigma F_x = 0 \qquad P_{AB} \cos 60° + P_{AC} \cos 30° + Q = 0$$
$$\Sigma F_y = 0 \qquad P_{AB} \sin 60° + P_{AC} \sin 30° - W = 0$$

the solution of which is

$$P_{AB} = 1.7321W + Q \qquad P_{AC} = -W - 1.7321Q$$

The strain energy of the truss is the sum of the strain energies of its members:

$$U = \sum \frac{P^2 L}{2EA}$$

According to Castigliano's theorem, Eq. (12.10b), the horizontal displacement of A is

$$\delta_A = \left.\frac{\partial U}{\partial Q}\right|_{Q=0} = \sum \left[\frac{PL}{EA}\frac{\partial P}{\partial Q}\right]_{Q=0} \qquad (a)$$

The computations are facilitated by the following table:

Member	P	L (in.)	A (in.²)	$\frac{\partial P}{\partial Q}$	$\left[\frac{PL}{A}\right]_{Q=0}$
AB	$1.7321W + Q$	144	4	1	$62.36W$
AC	$-W - 1.7321Q$	83.14	5	-1.7321	$-16.628W$

Substituting the data from the last two columns of the table into Eq. (a), we get

$$\delta_A = \frac{1}{E}[(62.36W)(1) + (-16.628W)(-1.7321)]$$

$$= 91.16\frac{W}{E} = 91.16\left(\frac{24 \times 10^3}{29 \times 10^6}\right) = 0.0754 \text{ in.} \qquad \text{Answer}$$

Because the answer is positive, the horizontal displacement of point A has the same direction as Q—that is, to the right.

Sample Problem 12.2

The round bar AB in Fig. (a) is formed into a quarter-circular arc of radius R that lies in the horizontal plane. The bar is built in at B and carries the vertical force P at end A. Find the vertical deflection at A.

Solution

Because the vertical deflection at A is the work-absorbing displacement of P, its computation is a straightforward application of Castigliano's theorem: $\delta_A = \partial U/\partial P$. We begin by deriving the expressions for the bending moment and torque in the bar. Examination of the free-body diagram in Fig. (b) shows that the moment arm of P is $R(1 - \cos\theta)$ about the torque axis and $R\sin\theta$ about the bending axis. Therefore,

$$T = PR(1 - \cos\theta) \qquad M = PR\sin\theta$$

$$\frac{\partial T}{\partial P} = R(1 - \cos\theta) \qquad \frac{\partial M}{\partial P} = R\sin\theta$$

Substituting the above expressions and $dx = R\,d\theta$ into Eq. (12.10a), we get

$$\delta_A = \frac{1}{GJ}\int_0^{\pi/2} T\frac{\partial T}{\partial P}R\,d\theta + \frac{1}{EI}\int_0^{\pi/2} M\frac{\partial M}{\partial P}R\,d\theta$$

$$= \frac{1}{GJ}\int_0^{\pi/2}[PR(1-\cos\theta)][R(1-\cos\theta)]R\,d\theta + \frac{1}{EI}\int_0^{\pi/2}(PR\sin\theta)(R\sin\theta)R\,d\theta$$

$$= \frac{PR^3}{GJ}\int_0^{\pi/2}(1-\cos\theta)^2\,d\theta + \frac{PR^3}{EI}\int_0^{\pi/2}\sin^2\theta\,d\theta$$

When we evaluate the integrals, the vertical deflection at A is found to be

$$\delta_A = \frac{PR^3}{GJ}\left(\frac{3\pi-8}{4}\right) + \frac{PR^3}{EI}\left(\frac{\pi}{4}\right) \qquad \text{Answer}$$

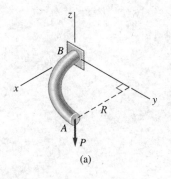

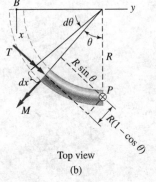

Top view
(b)

Sample Problem 12.3

The rigid frame in Fig. (a) is supported by a pin at A and a roller at D. (1) Find the value of $EI\delta_D$, where δ_D is the horizontal deflection at D due to the 800-N/m uniformly distributed load. (2) If the roller at D were replaced by a pin, determine the horizontal reaction at D. Assume that EI is constant throughout the frame, and consider only bending deformation.

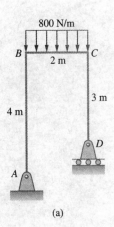

(a)

Solution

Part 1

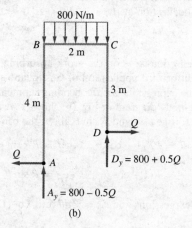

(b)

The free-body diagram of the frame is shown in Fig. (b). Because there is no horizontal force at D, we introduce the horizontal dummy load Q at that point. According to Castigliano's theorem, Eq. (12.10b), the horizontal deflection of D is

$$\delta_D = \left[\frac{\partial U}{\partial Q}\right]_{Q=0} = \int_L \left[\frac{M}{EI}\frac{\partial M}{\partial Q}\right]_{Q=0} dx \qquad \text{(a)}$$

where the integral extends over all members of the frame.

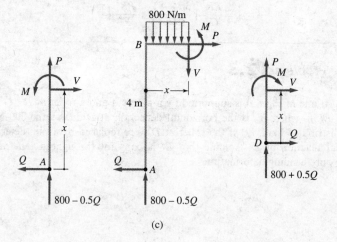

(c)

Figure (c) shows the free-body diagrams that can be used to derive the internal force system in each member of the frame. Our analysis requires only the bending moments, which are given in the table below. The table also lists the components of the integrand in Eq. (a). The origin of the x-coordinate used for each member is shown in Fig. (c).

	AB	BC	CD
M (N·m)	Qx	$4Q + (800 - 0.5Q)x - 400x^2$	Qx
$\dfrac{\partial M}{\partial Q}$ (m)	x	$4 - 0.5x$	x
$M\|_{Q=0}$ (N·m)	0	$800x - 400x^2$	0

Note that $M|_{Q=0} = 0$ for members AB and CD, so that only member BC contributes to the integral in Eq. (a). Therefore, the horizontal displacement of D is obtained from

$$EI\delta_D = \int_{L_{BC}} \left[M \frac{\partial M}{\partial Q}\right]_{Q=0} dx = \int_0^{2m} (800x - 400x^2)(4 - 0.5x)\, dx$$

$$= 1867 \text{ N·m}^3 \rightarrow \qquad \text{Answer}$$

Because the result is positive, the deflection at D is in the same direction as the dummy load Q—that is, to the right.

Part 2

The free-body diagram in Fig. (b) is still applicable, but now Q must be viewed as the horizontal (redundant) reaction at D. Because the displacement constraint at D is $\delta_Q = 0$ (δ_Q is the displacement in the direction of Q), Castigliano's theorem in Eq. (12.10c) becomes

$$0 = \frac{\partial U}{\partial Q} = \int_L \frac{M}{EI} \frac{\partial M}{\partial Q} dx$$

Substituting the data listed in the table and integrating along all three members, we get

$$0 = \int_0^{4m} (Qx)x\, dx + \int_0^{2m} [4Q + (800 - 0.5Q)x - 400x^2](4 - 0.5x)\, dx + \int_0^{3m} (Qx)x\, dx$$

$$= 21.33Q + (1866.7 + 24.67Q) + 9Q$$

which yields for the horizontal reaction at D

$$Q = -33.9 \text{ N} \qquad \text{Answer}$$

The negative sign means that the direction of Q is opposite to that shown in Fig. (b).

Problems

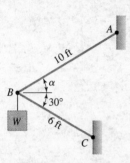

FIG. P12.1, P12.2

12.1 The aluminum truss carries the vertical load $W = 6000$ lb. Each member has a cross-sectional area of 0.6 in.^{2} and $E = 10 \times 10^6$ psi. Compute the horizontal and vertical displacements of joint B assuming $\alpha = 30°$.

12.2 Each member of the truss has a cross-sectional area of 0.6 in.^{2}. Member AB is made of steel, whereas member BC is made of aluminum. Assuming that $W = 6000$ lb and $\alpha = 45°$, determine the horizontal and vertical displacements of joint B. Use $E_{st} = 29 \times 10^6$ psi and $E_{al} = 10 \times 10^6$ psi.

12.3 The steel truss supports the load $F = 30$ kN. Determine the horizontal and vertical displacements of joint B. Use $E = 200$ GPa and the cross-sectional areas $A_{AB} = 300$ mm^2 and $A_{BC} = 500$ mm^2.

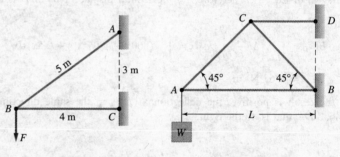

FIG. P12.3   FIG. P12.4

12.4 The members of the truss are made of the same material and have identical cross-sectional areas. Determine the vertical displacement of point A due to the applied load W.

12.5 Determine the vertical displacement at midspan for the simply supported beam.

12.6 For the simply supported beam, calculate the slope at point A due to the load P. (*Hint*: Introduce a dummy couple at A.)

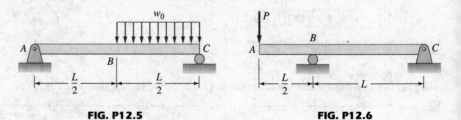

FIG. P12.5   FIG. P12.6

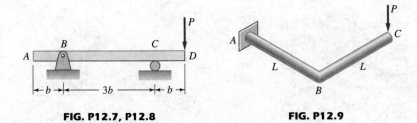

FIG. P12.7, P12.8   **FIG. P12.9**

12.7 For the overhanging beam, determine the vertical displacements at points A and D when the load P is applied.

12.8 Compute the slope of the overhanging beam at point C due to the load P. (*Hint*: Introduce a dummy couple at C.)

12.9 The bent circular rod AB is built in at A and carries the vertical load P at C. Determine the vertical displacement of point C.

12.10 The circular rod AB is bent into a semicircular arc of radius R. The rod is built in at A and carries the twisting moment T_0 at B. Determine the angle of twist at B.

12.11 For the bent rod described in Prob. 12.10, compute the vertical displacement of end B.

12.12 A vertical load P is applied to the cantilever frame. Assuming constant EI and considering only bending deformation, find the horizontal and vertical displacements of point C.

12.13 A circular rod is bent into a semicircular arc of radius R. When the horizontal load P is applied, determine the horizontal displacement of point C and the vertical displacement of point B. Consider only bending deformation.

12.14 Solve Prob. 12.13 assuming that the load P is applied vertically downward at C.

12.15 The frame shown in the figure, which has a constant bending rigidity EI, carries a 600-lb horizontal load at B. Considering only bending deformation, determine the value of $EI\delta_D$, where δ_D is the horizontal displacement of the roller support at D.

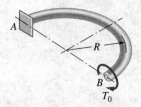

FIG. P12.10, P12.11

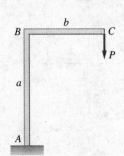

FIG. P12.12

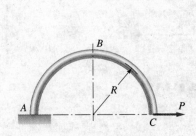

FIG. P12.13, P12.14

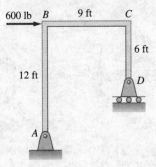

FIG. P12.15

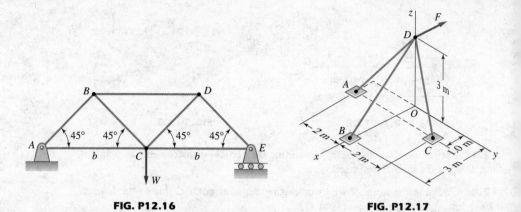

FIG. P12.16

FIG. P12.17

12.16 Find the vertical displacement at joint C of the steel truss. Assume that all members of the truss have the same cross-sectional area. (*Hint*: Use symmetry.)

12.17 Each member of the aluminum truss has a cross-sectional area of 500 mm^2. Determine the horizontal displacement of joint D caused by the force $F = 25$ kN acting in the negative x-direction. Use $E = 70$ GPa for aluminum.

12.18 The beam ABC of length $4b$ is built into the rigid wall at C and supported by a roller at B. Find the reaction at B when the load P is applied.

12.19 The cross-sectional area of each member of the steel truss is 4 in.^{2}. Find the force in member AD due to the 2.4-kip load.

12.20 Find the force in each member of the steel truss when the 3000-lb vertical load is applied to joint A. The area of each member is shown in the figure. Use $E = 29 \times 10^6$ psi for steel.

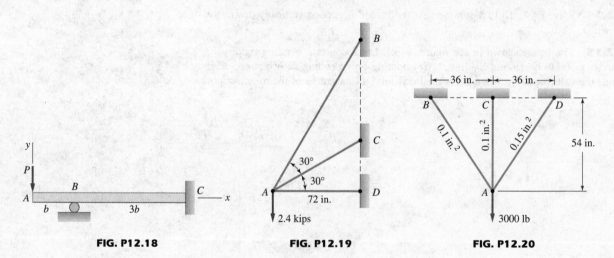

FIG. P12.18

FIG. P12.19

FIG. P12.20

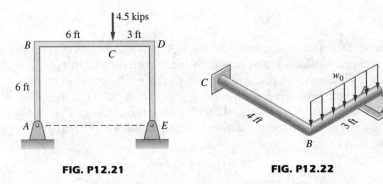
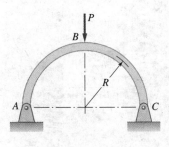

FIG. P12.21   **FIG. P12.22**   **FIG. P12.23**

12.21 The bending rigidity EI is the same for each member of the frame. Determine the horizontal component of the support reaction at E when the 4.5-kip vertical load is applied. Consider only bending deformations.

12.22 The bent steel rod ABC of diameter d is built into a rigid wall at C and rests on a support at A. Segment AB carries a uniformly distributed load $w_0 = 100$ lb/ft. Using $E = 30 \times 10^6$ psi and $G = 12 \times 10^6$ psi for steel, find the support reaction at A.

12.23 A uniform bar ABC is bent into a semicircular arc of radius R. The arc is supported by pins at A and C. Determine the magnitude of the horizontal component of the reaction at A when the vertical load P is applied at B. Consider only bending deformation.

12.24 Find the horizontal and vertical components of the reactions at A and D when the 600-N horizontal force is applied to the frame. The bending rigidity is EI for members AB and BC, and $2EI$ for member CD. Consider only deformation due to bending.

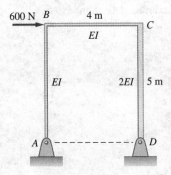

FIG. P12.24

12.3 Dynamic Loading

Static analysis implicitly assumes that the loads are applied so slowly that dynamic effects are negligible. Suddenly applied loading results in momentary displacements and stresses that can be much higher than those predicted by static analysis. In this article, we consider the effects of dynamic loading caused by a rigid mass colliding with a stationary, elastic body.

a. Assumptions

Our analysis of dynamic loading is based on the following simplifying assumptions:

- The stresses in the body remain below the proportional limit.
- The body and the impacting mass remain in contact during the collision (no rebound).
- No energy is lost during collision.

These assumptions lead to a highly idealized model of impact loading that gives only rough estimates for stresses and deformations. Because the loss of some energy to stress waves and heat is inevitable, the energy available to deform the body is less than what is predicted by the simplified theory. In other words, the actual stresses and displacements are less than the calculated values.

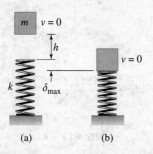

FIG. 12.6 Mass-spring model for dynamic loading: (a) before mass m is released; and (b) in the position of maximum deflection.

b. Mass-spring model

As a simple model of impact, consider the case in which a mass m is dropped onto a spring of stiffness k and negligible mass. As indicated in Fig. 12.6(a), the mass is released from rest and drops a distance h before making contact with the spring. Following the collision, the spring deforms, reaching its maximum deflection δ_{max} in Fig. 12.6(b) before rebounding. The displacement δ_{max} is known as the maximum *dynamic deflection*. The work-energy principle states that

$$U = \Delta T \qquad (12.11)$$

where U is the work done on the mass and ΔT represents the change in the kinetic energy of the mass. Applying this principle to the mass in Fig. 12.6(a), we conclude that the work done on the mass between the two positions shown is zero because there is no change in the kinetic energy ($T = 0$ in both positions).

The work done on the mass is

$$U = mg(h + \delta_{max}) - \frac{1}{2}k\delta_{max}^2$$

where the first term is the work done by gravity and the second term is the work done by the spring force (δ_{max} and the spring force have opposite directions, so the second term is negative). Because the total work is zero, we have

$$mg(h + \delta_{max}) - \frac{1}{2}k\delta_{max}^2 = 0 \qquad (a)$$

Substituting $mg/k = \delta_s$, where δ_s is the *static deformation* that would be produced by a gradual application of the weight mg, we can rearrange Eq. (a) in the form

$$\delta_{max}^2 - 2\delta_s\delta_{max} - 2\delta_s h = 0$$

which has the solution

$$\delta_{max} = \delta_s + \sqrt{\delta_s^2 + 2\delta_s h} \qquad (b)$$

Equation (b) can be written as

$$\boxed{\delta_{max} = n\delta_s} \qquad (12.12a)$$

where

$$\boxed{n = 1 + \sqrt{1 + \frac{2h}{\delta_s}}} \qquad (12.12b)$$

is called the *impact factor*. Note that the impact factor is a multiplier that converts the static deflection into the corresponding maximum dynamic deflection. Because the force in the spring is proportional to its deformation, the impact factor also applies to the spring force: $F_{max} = nF_s = n(mg)$.

Two special cases are of interest. If h is much larger than δ_s, we need keep only the term $2h/\delta_s$ in Eq. (12.12b), which then reduces to

$$n = \sqrt{\frac{2h}{\delta_s}} \qquad (h \gg \delta_s) \qquad (12.13)$$

The other case is $h = 0$ (the load is released when just touching the spring), for which Eq. (12.12b) yields $n = 2$. This result shows that the deflection caused by the sudden release of the load is twice as large as the deflection due to the same load when it is gradually applied.

c. Elastic bodies

The results obtained above for the mass-spring system remain valid if the spring is replaced by an elastic body of negligible mass because in both cases the load-displacement relationship is linear. As an example, consider the system shown in Fig. 12.7, where the mass m drops through the height h before striking a stop at the end of the rod. The static elongation of the rod would be $\delta_s = mgL/(EA)$. Assuming that $h \gg \delta_s$, we can use the impact factor Eq. (12.13), which gives for the maximum dynamic elongation of the rod

$$\delta_{\max} = n\delta_s = \sqrt{\frac{2h}{\delta_s}}\,\delta_s = \sqrt{2h\delta_s} = \sqrt{\frac{2hL}{EA}mg} \qquad \text{(c)}$$

Because the impact factor also applies to forces and stresses, the maximum dynamic stress in the rod can be obtained from

$$\sigma_{\max} = n\sigma_s = \sqrt{\frac{2h}{\delta_s}}\frac{mg}{A} = \sqrt{2h\left(\frac{EA}{mgL}\right)\left(\frac{mg}{A}\right)^2} = \sqrt{\frac{2Eh}{AL}mg} \qquad \text{(d)}$$

where σ_s is the static stress (stress caused by gradually applied loading). Equation (d) shows that the stress due to impact can be reduced by using a material with a smaller modulus of elasticity, or by increasing the area or the length of the rod. This is quite different from static tension, where the stress is independent of both E and L.

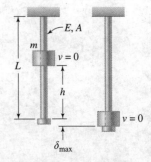

FIG. 12.7 The spring in the mass-spring model can be replaced by an elastic body, such as the vertical rod shown here.

d. Modulus of resilience; modulus of toughness

The modulus of resilience and the modulus of toughness are measures of the impact resistance of a material. These moduli are determined from the stress-strain diagram obtained from a simple tension test.

In simple tension, the work done by the stresses in deforming a *unit volume* of material is $U = \int_0^\epsilon \sigma\,d\epsilon$, which equals the area under the stress-strain diagram. The *modulus of resilience* U_r is defined as the maximum energy that the material can absorb per unit volume before it becomes permanently deformed. Therefore, U_r equals the area under the stress-strain diagram up to the yield point σ_{yp}. Referring to Fig. 12.8, we see that modulus of resilience is given by

$$U_r = \frac{1}{2}\sigma_{yp}\epsilon_{yp} = \frac{1}{2}\sigma_{yp}\left(\frac{\sigma_{yp}}{E}\right) = \frac{\sigma_{yp}^2}{2E} \qquad (12.14)$$

The *modulus of toughness*, denoted by U_t, is the energy absorbed by the material per unit volume up to rupture. Thus, U_t is equal to the area under the entire stress-strain diagram, as indicated in Fig. 12.8. It is evident that

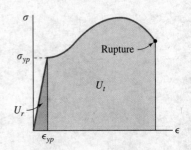

FIG. 12.8 The shaded areas equal the modulus of resilience (U_r) and the modulus of toughness (U_t).

the impact resistance of a material depends largely upon its ductility. Figure 12.9 shows the stress-strain diagrams for two materials. One material is a high-strength steel of low ductility; the other is a steel of lower strength but high ductility. In this case, the shaded area $(U_t)_1$ for the low-strength steel is larger than the area $(U_t)_2$ for the high-strength material. Thus, a ductile material can absorb more energy before rupture than can a stronger, but less ductile material. For this reason, ductile materials are usually selected for members subject to impact or shock loading.

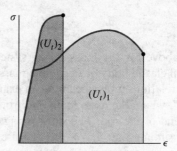

FIG. 12.9 Stress-strain diagrams for two grades of steel. The ductile, low-strength steel has greater toughness.

Sample Problem 12.4

An elevator that has a mass of 2000 kg is being lowered at the rate of 2 m/s. The hoisting drum is stopped suddenly when 30 m of cable has been unwound. If the cross-sectional area of the cable is 600 mm² and $E = 100$ GPa, compute the maximum force in the cable. Neglect the weight of the cable.

Solution

Figure (a) shows the elevator just before the hoisting drum is stopped. Because the speed of the elevator at that instant is constant, the force P in the cable equals the weight mg of the elevator. In the position of maximum dynamic elongation shown in Fig. (b), the speed of the elevator is zero and the cable force is $k\delta_{max} + mg$, where k is the effective spring stiffness of the cable. Between these two positions, P varies linearly with the elongation δ as shown in Fig. (c). Note that δ represents the dynamic elongation of the cable—that is, the elongation in addition to the static elongation $\delta_s = mg/k$.

Noting that P and δ have opposite directions, we see that the work of the cable force on the elevator is the negative of the area under the diagram in Fig. (c)—namely, $-(mg\delta_{max} + k\delta_{max}^2/2)$. Adding the work $mg\delta_{max}$ of the gravitational force, we obtain for the net work done on the elevator

$$U = -\frac{1}{2}k\delta_{max}^2$$

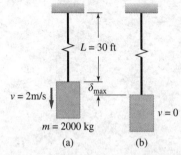

The elongation of the cable due to a static tensile force P is $\delta_s = PL/(EA)$, which yields for the effective spring stiffness of the cable

$$k = \frac{P}{\delta_s} = \frac{EA}{L} = \frac{(100 \times 10^9)(600 \times 10^{-6})}{30} = 2.0 \times 10^6 \text{ N/m}$$

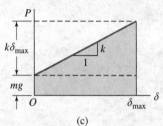

Therefore,

$$U = -\frac{1}{2}(2.0 \times 10^6)\delta_{max}^2 = -(1.0 \times 10^6)\delta_{max}^2$$

The change in the kinetic energy of the elevator between the positions shown in Figs. (a) and (b) is

$$\Delta T = T_{final} - T_{initial} = 0 - \frac{1}{2}mv^2 = -\frac{1}{2}(2000)(2)^2 = -4.0 \times 10^3 \text{ J}$$

Applying the work-energy principle $U = \Delta T$, we get

$$-(1.0 \times 10^6)\delta_{max}^2 = -4.0 \times 10^3$$

which gives

$$\delta_{max} = 63.25 \times 10^{-3} \text{ m}$$

The maximum force in the cable is

$$P_{max} = k\delta_{max} + mg = (2.0 \times 10^6)(63.25 \times 10^{-3}) + 2000(9.81)$$
$$= 146.1 \times 10^3 \text{ N} = 146.1 \text{ kN} \qquad \text{Answer}$$

Note that P_{max} is more than seven times the 19.62-kN weight of the elevator.

Sample Problem 12.5

The 80-kg block hits the simply supported beam at its midspan after a drop of 10 mm as shown in Fig. (a). Determine (1) the impact factor; and (2) the maximum dynamic bending stress in the beam. Use $E = 200$ GPa for the beam. Assume that the block and the beam stay in contact after the collision.

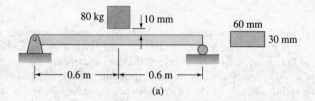

(a)

Solution

Part 1

The moment of inertia of the cross section of the beam about the neutral axis is

$$I = \frac{bh^3}{12} = \frac{60(30)^3}{12} = 135.0 \times 10^3 \text{ mm}^4 = 135.0 \times 10^{-9} \text{ m}^4$$

According to Table 6.3 on page 233, the static midspan deflection of the beam under the weight of the 80-kg mass is

$$\delta_{st} = \frac{(mg)L^3}{48EI} = \frac{(80 \times 9.81)(1.2)^3}{48(200 \times 10^9)(135.0 \times 10^{-9})} = 1.0464 \times 10^{-3} \text{ m}$$

From Eq. (12.12b), the impact factor is

$$n = 1 + \sqrt{1 + \frac{2h}{\delta_{st}}} = 1 + \sqrt{1 + \frac{2(0.010)}{1.0464 \times 10^{-3}}} = 5.485 \quad \text{Answer}$$

Part 2

The maximum dynamic load P_{max} at the midspan of the beam is obtained by multiplying the static load by the impact factor:

$$P_{max} = n(mg) = 5.485(80 \times 9.81) = 4305 \text{ N}$$

The maximum bending moment caused by this load occurs at the midspan, as shown in Fig. (b). Its value is

$$M_{max} = \frac{4305}{2}(0.6) = 1291.5 \text{ N} \cdot \text{m}$$

which results in the maximum dynamic bending stress

$$\sigma_{max} = \frac{M_{max} c}{I} = \frac{1291.5(0.015)}{135.0 \times 10^{-9}} = 143.5 \times 10^6 \text{ Pa} = 143.5 \text{ MPa} \quad \text{Answer}$$

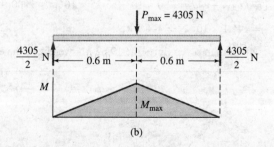

(b)

Problems

12.25 Calculate the modulus of resilience for the following three materials:

(a) Steel alloy $\quad E = 29 \times 10^3$ ksi $\quad \sigma_{yp} = 50$ ksi

(b) Brass $\quad E = 15 \times 10^3$ ksi $\quad \sigma_{yp} = 30$ ksi

(c) Aluminum alloy $\quad E = 10 \times 10^3$ ksi $\quad \sigma_{yp} = 40$ ksi

Which of the materials is best suited for absorbing impact without permanent deformation?

12.26 The 20-lb pendulum is released from rest in the position shown in the figure and strikes an aluminum bar 6 in. long. If the modulus of resilience of aluminum is 170 lb · in./in.^{3}, determine the smallest diameter d of the bar for which the impact can be absorbed without permanent deformation.

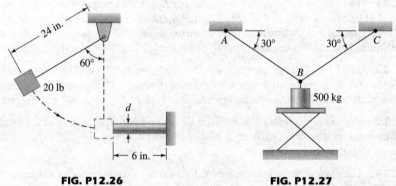

FIG. P12.26   FIG. P12.27

12.27 The 500-kg mass rests on a platform and is attached to the steel cable ABC. The platform is then gradually lowered until the slack in the cable is removed but all the weight of the mass is still supported by the platform. If the platform then suddenly collapses, find the maximum dynamic force in the cable.

12.28 A 100-lb weight falls through 6 ft and is then caught at the end of a wire rope 100 ft long having a cross-sectional area of 0.4 in.^{2}. Find the maximum dynamic stress in the rope, assuming $E = 15 \times 10^6$ psi.

12.29 The mass m attached to the end of a rope is lowered at the constant velocity v when the pulley at A suddenly jams. Show that the impact factor is

$$n = 1 + \sqrt{\frac{kv^2}{mg^2}}$$

where k is the effective spring stiffness of the rope.

FIG. P12.29

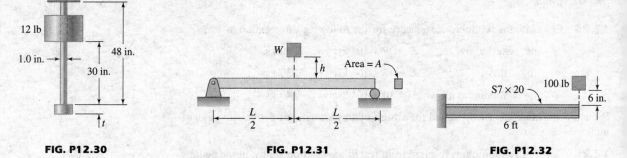

FIG. P12.30   FIG. P12.31   FIG. P12.32

12.30 The 12-lb weight falls 30 in. and strikes the head of a steel bolt 1.0 in. in diameter. Determine the smallest allowable thickness t of the head if the shear stress is limited to 12 ksi. Use $E = 29 \times 10^3$ ksi for the bolt.

12.31 The simply supported beam of length L has a rectangular cross section of area A. The beam is struck at its center by the weight W falling through the height h. Assuming that $h \gg \delta_s$, where δ_s is the static displacement of the beam, show that the maximum dynamic bending stress in the beam is $\sigma_{max} = \sqrt{18WhE/(AL)}$.

12.32 The S7 × 20 steel beam is used as a cantilever 6 ft long. The 100-lb weight falls through 6 in. before striking the free end of the beam. Determine the maximum dynamic stress and deflection caused by the impact. Use $E = 29 \times 10^6$ psi for steel.

12.33 The simply supported steel beam of rectangular cross section is hit by the 900-kg mass that is dropped from a height of 2.5 m. Using $E = 200$ GPa for the beam, compute the impact factor.

12.34 The mass m slides into the stepped elastic bar with the velocity v. Derive the expression for the maximum dynamic stress in the bar.

12.35 The 3-lb weight falls through the height h before striking the head of the stepped bolt. Calculate the largest value of h for which the maximum dynamic normal stress in the bolt does not exceed 36 ksi. Use $E = 29 \times 10^3$ ksi for the bolt.

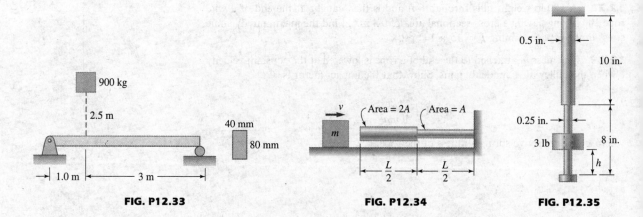

FIG. P12.33   FIG. P12.34   FIG. P12.35

12.4 Theories of Failure

Theories of failure, also called *failure criteria*, attempt to answer the question: Can data obtained from a uniaxial tension or compression test be used to predict failure under more complex loadings? By failure we mean yielding or rupture. So far, no universal method has been established that correlates failure in a uniaxial test with failure due to multiaxial loading. There are, however, several failure theories that work well enough for certain materials to be incorporated in design codes.

In general, there are two groups of failure criteria: one for brittle materials that fail by rupture, and the other for ductile materials that exhibit yielding. We limit our discussion to four simple, but popular, failure theories: two for brittle materials and two for ductile materials. Because all of these theories are expressed in terms of principal stresses, they presuppose that the three principal stresses at the critical points of the material have been determined. In this article, we consider only failure criteria for plane stress; that is, we assume that $\sigma_3 = 0$.

a. Brittle materials

Maximum Normal Stress Theory The *maximum normal stress theory* proposed by W. Rankine is the oldest, as well as the simplest, of all the theories of failure. This theory assumes that failure occurs when the largest principal stress in the material equals the ultimate stress σ_{ult} of the material in the uniaxial test. Obviously, this theory disregards the effect of the other two principal stresses. Nevertheless, the maximum normal stress theory does give results that agree well with test results on brittle materials that have about the same strength in tension and compression.

For plane stress, the maximum normal stress theory predicts failure when $|\sigma_1| = \sigma_{ult}$ or $|\sigma_2| = \sigma_{ult}$, where σ_1 and σ_2 are the principal stresses. In normalized form, the failure criterion is

$$\boxed{\frac{\sigma_1}{\sigma_{ult}} = \pm 1 \quad \text{or} \quad \frac{\sigma_2}{\sigma_{ult}} = \pm 1} \qquad (12.15)$$

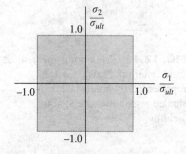

FIG. 12.10 Maximum normal stress failure criterion. States of stress represented by points inside the rectangle are safe against rupture.

The plots of Eqs. (12.15) are shown in Fig. 12.10. The outline of the square formed by the plots represents the failure criterion. If the points corresponding to the principal stresses in a material fall within the shaded area, the material will not fail. Points lying on or outside the shaded area represent stress states that will cause failure.

Mohr's Theory *Mohr's theory of failure* is used for materials that have different properties in tension and compression. To apply the theory, we must know the ultimate tensile stress $(\sigma_{ult})_t$ and the ultimate compressive stress $(\sigma_{ult})_c$ of the material, which are determined from uniaxial load tests. The Mohr's circles for these two states of stress are drawn on a single diagram, as shown in Fig. 12.11(a). The failure envelope for Mohr's theory is obtained by drawing two lines that are tangent to the circles. A given state of stress is considered safe if its Mohr's circle lies entirely within the failure envelope, which is the shaded area in Fig. 12.11(a). If any part of the circle is tangent to, or extends beyond the failure envelope, the theory predicts failure.

Mohr's theory can be refined if the ultimate shear stress τ_{ult}, obtainable from torsion tests, is also known. The Mohr's circle corresponding to this test can be added to the diagram, as shown in Fig. 12.11(b). The failure

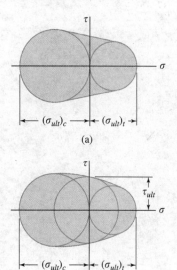

FIG. 12.11 Mohr's failure criterion. States of stress represented by points inside the shaded region are safe against rupture.

envelope is now obtained by drawing curves that are tangent to each of the three circles.

b. Ductile materials

Maximum Shear Stress Theory   The maximum shear stress theory, also known as *Tresca's yield criterion*, assumes that yielding occurs when the absolute maximum shear stress equals the maximum shear stress at yielding in the uniaxial tension test.

In Chapter 8, we found that for a plane state of stress the absolute maximum shear stress is the larger of $|\sigma_1|/2$, $|\sigma_2|/2$, or $|\sigma_1 - \sigma_2|/2$. In a uniaxial tension test, the maximum shear stress at yielding is $\sigma_{yp}/2$, so that Tresca's yield criterion can be written as

$$\boxed{\frac{\sigma_1}{\sigma_{yp}} = \pm 1 \quad \text{or} \quad \frac{\sigma_2}{\sigma_{yp}} = \pm 1 \quad \text{or} \quad \frac{\sigma_1}{\sigma_{yp}} - \frac{\sigma_2}{\sigma_{yp}} = \pm 1} \quad (12.16)$$

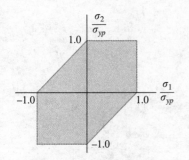

FIG. 12.12   Maximum shear stress yield criterion. States of stress represented by points inside the shaded region are safe against yielding.

The plot of Eqs. (12.16) in Fig. 12.12 forms a hexagon, called Tresca's hexagon. A state of stress is considered safe if its principal stresses are represented by a point within the hexagon. Points falling on or outside the hexagon represent stress states that will cause failure.

Maximum Distortion Energy Theory   The maximum distortion energy theory[1] is the most popular theory for predicting yielding in ductile materials. As pointed out previously, the work done by the forces that deform an elastic body is stored in the body as strain energy. A useful concept is the *strain energy density*, which is defined as the strain energy per unit volume at a point. The strain energy density can be divided into two parts: the *volumetric* strain energy density that is associated with a change in the volume of a material element, and the *distortion* strain energy density that changes the shape of the element without changing its volume. For an elastic body subjected to plane stress, the distortion strain energy density can be shown to be

$$U_d = \frac{1}{6G}(\sigma_1^2 + \sigma_2^2 - \sigma_1\sigma_2) \quad \text{(a)}$$

where G is the shear modulus.

The maximum distortion energy theory states that a material begins yielding when

$$U_d = (U_d)_{yp} \quad \text{(b)}$$

where $(U_d)_{yp}$ is the distortion strain energy density of the same material at the yield point in the uniaxial tension test. In uniaxial tension, the stresses at yielding are $\sigma_1 = \sigma_{yp}$ and $\sigma_2 = 0$, which upon substitution into Eq. (a) yield $(U_d)_{yp} = \sigma_{yp}^2/(6G)$. Substituting this result and Eq. (a) into Eq. (b), we obtain

---

[1] The maximum distortion enegy theory is also known as the Huber-Hencky-von Mises yield criterion, or the octahedral shear theory.

$$\sigma_1^2 + \sigma_2^2 - \sigma_1\sigma_2 = \sigma_{yp}^2$$

The normalized form of this yield criterion is

$$\left(\frac{\sigma_1}{\sigma_{yp}}\right)^2 + \left(\frac{\sigma_2}{\sigma_{yp}}\right)^2 - \left(\frac{\sigma_1}{\sigma_{yp}}\right)\left(\frac{\sigma_2}{\sigma_{yp}}\right) = 1 \qquad (12.17)$$

which is the equation of the ellipse shown in Fig. 12.13. A state of stress is considered safe if its principal stresses are represented by a point within the ellipse. Points falling on or outside the ellipse represent stress states that will cause failure. Because the ellipse encloses Tresca's hexagon (shown with dashed lines), the maximum distortion energy theory is less conservative than the maximum shear stress theory.

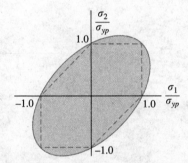

FIG. 12.13 Maximum distortion energy yield criterion. States of stress represented by points inside the elliptical region are safe against yielding. (Tresca's hexagon is shown with dashed lines.)

Sample Problem 12.6

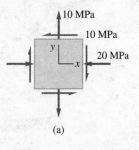

(a)

The ultimate strength of a brittle material is 40 MPa in tension and 50 MPa in compression. Use Mohr's failure criterion to determine whether the plane state of stress in Fig. (a) would result in rupture of this material.

Solution

We first draw the Mohr's circles representing the states of stress at rupture for uniaxial tension and for uniaxial compression, as shown in Fig. (b). We complete the failure envelope by drawing tangent lines to the two circles. Any state of stress with a Mohr's circle that lies entirely within the failure envelope (the shaded area in the figure) is deemed to be safe against rupture. Otherwise, failure is predicted.

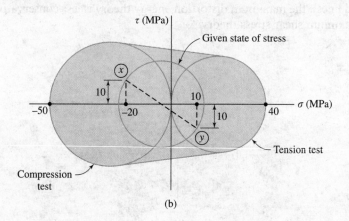

(b)

The Mohr's circle representing the given state of stress is also shown in Fig. (b). Because the circle lies within the failure envelope, this state of stress *would not cause rupture*.

Sample Problem 12.7

The 3-in.-diameter steel bar in Fig. (a) carries the bending moment $M = 2.21$ kip·ft and a torque T. If the yield strength of steel is 40 ksi, determine the largest torque T that can be applied without causing yielding. Use (1) the maximum shear stress theory; and (2) the maximum distortion energy theory.

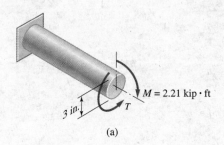

(a)

Solution

The maximum bending stress in the bar is

$$\sigma = \frac{32M}{\pi d^3} = \frac{32(2.21 \times 12)}{\pi(3)^3} = 10.00 \text{ ksi}$$

and the maximum shear stress due to torsion is given by

$$\tau = \frac{16T}{\pi d^3} = \frac{16T}{\pi(3)^3} = 0.188\,63T \text{ ksi}$$

where T is in kip·in. These stresses and the corresponding Mohr's circle are shown in Fig. (b). The radius of the circle is

$$R = \sqrt{(5)^2 + (0.188\,63T)^2} \text{ ksi} \qquad \text{(a)}$$

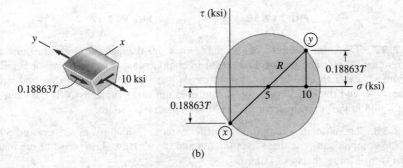

(b)

Part 1

Because the principal stresses have opposite signs, the absolute maximum shear stress is the radius R of the Mohr's circle. Therefore, the governing yield criterion, from Eq. (12.16), is $\sigma_1 - \sigma_2 = \sigma_{yp}$. Because $\sigma_1 - \sigma_2 = 2R$, the yield criterion becomes $2R = \sigma_{yp}$. Substituting R from Eq. (a), we get

$$2\sqrt{(5)^2 + (0.188\,63T)^2} = 40$$

which yields for the maximum safe torque

$$T = 102.67 \text{ kip} \cdot \text{in.} = 8.56 \text{ kip} \cdot \text{ft} \qquad \textit{Answer}$$

Part 2

Substituting $\sigma_1 = 5 + R$ ksi and $\sigma_2 = 5 - R$ ksi into the yield criterion $\sigma_1^2 + \sigma_2^2 - \sigma_1\sigma_2 = \sigma_{yp}^2$, we get

$$(5+R)^2 + (5-R)^2 - (5+R)(5-R) = 40^2$$

which yields

$$R = 22.91 \text{ ksi}$$

Therefore, Eq. (a) becomes

$$22.91 = \sqrt{(5)^2 + (0.188\,63T)^2}$$

from which the largest safe torque is

$$T = 118.5 \text{ kip} \cdot \text{in.} = 9.88 \text{ kip} \cdot \text{ft} \qquad \textit{Answer}$$

Problems

12.36 The ultimate strength of a brittle material is 30 MPa in tension and 40 MPa in compression. Using Mohr's failure criterion, determine whether the state of stress shown in the figure is safe against rupture.

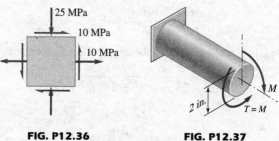

FIG. P12.36   FIG. P12.37

12.37 The 2-in.-diameter bar is made of a brittle material with the ultimate strengths of 20 ksi in tension and 30 ksi in compression. The bar carries a bending moment and a torque, both of magnitude M. (a) Use the maximum normal stress theory to find the largest value of M that does not cause rupture. (b) Is the value of M found in Part (a) safe according to Mohr's theory of failure?

12.38 The ultimate strength of a brittle material is 3000 psi in tension and 5000 psi in compression. Use these data to compute the ultimate shear stress of the material from Mohr's theory of failure.

12.39 The principal stresses at a point in a ductile material are $\sigma_1 = \sigma_0$, $\sigma_2 = 0.75\sigma_0$, and $\sigma_3 = 0$. If the yield strength of the material is 200 MPa, determine the value of σ_0 that initiates yielding using (a) the maximum shear stress theory; and (b) the maximum distortion energy theory.

12.40 Solve Prob. 12.39 if $\sigma_2 = -0.75\sigma_0$, all other data being unchanged.

12.41 The state of stress at the critical point in a ductile material is shown in the figure. If the yield strength of the material is 90 ksi, find the factor of safety against yielding using (a) the maximum shear stress theory; and (b) the maximum distortion energy theory.

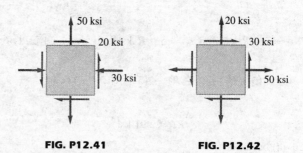

FIG. P12.41   FIG. P12.42

12.42 The state of stress at the critical point in a ductile material is shown. If the yield strength of the material is 90 ksi, find the factor of safety against yielding using (a) the maximum shear stress theory; and (b) the maximum distortion energy theory.

12.43 A solid shaft of diameter d transmits 80 kW of power at 60 Hz. Given that the yield strength of the material is 250 MPa, determine the value of d that provides a factor of safety of 3 against yielding. Use the maximum distortion energy theory.

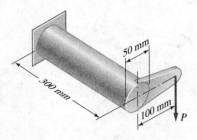

FIG. P12.44

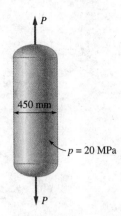

FIG. P12.45, P12.46

12.44 The 50-mm-diameter bar is made of steel with a yield strength of 200 MPa. Use the maximum distortion energy theory and a factor of safety of 2 against yielding to compute the largest allowable load P.

12.45 The thin-walled cylindrical vessel of 450-mm mean diameter and 20-mm wall thickness is pressurized internally to 20 MPa. The yield strength of the material is 320 MPa. Use the maximum shear stress theory to find the largest axial load P that can be applied in addition to the pressure without causing yielding.

12.46 Solve Prob. 12.45 using the maximum distortion energy theory.

12.47 A thin-walled cylindrical pressure vessel has a mean diameter of 11.75 in. and a wall thickness of 0.25 in. If the wall of the vessel yields at an internal pressure of 1750 psi, determine the yield strength of the material using the maximum distortion energy theory.

12.48 The cylindrical vessel with 11.75-in. mean diameter and 0.25-in. wall thickness carries an internal pressure of 750 psi and a torque of 650 kip·in. If the yield strength of the material is 36 ksi, compute the factor of safety against yielding using the maximum distortion energy theory.

12.49 The solid steel shaft 100 mm in diameter and 8 m long is subjected simultaneously to an axial compressive force P and the torque $T = 35$ kN·m. Determine the maximum safe value of P according to the maximum shear stress theory. Use $\sigma_{yp} = 200$ MPa.

12.50 The 8-m shaft, 100 mm in diameter, is made of a brittle material for which the ultimate stress in tension or compression is 40 MPa. The shaft carries simultaneously the axial compressive load $P = 200$ kN and the torque T. Calculate the maximum allowable value of T using the maximum normal stress theory.

12.51 The 3-in.-diameter rod carries the 4000-lb·ft bending moment and a torque T. The yield stress of the material is 60 ksi. Determine the largest allowable value of T using the maximum distortion energy theory and a factor of safety of 2.

12.52 Solve Prob. 12.51 using the maximum shear stress theory.

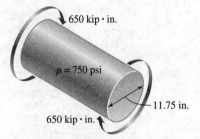

FIG. P12.48

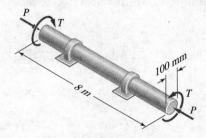

FIG. P12.49, P12.50

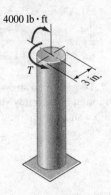

FIG. P12.51, P12.52

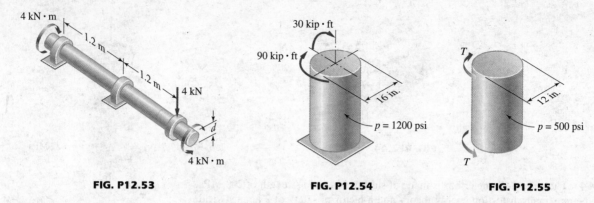

FIG. P12.53   **FIG. P12.54**   **FIG. P12.55**

12.53 The steel shaft carries simultaneously the 4-kN lateral force and the 4-kN·m torque. Find the smallest safe diameter d of the shaft by the maximum shear stress theory. Use $\sigma_{yp} = 240$ MPa and a factor of safety of 2.

12.54 The cylindrical pressure vessel with closed ends has a diameter of 16 in. and a wall thickness of 3/4 in. The vessel carries simultaneously an internal pressure of 1200 psi, a torque of 90 kip·ft, and a bending moment of 30 kip·ft. The yield strength of the material is 40 ksi. What is the factor of safety against yielding according to the maximum shear stress theory?

12.55 The cylindrical tank of 12-in. diameter is fabricated from 1/4-in. plate. The tank is subjected to an internal pressure of 500 psi and a torque T. Find the largest allowable value of T according to the maximum distortion energy theory. Use $\sigma_{yp} = 36$ ksi and a factor of safety of 2.

12.5 Stress Concentration

The elementary formulas for the computation of stresses that we have been using in previous chapters assumed that the cross sections of the members are either constant or change gradually. We now consider the effect of abrupt changes in cross section on the stress distribution. These changes produce localized regions of high stress known as *stress concentrations*. Stress concentrations are, in general, of more concern in brittle materials than in ductile materials. Ductile materials are able to yield locally, which redistributes the stress more evenly across the cross section. On the other hand, brittle materials, which cannot yield, are susceptible to abrupt failure caused by stress concentrations.

As an example of stress concentration, consider the plate with a small circular hole in Fig. 12.14. Although the plate is subjected to the uniform tensile stress σ_0, the stress distribution across the section through the center of the hole is nonuniform. From the theory of elasticity, this stress distribution is known to be[2]

$$\sigma = \frac{\sigma_0}{2}\left(2 + \frac{d^2}{4r^2} + \frac{3}{16}\frac{d^4}{r^4}\right) \qquad (a)$$

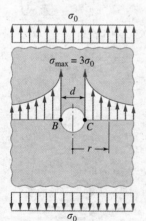

FIG. 12.14 Normal stress distribution in a uniformly stretched plate with a small round hole.

---

[2] See S. Timoshenko and J. N. Goodier, *Theory of Elasticity*, 3rd ed. (New York: McGraw-Hill, 1970).

where d is the diameter of the hole and r is the distance from the center of the hole. From Eq. (a), we find that the stresses at points B and C are $3\sigma_0$, which is three times the average stress.

A similar stress concentration is caused by the small elliptical hole shown in Fig. 12.15. The maximum stress at the ends of the horizontal axis of the hole is given by[3]

$$\sigma_{max} = \sigma_0\left(1 + 2\frac{b}{a}\right) \qquad \text{(b)}$$

Because this stress increases with the ratio b/a, a very high stress concentration is produced at the ends of a narrow crack ($b/a \gg 1$) that is perpendicular to the direction of the tensile stress. Therefore, such cracks tend to grow and may lead to catastrophic failure. The spreading of a crack may be stopped by drilling small holes at the ends of the crack, thus replacing a very large stress concentration by a smaller one. In ductile materials, localized yielding occurs at the crack tip, which has the same effect as a hole.

The *stress concentration factor k* is defined as

$$\boxed{k = \frac{\sigma_{max}}{\sigma_{nom}}} \qquad (12.18)$$

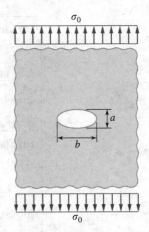

FIG. 12.15 Uniformly stretched plate with a small elliptical hole.

where σ_{max} is the maximum stress and σ_{nom} denotes the nominal stress (the stress calculated from an elementary formula). Thus, the maximum stresses for axial, torsional, and flexural loads on bars and beams are given by

$$\sigma_{max} = k\frac{P}{A} \qquad \tau_{max} = k\frac{Tr}{J} \qquad \sigma_{max} = k\frac{Mc}{I} \qquad (12.19)$$

Exact solutions for stresses, such as Eqs. (a) and (b) above, exist in only a few cases. However, various handbooks contain large numbers of stress concentration factors that have been determined either experimentally or by numerical solutions of the equations of elasticity theory.[4] A sample of available data is shown in Fig. 12.16.[5] The equations for the stress concentration factors were obtained by fitting cubic polynomials to data points computed for various values of r/D from elasticity theory. The graphs of these equations are shown in Fig. 12.17. The stress concentration factors in Fig. 12.16 assume that the nominal stress σ_{nom} is calculated using the *net cross-sectional area* passing through the point of maximum stress (the shaded areas in Fig. 12.16).

---

[3] Ibid.
[4] See, for example, Walter D. Pilkey, *Peterson's Stress Concentration Factors*, 2nd ed. (New York: John Wiley & Sons, 1977).
[5] The expressions in Fig. 12.16 were adapted from Warren C. Young, *Roark's Formulas for Stress and Strain*, 6th ed. (New York: McGraw-Hill, 1989).

(a) Circular hole in rectangular bar Axial tension $$k = 3.00 - 3.13\left(\frac{2r}{D}\right) + 3.66\left(\frac{2r}{D}\right)^2 - 1.53\left(\frac{2r}{D}\right)^3$$ In-plane bending $k = 2$ (independent of r/D)	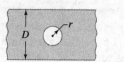
(b) Shoulder with circular fillet in rectangular bar Axial tension $$k = 1.976 - 0.385\left(\frac{2r}{D}\right) - 1.022\left(\frac{2r}{D}\right)^2 + 0.431\left(\frac{2r}{D}\right)^3$$ In-plane bending $$k = 1.976 - 1.925\left(\frac{2r}{D}\right) + 0.906\left(\frac{2r}{D}\right)^2 + 0.0430\left(\frac{2r}{D}\right)^3$$	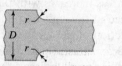
(c) Semicircular notches in rectangular bar Axial tension $$k = 3.065 - 3.370\left(\frac{2r}{D}\right) + 0.647\left(\frac{2r}{D}\right)^2 + 0.658\left(\frac{2r}{D}\right)^3$$ In-plane bending $$k = 3.065 - 6.269\left(\frac{2r}{D}\right) + 7.015\left(\frac{2r}{D}\right)^2 - 2.812\left(\frac{2r}{D}\right)^3$$	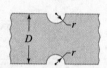
(d) Shoulder with circular fillet in circular shaft Axial tension $$k = 1.990 - 2.070\left(\frac{2r}{D}\right) + 1.938\left(\frac{2r}{D}\right)^2 - 0.857\left(\frac{2r}{D}\right)^3$$ Bending $$k = 1.990 - 2.429\left(\frac{2r}{D}\right) + 2.057\left(\frac{2r}{D}\right)^2 - 0.619\left(\frac{2r}{D}\right)^3$$ Torsion $$k = 1.580 - 1.796\left(\frac{2r}{D}\right) + 2.000\left(\frac{2r}{D}\right)^2 - 0.784\left(\frac{2r}{D}\right)^3$$	
(e) Semicircular groove in circular shaft Axial tension $$k = 3.04 - 5.42\left(\frac{2r}{D}\right) + 6.27\left(\frac{2r}{D}\right)^2 - 2.89\left(\frac{2r}{D}\right)^3$$ Bending $$k = 3.04 - 7.236\left(\frac{2r}{D}\right) + 9.375\left(\frac{2r}{D}\right)^2 - 4.179\left(\frac{2r}{D}\right)^3$$ Torsion $$k = 2.000 - 3.394\left(\frac{2r}{D}\right) + 4.231\left(\frac{2r}{D}\right)^2 - 1.837\left(\frac{2r}{D}\right)^3$$	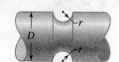

FIG. 12.16 Stress concentration factors for rectangular and circular bars.

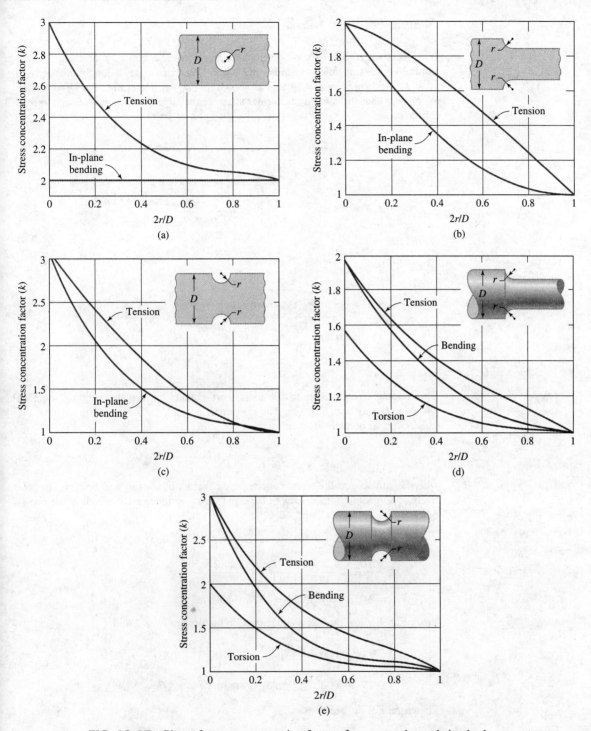

FIG. 12.17 Plots of stress concentration factors for rectangular and circular bars.

Sample Problem 12.8

The rectangular bar of cross-sectional dimensions $D = 250$ mm and $b = 20$ mm contains a central hole of radius $r = 50$ mm. The bar carries a longitudinal tensile load P of eccentricity $e = 50$ mm as shown in Fig. (a). Determine the largest value of P for which the maximum normal stress at the edge of the hole does not exceed 150 MPa.

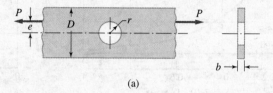

(a)

Solution

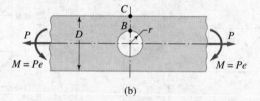

(b)

The loading is equivalent to the axial load P and the in-plane bending moment $M = Pe$, as shown in Fig. (b). The maximum normal stress σ_{max} at the edge of the hole occurs at point B; its magnitude is

$$\sigma_{max} = k_t \frac{P}{A} + k_b \frac{(Pe)r}{I} \qquad \text{(a)}$$

where k_t and k_b are the stress concentration factors in tension and bending, respectively. The geometric properties of the net cross-sectional area, which is the crosshatched area in Fig. (a), are

$$A = b(D - 2r) = 20[250 - 2(50)] = 3.0 \times 10^3 \text{ mm}^2 = 3.0 \times 10^{-3} \text{ m}^2$$

$$I = \frac{bD^3}{12} - \frac{b(2r)^3}{12} = \frac{20(250)^3}{12} - \frac{20(2 \times 50)^3}{12}$$

$$= 24.38 \times 10^6 \text{ mm}^4 = 24.38 \times 10^{-6} \text{ m}^4$$

From Fig. 12.6(a), the stress concentration factors are $k_b = 2$ and

$$k_t = 3.00 - 3.13\left(\frac{2r}{D}\right) + 3.66\left(\frac{2r}{D}\right)^2 - 1.53\left(\frac{2r}{D}\right)^3$$

$$= 3.00 - 3.13(0.4) + 3.66(0.4)^2 - 1.53(0.4)^3 = 2.236$$

Therefore, Eq. (a) becomes

$$\sigma_{max} = (2.236)\frac{P}{3.0 \times 10^{-3}} + (2)\frac{P(0.05)(0.05)}{24.38 \times 10^{-6}} = 950.4P$$

Substituting $\sigma_{max} = 150 \times 10^6$ Pa and solving for P yields

$$P = \frac{150 \times 10^6}{950.4} = 157.8 \times 10^3 \text{ N} = 157.8 \text{ kN} \qquad \textit{Answer}$$

Problems

12.56 The bar with a hole is subjected to the axial load P. Determine the maximum normal stress in the bar in terms of P, D, and b for (a) $r = 0.05D$; (b) $r = 0.1D$; and (c) $r = 0.2D$.

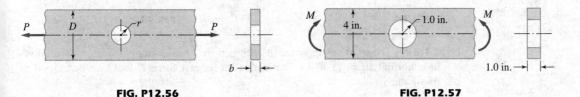

FIG. P12.56   **FIG. P12.57**

12.57 Determine the largest in-plane bending moment M that can be applied to the bar with a hole if the maximum normal stress at the edge of the hole is not to exceed 20 ksi.

12.58–12.61 Determine the maximum normal stress in the bar shown.

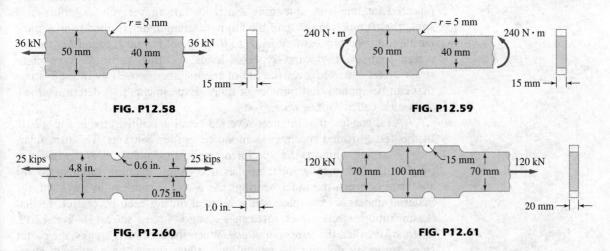

FIG. P12.58   **FIG. P12.59**

FIG. P12.60   **FIG. P12.61**

12.62 Determine the maximum shear stress in the stepped shaft due to the 7.5-kip·in. torque.

12.63 Find the largest axial load P that can be applied to the shaft if the maximum normal stress is limited to 200 MPa.

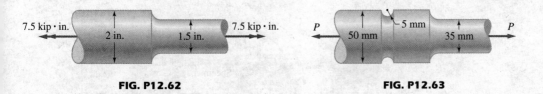

FIG. P12.62   **FIG. P12.63**

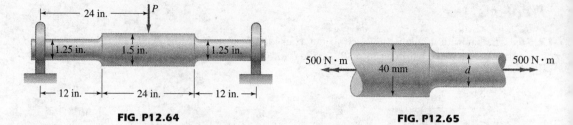

FIG. P12.64

FIG. P12.65

12.64 The shaft is supported at each end by a self-aligning bearing. If the allowable normal stress is 30 ksi, determine the largest force P that can be carried at midspan.

12.65 The stepped shaft carries a torque of 500 N · m. Determine the smallest allowable diameter d if the maximum shear stress is not to exceed 100 MPa.

12.6 Fatigue under Repeated Loading

Many machine parts and structures are subjected to cyclic stresses caused by repeated loading and unloading. Such loading may result in failure at a stress that is much lower than the ultimate strength determined from a static tensile test. The process that leads to this type of failure is called *fatigue*. To design members that carry repeated loads, we must know the number of stress cycles expected over the life of the member as well as the safe stress that can be applied that number of times. Experiments that determine these values are called *fatigue tests*.

A fatigue test that involves reversed bending is illustrated in Fig. 12.18. In this test, a round specimen is mounted in four bearings. The two outermost bearings support the specimen, while the middle bearings carry a weight W. This arrangement applies a constant bending moment to the specimen between the inner bearings. As a motor rotates the specimen, the material undergoes complete stress reversal during each revolution so that the maximum stress varies between $+\sigma_a$ and $-\sigma_a$ as shown in Fig. 12.19, where σ_a is called the *stress amplitude*. When the specimen breaks, the motor stops automatically and the revolution counter shows the number of stress cycles that produced the failure. A test program on a given material involves many identical specimens that are rotated under different values of W until they either fail or reach several million load cycles. The results are presented in the form of *S-N diagrams*, where the stress amplitude σ_a is plotted against

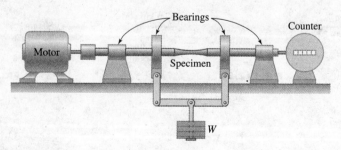

FIG. 12.18 Fatigue testing machine that applies reversed bending to the specimen.

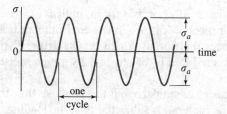

FIG. 12.19 Maximum normal stress versus time for the reversed bending test.

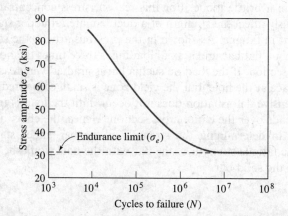

FIG. 12.20 S-N diagram for a steel alloy that has a finite endurance limit. (Some metals do not have an endurance limit.)

N, the number of cycles at failure. It is customary to use a logarithmic scale for the N-axis. Similar fatigue tests are used for axial loading and torsion.

A typical S-N diagram for a steel alloy is shown in Fig. 12.20. The stress at which the diagram becomes horizontal is called the *endurance limit*, or *fatigue limit*, and denoted by σ_e. Although no definite relationship exists between the endurance limit and the ultimate strength σ_{ult}, tests show that σ_e is usually between $0.4\sigma_{ult}$ and $0.5\sigma_{ult}$. Some metals, notably aluminum alloys and copper alloys, have no detectable endurance limit, in which case σ_e is taken to be the fatigue strength at a specific value of N, usually $N = 10^7$ or 10^6.

Many aspects of fatigue can be analyzed by *fracture mechanics*, which is a theory concerned with the propagation of cracks. Microscopic cracks either are initially present in a material or develop after a small number of load cycles due to stress concentrations near inclusions, grain boundaries, and other inhomogeneities. Under repeated loading, a crack tends to propagate until it reaches a critical size, at which time failure occurs. The task of fracture mechanics is to predict the rate of crack propagation under a given stress amplitude, which in turn determines the fatigue life of the component. In many critical applications, such as nuclear reactors, spacecraft, and submarines, fracture mechanics is a mandatory component of analysis. The detailed steps of such analyses are prescribed in the corresponding design codes.

The major problem in fatigue analysis is to account for all the factors that contribute to the fatigue life. It is known that fatigue strength is dependent not only on the metallurgical and structural aspects of the material, but also on the surface finish and the environmental conditions. Polishing the surface of a specimen after it has been milled can increase the fatigue strength by 30% or more. Moisture in the environment is another important factor that influences the rate at which cracks grow. Typically, the fatigue strength of metals is reduced by 5% to 15% by a high moisture content of the atmosphere.

Fatigue failures are catastrophic in the sense that they occur without warning. If a ductile material is subjected to static loading, extensive yielding of the material takes place before rupture takes place. Because fatigue occurs with very little yielding, even ductile materials fail by sudden fracture; that is, they fail in a brittle mode. For this reason, stress concentrations, which do not seriously affect the strength of a ductile material under static loading, are important in fatigue. As shown in the previous article, the stress around a small hole in a flat bar under axial loading is three times the nominal stress on the cross section. If the load on such a bar is gradually increased, yielding first takes place at the hole, but the yield zone is small compared to the cross section. Extensive deformation does not occur until the load is tripled, when yielding spreads over the entire cross section. Hence, the effect of the hole is insignificant in determining the strength of the bar under static loading. However, the fatigue strength of the bar with the hole would be almost one-third that of the solid bar.[6]

---

[6] The actual fatigue strength is usually somewhat higher than predicted from the stress concentration factor, an effect known as the *notch sensitivity*.

Computer Problems

C12.1 The truss in Fig. (a) consists of two members, denoted by ① and ②. The cross-sectional areas of the members are A_1 and A_2, and the modulus of elasticity E is the same for both members. If the junction O of the members is taken as the origin, the layout of the truss is specified by coordinates (x_1, y_1) and (x_2, y_2) of the support points. Given the coordinates of the support points, A_1, A_2, E, and the vertical load W, write an algorithm that uses Castigliano's theorem to compute the horizontal and vertical deflections of O. Apply the algorithm to the steel trusses (b); and (c). Use $E = 29 \times 10^6$ psi for steel.

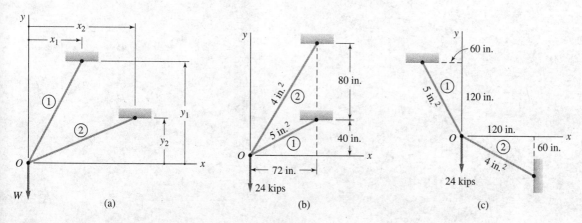

FIG. C12.1

C12.2 The uniform beam of length L in Fig. (a) caries a uniformly distributed load of intensity w_0. The beam has a built-in support at its right end and n roller supports located at distances $x_1, x_2, \ldots, x_n$ from the left end. Given L, w_0, n, and the coordinates of the roller supports, write an algorithm that computes the reactions at the roller supports using Castigliano's theorem. Apply the algorithm to the beams (b); and (c). (*Hint*: Consider the roller reactions $R_1, R_2, \ldots, R_n$ as redundant reactions and apply the displacement constraints $\delta_i = \partial U / \partial R_i = 0$, $i = 1, 2, \ldots, n$.)

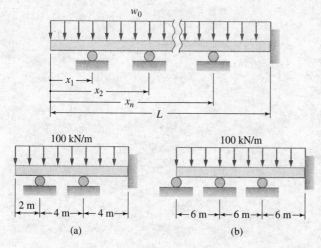

FIG. C12.2

C12.3 The stepped shaft carrying a torque T has two segments of diameters D and d. The radius of the circular fillet between the segments is $r = (D - d)/2$. No yielding is allowed anywhere in the shaft. Given T, D, and the yield stress σ_{yp} of the material, write an algorithm that computes the smallest allowable diameter d. Use the maximum distortion energy theory as the yield criterion. Run the algorithm with the following data: $T = 4$ kN·m, $D = 75$ mm, and $\sigma_{yp} = 480$ MPa.

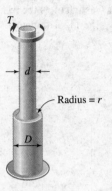

FIG. C12.3

13
Inelastic Action

Plastic deformation of a thin-walled aluminum channel section. In this case, the deformation was caused by inelastic buckling of the flanges.

13.1 Introduction

The analyses in the preceding chapters dealt almost exclusively with stresses that were below the proportional limit of the material. In other words, Hooke's law was assumed to apply. Analyses based on Hooke's law are entirely justified in most applications. Under normal service conditions, we want to prevent yielding because the resulting permanent deformation is generally undesirable. For this reason, the factor of safety for ductile materials was defined as $N = \sigma_w/\sigma_{yp}$, where σ_w is the maximum allowable stress due to the anticipated loading under *normal service conditions*. However, permanent deformation does not necessarily lead to catastrophic failure; it may only make the structure unusable. The implication here is that there are

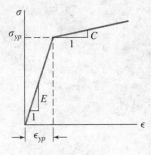

(a) Strain-hardening

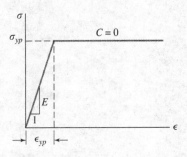

(b) Elastic-perfectly plastic

FIG. 13.1 Idealized stress-strain diagrams for strain-hardening and elastic-perfectly plastic materials.

two levels of safety: safety against permanent deformation and safety against catastrophic failure. The latter is very important in structures that may experience forces that greatly exceed the normal service loads, such as those imposed by earthquakes.

In this chapter, we consider the maximum loading that may be applied to a structure before it collapses. This loading is known as *limit loading*. With a ductile material, collapse does not occur until yielding has spread throughout the most highly stressed section in a statically determinate structure. The application of limit loads to indeterminate structures, called *limit analysis*, is discussed in Art. 13.5. We must emphasize that none of the concepts discussed in this chapter is applicable to brittle materials; we always assume that the material can undergo considerable plastic deformation before breaking.

The stress-strain relationship for ductile materials may be approximated by the idealized diagram shown in Fig. 13.1(a). The elastic portion of the diagram is a straight line with slope E, the modulus of elasticity of the material. The plastic portion is also a straight line beginning at the yield stress, σ_{yp}, and having a slope C. Because slope C is smaller than slope E, the increment of stress required to produce a specified increment of strain in the plastic range is less than it is in the elastic range. Such a material is said to be *strain-hardening*; it does not permit an increase in strain without an increase in stress. Figure 13.1(b) shows the idealized stress-strain diagram for an *elastic-perfectly plastic material*, for which $C = 0$. The diagram is assumed to be valid for both tension and compression. For materials of this type, flow can occur with no increase in stress beyond the proportional limit. Elastic-perfectly plastic behavior, which is a satisfactory model for low-carbon steels commonly used in building frames, is the only material model that we consider in this chapter.

13.2 Limit Torque

Torsion analysis of circular bars stressed into the plastic range is very similar to what we used in Art. 3.2 for elastic bars. As before, we assume that circular cross sections remain plane (do not warp) and perpendicular to the axis of the shaft. Consequently, the shear strain γ remains proportional to the radial distance from the center of the bar, but γ is now allowed to exceed the yield strain in shear.

Consider the response of a circular bar made of an elastic-perfectly plastic material that is twisted progressively through the elastic range into the fully plastic range. Until the yield stress τ_{yp} in shear is reached, the bar is elastic and has the stress distribution shown in Fig. 13.2(a). From the torsion formula $\tau_{max} = Tr/J$, the torque at the beginning of yielding (when $\tau_{max} = \tau_{yp}$) is

$$T_{yp} = \frac{J}{r}\tau_{yp} = \left(\frac{\pi r^4}{2}\right)\frac{\tau_{yp}}{r} = \frac{\pi r^3}{2}\tau_{yp} \qquad (a)$$

where T_{yp} is called the *yield torque*.

If we twist the bar beyond this point, the shear strains continue to increase but the shear stress is limited to the yield stress. Therefore, there is an intermediate radius r_i that forms the boundary between the elastic and plastic regions, as shown in Fig. 13.2(b). The stress in the outer, plastic portion

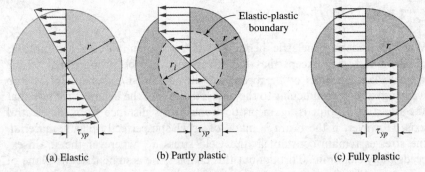

FIG. 13.2 Progression of the shear stress distribution in a shaft made of elastic-perfectly plastic material as the shaft is twisted into the fully plastic state.

equals the yield stress τ_{yp}, while the inner core remains elastic. The torque carried by the elastic (inner) region is

$$T_i = \frac{J_i}{r_i}\tau_{yp} = \left(\frac{\pi r_i^4}{2}\right)\frac{\tau_{yp}}{r_i} = \frac{\pi r_i^3}{2}\tau_{yp}$$

If we let ρ be the radial coordinate of an area element dA of the cross section, the torque carried by the plastic (outer) portion is

$$T_o = \int_{r_i}^{r} \rho(\tau_{yp}\, dA) = \tau_{yp}\int_{r_i}^{r} \rho(2\pi\rho\, d\rho) = \frac{2\pi}{3}(r^3 - r_i^3)\tau_{yp}$$

Therefore, the total torque carried by the cross section is

$$T = T_i + T_o = \frac{\pi r_i^3}{2}\tau_{yp} + \frac{2\pi}{3}(r^3 - r_i^3)\tau_{yp}$$

which reduces to

$$\boxed{T = \frac{\pi r^3}{6}\left(4 - \frac{r_i^3}{r^3}\right)\tau_{yp}} \qquad (13.1)$$

The torque required to produce the fully plastic state in Fig. 13.2(c) is called the *limit torque*, denoted by T_L. The limit torque can be found by setting $r_i = 0$ in Eq. (13.1), which gives

$$\boxed{T_L = \frac{2}{3}\pi r^3 \tau_{yp}} \qquad (13.2\text{a})$$

Note that the limit torque is one-third larger than the yield torque; that is,

$$\boxed{T_L = \frac{4}{3}T_{yp}} \qquad (13.2\text{b})$$

13.3 Limit Moment

When considering inelastic bending, we use the simplifying assumptions made in Art. 5.2, except that the stress need not be proportional to the strain. Plane sections of the beam are still assumed to remain plane (no warping) and perpendicular to the deformed axis of the beam. It follows that the strain at a point is again proportional to its distance from the neutral axis. However, if the beam is made of an elastic-perfectly plastic material, the stresses remain constant at the yield stress σ_{yp} wherever the strain exceeds the yield strain. Throughout this chapter, we assume that the plane of loading is also the plane of symmetry of the beam.

Consider the cantilever beam in Fig. 13.3(a) that carries a load P at its free end. We assume that P is large enough to cause yielding in the gray portion of the beam. At section a-a, the stresses on the outer fibers have just reached the yield stress, but the stress distribution is still elastic, as shown in Fig. 13.3(b). Applying the flexure formula $M_{max} = \sigma_{max} S = \sigma_{max}(bh^2/6)$, we find that the magnitude of the bending moment at this section is

$$M_{yp} = \sigma_{yp} \frac{bh^2}{6} \quad \text{(a)}$$

Because this moment initiates yielding, it is known as the *yield moment*.

At section b-b, the cross section is elastic over the depth $2y_i$ but plastic outside this depth, as shown by the stress distribution in Fig. 13.3(c). The stress is constant at σ_{yp} over the plastic portion and varies linearly over the elastic region. The bending moment carried by the elastic region, as determined by the flexure formula, is

$$M_i = \sigma_{yp} \frac{I_i}{y_i}$$

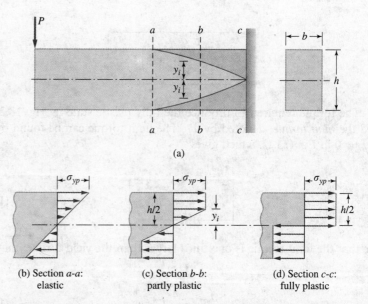

(b) Section a-a: elastic

(c) Section b-b: partly plastic

(d) Section c-c: fully plastic

FIG. 13.3 Normal stress distribution in a rectangular beam when section c-c has become fully plastic.

where I_i is the moment of inertia of the elastic region of the cross section about the neutral axis. For the plastic region, which here is symmetrical about the neutral axis, the bending moment is

$$M_o = 2 \int_{y_i}^{h/2} y(\sigma_{yp}\, dA) = 2\sigma_{yp} \int_{y_i}^{h/2} y\, dA = 2\sigma_{yp} Q_o$$

where Q_o is the first moment of the area of the plastic region that lies above (or below) the neutral axis. The total bending moment carried by the partly plastic cross section is

$$\boxed{M = M_i + M_o = \sigma_{yp}\left(\frac{I_i}{y_i} + 2Q_o\right)} \qquad (13.3)$$

At section c-c, the beam is fully plastic. As shown in Fig. 13.3(d), the stress is constant at σ_{yp} over the tensile and compressive portions of the cross section. The bending moment that causes this stress distribution, called the *limit moment* M_L, is given by

$$\boxed{M_L = 2\sigma_{yp} Q} \qquad (13.4a)$$

where Q is the first moment of the cross-sectional area above (or below) the neutral axis. The limit moment is the *largest bending moment that the cross section can carry*. Once a section has become fully plastic, it can bend further, but there will be no increase in the resisting moment. For this reason, fully plastic sections are known as *plastic hinges*.

For the rectangular cross section, the limit moment in Eq. (13.4a) becomes

$$M_L = 2\sigma_{yp}\left(\frac{bh}{2}\right)\left(\frac{h}{4}\right) = \sigma_{yp}\frac{bh^2}{4} \qquad (b)$$

Comparing Eqs. (a) and (b), we find that

$$\boxed{M_L = \frac{3}{2} M_{yp}} \qquad (13.4b)$$

Although the above discussion used a rectangular cross section, Eqs. (13.3) and (13.4) are valid for any section where the bending axis is also an axis of *symmetry*.

The ratio M_L/M_{yp} (equal to 3/2 for a rectangular section) varies with the shape of the cross section. Values of this ratio for several simple shapes are listed in Table 13.1. These ratios indicate that the limit moments for rectangular and circular sections are considerably larger than the yield moments, whereas the difference is much smaller for standard structural sections.

Cross section	M_L/M_{yp}
Solid rectangle	1.5
Solid circle	1.7
Thin-walled circular tube	1.27
Typical wide-flange beam	1.1

TABLE 13.1

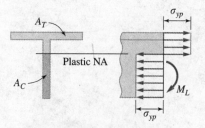

FIG. 13.4 If the bending axis is not an axis of symmetry of the cross section, the plastic neutral axis does not coincide with the elastic neutral axis.

If the bending axis is not an axis of symmetry, as is the case for the T-section shown in Fig. 13.4, the neutral axis changes its location as the section enters the plastic range. In the fully plastic case, the equilibrium condition of zero axial force is $\sigma_{yp}A_T = \sigma_{yp}A_C$, where A_T is the tension area of the cross section and A_C is the compression area, as indicated in the figure. Thus, the location of the plastic neutral axis is determined from

$$A_T = A_C \tag{13.5}$$

Once the plastic neutral axis has been located, the limit moment can be computed from

$$M_L = \sigma_{yp}(Q_T + Q_C) \tag{13.6}$$

where Q_T and Q_C are the first moments of the tension and compression areas, respectively, both computed about the plastic neutral axis. Note that because the plastic neutral axis does not pass through the centroid of the cross section, Q_T is not equal to Q_C.

Sample Problem 13.1

Determine the ratio M_L/M_{yp} for the T-section shown in Fig. (a).

Solution

Yield Moment From Fig. (b), the location of the elastic neutral axis (which coincides with the centroidal axis) is given by

$$\bar{y} = \frac{\sum A_i \bar{y}_i}{\sum A_i} = \frac{(100 \times 20)(160) + (20 \times 150)(75)}{(100 \times 20) + (20 \times 150)} = 109.0 \text{ mm} = 0.1090 \text{ m}$$

The moment of inertia of the cross-sectional area about the neutral axis is

$$I = \sum \left[\frac{b_i h_i^3}{12} + A_i(\bar{y}_i - \bar{y})^2 \right]$$

$$= \frac{100(20)^3}{12} + (100 \times 20)(160 - 109.0)^2 + \frac{20(150)^3}{12} + (20 \times 150)(75 - 109.0)^2$$

$$= 14.362 \times 10^6 \text{ mm}^4 = 14.362 \times 10^{-6} \text{ m}^4$$

Yielding will start at the bottom of the cross section when the bending moment reaches

$$M_{yp} = \frac{\sigma_{yp} I}{\bar{y}} = \frac{14.362 \times 10^{-6}}{0.1090} \sigma_{yp} = (131.76 \times 10^{-6})\sigma_{yp} \text{ N} \cdot \text{m}$$

Limit Moment The plastic neutral axis divides the cross section into the equal areas A_T and A_C, as indicated in Fig. (c). Denoting the distance of this axis from the bottom of the section by y_p, we get

$$20 y_p = 100(20) + 20(150 - y_p)$$

$$y_p = 125.0 \text{ mm}$$

The limit moment can be computed from Eq. (13.6):

$$M_L = \sigma_{yp}(Q_T + Q_C)$$

The sum of the first moments of the areas A_T and A_C about the plastic neutral axis is

$$Q_T + Q_C = (20 \times 125)\frac{125}{2} + \left[(100 \times 20)(35) + (20 \times 25)\frac{25}{2} \right]$$

$$= 232.5 \times 10^3 \text{ mm}^3 = 232.5 \times 10^{-6} \text{ m}^3$$

Therefore, the limit moment is

$$M_L = (232.5 \times 10^{-6})\sigma_{yp} \text{ N} \cdot \text{m}$$

The ratio of the limit moment to the yield moment is

$$\frac{M_L}{M_{yp}} = \frac{(232.5 \times 10^6)\sigma_{yp}}{(131.76 \times 10^6)\sigma_{yp}} = 1.765 \quad \quad \text{Answer}$$

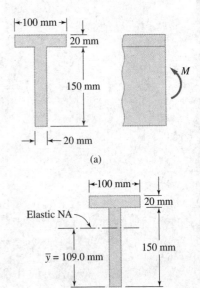

(a)

(b)

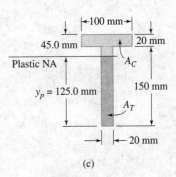

(c)

Problems

13.1 A solid circular shaft 3 in. in diameter is subjected to a torque T. If the yield stress in shear is $\tau_{yp} = 20$ ksi, determine (a) the yield torque T_{yp}; and (b) the limit torque T_L. (c) If $T = 140$ kip·in., to what radius does the elastic action extend?

13.2 Determine the ratio T_L/T_{yp} for a hollow circular shaft with an outer radius that is 1.5 times its inner radius.

13.3 When a torque T acts on a solid circular shaft, the elastic region extends to 80% of the outer radius. Determine the ratio T/T_{yp}.

13.4 Verify the ratio M_L/M_{yp} given in Table 13.1 for (a) a solid circle; and (b) a thin-walled circular tube.

13.5 Compute the ratio M_L/M_{yp} for a W250 × 80 beam.

13.6 Compute the ratio M_L/M_{yp} for a W8 × 40 beam.

13.7 For the T-section shown in the figure, determine the ratio M_L/M_{yp}.

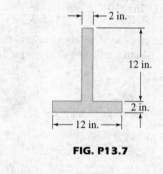

FIG. P13.7

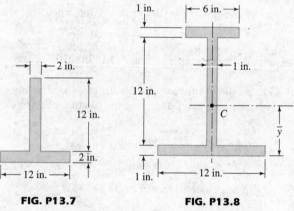

FIG. P13.8

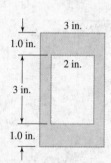

FIG. P13.11, P13.12

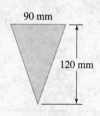

FIG. P13.13

13.8 The centroidal axis of the section shown is located at $\bar{y} = 5.7$ in., and the cross-sectional moment of inertia about this axis is $I = 855$ in.^{4}. Determine the ratio M_L/M_{yp} for a beam having this cross section.

13.9 A rectangular beam 60 mm wide and 140 mm deep is made of an elastic-perfectly plastic material for which $\sigma_{yp} = 240$ MPa. Compute the bending moment that causes one-half of the section to be in the plastic range.

13.10 Referring to the rectangular beam in Prob. 13.9, determine the bending moment that causes the middle three-fourths of the section to be in the elastic range.

13.11 If $\sigma_{yp} = 40$ ksi, compute the limit moment for a beam with the cross section shown in the figure.

13.12 If $\sigma_{yp} = 40$ ksi, determine the bending moment that causes the elastic region to extend 1.5 in. from the neutral axis for a beam with the cross section shown.

13.13 Compute the ratio M_L/M_{yp} for the triangular cross section.

13.4 Residual Stresses

a. Loading-unloading cycle

Experiments indicate that if a ductile material is loaded into the plastic range, as shown by curve OAB in Fig. 13.5(a), it unloads elastically following the path BC that is approximately parallel to the initial elastic path OA. Upon reloading, the material remains elastic up to the previously strained point B, after which it again becomes plastic (curve CBD). For an idealized elastic-perfectly plastic material, to which our analysis is limited, this loading, unloading, and reloading cycle is shown in Fig. 13.5(b).

The principal effect of unloading a material strained into the plastic range is to create a *permanent set*, such as the strain corresponding to $\overline{OC}$ in Fig. 13.5. This permanent set creates a system of self-balancing internal stresses, known as *residual stresses*. The magnitude and distribution of the residual stresses may be determined by superimposing the following two stress distributions: (1) the stress distribution (partly or fully plastic) caused by the given loading; and (2) the elastic stress distribution created by the unloading. However, this method of superposition cannot be used if the residual stresses thereby obtained would exceed the yield stress (an impossibility for an elastic-perfectly plastic material). It can be shown that this is indeed the case whenever the axis of bending is not an axis of symmetry of the cross section. Hence, the theory of elastic unloading is applicable only to bending about an axis of symmetry.

b. Torsion

As a first example of residual stress, we consider a circular bar strained into the fully plastic state by the limit torque T_L. As we saw in Art. 13.2, the limit torque is 4/3 times the yield torque, and the stress distribution is as shown in Fig. 13.6(a). To unload the bar, we now apply an opposite torque of magnitude T_L. We recall that the unloading is assumed to be elastic, so the result is the stress distribution shown in Fig. 13.6(b). Superimposing the loadings and the stress distributions of parts (a) and (b), we obtain an unloaded bar with the residual stress distribution shown in part (c).

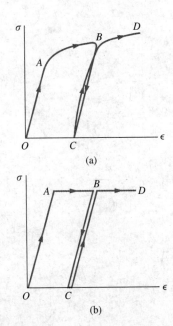

FIG. 13.5 (a) Stress-strain diagram for a ductile material showing loading-unloading; (b) the idealized diagram for an elastic-perfectly plastic material.

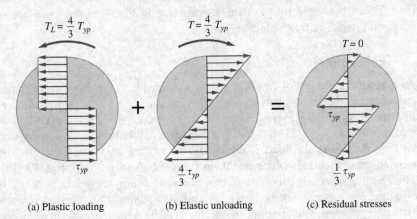

(a) Plastic loading   (b) Elastic unloading   (c) Residual stresses

FIG. 13.6 Determining the residual shear stress distribution in an elastic-perfectly plastic shaft after loading into the fully plastic state and then unloading.

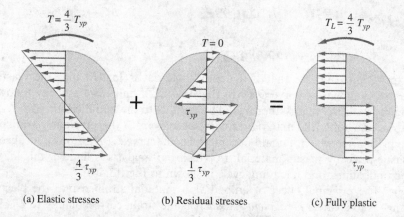

FIG. 13.7 If the limit torque is applied a second time, the fully plastic state is recovered.

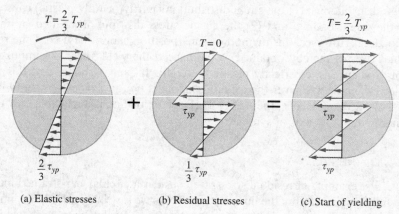

FIG. 13.8 A torque $T = (2/3)T_{yp}$ applied to a shaft that was previously twisted into the fully plastic range in the opposite direction will initiate additional yielding.

An interesting phenomenon of residual stresses is that the bar now behaves elastically if the original limit torque is reapplied, as shown in Fig. 13.7. Combining parts (a) and (b) of Fig. 13.7, we obtain the original plastic state shown in part (c). On the other hand, as shown in Fig. 13.8, no more than 2/3 of the yield torque can be reapplied in the *opposite sense* before additional yielding takes place.

c. Bending

As a second example of residual stress, we consider a beam of rectangular cross section that is strained into the fully plastic state by the limit moment. For the rectangular cross section, we know from Art. 13.3 that the limit moment is $M_L = (3/2)M_{yp}$, causing the stress distribution shown in Fig. 13.9(a). Releasing the load is equivalent to adding an equal but opposite moment to the section, which results in the stress distribution in Fig. 13.9(b). The residual stress distribution in the unloaded beam, shown in part (c), is obtained by superimposing parts (a) and (b). Again, the residual stresses are self-balancing, but if some of the material is removed, an unbalance is created. This explains why members that are cold-formed distort after machining.

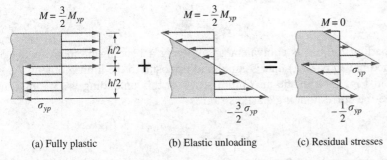

(a) Fully plastic (b) Elastic unloading (c) Residual stresses

FIG. 13.9 Determining the residual normal stress distribution in an elastic-perfectly plastic beam after loading into the fully plastic state and then unloading.

As we saw in the case of torsion, a beam that has been unloaded from the fully plastic state may be reloaded in the same sense, with the beam remaining elastic until the limit moment is reached. For reversed loading, the beam also remains elastic as long as the reversed moment does not exceed $(1/2)M_{yp}$. If this condition is not satisfied, further yielding will occur.

d. Elastic spring-back

As a final example of residual stress, we consider the inelastic bending of a straight rectangular bar around a 90° circular die, as shown in Fig. 13.10(a). When the bar is released, it springs back through the angle θ_s as shown in Fig. 13.10(b). This elastic spring-back is of great importance in metal-forming operations. The relationship between the radius R_0 of the die and the final radius of curvature R_f of the bar can be found by combining the plastic strain caused by loading with the elastic strain caused by unloading. The computations shown below are similar to those used to determine residual stresses.

Recalling from Art. 5.2 that the magnitude of the strain in bending is $\epsilon = y/\rho$, we see that the maximum strain in the bar in Fig. 13.10(a) is

$$\epsilon_0 = \frac{h/2}{R_0}$$

where h is the thickness of the bar. After spring-back in Fig. 13.10(b), the maximum strain is

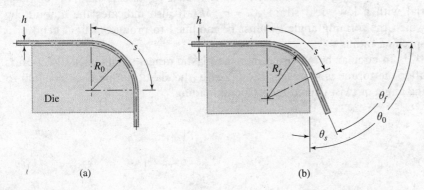

(a) (b)

FIG. 13.10 (a) Bar bent around a circular die; (b) elastic spring-back after unloading.

$$\epsilon_f = \frac{h/2}{R_f}$$

Unloading the bar is equivalent to applying a bending moment to the deformed bar in part (a) that is equal and opposite to the limit moment. As we saw in Fig. 13.9(b), the maximum stress in this unloading is $(3/2)\sigma_{yp}$, and hence the corresponding elastic strain is

$$\epsilon_e = \frac{\sigma}{E} = \frac{(3/2)\sigma_{yp}}{E}$$

Superimposing these strains, we obtain for the residual strain $\epsilon_f = \epsilon_0 - \epsilon_e$, or

$$\frac{h/2}{R_f} = \frac{h/2}{R_0} - \frac{(3/2)\sigma_{yp}}{E}$$

which can be rearranged in the form

$$\boxed{1 - \frac{R_0}{R_f} = \frac{3\sigma_{yp} R_0}{Eh}} \tag{13.7}$$

Letting s be the length of the bend, the angle θ_f associated with the final radius of R_f is found from $s = R_0 \theta_0 = R_f \theta_f$, or

$$\theta_f = \frac{R_0}{R_f} \theta_0$$

The spring-back angle may now be evaluated from

$$\theta_s = \theta_0 - \theta_f = \theta_0 \left(1 - \frac{R_0}{R_f}\right)$$

Substituting from Eq. (13.7), we finally obtain

$$\boxed{\theta_s = \theta_0 R_0 \left(\frac{3\sigma_{yp}}{Eh}\right)} \tag{13.8}$$

This result indicates that the relative amount of spring-back may be reduced by using (a) a smaller forming radius, (b) thicker bars, or (c) material with a low yield strain $\epsilon_{yp} = \sigma_{yp}/E$. It also indicates the amount by which the forming angle θ_0 must be modified to produce a final bend of a specified radius.

In circular bars twisted into the plastic range, spring-back also occurs after the torque is removed. In that case, the elastic spring-back is equal to the angle of twist caused by elastic unloading.

Sample Problem 13.2

The three equally spaced vertical rods in Fig. (a) are securely attached to a rigid horizontal bar. The two outer rods are made of an aluminum alloy, and the middle rod is steel. The cross-sectional areas of the rods are shown in the figure. Determine the residual stresses in the rods after the load P has been increased from zero to the limit load and then removed. The material properties are $E = 70$ GPa, $\sigma_{yp} = 330$ MPa for aluminum, and $E = 200$ GPa, $\sigma_{yp} = 290$ MPa for steel. Neglect the weight of the bar.

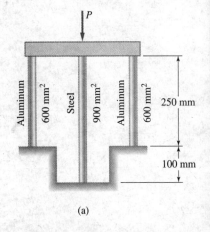

(a)

Solution

Limit Load The limit load P_L is the value of P at which all three rods yield. The axial forces in the bars at yielding are

$$P_{st} = (\sigma_{yp}A)_{st} = (290 \times 10^6)(900 \times 10^{-6}) = 261.0 \times 10^3 \text{ N} = 261.0 \text{ kN}$$

$$P_{al} = (\sigma_{yp}A)_{al} = (330 \times 10^6)(600 \times 10^{-6}) = 198.0 \times 10^3 \text{ N} = 198.0 \text{ kN}$$

Referring to the free-body diagram in Fig. (b), we obtain for the limit load

$$P_L = 261.0 + 2(198.0) = 657.0 \text{ kN}$$

Elastic Unloading Applying an upward load equal to P_L is equivalent to removing the limit load. Assuming the unloading is elastic, we apply the procedure discussed in Art. 2.5 for statically indeterminate problems. From the free-body diagram in Fig. (c), we obtain the equilibrium equation

$$P_{st} + 2P_{al} = 657.0 \text{ kN} \qquad \text{(a)}$$

Because the rigid bar imposes equal elongations on the three bars, the compatibility equation is

$$\left(\frac{PL}{EA}\right)_{st} = \left(\frac{PL}{EA}\right)_{al} \qquad \frac{P_{st}(350)}{(200)(900)} = \frac{P_{al}(250)}{(70)(600)}$$

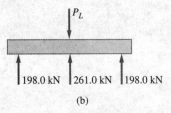

(b)

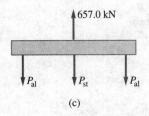

(c)

which yields

$$P_{st} = 3.061 P_{al} \qquad \text{(b)}$$

Solving Eqs. (a) and (b), we find that the unloading forces in the bars are

$$P_{st} = 397.4 \text{ kN} \quad \text{and} \quad P_{al} = 129.8 \text{ kN}$$

Residual Stresses The residual forces are obtained by superimposing the tensile forces caused by unloading and the compressive forces caused by loading. Considering tension as positive and compression as negative, we get

$$P_{st} = 397.4 - 261.0 = 137.4 \text{ kN}$$

$$P_{al} = 129.8 - 198.0 = -68.2 \text{ kN}$$

Thus, the residual stresses are

$$\sigma_{st} = \frac{P_{st}}{A_{st}} = \frac{137.4 \times 10^3}{900 \times 10^{-6}} = 152.7 \times 10^6 \text{ Pa} = 152.7 \text{ MPa} \qquad \textbf{Answer}$$

$$\sigma_{al} = -\frac{P_{al}}{A_{al}} = -\frac{68.2 \times 10^3}{600 \times 10^{-6}} = -113.7 \times 10^6 \text{ Pa} = -113.7 \text{ MPa} \qquad \textbf{Answer}$$

■

Problems

13.14 In Sample Problem 13.2, change the area of the steel bar to 1200 mm². If the load $P = 600$ kN is applied and then removed, determine the residual force in each bar.

13.15 The compound bar is firmly attached between rigid supports. The yield strengths for steel and the aluminum alloy are 42 ksi and 48 ksi, respectively. Determine the residual stresses if the limit load $P = P_L$ is applied and then removed.

13.16 Solve Prob. 13.15 if a load $P = 132$ kips is applied and then removed.

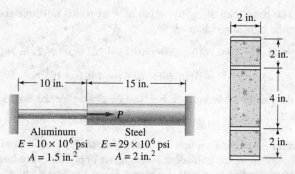

FIG. P13.15, P13.16   **FIG. P13.17**

13.17 A sandwich beam is made by bonding 1/8-in. strips of 2024-T3 aluminum alloy between layers of plastic to form the section shown in the figure. The plastic acts only to separate the aluminum strips; its bending rigidity is negligible. If a bending moment of 10 kip·ft is applied and then removed, determine the residual stresses in the inner and outer strips of aluminum. Use $\sigma_{yp} = 44$ ksi for aluminum.

13.18 A torque applied to a solid rod of radius r causes the elastic region to extend a distance $r/2$ from the center of the rod. If the torque is then removed, determine the residual shear stress distribution.

13.19 The outer diameter of a hollow shaft is twice its inner diameter. Determine the residual stress pattern after the limit torque has been applied and removed.

13.20 The bent rod is 10 mm in diameter. A torque causes arm CD to rotate through the angle θ relative to arm AB. Determine the value of θ if the two arms are to be 90° apart after the torque has been removed. Assume that $\tau_{yp} = 120$ MPa and $G = 80$ GPa.

13.21 A rectangular steel bar 50 mm wide by 90 mm deep is subjected to a bending moment that makes 80% of the bar plastic. Using $\sigma_{yp} = 260$ MPa, determine the residual stress distribution after the bending moment is removed.

13.22 A rectangular steel bar 30 mm wide by 60 mm deep is loaded by a bending moment of 6 kN·m, which is then removed. If $\sigma_{yp} = 280$ MPa, what is the residual stress at a point 20 mm from the neutral axis?

13.23 Consider a cross section of a beam, such as the T-section in Fig. 13.4, where the bending axis is not an axis of symmetry. Show that if this beam was loaded to a fully plastic state and then unloaded, the residual stress would be larger than the yield stress, which is impossible. (Recall that the theory of elastic unloading is applicable only when bending takes place about an axis of symmetry.)

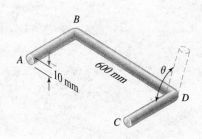

FIG. P13.20

13.24 A beam with the section shown in the figure carries a bending moment that causes the elastic region to extend 1.5 in. from the neutral axis. Determine the residual stress distribution after the bending moment has been removed. Use $\sigma_{yp} = 40$ ksi.

13.25 A strip of steel 1/4 in. thick is bent over a circular die of radius 3-7/8 in. Given that $\sigma_{yp} = 40$ ksi and $E = 29 \times 10^6$ psi, what will be the residual radius of curvature?

13.26 Referring to Prob. 13.25, determine the angle of contact with the circular die during bending so that the strip will have a permanent bend angle of 90°.

13.27 A circular die with radius 250 mm is used to bend a 2024-T4 aluminum alloy strip 10 mm thick. What is the required angle of contact if the strip is to have a permanent bend angle of 180°? Assume that $\sigma_{yp} = 330$ MPa and $E = 70$ GPa.

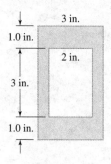

FIG. P13.24

13.5 Limit Analysis

Limit analysis is a method of determining the loading that causes a statically indeterminate structure to collapse. This method applies only to ductile materials, which in this simplified discussion are assumed to be elastic-perfectly plastic. The method is straightforward, consisting of two steps. The first step is a kinematic study of the structure to determine which parts must become fully plastic to permit the structure as a whole to undergo large deformations. The second step is an equilibrium analysis to determine the external loading that creates these fully plastic parts. We present limit analysis in the form of examples involving axial loading, torsion, and bending.[1]

a. Axial loading

Consider the rigid beam in Fig. 13.11, which is supported by a pin at O and the steel rods A and B of different lengths. In the elastic solution, the elongations of the rods are proportional to the distances of the rods from the pin. From Hooke's law, this condition gives us one relationship between axial forces P_A and P_B (compatibility equation). The second relationship is obtained from the equation of equilibrium $\Sigma M_O = 0$. The elastic analysis is valid as long as $W \leq W_{yp}$, where W_{yp} is the yield load (the magnitude of W at which one of the rods begins to yield).

When the load W is increased beyond W_{yp}, the rod that yielded first will keep yielding while maintaining its fully plastic axial load $P = \sigma_{yp} A$. The other rod, which is still elastic, will carry an increasing proportion of the loading until it also starts to yield. At that point, we will have reached the limit load W_L at which the structure collapses by rotation about the pin support.

Let us now find the limit load for the structure in Fig. 13.11 using the following properties of the rods:

Rod	A (mm²)	σ_{yp} (MPa)	E (GPa)
A	300	330	200
B	400	290	200

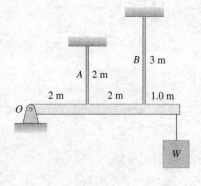

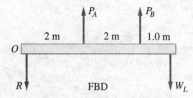

FIG. 13.11 Example of limit analysis. Both bars (A and B) are plastic at the limit value of the load W.

---

[1] One example of limit analysis involving axial loading was presented in Sample Problem 13.2.

Because both rods are in the plastic range at the limit load, the axial forces are given by

$$P_A = (\sigma_{yp}A)_A = (330 \times 10^6)(300 \times 10^{-6}) = 99.0 \times 10^3 \text{ N} = 99.0 \text{ kN}$$

$$P_B = (\sigma_{yp}A)_B = (290 \times 10^6)(400 \times 10^{-6}) = 116.0 \times 10^3 \text{ N} = 116.0 \text{ kN}$$

The limit load can now be obtained from the equilibrium equation $\Sigma M_O = 0$, which yields

$$5W_L = 2P_A + 4P_B = 2(99.0) + 4(116.0)$$

$$W_L = 132.4 \text{ kN}$$

For comparison, the yield load for this structure can be shown to be $W_{yp} = 118.9$ kN, with bar A yielding.

b. Torsion

As an example of limit analysis in torsion, consider the compound shaft shown in Fig. 13.12. The ends of the shaft are attached to rigid supports. The problem is to determine the limit torque T_L (the largest torque that the shaft can carry at the junction of the steel and aluminum segments).

If we assume the materials are elastic-perfectly plastic, the limit torque of the shaft occurs when both segments have reached their limit torques. From Eq. (b) in Art. 13.2, the limit torques in the segments are

$$(T_L)_{\text{al}} = \frac{2}{3}\pi(r^3\tau_{yp})_{\text{al}} = \frac{2}{3}\pi(0.035)^3(160 \times 10^6) = 14.368 \times 10^3 \text{ N} \cdot \text{m}$$

$$(T_L)_{\text{st}} = \frac{2}{3}\pi(r^3\tau_{yp})_{\text{st}} = \frac{2}{3}\pi(0.025)^3(140 \times 10^6) = 4.581 \times 10^3 \text{ N} \cdot \text{m}$$

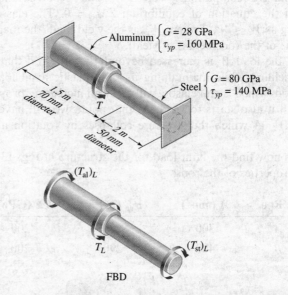

FIG. 13.12 Example of limit analysis. Both segments of the shaft are fully plastic at the limit value of the applied torque T.

The limit torque of the shaft is obtained from the equilibrium of moments about the axis of the shaft, which yields

$$T_L = (T_L)_{\text{al}} + (T_L)_{\text{st}} = (14.368 + 4.581) \times 10^3$$
$$= 18.95 \times 10^3 \text{ N} \cdot \text{m} = 18.95 \text{ kN} \cdot \text{m}$$

In contrast, the yield torque can be shown to be $T_{yp} = 9.60$ kN·m, which occurs when the maximum shear stress in steel reaches its yield value.

c. Bending

Let us return to the cantilever beam shown in Fig. 13.3. As the load P is increased, section c-c at the support goes successively through elastic and partly plastic states until it becomes fully plastic. Sections between a-a and b-b are then partly plastic, whereas the rest of the beam remains elastic. The fully plastic section is called a *plastic hinge* because it allows the beam to rotate about the support without an increase in the bending moment. The bending moment at the plastic hinge is, of course, the limit moment M_L. Once the plastic hinge has formed, the beam in Fig. 13.3 would collapse.

The collapse of a statically determinate beam is synonymous with the formation of a plastic hinge. Statically indeterminate beams do not collapse until enough plastic hinges have formed to make the collapse kinematically possible. The kinematic representation of the collapse is called the *collapse mechanism*. Three examples of collapse mechanisms are shown in Fig. 13.13. Note that the number of plastic hinges (shown by solid circles) required in the collapse mechanism increases with the number of support redundancies. Thus, the simply supported beam requires only one plastic hinge, whereas three plastic hinges are required for the beam with built-in ends.

In general, plastic hinges form where the bending moment is a maximum, which includes built-in supports and sections with zero shear force (since $V = dM/dx$ is an equilibrium equation, its validity is not limited to elastic beams). Thus, the location of plastic hinges is usually obvious for beams subjected to concentrated loads. The task is more difficult for statically indeterminate beams carrying distributed loads. Sometimes more than one collapse mechanism is possible, in which case we must compute the collapse load for each mechanism and choose the smallest collapse load as the actual limit load. These concepts are discussed in the sample problems that follow.

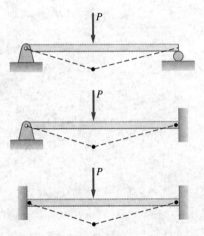

FIG. 13.13 The collapse mechanism of a beam depends on the supports. Each extra support constraint requires an additional plastic hinge in the collapse mechanism.

Sample Problem 13.3

The beam in Fig. (a) has a built-in support at each end and carries a uniformly distributed load of intensity w. Knowing that moments at the supports in the elastic state are

$$M_A = M_B = \frac{wL^2}{12} \tag{a}$$

determine the ratio of the limit load (the value of w at collapse) to the yield load (the value of w when yielding begins).

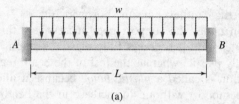

(a)

Solution

The collapse mechanism requires the formation of three plastic hinges. Due to symmetry of the beam and the loading, these hinges are located at A, B, and C, giving rise to the collapse mechanism shown in Fig. (b). The bending moments at the plastic hinges are equal to the limit moment M_L. The relationship between w_L (the limit value of w) and M_L can now be determined from equilibrium using the free-body diagrams of the two beam segments shown in Fig. (c). Note that as a consequence of symmetry, the shear force at C is zero. Applying the equilibrium equation $\Sigma M_A = 0$ to the left half of the beam, we get

$$2M_L = \frac{w_L L}{2}\left(\frac{L}{4}\right) \qquad M_L = \frac{w_L L^2}{16} \tag{b}$$

Substituting $M_A = M_B = M_{yp}$ and $w = w_{yp}$ into Eq. (a), we obtain for the relationship between the yield moment and the yield load

$$M_{yp} = \frac{w_{yp} L^2}{12} \tag{c}$$

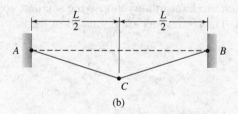

(b)

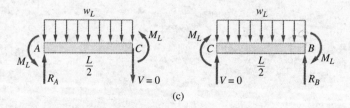

(c)

Dividing Eq. (b) by Eq. (c) gives us the ratio of the limit load to the yield load:

$$\frac{w_L}{w_{yp}} = \frac{4}{3}\frac{M_L}{M_{yp}} \qquad \text{Answer}$$

The ratios M_L/M_{yp} listed in Table 13.1 are appreciable for rectangular or circular cross sections, but for standard structural shapes they are so close to unity that M_L can be taken to be equal to M_{yp} without appreciable error.

Sample Problem 13.4

The propped cantilever beam in Fig. (a) carries a uniformly distributed load of intensity w. Determine the limit moment M_L in terms of the limit load w_L.

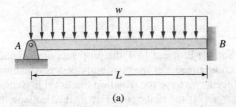

(a)

Solution

The collapse mechanism, which requires two plastic hinges, is shown in Fig. (b). The location of the plastic hinge at C is unknown, but it may be found from the condition that the vertical shear force is zero where the maximum bending moment occurs ($V = dM/dx$ is an equilibrium equation that is valid whether or not the stresses are in the elastic range). The free-body diagrams of the beam segments on either side of C are shown in Fig. (c). Note that the moments at B and C are equal to the limit moment M_L. The equilibrium condition $\Sigma M_A = 0$ applied to segment AC yields

$$M_L = \frac{w_L x^2}{2} \qquad \text{(a)}$$

and the equation $\Sigma M_B = 0$ for segment CB becomes

$$2M_L = \frac{w_L(L-x)^2}{2} \qquad \text{(b)}$$

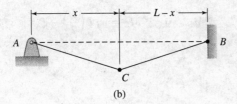

(b)

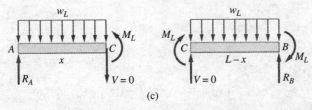

(c)

The solution of Eqs. (a) and (b) is $x = 0.414L$, which locates the plastic hinge C, and

$$M_L = 0.0858 w_L L^2 \qquad \text{Answer}$$

Sample Problem 13.5

The two cantilever beams in Fig. (a), separated by a roller at B, jointly support the uniformly distributed load of intensity w. Determine the limit load w_L in terms of the limit moment M_L. Assume that the beams have identical cross sections.

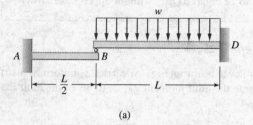

(a)

Solution

In this variation of Sample Problem 13.4, the prop support is replaced by a cantilever beam. This modification introduces the possibility of a plastic hinge forming at A. If that hinge does not form, we have the situation discussed in Sample Problem 13.4 in which $M_L = 0.0858 w_L L^2$. If the hinge forms at A, then collapse is possible by rotation about plastic hinges at A and D as shown in Fig. (b).

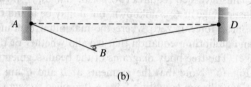

(b)

To determine the limit load for the collapse mechanism in Fig. (b), we draw the free-body diagrams for the two beams, as shown in Fig. (c). Note that the moments at A and D have been set equal to the limit moment M_L. The equilibrium equations for the two beams are

$$\Sigma M_A = 0 \qquad M_L = P\left(\frac{L}{2}\right)$$

$$\Sigma M_D = 0 \qquad M_L = w_L L\left(\frac{L}{2}\right) - P(L)$$

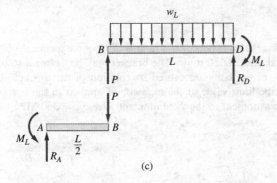

(c)

Eliminating the contact force P from these two equations, we obtain

$$M_L = \frac{w_L L^2}{6} = 0.1667 w_L L^2$$

Because this expression gives a smaller w_L than the one found in Sample Problem 13.4, the collapse mechanism is as shown in Fig. (b), the limit load being

$$w_L = \frac{6M_L}{L^2}. \qquad \text{Answer}$$

Problems

13.28 The bracket is fastened to a rigid wall by three identical bolts, each having a cross-sectional area of 150 mm². The bracket may be assumed to be rigid, so that the elongations of the bolts are caused by rotation of the bracket about corner O. Determine (a) the limit value of the moment M applied to the bracket, and (b) the ratio of the limit moment to the yield moment. Use $\sigma_{yp} = 300$ MPa for the bolts.

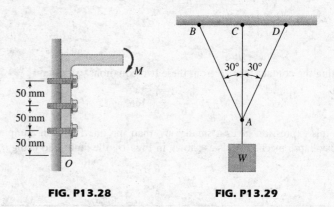

FIG. P13.28   **FIG. P13.29**

13.29 The three steel rods, each 0.5 in.² in cross-sectional area, jointly support the load W. Assuming that there is no slack or stress in the rods before the load is applied, determine the ratio of the limit load to the yield load. Use $\sigma_{yp} = 35$ ksi for steel.

13.30 Determine the limit value of the torque T in Fig. 13.12 if the torque acts at 1.0 m from the right support.

13.31 Referring to Sample Problem 13.5, let the length of beam AB be L_1. Show that Fig. (b) represents the collapse mechanism only if $L_1 > 0.207L$.

13.32 Determine the limit load P_L for the propped cantilever beam in terms of the limit moment M_L.

13.33 If the two cantilever beams have the same limit moment M_L, determine the limit value of the load P.

13.34 The two cantilever beams have rectangular cross sections with the dimensions shown in the figure. Determine the limit value of the load P if $\sigma_{yp} = 40$ ksi.

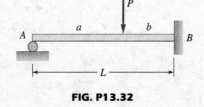

FIG. P13.32

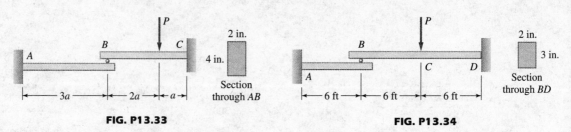

FIG. P13.33   **FIG. P13.34**

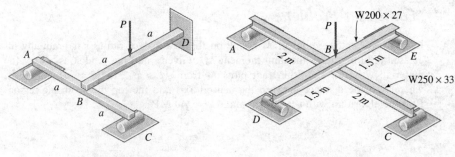

FIG. P13.35   **FIG. P13.36**

13.35 The load P is supported by a cantilever beam resting on a simply supported beam. If the limit moments are M_L for the cantilever beam and $0.75M_L$ for the simply supported beam, determine the limit value of P.

13.36 The two simply supported, wide-flange beams are mounted at right angles and are in contact with each other at their midpoints. At the crossover point, the beams jointly support a load P. Determine the limit value of P. Use $\sigma_{yp} = 290$ MPa and assume $M_L = M_{yp}$.

13.37 The restrained beam carries two loads of magnitude P. Denoting the ratio of the limit moment to the yield moment by K, determine the ratio P_L/P_{yp} in terms of K. (*Note*: If the beam is elastic, the magnitude of the bending moment at each support is $2Pa/3$.)

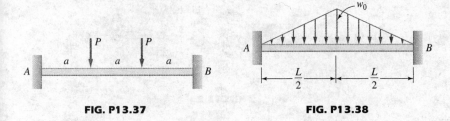

FIG. P13.37   **FIG. P13.38**

13.38 Letting K denote the ratio of the limit moment to the yield moment, determine the ratio $(w_0)_L/(w_0)_{yp}$ in terms of K for the restrained beam. (*Note*: Given elastic behavior, the magnitude of the bending moment at each support is $5w_0L^2/96$.)

13.39 Determine the limit value of P in terms of the limit moment M_L for the restrained beam in the figure.

13.40 The continuous beam carries a uniformly distributed load of intensity w over its entire length. Determine the limit value of w in terms of the limit moment M_L.

13.41 Repeat Prob. 13.40 assuming that both ends of the beam are built into rigid walls.

13.42 Determine the limit value of P in terms of the limit moment M_L for the beam shown in the figure.

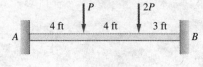

FIG. P13.39

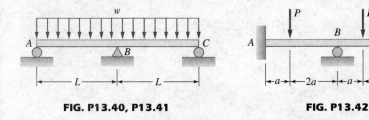

FIG. P13.40, P13.41   **FIG. P13.42**

Computer Problems

C13.1 A bending moment M acting on the steel bar of radius r is gradually increased until it reaches the limit moment M_L. Given r and the yield stress σ_{yp} of the material, write an algorithm that plots M from $M = M_{yp}$ to $M = M_L$ versus y_i, where y_i is the distance between the neutral axis and the top of the elastic region. Run the algorithm with $r = 40$ mm and $\sigma_{yp} = 240$ MPa.

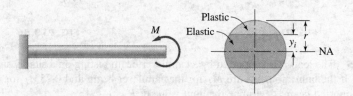

FIG. C13.1

C13.2 The steel shaft AB of length L and radius r carries a torque T that is gradually increased until it reaches the limit torque T_L. Given L, r, and the material properties G and τ_{yp}, write an algorithm that plots the angle of rotation of end B versus T from $T = T_{yp}$ to $T = 0.99T_L$. Run the algorithm with the data $L = 1.0$ m, $r = 50$ mm, $G = 80$ GPa, and $\tau_{yp} = 150$ MPa.

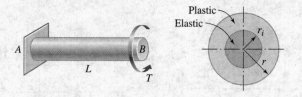

FIG. C13.2

Appendix A
Review of Properties of Plane Areas

A.1 First Moments of Area; Centroid

The first moments of a plane area A about the x- and y-axes are defined as

$$Q_x = \int_A y\,dA \qquad Q_y = \int_A x\,dA \qquad \text{(A.1)}$$

where dA is an infinitesimal element of A located at (x, y), as shown in Fig. A.1. The values of Q_x and Q_y may be positive, negative, or zero, depending on the location of the origin O of the coordinate axes. The dimension of the first moment of area is $[L^3]$; hence, its units are mm^3, in.^{3}, and so on.

The centroid C of the area is defined as the point in the xy-plane that has the coordinates (see Fig. A.1)

$$\bar{x} = \frac{Q_y}{A} \qquad \bar{y} = \frac{Q_x}{A} \qquad \text{(A.2)}$$

It follows that if A and $(\bar{x}, \bar{y})$ are known, the first moments of the area can be computed from $Q_x = A\bar{y}$ and $Q_y = A\bar{x}$. The following are useful properties of the first moments of area:

- If the origin of the xy-coordinate system is the centroid of the area (in which case $\bar{x} = \bar{y} = 0$), then $Q_x = Q_y = 0$.
- Whenever the area has an axis of symmetry, the centroid of the area will lie on that axis.

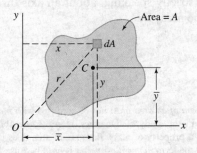

FIG. A.1

If the area can be subdivided into simple geometric shapes (rectangles, circles, etc.) of areas $A_1, A_2, A_2, \ldots$, it is convenient to use Eqs. (A.1) in the form

$$Q_x = \sum \left(\int_{A_i} y\, dA \right) = \sum (Q_x)_i = \sum A_i \bar{y}_i \qquad \text{(A.3a)}$$

$$Q_y = \sum \left(\int_{A_i} x\, dA \right) = \sum (Q_y)_i = \sum A_i \bar{x}_i \qquad \text{(A.3b)}$$

where $(\bar{x}_i, \bar{y}_i)$ are the coordinates of the centroid of area A_i. Because the centroidal coordinates of simple shapes are known, Eqs. (A.3) allow us to compute the first moments without using integration.

A.2 Second Moments of Area

a. Moments and product of inertia

Referring again to Fig. A.1, we can define the second moments of a plane area A with respect to the xy-axes by

$$I_x = \int_A y^2\, dA \qquad I_y = \int_A x^2\, dA \qquad I_{xy} = \int_A xy\, dA \qquad \text{(A.4)}$$

The integrals I_x and I_y are commonly called the *moments of inertia*,[1] whereas I_{xy} is known as the *product of inertia*. The moments of inertia are always positive, but the product of inertia can be positive, negative, or zero. The dimension of the second moment of area is $[L^4]$. Therefore, the units are mm^4, in.^{4}, and so forth.

> **Caution** Recall that the first moment of an area about the x-axis can be evaluated from $Q_x = A\bar{y}$, where $\bar{y}$ is the centroidal coordinate of the area. A common mistake is to extrapolate this formula to the moment of inertia by wrongly assuming that $I_x = A\bar{y}^2$.

If either the x- or y-axis is an axis of symmetry of the area, then $I_{xy} = 0$. This result can be deduced by inspection of the area shown in Fig. A.2. Because the y-axis is an axis of symmetry, for every dA with coordinates (x, y), there is a dA with coordinates $(-x, y)$. It follows that $I_{xy} = \int_A xy\, dA = 0$ when the integration is performed over the entire region.

Referring to Fig. A.1, we can define the *polar moment of inertia of an area* about point O (strictly speaking, about an axis through O, perpendicular to the plane of the area) by

---

[1] The term *moment of inertia of an area* must not be confused with *moment of inertia of a body*. The latter, which occurs in the study of dynamics, refers to the ability of a body to resist a change in rotation and is a property of mass. Because an area does not have mass, it does not possess inertia. However, the term *moment of inertia* is used because the integrals in Eqs. (A.4) are similar to the expression $\int r^2\, dm$ that defines the moment of inertia of a body.

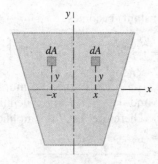

FIG. A.2

$$J_O = \int_A r^2 \, dA \qquad (A.5)$$

where r is the distance from O to the area element dA. Substituting $r^2 = y^2 + x^2$, we obtain

$$J_O = \int_A r^2 \, dA = \int_A (y^2 + x^2) \, dA = \int_A y^2 \, dA + \int_A x^2 \, dA$$

or

$$J_O = I_x + I_y \qquad (A.6)$$

This relationship states that the polar moment of inertia of an area about a point O equals the sum of the moments of inertia of the area about two perpendicular axes that intersect at O.

b. Parallel-axis theorems

In Fig. A.3, let C be the centroid of the area A, and let the x'-axis be the centroidal x-axis (the axis passing through C that is parallel to the x-axis). We denote the moment of inertia about the x'-axis by $\bar{I}_x$. The y-coordinate of the area element dA can be written as $y = \bar{y} + y'$, where $\bar{y}$ is the distance between the two axes. From Eq. (A.4), the moment of inertia of the area

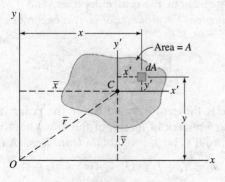

FIG. A.3

about the x-axis ($\bar{y}$ is constant) becomes

$$I_x = \int_A y^2 \, dA = \int_A (\bar{y} + y')^2 \, dA = \bar{y}^2 \int_A dA + 2\bar{y} \int_A y' \, dA + \int_A (y')^2 \, dA \quad \text{(a)}$$

We note that $\int_A dA = A$, $\int_A y' \, dA = 0$ (the first moment of the area about a centroidal axis is zero), and $\int_A (y')^2 \, dA = \bar{I}_x$ (definition of the moment of inertia about the x'-axis). Therefore, Eq. (a) simplifies to

$$\boxed{I_x = \bar{I}_x + A\bar{y}^2} \quad \text{(A.7a)}$$

which is known as the *parallel-axis theorem* for the moment of inertia of an area. The distance $\bar{x}$ is sometimes called the *transfer distance* (the distance through which the moment of inertia is to be "transferred").

Caution The parallel-axis theorem is valid only if $\bar{I}_x$ is the moment of inertia about the centroidal x-axis. If this is not the case, the integral $\int_A y' \, dA$ in Eq. (a) does not vanish, giving rise to an additional term in Eq. (A.7a).

Because the direction of the x-axis can be chosen arbitrarily, the parallel-axis theorem applies to axes that have any orientation. For example, applying the theorem to the y-axis yields

$$\boxed{I_y = \bar{I}_y + A\bar{x}^2} \quad \text{(A.7b)}$$

where $\bar{I}_y$ is the moment of inertia about the centroidal y-axis (that is, the y'-axis in Fig. A.3) and $\bar{x}$ is the x-coordinate of the centroid.

By substituting $x = x' + \bar{x}$ and $y = y' + \bar{y}$ into the expression for I_{xy} in Eq. (A.4), we obtain the parallel-axis theorem for the product of inertia:

$$I_{xy} = \int_A xy \, dA = \int_A (x' + \bar{x})(y' + \bar{y}) \, dA$$

$$= \int_A x'y' \, dA + \bar{x} \int_A y' \, dA + \bar{y} \int_A x' \, dA + \bar{x}\bar{y} \int_A dA$$

We note that $\int_A x'y' \, dA = \bar{I}_{xy}$ is the product of inertia with respect to the centroidal axes. Also, $\int_A y' \, dA = \int_A x' \, dA = 0$ because they represent the first moments of the area about the centroidal axes, and $\int_A dA = A$. Therefore, the *parallel-axis theorem for products of inertia* becomes

$$\boxed{I_{xy} = \bar{I}_{xy} + A\bar{x}\bar{y}} \quad \text{(A.8)}$$

A parallel-axis theorem also exists for the polar moment of inertia. Denoting the polar moment of inertia of the area about the origin O by J_O and about the centroid C by $\bar{J}_C$, we obtain from Eqs. (A.6) and (A.7)

$$J_O = I_x + I_y = (\bar{I}_x + A\bar{y}^2) + (\bar{I}_y + A\bar{x}^2)$$

Because $\bar{I}_x + \bar{I}_y = \bar{J}_C$, this equation becomes

$$\boxed{J_O = \bar{J}_C + A\bar{r}^2} \qquad (A.9)$$

where $\bar{r} = \sqrt{\bar{x}^2 + \bar{y}^2}$ is the distance between O and C, as shown in Fig. A.3.

c. Radii of gyration

In some structural engineering applications, it is common practice to introduce the *radius of gyration of an area*. The radii of gyration about the x-axis, the y-axis, and the point O are defined as

$$\boxed{k_x = \sqrt{\frac{I_x}{A}} \qquad k_y = \sqrt{\frac{I_y}{A}} \qquad k_O = \sqrt{\frac{J_O}{A}}} \qquad (A.10)$$

The dimension of the radius of gyration is $[L]$. However, the radius of gyration is not a distance that has a clear-cut physical meaning, nor can it be determined by direct measurement. It can be determined only by computation using Eq. (A.10).

The radii of gyration are related by

$$\boxed{k_O^2 = k_x^2 + k_y^2} \qquad (A.11)$$

which can be obtained by substituting Eqs. (A.10) into Eq. (A.6).

d. Method of composite areas

Consider a plane area A that has been divided into the subareas $A_1, A_2, A_3, \ldots$. The moment of inertia of the area A about an axis can be computed by summing the moments of inertia of the subareas about the same axis. This technique, known as the *method of composite areas*, follows directly from the property of definite integrals: The integral of a sum equals the sum of the integrals. For example, the moment of inertia I_x about the x-axis is

$$I_x = \int_A y^2\, dA = \int_{A_1} y^2\, dA + \int_{A_2} y^2\, dA + \int_{A_3} y^2\, dA + \cdots$$

which can be written as

$$I_x = (I_x)_1 + (I_x)_2 + (I_x)_3 + \cdots \qquad (A.12a)$$

where $(I_x)_i$ is the moment of inertia of subarea A_i about the x-axis. The method of composite areas also applies to the computation of the polar moment of inertia:

$$J_O = (J_O)_1 + (J_O)_2 + (J_O)_3 + \cdots \qquad (A.12b)$$

where $(J_O)_i$ is the polar moment of inertia of subarea A_i with respect to point O.

Before the moments or products of inertia can be summed, they must be transferred to common axes using the parallel-axis theorems. The table in Fig. A.4 lists the area properties of simple shapes that can be used for the method of composite areas.

APPENDIX A Review of Properties of Plane Areas

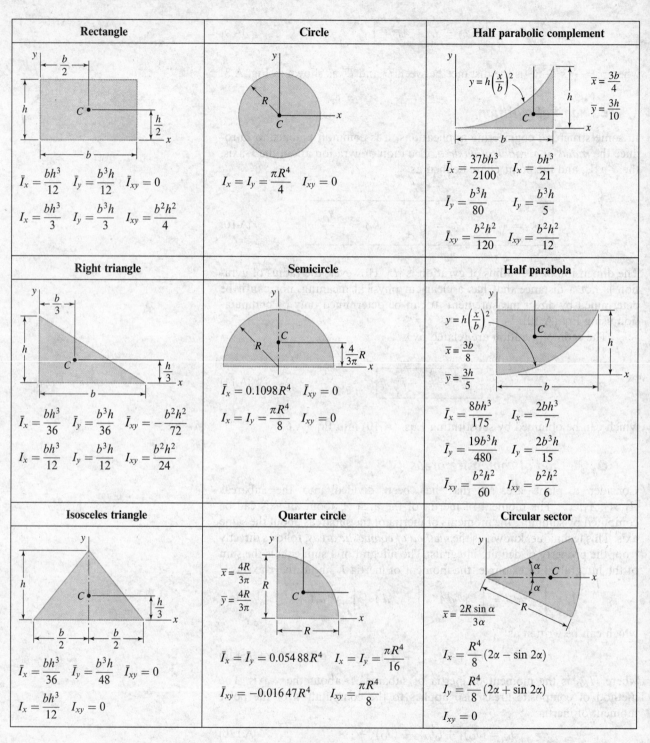

FIG. A.4

A.2 Second Moments of Area

Triangle	Quarter ellipse
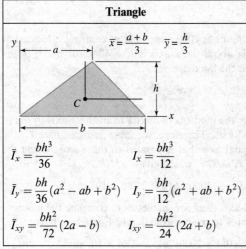	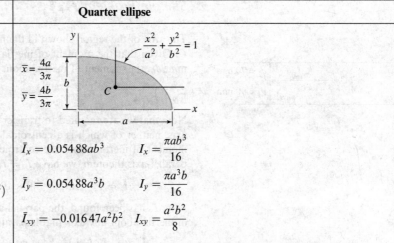
$\bar{I}_x = \dfrac{bh^3}{36}$ $\quad I_x = \dfrac{bh^3}{12}$	$\bar{I}_x = 0.05488ab^3$ $\quad I_x = \dfrac{\pi ab^3}{16}$
$\bar{I}_y = \dfrac{bh}{36}(a^2 - ab + b^2)$ $\quad I_y = \dfrac{bh}{12}(a^2 + ab + b^2)$	$\bar{I}_y = 0.05488a^3 b$ $\quad I_y = \dfrac{\pi a^3 b}{16}$
$\bar{I}_{xy} = \dfrac{bh^2}{72}(2a - b)$ $\quad I_{xy} = \dfrac{bh^2}{24}(2a + b)$	$\bar{I}_{xy} = -0.01647a^2 b^2$ $\quad I_{xy} = \dfrac{a^2 b^2}{8}$

FIG. A.4 (continued)

Sample Problem A.1

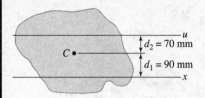

The area of the region shown in the figure is 2000 mm^2, and its centroid is located at C. Given that the moment of inertia about the x-axis is $I_x = 40 \times 10^6$ mm^4, determine I_u, the moment of inertia about the u-axis.

Solution

Note that we are required to transfer the moment of inertia from the x-axis to the u-axis, neither of which is a centroidal axis. Therefore, we must first calculate $\bar{I}_x$, the moment of inertia about the centroidal axis that is parallel to the x-axis. Using the parallel-axis theorem, we have $I_x = \bar{I}_x + Ad_1^2$, which gives

$$\bar{I}_x = I_x - Ad_1^2 = (40 \times 10^6) - 2000(90)^2 = 23.80 \times 10^6 \text{ mm}^4$$

After $\bar{I}_x$ has been found, the parallel-axis theorem enables us to compute the moment of inertia about any axis that is parallel to the centroidal axis. For I_u, we have

$$I_u = \bar{I}_x + Ad_2^2 = (23.80 \times 10^6) + 2000(70)^2 = 33.6 \times 10^6 \text{ mm}^4 \quad \textbf{Answer}$$

A common error is to use the parallel-axis theorem to transfer the moment of inertia between two axes, neither of which is a centroidal axis. In this problem, for example, it is tempting to write $I_u = I_x + A(d_1 + d_2)^2$, which would result in an incorrect answer for I_u.

Sample Problem A.2

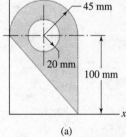

For the plane area shown in Fig. (a), calculate the moment of inertia about (1) the x-axis; and (2) the centroidal x-axis.

Solution

Part 1

We consider the area to be composed of the three parts shown in Figs. (b)–(d): a triangle, plus a semicircle, minus a circle. The moment of inertia for each part is obtained in three steps. First, we compute the moment of inertia of each part about its own centroidal axes using the table in Fig. A.4. Each of the moments of inertia is then transferred to the x-axis using the parallel-axis theorem. Finally, we obtain the moment of inertia of the composite area by combining the moments of inertia of the parts.

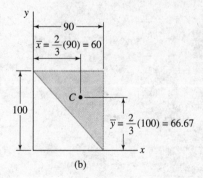
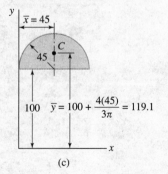
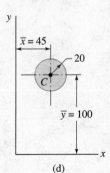

Triangle

$$A = \frac{bh}{2} = \frac{90(100)}{2} = 4500 \text{ mm}^2$$

$$\bar{I}_x = \frac{bh^3}{36} = \frac{90(100)^3}{36} = 2.500 \times 10^6 \text{ mm}^4$$

$$I_x = \bar{I}_x + A\bar{y}^2 = (2.500 \times 10^6) + 4500(66.67)^2 = 22.50 \times 10^6 \text{ mm}^4$$

Semicircle

$$A = \frac{\pi R^2}{2} = \frac{\pi(45)^2}{2} = 3181 \text{ mm}^2$$

$$\bar{I}_x = 0.1098 R^4 = 0.1098(45)^4 = 0.4503 \times 10^6 \text{ mm}^4$$

$$I_x = \bar{I}_x + A\bar{y}^2 = (0.4503 \times 10^6) + 3181(119.1)^2 = 45.57 \times 10^6 \text{ mm}^4$$

Circle (to be removed)

$$A = \pi R^2 = \pi(20)^2 = 1257 \text{ mm}^2$$

$$\bar{I}_x = \frac{\pi R^4}{4} = \frac{\pi(20)^4}{4} = 0.1257 \times 10^6 \text{ mm}^4$$

$$I_x = \bar{I}_x + A\bar{y}^2 = (0.1257 \times 10^6) + 1257(100)^2 = 12.70 \times 10^6 \text{ mm}^4$$

Composite Area

$$A = \Sigma A_i = 4500 + 3181 - 1257 = 6424 \text{ mm}^2$$

$$I_x = \Sigma(I_x)_i = (22.50 + 45.57 - 12.70) \times 10^6 = 55.37 \times 10^6 \text{ mm}^4 \quad \text{Answer}$$

Part 2

Before we can transfer the moment of inertia computed in Part 1 to the centroidal x-axis, we must find $\bar{y}$, the y-coordinate of the centroid of the composite area. Combining Eqs. (A.2) and (A.3a), we get

$$\bar{y} = \frac{Q_x}{A} = \frac{\Sigma A_i \bar{y}_i}{A} = \frac{4500(66.7) + 3181(119.1) - 1257(100)}{6424} = 86.13 \text{ mm}$$

From the parallel-axis theorem, the moment of inertia about the centroidal x-axis is

$$\bar{I}_x = I_x - A\bar{y}^2 = (55.37 \times 10^6) - 6424(86.13)^2 = 7.71 \times 10^6 \text{ mm}^4 \quad \text{Answer}$$

Sample Problem A.3

Calculate the product of inertia I_{xy} for the area shown in Fig. (a).

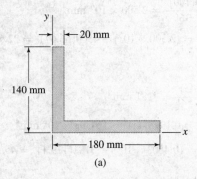

(a)

Solution

We may view the area as the composite of the two rectangles shown in Fig. (b). We can compute I_{xy} of each rectangle using the parallel-axis theorem for the product of inertia, and then add the results. Noting that due to symmetry $\bar{I}_{xy} = 0$ for each rectangle, we make the following calculations.

20-mm by 140-mm Rectangle

$$I_{xy} = \bar{I}_{xy} + A\bar{x}\bar{y} = 0 + (140 \times 20)(10)(70) = 1.960 \times 10^6 \text{ mm}^4$$

160-mm by 20-mm Rectangle

$$I_{xy} = \bar{I}_{xy} + A\bar{x}\bar{y} = 0 + (160 \times 20)(100)(10) = 3.200 \times 10^6 \text{ mm}^4$$

Composite Area

$$I_{xy} = \Sigma(I_{xy})_i = (1.960 + 3.200) \times 10^6 = 5.16 \times 10^6 \text{ mm}^4 \qquad \textit{Answer}$$

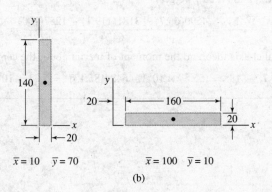

(b)

Problems

A.1 The properties of the area shown in the figure are $\bar{J}_C = 50 \times 10^3$ mm^4, $I_x = 600 \times 10^3$ mm^4, and $I_y = 350 \times 10^3$ mm^4. Calculate the area A, $\bar{I}_x$, and $\bar{I}_y$.

A.2 The moments of inertia of the trapezoid about the x- and u-axes are $I_x = 14 \times 10^9$ mm^4 and $I_u = 38 \times 10^9$ mm^4, respectively. Given that $h = 200$ mm, determine the area A and the radius of gyration about the centroidal x-axis.

A.3 Find the distance h for which the moment of inertia of the trapezoid about the u-axis is $I_u = 120 \times 10^9$ mm^4, given that $A = 90 \times 10^3$ mm^2 and $I_x = 14 \times 10^9$ mm^4.

A.4 Calculate $\bar{y}$ and the moment of inertia of the T-section about the centroidal x-axis.

A.5 Find $\bar{y}$ and the moments of inertia about the centroidal x- and y-axes for the area shown in the figure.

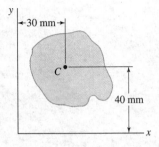

FIG. PA.1

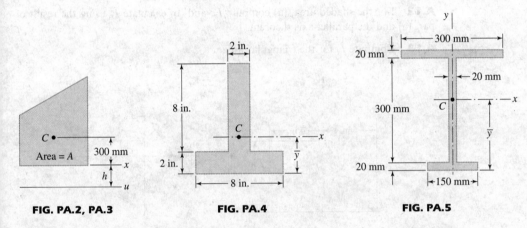

FIG. PA.2, PA.3   FIG. PA.4   FIG. PA.5

A.6 Calculate $\bar{I}_x$ for the shaded area given that $\bar{y} = 68.54$ mm.

A.7 Compute $\bar{I}_y$ for the shaded area given that $\bar{x} = 25.86$ mm.

A.8 Calculate $\bar{I}_x$ for the Z-section.

A.9 The section shown is formed by lacing together two C200 × 28 channel sections. Determine the distance d for which the moments of inertia of the section about the x- and y-axes are equal. Neglect the moment of inertia of the lattice bars that are indicated by the dashed lines.

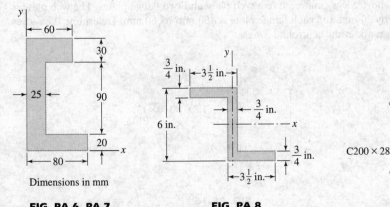

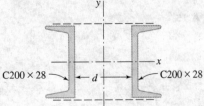

FIG. PA.6, PA.7   FIG. PA.8   FIG. PA.9

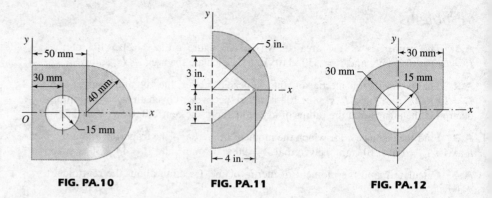

FIG. PA.10   **FIG. PA.11**   **FIG. PA.12**

A.10 Compute the polar moment of inertia of the shaded area about point O.

A.11 Calculate the moment of inertia about the x-axis for the shaded area.

A.12 For the shaded area, (a) compute I_x; and (b) calculate $\bar{I}_x$ using the result of part (a) and the parallel-axis theorem.

A.13 Compute $\bar{I}_x$ for the triangular area.

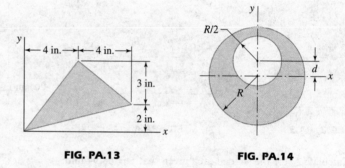

FIG. PA.13   **FIG. PA.14**

A.14 The shaded area consists of a circle of radius R from which a circle of radius $R/2$ has been removed. For what distance d will k_x for the shaded area be the same as k_x for the larger circle before the removal of the smaller circle?

A.15 The short legs of four L6 × 4 × 1/2 angle sections are connected to a 23-1/2-in. by 5/16-in. web plate to form the plate and angle girder. Compute the radius of gyration of the section about the centroidal x-axis.

A.16 A plate and angle column is made of four L200 × 100 × 13 angle sections with the shorter legs connected to a web plate and two flange plates. The web plate is 350 mm by 20 mm and each flange plate is 460 mm by 60 mm. Determine the radius of gyration about the centroidal x-axis.

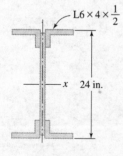

FIG. PA.15

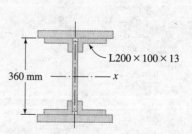

FIG. PA.16

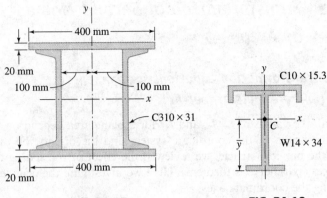

FIG. PA.17

FIG. PA.18

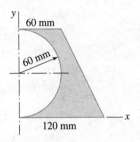

FIG. PA.19

A.17 Calculate $\bar{I}_x$ and $\bar{I}_y$ for the built-up column section that is composed of two 400-mm by 20-mm plates connected to two C310 × 31 channels.

A.18 A C10 × 15.3 channel is welded to the top of a W14 × 34 as shown. Determine $\bar{y}$ and the moment of inertia of the composite section about the centroidal x-axis.

A.19 Compute the product of inertia of the area shown with respect to the xy-axes.

A.20 Calculate the product of inertia of the area shown with respect to the xy-axes.

A.21 Find $\bar{I}_{xy}$ for the area shown in the figure.

A.22 The figure shows the cross section of a standard L80 × 60 × 10 angle section. Compute I_{xy} of the cross-sectional area.

A.23 Calculate $\bar{I}_{xy}$ for the area shown given that $\bar{x} = 25.86$ mm and $\bar{y} = 68.54$ mm.

A.24 Compute $\bar{I}_{xy}$ for the area shown in the figure.

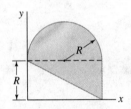

FIG. PA.20

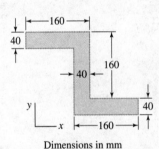

Dimensions in mm

FIG. PA.21

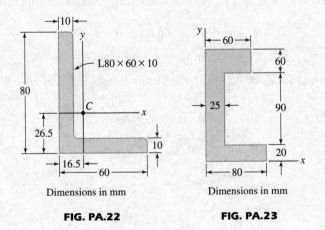

FIG. PA.22

FIG. PA.23

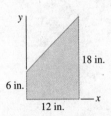

FIG. PA.24

A.3 Transformation of Second Moments of Area

a. Transformation equations for moments and products of inertia

In general, the values of I_x, I_y, and I_{xy} for a plane area depend on the location of the origin of the coordinate system and the orientation of the xy-axes. In the previous article, we reviewed the effect of translating the coordinate axes (parallel-axis theorem). Here we investigate the changes caused by rotating the coordinate axes.

Consider the plane area A and the two coordinate systems shown in Fig. A.5. The coordinate systems have the same origin O, but the uv-axes are inclined at the angle θ to the xy-axes. We now derive the *transformation equations* that enable us to compute I_u, I_v, and I_{uv} for the area in terms of I_x, I_y, I_{xy}, and θ. We start with the transformation equations for the position coordinates, which can be obtained from Fig. A.5:

$$u = y \sin \theta + x \cos \theta \qquad v = y \cos \theta - x \sin \theta \qquad \text{(A.13)}$$

By definition, the moment of inertia about the u-axis is $I_u = \int_A v^2 \, dA$. Substituting v from Eq. (A.13), we get

$$I_u = \int_A (y \cos \theta - x \sin \theta)^2 \, dA$$

$$= \cos^2 \theta \int_A y^2 \, dA - 2 \sin \theta \cos \theta \int_A xy \, dA + \sin^2 \theta \int_A x^2 \, dA$$

Because the integrals represent the second moments of the area with respect to the xy-axes, we have

$$I_u = I_x \cos^2 \theta - 2 I_{xy} \sin \theta \cos \theta + I_y \sin^2 \theta \qquad \text{(A.14a)}$$

The equations for I_v and I_{uv} may be derived in a similar manner; the results are

$$I_v = I_x \sin^2 \theta + 2 I_{xy} \sin \theta \cos \theta + I_y \cos^2 \theta \qquad \text{(A.14b)}$$

$$I_{uv} = (I_x - I_y) \sin \theta \cos \theta + I_{xy}(\cos^2 \theta - \sin^2 \theta) \qquad \text{(A.14c)}$$

The equation for I_v could also be derived by replacing θ by $(\theta + 90°)$ in Eq. (A.14a).

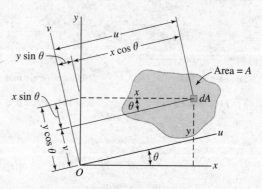

FIG. A.5

Using the trigonometric identities

$$\sin 2\theta = 2\sin\theta\cos\theta \qquad \cos 2\theta = \cos^2\theta - \sin^2\theta$$

$$\cos^2\theta = \frac{1}{2}(1+\cos 2\theta) \qquad \sin^2\theta = \frac{1}{2}(1-\cos 2\theta)$$

we can write Eqs. (A.14) in the form

$$I_u = \frac{I_x + I_y}{2} + \frac{I_x - I_y}{2}\cos 2\theta - I_{xy}\sin 2\theta \qquad \text{(A.15a)}$$

$$I_v = \frac{I_x + I_y}{2} - \frac{I_x - I_y}{2}\cos 2\theta + I_{xy}\sin 2\theta \qquad \text{(A.15b)}$$

$$I_{uv} = \frac{I_x - I_y}{2}\sin 2\theta + I_{xy}\cos 2\theta \qquad \text{(A.15c)}$$

b. Comparison with stress transformation equations

If we replace x' by u and y' by v, the transformation equations for plane stress derived in Chapter 8 are

$$\sigma_u = \frac{\sigma_x + \sigma_y}{2} + \frac{\sigma_x - \sigma_y}{2}\cos 2\theta + \tau_{xy}\sin 2\theta \qquad \text{(8.5a, repeated)}$$

$$\sigma_v = \frac{\sigma_x + \sigma_y}{2} - \frac{\sigma_x - \sigma_y}{2}\cos 2\theta - \tau_{xy}\sin 2\theta \qquad \text{(8.5b, repeated)}$$

$$\tau_{uv} = -\frac{\sigma_x - \sigma_y}{2}\sin 2\theta + \tau_{xy}\cos 2\theta \qquad \text{(8.5c, repeated)}$$

Comparing Eqs. (8.5) with Eqs. (A.15), we see that the transformation equations for plane stress and for second moments of area are identical if we make the following associations:

Plane stress	σ_x	σ_y	σ_u	σ_v	τ_{xy}	τ_{uv}
Second moments of area	I_x	I_y	I_u	I_v	$-I_{xy}$	$-I_{uv}$

(A.16)

Therefore, we can use the plane stress equations for second moments of area by simply switching the symbols as indicated in Eq. (A.16).

c. Principal moments of inertia and principal axes

The expression for the principal stresses derived in Art. 8.5 is

$$\left.\begin{array}{c}\sigma_1\\\sigma_2\end{array}\right\} = \frac{\sigma_x + \sigma_y}{2} \pm \sqrt{\left(\frac{\sigma_x - \sigma_y}{2}\right)^2 + \tau_{xy}^2} \qquad \text{(8.10, repeated)}$$

Changing the symbols according to the associations in Eq. (A.16), we obtain the expression for the principal moments of inertia:

$$\left.\begin{array}{c} I_1 \\ I_2 \end{array}\right\} = \frac{I_x + I_y}{2} \pm \sqrt{\left(\frac{I_x - I_y}{2}\right)^2 + I_{xy}^2} \qquad \text{(A.17)}$$

Similar modification of Eq. (8.7): $\tan 2\theta = 2\tau_{xy}/(\sigma_x - \sigma_y)$ yields

$$\tan 2\theta = -\frac{2I_{xy}}{I_x - I_y} \qquad \text{(A.18)}$$

where θ defines the orientation of the principal axes. As in the case of stress, Eq. (A.18) yields two solutions for 2θ that differ by 180°. If we denote one solution by $2\theta_1$, the second solution is $2\theta_2 = 2\theta_1 + 180°$. Hence, the two principal directions differ by 90°.

In Chapter 8, we showed that there are no shear stresses on the principal planes. By analogy, we conclude that the product of inertia is zero with respect to the principal axes.

d. Mohr's circle for second moments of area

We saw in Chapter 8 that the transformation equations for plane stress can be represented by the Mohr's circle shown in Fig. A.6(a). The coordinates of each point on the circle correspond to the normal and shear stresses acting on a specific set of perpendicular planes. Mohr's circle also applies to second moments of area, where the coordinates of each point represent the moment and the product of inertia with respect to a specific set of perpendicular axes. Because τ_{xy} is associated with $-I_{xy}$, the circle for the second moments of area is "flipped" about the horizontal axis, as shown in Fig. A.6(b). That is, the point (I_x, I_{xy}) is plotted *above* the horizontal axis if I_{xy} is positive. In contrast, (σ_x, τ_{xy}) is plotted *below* the axis for positive τ_{xy}. Otherwise, the properties of the two circles are identical.

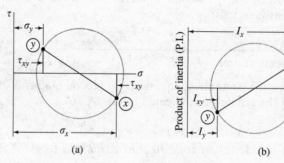

FIG. A.6

Sample Problem A.4

For the area shown in Fig. (a), use the transformation equations to calculate (1) the principal moments of inertia at the centroid C of the area and the corresponding principal directions; and (2) the moments and products of inertia with respect to the uv-axes.

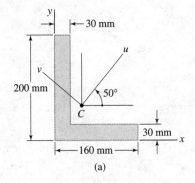
(a)

Solution

Preliminary Calculations

We consider the area to be a composite of the two rectangles shown in Fig. (b). The coordinates of C are obtained from

$$\bar{x} = \frac{\Sigma A_i \bar{x}_i}{\Sigma A_i} = \frac{6000(15) + 3900(30 + 65)}{6000 + 3900} = 46.52 \text{ mm}$$

$$\bar{y} = \frac{\Sigma A_i \bar{y}_i}{\Sigma A_i} = \frac{6000(100) + 3900(15)}{6000 + 3900} = 66.52 \text{ mm}$$

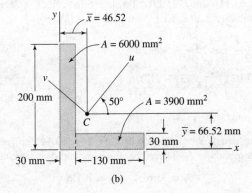
(b)

The second moments of the composite area with respect to the centroidal xy-axes are

$$\bar{I}_x = \sum \left[\frac{b_i h_i^3}{12} + A_i(\bar{y}_i - \bar{y})^2 \right]$$

$$= \frac{30(200)^3}{12} + 6000(100 - 66.52)^2 + \frac{130(30)^3}{12} + 3900(15 - 66.52)^2$$

$$= 37.37 \times 10^6 \text{ mm}^4$$

$$\bar{I}_y = \sum \left[\frac{b_i h_i^3}{12} + A_i(\bar{x}_i - \bar{x})^2 \right]$$

$$= \frac{200(30)^3}{12} + 6000(15 - 46.52)^2 + \frac{30(130)^3}{12} + 3900(95 - 46.52)^2$$

$$= 21.07 \times 10^6 \text{ mm}^4$$

$$\bar{I}_{xy} = \sum A_i(\bar{x}_i - \bar{x})(\bar{y}_i - \bar{y})$$

$$= 6000(15 - 46.52)(100 - 66.52) + 3900(15 - 66.52)(95 - 46.52)$$

$$= -16.073 \times 10^6 \text{ mm}^4$$

Part 1

Substituting the values for $\bar{I}_x, \bar{I}_y,$ and $\bar{I}_{xy}$ into Eq. (A.17) yields

$$\left.\begin{array}{c}I_1\\I_2\end{array}\right\} = \frac{\bar{I}_x + \bar{I}_y}{2} \pm \sqrt{\left(\frac{\bar{I}_x - \bar{I}_y}{2}\right)^2 + \bar{I}_{xy}^2}$$

$$= \left[\frac{37.37 + 21.07}{2} \pm \sqrt{\left(\frac{37.37 - 21.07}{2}\right)^2 + (-16.073)^2}\right] \times 10^6$$

$$= (29.22 \pm 18.02) \times 10^6 \text{ mm}^4$$

which yields for the principal moments of inertia at the centroid

$$I_1 = 47.2 \times 10^6 \text{ mm}^4 \qquad I_2 = 11.20 \times 10^6 \text{ mm}^4 \qquad \textit{Answer}$$

For the principal directions, Eq. (A.18) yields

$$\tan 2\theta = -\frac{2\bar{I}_{xy}}{\bar{I}_x - \bar{I}_y} = -\frac{2(-16.073)}{37.37 - 21.07} = 1.9722$$

which gives $2\theta = 63.11°$ and $243.11°$. To determine which of these angles corresponds to I_1, we substitute $2\theta = 63.11°$ into Eq. (A.15a):

$$I_u = \frac{\bar{I}_x + \bar{I}_y}{2} + \frac{\bar{I}_x - \bar{I}_y}{2}\cos 2\theta - \bar{I}_{xy}\sin 2\theta$$

$$= \left[\left(\frac{37.37 + 21.07}{2}\right) + \left(\frac{37.37 - 21.07}{2}\right)\cos 63.11° - (-16.073)\sin 63.11°\right] \times 10^{-6}$$

$$= 47.2 \times 10^{-6} \text{ mm}^4$$

Because this value equals I_1, we conclude that $2\theta = 63.11°$ corresponds to I_1 and $2\theta = 243.11°$ corresponds to I_2. Therefore, the principal directions are

$$\theta_1 = 31.6° \qquad \theta_2 = 121.6° \qquad \textit{Answer}$$

The principal axes, labeled 1 and 2, are shown in Fig. (c).

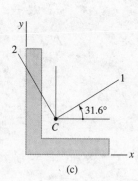

(c)

Part 2

To compute the moments and product of inertia relative to the uv-axes shown in Fig. (a), we substitute $\theta = 50°$ into the transformation equations, Eqs. (A.15):

$$I_u = \left[\frac{37.37 + 21.07}{2} + \frac{37.37 - 21.07}{2}\cos 100° - (-16.073)\sin 100°\right] \times 10^{-6}$$

$$= 43.6 \times 10^6 \text{ mm}^4 \qquad \textit{Answer}$$

$$I_v = \left[\frac{37.37 + 21.07}{2} - \frac{37.37 - 21.07}{2}\cos 100° + (-16.073)\sin 100°\right] \times 10^{-6}$$

$$= 14.81 \times 10^6 \text{ mm}^4 \qquad \textit{Answer}$$

$$I_{uv} = \left[\frac{37.37 - 21.07}{2}\sin 100° + (-16.073)\cos 100°\right] \times 10^{-6}$$

$$= 10.82 \times 10^6 \text{ mm}^4 \qquad \textit{Answer}$$

Sample Problem A.5

Solve Sample Problem A.4 using Mohr's circle instead of the transformation equations.

Solution

From the solution of Sample Problem A.4, we have $\bar{I}_x = 37.37 \times 10^6$ mm^4, $\bar{I}_y = 21.07 \times 10^6$ mm^4, and $\bar{I}_{xy} = -16.073 \times 10^6$ mm^4. Using these values, we plot the Mohr's circle shown in the figure. The points on the circle that correspond to the second moments of area about the centroidal xy-axes are labeled $\textcircled{x}$ and $\textcircled{y}$, respectively. Because $\bar{I}_{xy}$ is negative, $\textcircled{x}$ is plotted below the abscissa. The parameters of the circle, which can be obtained from geometry, are

$$b = \frac{37.37 + 21.07}{2} \times 10^6 = 29.22 \times 10^6 \text{ mm}^4$$

$$R = \sqrt{\left(\frac{37.37 - 21.07}{2}\right)^2 + (16.073)^2} \times 10^6 = 18.021 \times 10^6 \text{ mm}^4$$

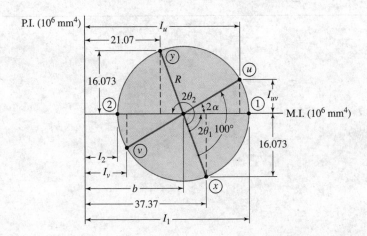

Part 1

The points labeled $\textcircled{1}$ and $\textcircled{2}$ on the Mohr's circle correspond to the principal moments of inertia. From the figure, we see that

$$I_1 = b + R = (29.22 + 18.021) \times 10^6 = 47.2 \times 10^6 \text{ mm}^4 \quad \text{Answer}$$

$$I_2 = b - R = (29.22 - 18.021) \times 10^6 = 11.20 \times 10^6 \text{ mm}^4 \quad \text{Answer}$$

The principal directions are found by calculating the angles θ_1 and θ_2 shown on the circle. Using trigonometry, we get

$$2\theta_1 = \sin^{-1}\frac{16.073}{R} = \sin^{-1}\frac{16.073}{18.021} = 63.11°$$

Therefore,

$$\theta_1 = 31.6° \qquad \theta_2 = 31.6° + 90° = 121.6° \quad \text{Answer}$$

Note that on the circle the central angle from point $\textcircled{x}$ to point $\textcircled{1}$ is $2\theta_1$ in the counterclockwise direction. Therefore, the principal direction corresponding to I_1 is $\theta_1 = 31.6°$, measured counterclockwise from the centroidal x-axis. (Recall that the angles on Mohr's circle are twice the angles between axes, measured in the same direction.) This leads to the same result as shown in Fig. (c) of Sample Problem A.4.

Part 2

The coordinates of (u) and (v) on the Mohr's circle are the second moments of area with respect to the uv-axes. Because the angle measured from the x-axis to the u-axis is 50° counterclockwise, the angle from (x) to (u) is 100°, also counterclockwise. Of course, (v) is located at the opposite end of the diameter from (u). To facilitate our computations, we have introduced the central angle 2α between (1) and (u), where $2\alpha = 100° - 2\theta_1 = 100° - 63.11° = 36.89°$. Referring to the circle, we find that

$$I_u = b + R \cos 2\alpha = (29.22 + 18.021 \cos 36.89°) \times 10^6$$

$$= 43.6 \times 10^6 \text{ mm}^4 \qquad \textit{Answer}$$

$$I_v = b - R \cos 2\alpha = (29.22 - 18.021 \cos 36.89°) \times 10^6$$

$$= 14.81 \times 10^6 \text{ mm}^4 \qquad \textit{Answer}$$

$$I_{uv} = R \sin 2\alpha = (18.021 \sin 36.89°) \times 10^6$$

$$= 10.82 \times 10^6 \text{ mm}^4 \qquad \textit{Answer}$$

Note that because (u) is above the abscissa, I_{uv} is positive.

Problems

Solve Problems A.25–A.29 using the transformation equations for moments and products of inertia.

A.25 The properties of the area shown in the figure are $I_x = 4000$ in.^{4}, $I_y = 1000$ in.^{4}, and $I_{xy} = -800$ in.^{4}. Determine I_u, I_v, and I_{uv} for $\theta = 120°$.

A.26 The properties of the area shown are $I_x = 10 \times 10^6$ mm^4, $I_y = 20 \times 10^6$ mm^4, and $I_{xy} = 12 \times 10^6$ mm^4. Compute I_u, I_v, and I_{uv} for $\theta = 33.7°$.

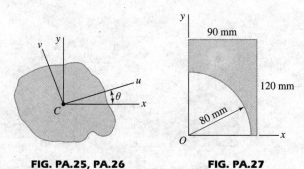

FIG. PA.25, PA.26   **FIG. PA.27**

A.27 For the area shown, determine the principal moments of inertia at point O and the corresponding principal directions.

A.28 The L80 × 60 × 10 angle section has the cross-sectional properties $I_x = 0.808 \times 10^6$ mm^4, $I_y = 0.388 \times 10^6$ mm^4, and $I_2 = 0.213 \times 10^6$ mm^4, where I_2 is a principal centroidal moment of inertia. Assuming I_{xy} is negative, use the transformation equations to compute (a) I_1, the other principal centroidal moment of inertia; and (b) the principal directions.

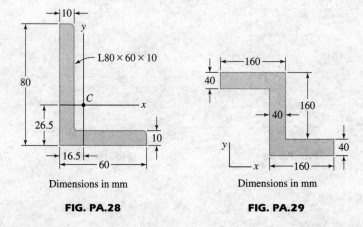

FIG. PA.28   **FIG. PA.29**

A.29 Compute the principal centroidal moments of inertia and the corresponding principal directions for the area shown in the figure.

Solve Problems A.30–A.39 using Mohr's circle.

A.30 See Prob. A.25.

A.31 See Prob. A.26.

A.32 See Prob. A.27.

A.33 See Prob. A.28.

A.34 See Prob. A.29.

A.35 Find the moments and the product of inertia of the rectangle about the uv-axes at the centroid C.

A.36 Calculate I_u, I_v, and I_{uv} for the area shown in the figure.

A.37 Determine the moments and product of inertia for the triangle about the uv-axes.

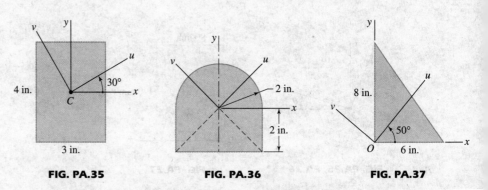

FIG. PA.35 FIG. PA.36 FIG. PA.37

A.38 Calculate the principal moments of inertia and the principal directions at point O for the area shown.

A.39 The properties of the area shown in the figure are $I_x = 140$ in.^{4}, $I_y = 264$ in.^{4}, and $I_{xy} = -116$ in.^{4}. Determine I_u, I_v, and I_{uv}. Note that the u-axis passes through point B.

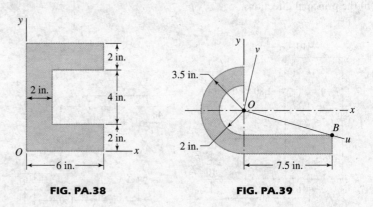

FIG. PA.38 FIG. PA.39

Appendix B
Tables

B-1   Average Physical Properties of Common Metals
B-2   Properties of Wide-Flange Sections (W-Shapes): SI Units
B-3   Properties of I-Beam Sections (S-Shapes): SI Units
B-4   Properties of Channel Sections: SI Units
B-5   Properties of Equal Angle Sections: SI Units
B-6   Properties of Unequal Angle Sections: SI Units
B-7   Properties of Wide-Flange Sections (W-Shapes): U.S. Customary Units
B-8   Properties of I-Beam Sections (S-Shapes): U.S. Customary Units
B-9   Properties of Channel Sections: U.S. Customary Units
B-10   Properties of Equal and Unequal Angle Sections: U.S. Customary Units

Acknowledgments

Data for Tables B-2 through B-6 are taken from *Handbook of Steel Construction*, 4th ed., 1985, Canadian Institute of Steel Construction.

Data for Tables B-7 through B-10 are taken from the AISC *Manual of Steel Construction*, 8th ed., 1980, American Institute of Steel Construction, Inc.

APPENDIX B Tables

TABLE B-1 *Average Physical Properties of Common Metals: SI Units*

Metal	Density (kg/m³)	Temp. coeff. of linear expansion [μm/(m·°C)]	Proportional limit (MPa)[a] Tension	Proportional limit (MPa)[a] Shear	Ultimate strength (MPa) Tension	Ultimate strength (MPa) Comp.	Ultimate strength (MPa) Shear	Modulus of elasticity (GPa)[a] Tension, E	Modulus of elasticity (GPa)[a] Shear, G	Percentage of elongation (in 50 mm)
Steel, 0.2% carbon, hot rolled	7850	Varies from 11.0 to 13.2. Average is 11.7	240	150	410	[b]	310	200	80	35
0.2% carbon, cold rolled	7850		420	250	550	[b]	420	200	80	18
0.6% carbon, hot rolled	7850		420	250	690	[b]	550	200	80	15
0.8% carbon, hot rolled	7850		480	290	830	[b]	730	200	80	10
Gray cast iron	7200	10.8	[c]	[d]	140	520	[d]	100	40	Slight
Malleable cast iron	7200	11.9	250	160	370	[b]	330	170	90	18
Wrought iron	7700	12.1	210	130	350	[b]	240	190	70	35
Aluminum, cast	2650	23.1	60	—	90	[b]	70	70	30	20
Aluminum alloy 17ST	2700	23.1	220	150	390	[b]	220	71	30	—
Brass, rolled (70% Cu, 30% Zn)	8500	18.7	170	110	380	[b]	330	100	40	30
Bronze, cast	8200	18.0	140	—	230	390	—	80	35	10
Copper, hard-drawn	8800	16.8	260	160	380	[b]	—	120	40	4

[a] The proportional limit and modulus of elasticity for compression may be assumed equal to these values for tension except for cast iron where the proportional limit is approximately 180 MPa.
[b] The ultimate compressive strength for ductile materials may be taken as the yield point, which is slightly greater than the proportional limit in tension.
[c] Not well defined; approximately 40 MPa.
[d] Cast iron fails by diagonal tension.

TABLE B-1 Average Physical Properties of Common Metals: U.S. Customary Units (continued)

Metal	Weight (lb/ft³)	Temp. coeff. of linear expansion [μ in./(in.·°F)]	Proportional limit (ksi)[a] Tension	Proportional limit (ksi)[a] Shear	Ultimate strength (ksi) Tension	Ultimate strength (ksi) Comp.	Ultimate strength (ksi) Shear	Modulus of elasticity (psi)[a] Tension, E	Modulus of elasticity (psi)[a] Shear, G	Percentage of elongation (in 2 in.)
Steel, 0.2% carbon hot rolled	490	Varies from 6.1 to 7.3 Average is 6.5	35	21	60	[b]	45	29×10^6	12×10^6	35
0.2% carbon cold rolled	490		60	36	80	[b]	60	29×10^6	12×10^6	18
0.6% carbon hot rolled	490		60	36	100	[b]	80	29×10^6	12×10^6	15
0.8% carbon hot rolled	490		70	42	120	[b]	105	29×10^6	12×10^6	10
Gray cast iron	450	6	[c]	[d]	20	75	[d]	15×10^6	6×10^6	Slight
Malleable cast iron	450	6.6	36	23	54	—	48	25×10^6	12.5×10^6	18
Wrought iron	480	6.7	30	18	50	[b]	35	27×10^6	10×10^6	35
Aluminum, cast	165	12.8	9	—	13	[b]	10.5	10×10^6	4×10^6	20
Aluminum alloy 17ST	168	12.8	32	21	56	[b]	32	10.3×10^6	4×10^6	—
Brass, rolled (70% Cu, 30% Zn)	530	10.4	25	15	55	[b]	48	14×10^6	6×10^6	30
Bronze, cast	510	10	20	—	33	56	—	12×10^6	5×10^6	10
Copper, hard-drawn	550	9.3	38	23	55	[b]	—	17×10^6	6×10^6	4

[a] The proportional limit and modulus of elasticity for compression may be assumed to equal these values for tension except for cast iron where the proportional limit is 26 ksi. The ultimate compressive strength for ductile materials may be taken as the yield point, which is slightly greater than the proportional limit in tension.
[b] Not well defined; approximately 6 ksi.
[c] Cast iron fails by diagonal tension.

TABLE B-2 Properties of Wide-Flange Sections (W-Shapes): SI Units

Designation	Theoretical mass (kg/m)	Area (mm²)	Depth (mm)	Flange Width (mm)	Flange Thickness (mm)	Web thickness (mm)	Axis X-X I (10^6 mm^4)	Axis X-X $S = I/c$ (10^3 mm^3)	Axis X-X $r = \sqrt{I/A}$ (mm)	Axis Y-Y I (10^6 mm^4)	Axis Y-Y $S = I/c$ (10^3 mm^3)	Axis Y-Y $r = \sqrt{I/A}$ (mm)
W920 × 446	447.2	57 000	933	423	42.7	24.0	8470	18 200	385	540	2550	97.3
× 417	418.1	53 300	928	422	39.9	22.5	7880	17 000	385	501	2370	97.0
× 387	387.0	49 300	921	420	36.6	21.3	7180	15 600	382	453	2160	95.9
× 365	364.6	46 400	916	419	34.3	20.3	6710	14 600	380	421	2010	95.3
× 342	342.4	43 600	912	418	32.0	19.3	6250	13 700	379	390	1870	94.6
× 313	312.7	39 800	932	309	34.5	21.1	5480	11 800	371	170	1100	65.4
× 289	288.6	36 800	927	308	32.0	19.4	5040	10 900	370	156	1020	65.1
× 271	271.7	34 600	923	307	30.0	18.4	4720	10 200	369	145	946	64.7
× 253	253.7	32 300	919	306	27.9	17.3	4370	9520	368	134	874	64.4
× 238	238.3	30 400	915	305	25.9	16.5	4060	8880	365	123	806	63.6
× 223	224.2	28 600	911	304	23.9	15.9	3770	8270	363	112	738	62.6
× 201	201.3	25 600	903	304	20.1	15.2	3250	7200	356	94.4	621	60.7
W840 × 359	359.4	45 800	868	403	35.6	21.1	5910	13 600	359	389	1930	92.2
× 329	329.4	42 000	862	401	32.4	19.7	5350	12 400	357	349	1740	91.2
× 299	299.3	38 100	855	400	29.2	18.2	4790	11 200	355	312	1560	90.5
× 226	226.6	28 900	851	294	26.8	16.1	3400	7990	343	114	774	62.8
× 210	210.8	26 800	846	293	24.4	15.4	3110	7340	341	103	700	62.0
× 193	193.5	24 700	840	292	21.7	14.7	2780	6630	335	90.3	618	60.5
× 176	176.0	22 400	835	292	18.8	14.0	2460	5900	331	78.2	536	59.1
W760 × 314	314.4	40 100	786	384	33.4	19.7	4270	10 900	326	316	1640	88.8
× 284	283.9	36 200	779	382	30.1	18.0	3810	9790	324	280	1470	87.9
× 257	257.6	32 800	773	381	27.1	16.6	3420	8840	323	250	1310	87.3
× 196	196.8	25 100	770	268	25.4	15.6	2400	6240	309	81.7	610	57.1
× 185	184.8	23 500	766	267	23.6	14.9	2230	5820	308	75.1	563	56.5
× 173	173.6	22 100	762	267	21.6	14.4	2060	5400	305	68.7	515	55.8
× 161	160.4	20 400	758	266	19.3	13.8	1860	4900	302	60.7	457	54.5
× 147	147.1	18 700	753	265	17.0	13.2	1660	4410	298	52.9	399	53.2

Designation										
W690 × 265	264.5	33 700	706	358	30.2	18.4	2900	8220	293	231	1290	82.8
× 240	239.9	30 600	701	356	27.4	16.8	2610	7450	292	206	1160	82.0
× 217	217.8	27 700	695	355	24.8	15.4	2340	6740	291	185	1040	81.7
× 170	169.9	21 600	693	256	23.6	14.5	1700	4910	281	66.2	517	55.4
× 152	152.1	19 400	688	254	21.1	13.1	1510	4380	279	57.8	455	54.6
× 140	139.8	17 800	684	254	18.9	12.4	1360	3980	276	51.7	407	53.9
× 125	125.6	16 000	678	253	16.3	11.7	1190	3500	273	44.1	349	52.5
W610 × 241	241.7	30 800	635	329	31.0	17.9	2150	6780	264	184	1120	77.3
× 217	217.9	27 800	628	328	27.7	16.5	1910	6070	262	163	995	76.6
× 195	195.6	24 900	622	327	24.4	15.4	1680	5400	260	142	871	75.5
× 174	174.3	22 200	616	325	21.6	14.0	1470	4780	257	124	761	74.7
× 155	154.9	19 700	611	324	19.0	12.7	1290	4220	256	108	666	74.0
× 140	140.1	17 900	617	230	22.2	13.1	1120	3630	250	45.1	392	50.2
× 125	125.1	15 900	612	229	19.6	11.9	985	3220	249	39.3	343	49.7
× 113	113.4	14 400	608	228	17.3	11.2	875	2880	247	34.3	300	48.8
× 101	101.7	13 000	603	228	14.9	10.5	764	2530	242	29.5	259	47.6
× 92	92.3	11 800	603	179	15.0	10.9	646	2140	234	14.4	161	34.9
× 82	81.9	10 400	599	178	12.8	10.0	560	1870	232	12.1	136	34.1
W530 × 219	218.9	27 900	560	318	29.2	18.3	1510	5390	233	157	986	75.0
× 196	196.5	25 000	554	316	26.3	16.5	1340	4840	232	139	877	74.6
× 182	181.7	23 100	551	315	24.4	15.2	1240	4480	232	127	808	74.1
× 165	165.3	21 100	546	313	22.2	14.0	1110	4060	229	114	726	73.5
× 150	150.6	19 200	543	312	20.3	12.7	1010	3710	229	103	659	73.2
× 138	138.3	17 600	549	214	23.6	14.7	861	3140	221	38.7	362	46.9
× 123	123.2	15 700	544	212	21.2	13.1	761	2800	220	33.8	319	46.4
× 109	109.0	13 900	539	211	18.8	11.6	667	2480	219	29.5	280	46.1
× 101	101.4	12 900	537	210	17.4	10.9	617	2300	219	26.9	256	45.7
× 92	92.5	11 800	533	209	15.6	10.2	552	2070	216	23.8	228	44.9
× 82[a]	82.4	10 500	528	209	13.3	9.5	479	1810	214	20.3	194	44.0
× 85	84.7	10 800	535	166	16.5	10.3	485	1810	212	12.6	152	34.2
× 74	74.7	9 520	529	166	13.6	9.7	411	1550	208	10.4	125	33.1
× 66	65.7	8 370	525	165	11.4	8.9	351	1340	205	8.57	104	32.0

(continues)

TABLE B-2 Properties of Wide-Flange Sections (W-Shapes): SI Units (continued)

Designation	Theoretical mass (kg/m)	Area (mm²)	Depth (mm)	Flange Width (mm)	Flange Thickness (mm)	Web thickness (mm)	Axis X-X I (10^6 mm⁴)	Axis X-X $S = I/c$ (10^3 mm³)	Axis X-X $r = \sqrt{I/A}$ (mm)	Axis Y-Y I (10^6 mm⁴)	Axis Y-Y $S = I/c$ (10^3 mm³)	Axis Y-Y $r = \sqrt{I/A}$ (mm)
W460 × 177	177.3	22 600	482	286	26.9	16.6	910	3 780	201	105	735	68.2
× 158	157.7	20 100	476	284	23.9	15.0	796	3 350	199	91.4	643	67.4
× 144	144.6	18 400	472	283	22.1	13.6	726	3 080	199	83.6	591	67.4
× 128	128.4	16 400	467	282	19.6	12.2	637	2 730	197	73.3	520	66.9
× 113	113.1	14 400	463	280	17.3	10.8	556	2 400	196	63.3	452	66.3
× 106	105.8	13 500	469	194	20.6	12.6	488	2 080	190	25.1	259	43.1
× 97	96.6	12 300	466	193	19.0	11.4	445	1 910	190	22.8	237	43.1
× 89	89.3	11 400	463	192	17.7	10.5	410	1 770	190	20.9	218	42.8
× 82	81.9	10 400	460	191	16.0	9.9	370	1 610	189	18.6	195	42.3
× 74	74.2	9 450	457	190	14.5	9.0	333	1 460	188	16.6	175	41.9
× 67[a]	68.1	8 680	454	190	12.7	8.5	300	1 320	186	14.6	153	41.0
× 61[a]	60.9	7 760	450	189	10.8	8.1	259	1 150	183	12.2	129	39.7
× 68	68.5	8 730	459	154	15.4	9.1	297	1 290	184	9.41	122	32.8
× 60	59.6	7 590	455	153	13.3	8.0	255	1 120	183	7.96	104	32.4
× 52	52.0	6 630	450	152	10.8	7.6	212	943	179	6.34	83.4	30.9
W410 × 149	149.3	19 000	431	265	25.0	14.9	619	2 870	180	77.7	586	63.9
× 132	132.1	16 800	425	263	22.2	13.3	538	2 530	179	67.4	512	63.3
× 114	114.5	14 600	420	261	19.3	11.6	462	2 200	178	57.2	439	62.6
× 100	99.6	12 700	415	260	16.9	10.0	398	1 920	177	49.5	381	62.4
× 85	85.0	10 800	417	181	18.2	10.9	315	1 510	171	18.0	199	40.8
× 74	74.9	9 550	413	180	16.0	9.7	275	1 330	170	15.6	173	40.4
× 67	67.5	8 600	410	179	14.4	8.8	246	1 200	169	13.8	154	40.1
× 60	59.5	7 580	407	178	12.8	7.7	216	1 060	169	12.0	135	39.8
× 54	53.4	6 810	403	177	10.9	7.5	186	924	165	10.1	114	38.5
× 46	46.2	5 890	403	140	11.2	7.0	156	773	163	5.14	73.4	29.5
× 39	39.2	4 990	399	140	8.8	6.4	127	634	160	4.04	57.7	28.5
W360 × 1086	1087.9	139 000	569	454	125	78.0	5960	20 900	207	1960	8 650	119
× 990	991.0	126 000	550	448	115	71.9	5190	18 900	203	1730	7 740	117
× 900	902.2	115 000	531	442	106	65.9	4500	17 000	198	1530	6 940	115
× 818	819.0	104 000	514	437	97.0	60.5	3920	15 300	194	1360	6 200	114
× 744	744.3	94 800	498	432	88.9	55.6	3420	13 700	190	1200	5 550	113
× 677	677.8	86 300	483	428	81.5	51.2	2990	12 400	186	1070	4 990	111
× 634	634.3	80 800	474	424	77.1	47.6	2740	11 600	184	983	4 630	110
× 592	592.6	75 500	465	421	72.3	45.0	2500	10 800	182	902	4 280	109
× 551	550.6	70 100	455	418	67.6	42.0	2260	9 940	180	825	3 950	108
× 509	509.5	64 900	446	416	62.7	39.1	2050	9 170	178	754	3 630	108

Designation											
W360 × 463	462.8	59 000	435	412	57.4	35.8	1800	8280	175	670	3250	107
× 421	421.7	53 700	425	409	52.6	32.8	1600	7510	173	601	2940	106
× 382	382.4	48 700	416	406	48.0	29.8	1410	6790	170	536	2640	105
× 347	347.0	44 200	407	404	43.7	27.2	1250	6140	168	481	2380	104
× 314	313.4	39 900	399	401	39.6	24.9	1100	5530	166	426	2120	103
× 287	287.6	36 600	393	399	36.6	22.6	997	5070	165	388	1940	103
W360 × 262	262.7	33 500	387	398	33.3	21.1	894	4620	163	350	1760	102
× 237	236.3	30 100	380	395	30.2	18.9	788	4150	162	310	1570	101
× 216	216.3	27 600	375	394	27.7	17.3	712	3790	161	283	1430	101
× 196	196.5	25 000	372	374	26.2	16.4	636	3420	159	229	1220	95.7
× 179	179.2	22 800	368	373	23.9	15.0	575	3120	159	207	1110	95.3
× 162	162.0	20 600	364	371	21.8	13.3	516	2830	158	186	1000	95.0
× 147	147.5	18 800	360	370	19.8	12.3	463	2570	157	167	904	94.2
× 134	134.0	17 100	356	369	18.0	11.2	415	2330	156	151	817	94.0
× 122	121.7	15 500	363	257	21.7	13.0	365	2010	153	61.5	478	63.0
× 110	110.2	14 000	360	256	19.9	11.4	331	1840	154	55.7	435	63.1
× 101	101.2	12 900	357	255	18.3	10.5	302	1690	153	50.6	397	62.6
× 91	90.8	11 600	353	254	16.4	9.5	267	1510	152	44.8	353	62.1
× 79	79.3	10 100	354	205	16.8	9.4	227	1280	150	24.2	236	48.9
× 72	71.5	9 110	350	204	15.1	8.6	201	1150	149	21.4	210	48.5
× 64	63.9	8 140	347	203	13.5	7.7	178	1030	148	18.8	186	48.1
× 57	56.7	7 220	358	172	13.1	7.9	161	897	149	11.1	129	39.2
× 51	50.6	6 450	355	171	11.6	7.2	141	796	148	9.68	113	38.7
× 45	45.0	5 730	352	171	9.8	6.9	122	691	146	8.18	95.7	37.8
× 39	39.1	4 980	353	128	10.7	6.5	102	580	143	3.75	58.6	27.4
× 33	32.8	4 170	349	127	8.5	5.8	82.7	474	141	2.91	45.8	26.4

(continues)

TABLE B-2 Properties of Wide-Flange Sections (W-Shapes): SI Units (continued)

Designation	Theoretical mass (kg/m)	Area (mm^2)	Depth (mm)	Flange Width (mm)	Flange Thickness (mm)	Web thickness (mm)	Axis X-X I (10^6 mm^4)	Axis X-X $S = I/c$ (10^3 mm^3)	Axis X-X $r = \sqrt{I/A}$ (mm)	Axis Y-Y I (10^6 mm^4)	Axis Y-Y $S = I/c$ (10^3 mm^3)	Axis Y-Y $r = \sqrt{I/A}$ (mm)
W310 × 500	500.4	63 700	427	340	75.1	45.1	1690	7910	163	494	2910	88.1
× 454	454.0	57 800	415	336	68.7	41.3	1480	7130	160	436	2600	86.9
× 415	415.1	52 900	403	334	62.7	38.9	1300	6450	157	391	2340	86.0
× 375	374.3	47 700	391	330	57.1	35.4	1130	5760	154	343	2080	84.8
× 342	343.3	43 700	382	328	52.6	32.6	1010	5260	152	310	1890	84.2
× 313	313.3	39 900	374	325	48.3	30.0	896	4790	150	277	1700	83.3
× 283	283.0	36 000	365	322	44.1	26.9	787	4310	148	246	1530	82.7
× 253	252.9	32 200	356	319	39.6	24.4	682	3830	146	215	1350	81.7
× 226	226.8	28 900	348	317	35.6	22.1	596	3420	144	189	1190	80.9
× 202	202.6	25 800	341	315	31.8	20.1	520	3050	142	166	1050	80.2
× 179	178.8	22 800	333	313	28.1	18.0	445	2680	140	144	919	79.5
× 158	157.4	20 100	327	310	25.1	15.5	386	2360	139	125	805	78.9
× 143	143.1	18 200	323	309	22.9	14.0	348	2150	138	113	729	78.8
× 129	129.6	16 500	318	308	20.6	13.1	308	1940	137	100	652	77.8
× 118	117.5	15 000	314	307	18.7	11.9	275	1750	135	90.2	588	77.5
× 107	106.9	13 600	311	306	17.0	10.9	248	1590	135	81.2	531	77.3
× 97	96.8	12 300	308	305	15.4	9.9	222	1440	134	72.9	478	77.0
× 86	86.4	11 000	310	254	16.3	9.1	199	1280	135	44.5	351	63.6
× 79	78.9	10 100	306	254	14.6	8.8	177	1160	132	39.9	314	62.9
× 74	74.5	9490	310	205	16.3	9.4	165	1060	132	23.4	229	49.7
× 67	66.8	8510	306	204	14.6	8.5	145	949	131	20.7	203	49.3
× 60	59.6	7590	303	203	13.1	7.5	129	849	130	18.3	180	49.1
× 52	52.3	6670	317	167	13.2	7.6	118	747	133	10.3	123	39.3
× 45	44.6	5690	313	166	11.2	6.6	99.2	634	132	8.55	103	38.8
W310 × 39	38.7	4940	310	165	9.7	5.8	85.1	549	131	7.27	88.1	38.4
× 33	32.8	4180	313	102	10.8	6.6	65.0	415	125	1.92	37.6	21.4
× 28	28.4	3610	309	102	8.9	6.0	54.3	351	123	1.58	31.0	20.9
× 24	23.8	3040	305	101	6.7	5.6	42.7	280	119	1.16	22.9	19.5
× 21	21.1	2690	303	101	5.7	5.1	37.0	244	117	0.983	19.5	19.1

W250 × 167	167.4	21300	289	265	31.8	19.2	300	2080	119	98.8	746	68.1
× 149	148.9	19000	282	263	28.4	17.3	259	1840	117	86.2	656	67.4
× 131	131.1	16700	275	261	25.1	15.4	221	1610	115	74.5	571	66.8
× 115	114.8	14600	269	259	22.1	13.5	189	1410	114	64.1	495	66.3
× 101	101.2	12900	264	257	19.6	11.9	164	1240	113	55.5	432	65.6
× 89	89.6	11400	260	256	17.3	10.7	143	1100	112	48.4	378	65.2
× 80	80.1	10200	256	255	15.6	9.4	126	982	111	43.1	338	65.0
× 73	72.9	9280	253	254	14.2	8.6	113	891	110	38.8	306	64.7
× 67	67.1	8550	257	204	15.7	8.9	104	806	110	22.2	218	51.0
× 58	58.2	7420	252	203	13.5	8.0	87.3	693	108	18.8	186	50.3
× 49	49.0	6250	247	202	11.0	7.4	70.6	572	106	15.1	150	49.2
× 45	44.9	5720	266	148	13.0	7.6	71.1	534	111	7.03	95.1	35.1
× 39	38.7	4920	262	147	11.2	6.6	60.1	459	111	5.94	80.8	34.7
× 33	32.7	4170	258	146	9.1	6.1	48.9	379	108	4.73	64.7	33.7
× 28	28.5	3630	260	102	10.0	6.4	40.0	307	105	1.78	34.8	22.1
× 25	25.3	3230	257	102	8.4	6.1	34.2	266	103	1.49	29.2	21.5
× 22	22.4	2850	254	102	6.9	5.8	28.9	227	101	1.23	24.0	20.8
× 18	17.9	2270	251	101	5.3	4.8	22.4	179	99.3	0.913	18.1	20.1
W200 × 100	99.5	12700	229	210	23.7	14.5	113	989	94.3	36.6	349	53.7
× 86	86.7	11100	222	209	20.6	13.0	94.7	853	92.4	31.4	300	53.2
× 71	71.5	9110	216	206	17.4	10.2	76.6	709	91.7	25.4	246	52.8
× 59	59.4	7560	210	205	14.2	9.1	61.1	582	89.9	20.4	199	51.9
× 52	52.3	6660	206	204	12.6	7.9	52.7	512	89.0	17.8	175	51.7
× 46	46.0	5860	203	203	11.0	7.2	45.5	448	88.1	15.3	151	51.1
× 42	41.7	5310	205	166	11.8	7.2	40.9	399	87.7	9.00	108	41.2
× 36	35.9	4580	201	165	10.2	6.2	34.4	342	86.7	7.64	92.6	40.8
× 31	31.4	4000	210	134	10.2	6.4	31.4	299	88.6	4.10	61.1	32.0
× 27	26.6	3390	207	133	8.4	5.8	25.8	249	87.2	3.30	49.6	31.2
× 22	22.4	2860	206	102	8.0	6.2	20.0	194	83.6	1.42	27.8	22.3
× 19	19.4	2480	203	102	6.5	5.8	16.6	163	81.8	1.15	22.6	21.5
× 15	15.0	1900	200	100	5.2	4.3	12.7	127	81.8	0.869	17.4	21.4
W150 × 37	37.1	4730	162	154	11.6	8.1	22.2	274	68.5	7.07	91.8	38.7
× 30	29.8	3790	157	153	9.3	6.6	17.2	219	67.4	5.56	72.6	38.3
× 22	22.3	2850	152	152	6.6	5.8	12.1	159	65.2	3.87	50.9	36.8
× 24	24.0	3060	160	102	10.3	6.6	13.4	168	66.2	1.83	35.8	24.5
× 18	18.0	2290	153	102	7.1	5.8	9.16	120	63.2	1.26	24.7	23.5
× 14	13.6	1730	150	100	5.5	4.3	6.87	91.5	63.0	0.918	18.4	23.0
W130 × 28	28.1	3580	131	128	10.9	6.9	10.9	167	55.2	3.81	59.6	32.6
× 24	23.6	3010	127	127	9.1	6.1	8.80	139	54.1	3.11	49.0	32.1
W100 × 19	19.4	2470	106	103	8.8	7.1	4.76	89.9	43.9	1.61	31.2	25.5

[a]Produced exclusively by Algoma Steel (Canada).

APPENDIX B Tables

TABLE B-3 Properties of I-Beam Sections (S-Shapes): SI Units

Designation	Theoretical mass (kg/m)	Area (mm²)	Depth (mm)	Flange Width (mm)	Flange Thickness (mm)	Web thickness (mm)	Axis X-X I (10^6 mm⁴)	Axis X-X $S = I/c$ (10^3 mm³)	Axis X-X $r = \sqrt{I/A}$ (mm)	Axis Y-Y I (10^6 mm⁴)	Axis Y-Y $S = I/c$ (10^3 mm³)	Axis Y-Y $r = \sqrt{I/A}$ (mm)
S610 × 180	180.0	22 900	622	204	27.7	20.3	1310	4220	239	34.7	340	38.9
× 158	157.8	20 100	622	200	27.7	15.7	1220	3940	246	32.4	324	40.1
× 149	148.7	18 900	610	184	22.1	18.9	996	3270	230	20.1	218	32.6
× 134	134.4	17 100	610	181	22.1	15.9	939	3080	234	18.9	209	33.2
× 119	119.1	15 200	610	178	22.1	12.7	879	2880	240	17.9	201	34.3
S510 × 143	143.3	18 300	516	183	23.4	20.3	702	2720	196	21.1	231	34.0
× 128	128.9	16 400	516	179	23.4	16.8	660	2560	201	19.6	219	34.6
× 112	111.4	14 200	508	162	20.2	16.1	532	2090	194	12.5	155	29.7
× 98.2	98.4	12 500	508	159	20.2	12.8	497	1960	199	11.7	148	30.6
S460 × 104	104.7	13 300	457	159	17.6	18.1	387	1690	171	10.3	129	27.8
× 81.4	81.6	10 400	457	152	17.6	11.7	335	1470	179	8.77	115	29.0
S380 × 74	74.6	9500	381	143	15.8	14.0	203	1060	146	6.60	92.3	26.4
× 64	63.9	8150	381	140	15.8	10.4	187	980	151	6.11	87.3	27.4
S310 × 74	74.4	9470	305	139	16.7	17.4	127	833	116	6.60	94.9	26.4
× 60.7	60.6	7730	305	133	16.7	11.7	113	744	121	5.67	85.3	27.1
× 52	52.2	6650	305	129	13.8	10.9	95.8	629	120	4.16	64.5	25.0
× 47	47.4	6040	305	127	13.8	8.9	91.1	597	123	3.94	62.1	25.5
S250 × 52	52.3	6660	254	126	12.5	15.1	61.6	485	96.2	3.56	56.5	23.1
× 38	37.8	4820	254	118	12.5	7.9	51.4	405	103	2.84	48.2	24.3
S200 × 34	34.3	4370	203	106	10.8	11.2	27.0	266	78.6	1.81	34.2	20.4
× 27	27.5	3500	203	102	10.8	6.9	24.0	237	82.8	1.59	31.1	21.3
S180 × 30	29.9	3800	178	98	10.0	11.4	17.8	200	68.4	1.34	27.3	18.8
× 22.8	22.9	2910	178	93	10.0	6.4	15.4	173	72.7	1.12	24.0	19.6
S150 × 26	25.7	3270	152	91	9.1	11.8	10.9	144	57.7	0.981	21.6	17.3
× 19	18.6	2370	152	85	9.1	5.9	9.19	121	62.3	0.776	18.2	18.1
S130 × 22	21.9	2790	127	83	8.3	12.5	6.33	99.6	47.6	0.690	16.6	15.7
× 15	14.8	1890	127	76	8.3	5.4	5.12	80.6	52.0	0.508	13.4	16.4
S100 × 14.1	14.2	1800	102	71	7.4	8.3	2.85	55.8	39.8	0.376	10.6	14.5
× 11	11.4	1450	102	68	7.4	4.8	2.55	50.1	41.9	0.324	9.52	14.9
S75 × 11	11.2	1430	76	64	6.6	8.9	1.22	32.0	29.2	0.249	7.77	13.2
× 8	8.4	1070	76	59	6.6	4.3	1.04	27.4	31.2	0.190	6.43	13.3

TABLE B-4 Properties of Channel Sections: SI Units

Designation	Theoretical mass (kg/m)	Area (mm²)	Depth (mm)	Flange Width (mm)	Flange Thickness (mm)	Web thickness (mm)	Axis X-X I (10⁶ mm⁴)	Axis X-X $S = I/c$ (10³ mm³)	Axis X-X $r = \sqrt{I/A}$ (mm)	Axis Y-Y I (10⁶ mm⁴)	Axis Y-Y $S = I/c$ (10³ mm³)	Axis Y-Y $r = \sqrt{I/A}$ (mm)	x (mm)
C380 × 74	74.4	9480	381	94	16.5	18.2	168	881	133	4.60	62.4	22.0	20.3
× 60	59.4	7570	381	89	16.5	13.2	145	760	138	3.84	55.5	22.5	19.7
× 50	50.5	6430	381	86	16.5	10.2	131	687	143	3.39	51.4	23.0	20.0
C310 × 45	44.7	5690	305	80	12.7	13.0	67.3	442	109	2.12	33.6	19.3	17.0
× 37	37.1	4720	305	77	12.7	9.8	59.9	393	113	1.85	30.9	19.8	17.1
× 31	30.8	3920	305	74	12.7	7.2	53.5	351	117	1.59	28.2	20.1	17.5
C250 × 45	44.5	5670	254	76	11.1	17.1	42.8	337	86.9	1.60	26.8	16.8	16.3
× 37	37.3	4750	254	73	11.1	13.4	37.9	299	89.3	1.40	24.3	17.2	15.7
× 30	29.6	3780	254	69	11.1	9.6	32.7	257	93.0	1.16	21.5	17.5	15.3
× 23	22.6	2880	254	65	11.1	6.1	27.8	219	98.2	0.922	18.8	17.9	15.9
C230 × 30	29.8	3800	229	67	10.5	11.4	25.5	222	81.9	1.01	19.3	16.3	14.8
× 22	22.3	2840	229	63	10.5	7.2	21.3	186	86.6	0.806	16.8	16.8	14.9
× 20	19.8	2530	229	61	10.5	5.9	19.8	173	88.5	0.716	15.6	16.8	15.1
C200 × 28	27.9	3560	203	64	9.9	12.4	18.2	180	71.5	0.825	16.6	15.2	14.4
× 21	20.4	2600	203	59	9.9	7.7	14.9	147	75.7	0.627	13.9	15.5	14.0
× 17	17.0	2170	203	57	9.9	5.6	13.5	133	78.9	0.544	12.8	15.8	14.5
C180 × 22	21.9	2780	178	58	9.3	10.6	11.3	127	63.8	0.568	12.8	14.3	13.5
× 18	18.2	2310	178	55	9.3	8.0	10.0	113	65.8	0.476	11.4	14.4	13.2
× 15	14.5	1850	178	53	9.3	5.3	8.86	99.6	69.2	0.405	10.3	14.8	13.8
C150 × 19	19.2	2450	152	54	8.7	11.1	7.12	93.7	53.9	0.425	10.3	13.2	12.9
× 16	15.5	1980	152	51	8.7	8.0	6.22	81.9	56.0	0.351	9.13	13.3	12.6
× 12	12.1	1540	152	48	8.7	5.1	5.36	70.6	59.0	0.279	7.93	13.5	12.8
C130 × 17	17.2	2190	127	52	8.1	12.0	4.36	68.7	44.6	0.346	8.85	12.6	12.9
× 13	13.3	1700	127	47	8.1	8.3	3.66	57.6	46.4	0.252	7.20	12.2	11.9
× 10	9.9	1260	127	44	8.1	4.8	3.09	48.6	49.5	0.195	6.14	12.4	12.2
C100 × 11	10.8	1370	102	43	7.5	8.2	1.91	37.4	37.3	0.174	5.52	11.3	11.5
× 9	9.4	1190	102	42	7.5	6.3	1.77	34.6	38.6	0.158	5.18	11.5	11.6
× 8	8.0	1020	102	40	7.5	4.7	1.61	31.6	39.7	0.132	4.65	11.4	11.6
C75 × 9	8.8	1120	76	40	6.9	9.0	0.85	22.3	27.5	0.123	4.31	10.5	11.4
× 7	7.3	933	76	37	6.9	6.6	0.75	19.7	28.3	0.096	3.67	10.1	10.8
× 6	6.0	763	76	35	6.9	4.3	0.67	17.6	29.6	0.077	3.21	10.1	10.9

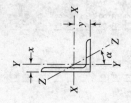

TABLE B-5 Properties of Equal Angle Sections: SI Units

Size and thickness (mm)	Theoretical mass (kg/m)	Area (mm²)	Axis X-X and Axis Y-Y				Axis Z-Z	
			I (10^6 mm⁴)	$S = I/c$ (10^3 mm³)	$r = \sqrt{I/A}$ (mm)	x or y (mm)	$r = \sqrt{I/A}$ (mm)	
200 × 200 × 30	87.1	11 100	40.3	290	60.3	60.9	39.0	
× 25	73.6	9 380	34.8	247	60.9	59.2	39.1	
× 20	59.7	7 600	28.8	202	61.6	57.4	39.3	
× 16	48.2	6 140	23.7	165	62.1	55.9	39.5	
× 13	39.5	5 030	19.7	136	62.6	54.8	39.7	
× 10	30.6	3 900	15.5	106	63.0	53.7	39.9	
150 × 150 × 20	44.0	5 600	11.6	110	45.5	44.8	29.3	
× 16	35.7	4 540	9.63	90.3	46.0	43.4	29.4	
× 13	29.3	3 730	8.05	74.7	46.4	42.3	29.6	
× 10	22.8	2 900	6.37	58.6	46.9	41.2	29.8	
125 × 125 × 16	29.4	3 740	5.41	61.5	38.0	37.1	24.4	
× 13	24.2	3 080	4.54	51.1	38.4	36.0	24.5	
× 10	18.8	2 400	3.62	40.2	38.8	34.9	24.7	
× 8	15.2	1 940	2.96	32.6	39.1	34.2	24.8	
100 × 100 × 16	23.1	2 940	2.65	38.3	30.0	30.8	19.5	
× 13	19.1	2 430	2.24	31.9	30.4	29.8	19.5	
× 10	14.9	1 900	1.80	25.2	30.8	28.7	19.7	
× 8	12.1	1 540	1.48	20.6	31.1	28.0	19.8	
× 6	9.14	1 160	1.14	15.7	31.3	27.2	19.9	
90 × 90 × 13	17.0	2 170	1.60	25.6	27.2	27.2	17.6	
× 10	13.3	1 700	1.29	20.2	27.6	26.2	17.6	
× 8	10.8	1 380	1.07	16.5	27.8	25.5	17.7	
× 6	8.20	1 040	0.826	12.7	28.1	24.7	17.9	

Size							
75 × 75 × 13	14.0	1780	0.892	17.3	22.4	23.5	14.6
× 10	11.0	1400	0.725	13.8	22.8	22.4	14.6
× 8	8.92	1140	0.602	11.3	23.0	21.7	14.7
× 6	6.78	864	0.469	8.68	23.3	21.0	14.8
× 5	5.69	725	0.398	7.32	23.4	20.6	14.9
65 × 65 × 10	9.42	1200	0.459	10.2	19.6	19.9	12.7
× 8	7.66	976	0.383	8.36	19.8	19.2	12.7
× 6	5.84	744	0.300	6.44	20.1	18.5	12.8
× 5	4.91	625	0.255	5.45	20.2	18.1	12.9
55 × 55 × 10	7.85	1000	0.268	7.11	16.4	17.4	10.7
× 8	6.41	816	0.225	5.87	16.6	16.7	10.7
× 6	4.90	624	0.177	4.54	16.9	16.0	10.8
× 5	4.12	525	0.152	3.85	17.0	15.6	10.8
× 4	3.33	424	0.125	3.13	17.1	15.2	10.9
× 3	2.52	321	0.096	2.39	17.3	14.9	11.0
45 × 45 × 8	5.15	656	0.118	3.82	13.4	14.2	8.76
× 6	3.96	504	0.094	2.98	13.7	13.4	8.79
× 5	3.34	425	0.081	2.53	13.8	13.1	8.82
× 4	2.70	344	0.067	2.07	13.9	12.7	8.87
× 3	2.05	261	0.052	1.58	14.1	12.4	8.93
35 × 35 × 6	3.01	384	0.042	1.74	10.5	10.9	6.81
× 5	2.55	325	0.036	1.49	10.6	10.6	6.83
× 4	2.07	264	0.030	1.22	10.7	10.2	6.86
× 3	1.58	201	0.024	0.940	10.8	9.86	6.91
25 × 25 × 5	1.77	225	0.012	0.724	7.39	8.06	4.87
× 4	1.44	184	0.010	0.599	7.50	7.71	4.87
× 3	1.11	141	0.008	0.465	7.63	7.35	4.89

APPENDIX B Tables

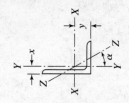

TABLE B-6 *Properties of Unequal Angle Sections: SI Units*

Size and thickness (mm)	Theoretical mass (kg/m)	Area (mm^2)	Axis X-X				Axis Y-Y				Axis Z-Z	
			I (10^6 mm^4)	$S = I/c$ (10^3 mm^3)	$r = \sqrt{I/A}$ (mm)	y (mm)	I (10^6 mm^4)	$S = I/c$ (10^3 mm^3)	$r = \sqrt{I/A}$ (mm)	x (mm)	$r = \sqrt{I/A}$ (mm)	$\tan \alpha$
200 × 150 × 25	63.8	8120	31.6	236	62.3	66.3	15.1	139	43.2	41.3	32.0	0.543
× 20	51.8	6600	26.2	193	63.0	64.5	12.7	115	43.8	39.5	32.1	0.549
× 16	42.0	5340	21.6	158	63.5	63.1	10.5	93.8	44.3	38.1	32.3	0.554
× 13	34.4	4380	17.9	130	64.0	62.0	8.77	77.6	44.7	37.0	32.5	0.557
200 × 100 × 20	44.0	5600	22.6	180	63.6	74.3	3.84	50.8	26.2	24.3	21.3	0.256
× 16	35.7	4540	18.7	147	64.2	72.8	3.22	41.8	26.6	22.8	21.4	0.262
× 13	29.3	3730	15.6	121	64.6	71.7	2.72	34.7	27.0	21.7	21.6	0.266
× 10	22.8	2900	12.3	94.8	65.1	70.5	2.18	27.4	27.4	20.5	21.8	0.271
150 × 100 × 16	29.4	3740	8.40	84.8	47.4	50.9	3.00	40.4	28.3	25.9	21.6	0.434
× 13	24.2	3080	7.03	70.2	47.8	49.9	2.53	33.7	28.7	24.9	21.7	0.440
× 10	18.8	2400	5.58	55.1	48.2	48.8	2.03	26.6	29.1	23.8	21.9	0.445
× 8	15.2	1940	4.55	44.6	48.5	48.0	1.67	21.6	29.3	23.0	22.0	0.448
125 × 90 × 16	25.0	3180	4.84	58.5	39.0	42.2	2.09	32.0	25.6	24.7	19.2	0.499
× 13	20.6	2630	4.07	48.6	39.4	41.2	1.77	26.7	26.0	23.7	19.3	0.505
× 10	16.1	2050	3.25	38.2	39.8	40.1	1.42	21.1	26.4	22.6	19.5	0.511
× 8	13.0	1660	2.66	31.1	40.1	39.3	1.18	17.2	26.6	21.8	19.6	0.515
125 × 75 × 13	19.1	2430	3.82	47.1	39.6	43.9	1.04	18.5	20.7	18.9	16.2	0.356
× 10	14.9	1900	3.05	37.1	40.0	42.8	0.841	14.7	21.0	17.8	16.3	0.363
× 8	12.1	1540	2.50	30.1	40.3	42.1	0.697	12.0	21.3	17.1	16.4	0.367
× 6	9.14	1160	1.92	23.0	40.6	41.3	0.542	9.23	21.6	16.3	16.6	0.372
100 × 90 × 13	18.1	2300	2.17	31.4	30.7	31.1	1.66	25.9	26.8	26.1	18.4	0.796
× 10	14.1	1800	1.74	24.9	31.1	30.0	1.33	20.5	27.2	25.0	18.5	0.800
× 8	11.4	1460	1.43	20.3	31.4	29.3	1.10	16.8	27.5	24.3	18.6	0.802
× 6	8.67	1100	1.11	15.5	31.7	28.5	0.853	12.8	27.8	23.5	18.7	0.805

100 × 75 × 13	16.5	2110	2.04	30.6	31.1	33.4	0.976	18.0	21.5	20.9	16.0	0.541
× 10	13.0	1650	1.64	24.2	31.5	32.3	0.791	14.3	21.9	19.8	16.1	0.549
× 8	10.5	1340	1.35	19.7	31.8	31.5	0.656	11.7	22.2	19.0	16.2	0.554
× 6	7.96	1010	1.04	15.1	32.1	30.8	0.511	9.01	22.4	18.3	16.3	0.559
90 × 75 × 13	15.5	1980	1.51	24.8	27.6	29.3	0.946	17.8	21.9	21.8	15.6	0.672
× 10	12.2	1550	1.22	19.7	28.0	28.2	0.767	14.1	22.2	20.7	15.7	0.679
× 8	9.86	1260	1.01	16.1	28.3	27.5	0.636	11.6	22.5	20.0	15.8	0.683
× 6	7.49	954	0.779	12.3	28.6	26.8	0.495	8.89	22.8	19.3	15.9	0.687
× 5	6.28	800	0.660	10.4	28.7	26.4	0.421	7.50	22.9	18.9	16.0	0.689
90 × 65 × 10	11.4	1450	1.16	19.2	28.3	29.8	0.507	10.6	18.7	17.3	13.9	0.506
× 8	9.23	1180	0.958	15.7	28.5	29.1	0.422	8.72	18.9	16.6	14.0	0.512
× 6	7.02	894	0.743	12.1	28.8	28.4	0.330	6.72	19.2	15.9	14.2	0.518
× 5	5.89	750	0.629	10.2	29.0	28.0	0.281	5.68	19.4	15.5	14.2	0.520
80 × 60 × 10	10.2	1300	0.808	15.1	24.9	26.5	0.388	8.92	17.3	16.5	12.8	0.543
× 8	8.29	1060	0.670	12.4	25.2	25.8	0.324	7.33	17.5	15.8	12.9	0.549
× 6	6.31	804	0.522	9.50	25.5	25.1	0.254	5.66	17.8	15.1	13.0	0.555
× 5	5.30	675	0.443	8.02	25.6	24.7	0.217	4.79	17.9	14.7	13.0	0.558
70 × 50 × 8	7.35	936	0.525	10.6	23.7	25.5	0.187	5.06	14.1	13.0	10.8	0.434
× 6	5.60	714	0.410	8.15	24.0	24.7	0.148	3.92	14.4	12.2	10.9	0.441
× 5	4.71	600	0.349	6.88	24.1	24.4	0.127	3.32	14.5	11.9	10.9	0.445
65 × 50 × 8	6.72	856	0.351	8.03	20.2	21.3	0.180	4.97	14.5	13.8	10.6	0.572
× 6	5.13	654	0.275	6.19	20.5	20.6	0.142	3.85	14.7	13.1	10.7	0.580
× 5	4.32	550	0.235	5.24	20.7	20.2	0.122	3.27	14.9	12.7	10.8	0.583
× 4	3.49	444	0.192	4.25	20.8	19.9	0.100	2.66	15.0	12.4	10.8	0.587
55 × 35 × 6	3.96	504	0.152	4.23	17.4	19.0	0.048	1.85	9.77	9.04	7.55	0.396
× 5	3.34	425	0.130	3.59	17.5	18.7	0.041	1.58	9.89	8.68	7.59	0.401
× 4	2.70	344	0.107	2.92	17.7	18.3	0.034	1.29	10.0	8.31	7.65	0.406
× 3	2.05	261	0.083	2.23	17.8	17.9	0.027	0.994	10.2	7.94	7.72	0.411
45 × 30 × 6	3.25	414	0.082	2.79	14.0	15.7	0.029	1.32	8.35	8.22	6.44	0.426
× 5	2.75	350	0.070	2.37	14.2	15.4	0.025	1.13	8.46	7.86	6.47	0.433
× 4	2.23	284	0.058	1.94	14.3	15.0	0.021	0.930	8.58	7.49	6.51	0.439
× 3	1.70	216	0.045	1.49	14.5	14.6	0.016	0.717	8.72	7.12	6.57	0.445

TABLE B-7 Properties of Wide-Flange Sections (W-Shapes): U.S. Customary Units

Designation	Area (in.²)	Depth (in.)	Web thickness (in.)	Flange Width (in.)	Flange Thickness (in.)	Axis X-X I (in.⁴)	Axis X-X $S = I/c$ (in.³)	Axis X-X $r = \sqrt{I/A}$ (in.)	Axis Y-Y I (in.⁴)	Axis Y-Y $S = I/c$ (in.³)	Axis Y-Y $r = \sqrt{I/A}$ (in.)
W36 × 300	88.3	36.74	0.945	16.655	1.680	20 300	1110	15.2	1300	156	3.83
× 280	82.4	36.52	0.885	16.595	1.570	18 900	1030	15.1	1200	144	3.81
× 260	76.5	36.26	0.840	16.550	1.440	17 300	953	15.0	1090	132	3.78
× 245	72.1	36.08	0.800	16.510	1.350	16 100	895	15.0	1010	123	3.75
× 230	67.6	35.90	0.760	16.470	1.260	15 000	837	14.9	940	114	3.73
W36 × 210	61.8	36.69	0.830	12.180	1.360	13 200	719	14.6	411	67.5	2.58
× 194	57.0	36.49	0.765	12.115	1.260	12 100	664	14.6	375	61.9	2.56
× 182	53.6	36.33	0.725	12.075	1.180	11 300	623	14.5	347	57.6	2.55
× 170	50.0	36.17	0.680	12.030	1.100	10 500	580	14.5	320	53.2	2.53
× 160	47.0	36.01	0.650	12.000	1.020	9 750	542	14.4	295	49.1	2.50
× 150	44.2	35.85	0.625	11.975	0.940	9 040	504	14.3	270	45.1	2.47
× 135	39.7	35.55	0.600	11.950	0.790	7 800	439	14.0	225	37.7	2.38
W33 × 241	70.9	34.18	0.830	15.860	1.400	14 200	829	14.1	932	118	3.63
× 221	65.0	33.93	0.775	15.805	1.275	12 800	757	14.1	840	106	3.59
× 201	59.1	33.68	0.715	15.745	1.150	11 500	684	14.0	749	95.2	3.56

W33 × 152	44.7	33.49	0.635	11.565	1.055	8160	487	13.5	273	47.2	2.47	
× 141	41.6	33.30	0.605	11.535	0.960	7450	448	13.4	246	42.7	2.43	
× 130	38.3	33.09	0.580	11.510	0.855	6710	406	13.2	218	37.9	2.39	
× 118	34.7	32.86	0.550	11.480	0.740	5900	359	13.0	187	32.6	2.32	
W30 × 211	62.0	30.94	0.775	15.105	1.315	10300	663	12.9	757	100	3.49	
× 191	56.1	30.68	0.710	15.040	1.185	9170	598	12.8	673	89.5	3.46	
× 173	50.8	30.44	0.655	14.985	1.065	8200	539	12.7	598	79.8	3.43	
W30 × 132	38.9	30.31	0.615	10.545	1.000	5770	380	12.2	196	37.2	2.25	
× 124	36.5	30.17	0.585	10.515	0.930	5360	355	12.1	181	34.4	2.23	
× 116	34.2	30.01	0.565	10.495	0.850	4930	329	12.0	164	31.3	2.19	
× 108	31.7	29.83	0.545	10.475	0.760	4470	299	11.9	146	27.9	2.15	
× 99	29.1	29.65	0.520	10.450	0.670	3990	269	11.7	128	24.5	2.10	
W27 × 178	52.3	27.81	0.725	14.085	1.190	6990	502	11.6	555	78.8	3.26	
× 161	47.4	27.59	0.660	14.020	1.080	6280	455	11.5	497	70.9	3.24	
× 146	42.9	27.38	0.605	13.965	0.975	5630	411	11.4	443	63.5	3.21	
W27 × 114	33.5	27.29	0.570	10.070	0.930	4090	299	11.0	159	31.5	2.18	
× 102	30.0	27.09	0.515	10.015	0.830	3620	267	11.0	139	27.8	2.15	
× 94	27.7	26.92	0.490	9.990	0.745	3270	243	10.9	124	24.8	2.12	
× 84	24.8	26.71	0.460	9.960	0.640	2850	213	10.7	106	21.2	2.07	
W24 × 162	47.7	25.00	0.705	12.955	1.220	5170	414	10.4	443	68.4	3.05	
× 146	43.0	24.74	0.650	12.900	1.090	4580	371	10.3	391	60.5	3.01	
× 131	38.5	24.48	0.605	12.855	0.960	4020	329	10.2	340	53.0	2.97	
× 117	34.4	24.26	0.550	12.800	0.850	3540	291	10.1	297	46.5	2.94	
× 104	30.6	24.06	0.500	12.750	0.750	3100	258	10.1	259	40.7	2.91	

(continues)

TABLE B-7 Properties of Wide-Flange Sections (W-Shapes): U.S. Customary Units (continued)

Designation	Area (in.²)	Depth (in.)	Web thickness (in.)	Flange Width (in.)	Flange Thickness (in.)	Axis X-X I (in.⁴)	Axis X-X $S = I/c$ (in.³)	Axis X-X $r = \sqrt{I/A}$ (in.)	Axis Y-Y I (in.⁴)	Axis Y-Y $S = I/c$ (in.³)	Axis Y-Y $r = \sqrt{I/A}$ (in.)
W24 × 94	27.7	24.31	0.515	9.065	0.875	2700	222	9.87	109	24.0	1.98
× 84	24.7	24.10	0.470	9.020	0.770	2370	196	9.79	94.4	20.9	1.95
× 76	22.4	23.92	0.440	8.990	0.680	2100	176	9.69	82.5	18.4	1.92
× 68	20.1	23.73	0.415	8.965	0.585	1830	154	9.55	70.4	15.7	1.87
W24 × 62	18.2	23.74	0.430	7.040	0.590	1550	131	9.23	34.5	9.80	1.38
× 55	16.2	23.57	0.395	7.005	0.505	1350	114	9.11	29.1	8.30	1.34
W21 × 147	43.2	22.06	0.720	12.510	1.150	3630	329	9.17	376	60.1	2.95
× 132	38.8	21.83	0.650	12.440	1.035	3220	295	9.12	333	53.5	2.93
× 122	35.9	21.68	0.600	12.390	0.960	2960	273	9.09	305	49.2	2.92
× 111	32.7	21.51	0.550	12.340	0.875	2670	249	9.05	274	44.5	2.90
× 101	29.8	21.36	0.500	12.290	0.800	2420	227	9.02	248	40.3	2.89
W21 × 93	27.3	21.62	0.580	8.420	0.930	2070	192	8.70	92.9	22.1	1.84
× 83	24.3	21.43	0.515	8.355	0.835	1830	171	8.67	81.4	19.5	1.83
× 73	21.5	21.24	0.455	8.295	0.740	1600	151	8.64	70.6	17.0	1.81
× 68	20.0	21.13	0.430	8.270	0.685	1480	140	8.60	64.7	15.7	1.80
× 62	18.3	20.99	0.400	8.240	0.615	1330	127	8.54	57.5	13.9	1.77
W21 × 57	16.7	21.06	0.405	6.555	0.650	1170	111	8.36	30.6	9.35	1.35
× 50	14.7	20.83	0.380	6.530	0.535	984	94.5	8.18	24.9	7.64	1.30
× 44	13.0	20.66	0.350	6.500	0.450	843	81.6	8.06	20.7	6.36	1.26
W18 × 119	35.1	18.97	0.655	11.265	1.060	2190	231	7.90	253	44.9	2.69
× 106	31.1	18.73	0.590	11.200	0.940	1910	204	7.84	220	39.4	2.66
× 97	28.5	18.59	0.535	11.145	0.870	1750	188	7.82	201	36.1	2.65
× 86	25.3	18.39	0.480	11.090	0.770	1530	166	7.77	175	31.6	2.63
× 76	22.3	18.21	0.425	11.035	0.680	1330	146	7.73	152	27.6	2.61
W18 × 71	20.8	18.47	0.495	7.635	0.810	1170	127	7.50	60.3	15.8	1.70
× 65	19.1	18.35	0.450	7.590	0.750	1070	117	7.49	54.8	14.4	1.69
× 60	17.6	18.24	0.415	7.555	0.695	984	108	7.47	50.1	13.3	1.69
× 55	16.2	18.11	0.390	7.530	0.630	890	98.3	7.41	44.9	11.9	1.67
× 50	14.7	17.99	0.355	7.495	0.570	800	88.9	7.38	40.1	10.7	1.65

Designation											
W18 × 46	13.5	18.06	0.360	6.060	0.605	712	78.8	7.25	22.5	7.43	1.29
× 40	11.8	17.90	0.315	6.015	0.525	612	68.4	7.21	19.1	6.35	1.27
× 35	10.3	17.70	0.300	6.000	0.425	510	57.6	7.04	15.3	5.12	1.22
W16 × 100	29.4	16.97	0.585	10.425	0.985	1490	175	7.10	186	35.7	2.51
× 89	26.2	16.75	0.525	10.365	0.875	1300	155	7.05	163	31.4	2.49
× 77	22.6	16.52	0.455	10.295	0.760	1110	134	7.00	138	26.9	2.47
× 67	19.7	16.33	0.395	10.235	0.665	954	117	6.96	119	23.2	2.46
W16 × 57	16.8	16.43	0.430	7.120	0.715	758	92.2	6.72	43.1	12.1	1.60
× 50	14.7	16.26	0.380	7.070	0.630	659	81.0	6.68	37.2	10.5	1.59
× 45	13.3	16.13	0.345	7.035	0.565	586	72.7	6.65	32.8	9.34	1.57
× 40	11.8	16.01	0.305	6.995	0.505	518	64.7	6.63	28.9	8.25	1.57
× 36	10.6	15.86	0.295	6.985	0.430	448	56.5	6.51	24.5	7.00	1.52
W16 × 31	9.12	15.88	0.275	5.525	0.440	375	47.2	6.41	12.4	4.49	1.17
× 26	7.68	15.69	0.250	5.500	0.345	301	38.4	6.26	9.59	3.49	1.12
W14 × 730	215.0	22.42	3.070	17.890	4.910	14300	1280	8.17	4720	527	4.69
× 665	196.0	21.64	2.830	17.650	4.520	12400	1150	7.98	4170	472	4.62
× 605	178.0	20.92	2.595	17.415	4.160	10800	1040	7.80	3680	423	4.55
× 550	162.0	20.24	2.380	17.200	3.820	9430	931	7.63	3250	378	4.49
× 500	147.0	19.60	2.190	17.010	3.500	8210	838	7.48	2880	339	4.43
× 455	134.0	19.02	2.015	16.835	3.210	7190	756	7.33	2560	304	4.38
W14 × 426	125.0	18.67	1.875	16.695	3.035	6600	707	7.26	2360	283	4.34
× 398	117.0	18.29	1.770	16.590	2.845	6000	656	7.16	2170	262	4.31
× 370	109.0	17.92	1.655	16.475	2.660	5440	607	7.07	1990	241	4.27
× 342	101.0	17.54	1.540	16.360	2.470	4900	559	6.98	1810	221	4.24

(continues)

TABLE B-7 Properties of Wide-Flange Sections (W-Shapes): U.S. Customary Units (continued)

Designation	Area (in.²)	Depth (in.)	Web thickness (in.)	Flange Width (in.)	Flange Thickness (in.)	Axis X-X I (in.⁴)	Axis X-X $S=I/c$ (in.³)	Axis X-X $r=\sqrt{I/A}$ (in.)	Axis Y-Y I (in.⁴)	Axis Y-Y $S=I/c$ (in.³)	Axis Y-Y $r=\sqrt{I/A}$ (in.)
W14 × 311	91.4	17.12	1.410	16.230	2.260	4330	506	6.88	1610	199	4.20
× 283	83.3	16.74	1.290	16.110	2.070	3840	459	6.79	1440	179	4.17
× 257	75.6	16.38	1.175	15.995	1.890	3400	415	6.71	1290	161	4.13
× 233	68.5	16.04	1.070	15.890	1.720	3010	375	6.63	1150	145	4.10
× 211	62.0	15.72	0.980	15.800	1.560	2660	338	6.55	1030	130	4.07
× 193	56.8	15.48	0.890	15.710	1.440	2400	310	6.50	931	119	4.05
× 176	51.8	15.22	0.830	15.650	1.310	2140	281	6.43	838	107	4.02
× 159	46.7	14.98	0.745	15.565	1.190	1900	254	6.38	748	96.2	4.00
× 145	42.7	14.78	0.680	15.500	1.090	1710	232	6.33	677	87.3	3.98
W14 × 132	38.8	14.66	0.645	14.725	1.030	1530	209	6.28	548	74.5	3.76
× 120	35.3	14.48	0.590	14.670	0.940	1380	190	6.24	495	67.5	3.74
× 109	32.0	14.32	0.525	14.605	0.860	1240	173	6.22	447	61.2	3.73
× 99	29.1	14.16	0.485	14.565	0.780	1110	157	6.17	402	55.2	3.71
× 90	26.5	14.02	0.440	14.520	0.710	999	143	6.14	362	49.9	3.70
W14 × 82	24.1	14.31	0.510	10.130	0.855	882	123	6.05	148	29.3	2.48
× 74	21.8	14.17	0.450	10.070	0.785	796	112	6.04	134	26.6	2.48
× 68	20.0	14.04	0.415	10.035	0.720	723	103	6.01	121	24.2	2.46
× 61	17.9	13.89	0.375	9.995	0.645	640	92.2	5.98	107	21.5	2.45
W14 × 53	15.6	13.92	0.370	8.060	0.660	541	77.8	5.89	57.7	14.3	1.92
× 48	14.1	13.79	0.340	8.030	0.595	485	70.3	5.85	51.4	12.8	1.91
× 43	12.6	13.66	0.305	7.995	0.530	428	62.7	5.82	45.2	11.3	1.89
W14 × 38	11.2	14.10	0.310	6.770	0.515	385	54.6	5.87	26.7	7.88	1.55
× 34	10.0	13.98	0.285	6.745	0.455	340	48.6	5.83	23.3	6.91	1.53
× 30	8.85	13.84	0.270	6.730	0.385	291	42.0	5.73	19.6	5.82	1.49
W14 × 26	7.69	13.91	0.255	5.025	0.420	245	35.3	5.65	8.91	3.54	1.08
× 22	6.49	13.74	0.230	5.000	0.335	199	29.0	5.54	7.00	2.80	1.04

W12 × 336	98.8	16.82	1.775	13.385	2.955	4060	483	6.41	1190	177	3.47
× 305	89.6	16.32	1.625	13.235	2.705	3550	435	6.29	1050	159	3.42
× 279	81.9	15.85	1.530	13.140	2.470	3110	393	6.16	937	143	3.38
× 252	74.1	15.41	1.395	13.005	2.250	2720	353	6.06	828	127	3.34
× 230	67.7	15.05	1.285	12.895	2.070	2420	321	5.97	742	115	3.31
× 210	61.8	14.71	1.180	12.790	1.900	2140	292	5.89	664	104	3.28
× 190	55.8	14.38	1.060	12.670	1.735	1890	263	5.82	589	93.0	3.25
× 170	50.0	14.03	0.960	12.570	1.560	1650	235	5.74	517	82.3	3.22
× 152	44.7	13.71	0.870	12.480	1.400	1430	209	5.66	454	72.8	3.19
× 136	39.9	13.41	0.790	12.400	1.250	1240	186	5.58	398	64.2	3.16
× 120	35.3	13.12	0.710	12.320	1.105	1070	163	5.51	345	56.0	3.13
× 106	31.2	12.89	0.610	12.220	0.990	933	145	5.47	301	49.3	3.11
× 96	28.2	12.71	0.550	12.160	0.900	833	131	5.44	270	44.4	3.09
× 87	25.6	12.53	0.515	12.125	0.810	740	118	5.38	241	39.7	3.07
× 79	23.2	12.38	0.470	12.080	0.735	662	107	5.34	216	35.8	3.05
× 72	21.1	12.25	0.430	12.040	0.670	597	97.4	5.31	195	32.4	3.04
× 65	19.1	12.12	0.390	12.000	0.605	533	87.9	5.28	174	29.1	3.02
W12 × 58	17.0	12.19	0.360	10.010	0.640	475	78.0	5.28	107	21.4	2.51
× 53	15.6	12.06	0.345	9.995	0.575	425	70.6	5.23	95.8	19.2	2.48
W12 × 50	14.7	12.19	0.370	8.080	0.640	394	64.7	5.18	56.3	13.9	1.96
× 45	13.2	12.06	0.335	8.045	0.575	350	58.1	5.15	50.0	12.4	1.94
× 40	11.8	11.94	0.295	8.005	0.515	310	51.9	5.13	44.1	11.0	1.93
W12 × 35	10.3	12.50	0.300	6.560	0.520	285	45.6	5.25	24.5	7.47	1.54
× 30	8.79	12.34	0.260	6.520	0.440	238	38.6	5.21	20.3	6.24	1.52
× 26	7.65	12.22	0.230	6.490	0.380	204	33.4	5.17	17.3	5.34	1.51

(continues)

TABLE B-7 Properties of Wide-Flange Sections (W-Shapes): U.S. Customary Units (continued)

Designation	Area (in.²)	Depth (in.)	Web thickness (in.)	Flange Width (in.)	Flange Thickness (in.)	Axis X-X I (in.⁴)	Axis X-X $S=I/c$ (in.³)	Axis X-X $r=\sqrt{I/A}$ (in.)	Axis Y-Y I (in.⁴)	Axis Y-Y $S=I/c$ (in.³)	Axis Y-Y $r=\sqrt{I/A}$ (in.)
W12 × 22	6.48	12.31	0.260	4.030	0.425	156	25.4	4.91	4.66	2.31	0.847
× 19	5.57	12.16	0.235	4.005	0.350	130	21.3	4.82	3.76	1.88	0.822
× 16	4.71	11.99	0.220	3.990	0.265	103	17.1	4.67	2.82	1.41	0.773
× 14	4.16	11.91	0.200	3.970	0.225	88.6	14.9	4.62	2.36	1.19	0.753
W10 × 112	32.9	11.36	0.755	10.415	1.250	716	126	4.66	236	45.3	2.68
× 100	29.4	11.10	0.680	10.340	1.120	623	112	4.60	207	40.0	2.65
× 88	25.9	10.84	0.605	10.265	0.990	534	98.5	4.54	179	34.8	2.63
× 77	22.6	10.60	0.530	10.190	0.870	455	85.9	4.49	154	30.1	2.60
× 68	20.0	10.40	0.470	10.130	0.770	394	75.7	4.44	134	26.4	2.59
× 60	17.6	10.22	0.420	10.080	0.680	341	66.7	4.39	116	23.0	2.57
× 54	15.8	10.09	0.370	10.030	0.615	303	60.0	4.37	103	20.6	2.56
× 49	14.4	9.98	0.340	10.000	0.560	272	54.6	4.35	93.4	18.7	2.54
W10 × 45	13.3	10.10	0.350	8.020	0.620	248	49.1	4.32	53.4	13.3	2.01
× 39	11.5	9.92	0.315	7.985	0.530	209	42.1	4.27	45.0	11.3	1.98
× 33	9.71	9.73	0.290	7.960	0.435	170	35.0	4.19	36.6	9.20	1.94
W10 × 30	8.84	10.47	0.300	5.810	0.510	170	32.4	4.38	16.7	5.75	1.37
× 26	7.61	10.33	0.260	5.770	0.440	144	27.9	4.35	14.1	4.89	1.36
× 22	6.49	10.17	0.240	5.750	0.360	118	23.2	4.27	11.4	3.97	1.33
W10 × 19	5.62	10.24	0.250	4.020	0.395	96.3	18.8	4.14	4.29	2.14	0.874
× 17	4.99	10.11	0.240	4.010	0.330	81.9	16.2	4.05	3.56	1.78	0.844
× 15	4.41	9.99	0.230	4.000	0.270	68.9	13.8	3.95	2.89	1.45	0.810
× 12	3.54	9.87	0.190	3.960	0.210	53.8	10.9	3.90	2.18	1.10	0.785
W8 × 67	19.7	9.00	0.570	8.280	0.935	272	60.4	3.72	88.6	21.4	2.12
× 58	17.1	8.75	0.510	8.220	0.810	228	52.0	3.65	75.1	18.3	2.10
× 48	14.1	8.50	0.400	8.110	0.685	184	43.3	3.61	60.9	15.0	2.08
× 40	11.7	8.25	0.360	8.070	0.560	146	35.5	3.53	49.1	12.2	2.04
× 35	10.3	8.12	0.310	8.020	0.495	127	31.2	3.51	42.6	10.6	2.03
× 31	9.13	8.00	0.285	7.995	0.435	110	27.5	3.47	37.1	9.27	2.02

W8 × 28	8.25	8.06	0.285	6.535	0.465	98.0	24.3	3.45	21.7	6.63	1.62
× 24	7.08	7.93	0.245	6.495	0.400	82.8	20.9	3.42	18.3	5.63	1.61
W8 × 21	6.16	8.28	0.250	5.270	0.400	75.3	18.2	3.49	9.77	3.71	1.26
× 18	5.26	8.14	0.230	5.250	0.330	61.9	15.2	3.43	7.97	3.04	1.23
W8 × 15	4.44	8.11	0.245	4.015	0.315	48.0	11.8	3.29	3.41	1.70	0.876
× 13	3.84	7.99	0.230	4.000	0.255	39.6	9.91	3.21	2.73	1.37	0.843
× 10	2.96	7.89	0.170	3.940	0.205	30.8	7.81	3.22	2.09	1.06	0.841
W6 × 25	7.34	6.38	0.320	6.080	0.455	53.4	16.7	2.70	17.1	5.61	1.52
× 20	5.87	6.20	0.260	6.020	0.365	41.4	13.4	2.66	13.3	4.41	1.50
× 15	4.43	5.99	0.230	5.990	0.260	29.1	9.72	2.56	9.32	3.11	1.46
W6 × 16	4.74	6.28	0.260	4.030	0.405	32.1	10.2	2.60	4.43	2.20	0.966
× 12	3.55	6.03	0.230	4.000	0.280	22.1	7.31	2.49	2.99	1.50	0.918
× 9	2.68	5.90	0.170	3.940	0.215	16.4	5.56	2.47	2.19	1.11	0.905
W5 × 19	5.54	5.15	0.270	5.030	0.430	26.2	10.2	2.17	9.13	3.63	1.28
× 16	4.68	5.01	0.240	5.000	0.360	21.3	8.51	2.13	7.51	3.00	1.27
W4 × 13	3.83	4.16	0.280	4.060	0.345	11.3	5.46	1.72	3.86	1.90	1.00

TABLE B-8 Properties of I-Beam Sections (S-Shapes): U.S. Customary Units

Designation	Area (in.²)	Depth (in.)	Web thickness (in.)	Flange Width (in.)	Flange Thickness (in.)	Axis X-X I (in.⁴)	Axis X-X $S = I/c$ (in.³)	Axis X-X $r = \sqrt{I/A}$ (in.)	Axis Y-Y I (in.⁴)	Axis Y-Y $S = I/c$ (in.³)	Axis Y-Y $r = \sqrt{I/A}$ (in.)
S24 × 121	35.6	24.50	0.800	8.050	1.090	3160	258	9.43	83.3	20.7	1.53
× 106	31.2	24.50	0.620	7.870	1.090	2940	240	9.71	77.1	19.6	1.57
S24 × 100	29.3	24.00	0.745	7.245	0.870	2390	199	9.02	47.7	13.2	1.27
× 90	26.5	24.00	0.625	7.125	0.870	2250	187	9.21	44.9	12.6	1.30
× 80	23.5	24.00	0.500	7.000	0.870	2100	175	9.47	42.2	12.1	1.34
S20 × 96	28.2	20.30	0.800	7.200	0.920	1670	165	7.71	50.2	13.9	1.33
× 86	25.3	20.30	0.660	7.060	0.920	1580	155	7.89	46.8	13.3	1.36
S20 × 75	22.0	20.00	0.635	6.385	0.795	1280	128	7.62	29.8	9.32	1.16
× 66	19.4	20.00	0.505	6.255	0.795	1190	119	7.83	27.7	8.85	1.19
S18 × 70	20.6	18.00	0.711	6.251	0.691	926	103	6.71	24.1	7.72	1.08
× 54.7	16.1	18.00	0.461	6.001	0.691	804	89.4	7.07	20.8	6.94	1.14
S15 × 50	14.7	15.00	0.550	5.640	0.622	486	64.8	5.75	15.7	5.57	1.03
× 42.9	12.6	15.00	0.411	5.501	0.622	447	59.6	5.95	14.4	5.23	1.07
S12 × 50	14.7	12.00	0.687	5.477	0.659	305	50.8	4.55	15.7	5.74	1.03
× 40.8	12.0	12.00	0.462	5.252	0.659	272	45.4	4.77	13.6	5.16	1.06

S12 × 35	10.3	12.00	0.428	5.078	0.544	229	38.2	4.72	9.87	3.89	0.980
× 31.8	9.35	12.00	0.350	5.000	0.544	218	36.4	4.83	9.36	3.74	1.00
S10 × 35	10.3	10.00	0.594	4.944	0.491	147	29.4	3.78	8.36	3.38	0.901
× 25.4	7.46	10.00	0.311	4.661	0.491	124	24.7	4.07	6.79	2.91	0.954
S8 × 23	6.77	8.00	0.441	4.171	0.426	64.9	16.2	3.10	4.31	2.07	0.798
× 18.4	5.41	8.00	0.271	4.001	0.426	57.6	14.4	3.26	3.73	1.86	0.831
S7 × 20	5.88	7.00	0.450	3.860	0.392	42.4	12.1	2.69	3.17	1.64	0.734
× 15.3	4.50	7.00	0.252	3.662	0.392	36.7	10.5	2.86	2.64	1.44	0.766
S6 × 17.25	5.07	6.00	0.465	3.565	0.359	26.3	8.77	2.28	2.31	1.30	0.675
× 12.5	3.67	6.00	0.232	3.332	0.359	22.1	7.37	2.45	1.82	1.09	0.705
S5 × 14.75	4.34	5.00	0.494	3.284	0.326	15.2	6.09	1.87	1.67	1.01	0.620
× 10	2.94	5.00	0.214	3.004	0.326	12.3	4.92	2.05	1.22	0.809	0.643
S4 × 9.5	2.79	4.00	0.326	2.796	0.293	6.79	3.39	1.56	0.903	0.646	0.569
× 7.7	2.26	4.00	0.193	2.663	0.293	6.08	3.04	1.64	0.764	0.574	0.581
S3 × 7.5	2.21	3.00	0.349	2.509	0.260	2.93	1.95	1.15	0.586	0.468	0.516
× 5.7	1.67	3.00	0.170	2.330	0.260	2.52	1.68	1.23	0.455	0.390	0.522

TABLE B-9 Properties of Channel Sections: U.S. Customary Units

Designation	Area (in.²)	Depth (in.)	Web thickness (in.)	Flange Width (in.)	Flange Average thickness (in.)	Axis X-X I (in.⁴)	Axis X-X $S = I/c$ (in.³)	Axis X-X $r = \sqrt{I/A}$ (in.)	Axis Y-Y I (in.⁴)	Axis Y-Y $S = I/c$ (in.³)	Axis Y-Y $r = \sqrt{I/A}$ (in.)	x (in.)
C15 × 50	14.7	15.00	0.716	3.716	0.650	404	53.8	5.24	11.0	3.78	0.867	0.798
× 40	11.8	15.00	0.520	3.520	0.650	349	46.5	5.44	9.23	3.37	0.886	0.777
× 33.9	9.96	15.00	0.400	3.400	0.650	315	42.0	5.62	8.13	3.11	0.904	0.787
C12 × 30	8.82	12.00	0.510	3.170	0.501	162	27.0	4.29	5.14	2.06	0.763	0.674
× 25	7.35	12.00	0.387	3.047	0.501	144	24.1	4.43	4.47	1.88	0.780	0.674
× 20.7	6.09	12.00	0.282	2.942	0.501	129	21.5	4.61	3.88	1.73	0.799	0.698
C10 × 30	8.82	10.00	0.673	3.033	0.436	103	20.7	3.42	3.94	1.65	0.669	0.649
× 25	7.35	10.00	0.526	2.886	0.436	91.2	18.2	3.52	3.36	1.48	0.676	0.617
× 20	5.88	10.00	0.379	2.739	0.436	78.9	15.8	3.66	2.81	1.32	0.692	0.606
× 15.3	4.49	10.00	0.240	2.600	0.436	67.4	13.5	3.87	2.28	1.16	0.713	0.634
C9 × 20	5.88	9.00	0.448	2.648	0.413	60.9	13.5	3.22	2.42	1.17	0.642	0.583
× 15	4.41	9.00	0.285	2.485	0.413	51.0	11.3	3.40	1.93	1.01	0.661	0.586
× 13.4	3.94	9.00	0.233	2.433	0.413	47.9	10.6	3.48	1.76	0.962	0.669	0.601
C8 × 18.75	5.51	8.00	0.487	2.527	0.390	44.0	11.0	2.82	1.98	1.01	0.599	0.565
× 13.75	4.04	8.00	0.303	2.343	0.390	36.1	9.03	2.99	1.53	0.854	0.615	0.553
× 11.5	3.38	8.00	0.220	2.260	0.390	32.6	8.14	3.11	1.32	0.781	0.625	0.571
C7 × 14.75	4.33	7.00	0.419	2.299	0.366	27.2	7.78	2.51	1.38	0.779	0.564	0.532
× 12.25	3.60	7.00	0.314	2.194	0.366	24.2	6.93	2.60	1.17	0.703	0.571	0.525
× 9.8	2.87	7.00	0.210	2.090	0.366	21.3	6.08	2.72	0.968	0.625	0.581	0.540
C6 × 13	3.83	6.00	0.437	2.157	0.343	17.4	5.80	2.13	1.05	0.642	0.525	0.514
× 10.5	3.09	6.00	0.314	2.034	0.343	15.2	5.06	2.22	0.866	0.564	0.529	0.499
× 8.2	2.40	6.00	0.200	1.920	0.343	13.1	4.38	2.34	0.693	0.492	0.537	0.511
C5 × 9	2.64	5.00	0.325	1.885	0.320	8.90	3.56	1.83	0.632	0.450	0.489	0.478
× 6.7	1.97	5.00	0.190	1.750	0.320	7.49	3.00	1.95	0.479	0.378	0.493	0.484
C4 × 7.25	2.13	4.00	0.321	1.721	0.296	4.59	2.29	1.47	0.433	0.343	0.450	0.459
× 5.4	1.59	4.00	0.184	1.584	0.296	3.85	1.93	1.56	0.319	0.283	0.449	0.457
C3 × 6	1.76	3.00	0.356	1.596	0.273	2.07	1.38	1.08	0.305	0.268	0.416	0.455
× 5	1.47	3.00	0.258	1.498	0.273	1.85	1.24	1.12	0.247	0.233	0.410	0.438
× 4.1	1.21	3.00	0.170	1.410	0.273	1.66	1.10	1.17	0.197	0.202	0.404	0.436

TABLE B-10 Properties of Equal and Unequal Angle Sections: U.S. Customary Units

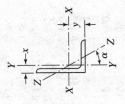

Size and thickness (in.)	Weight per foot (lb)	Area (in.²)	Axis X-X				Axis Y-Y				Axis Z-Z	
			I (in.⁴)	$S = I/c$ (in.³)	$r = \sqrt{I/A}$ (in.)	y (in.)	I (in.⁴)	$S = I/c$ (in.³)	$r = \sqrt{I/A}$ (in.)	x (in.)	$r = \sqrt{I/A}$ (in.)	$\tan \alpha$
L9 × 4 × 5/8	26.3	7.73	64.9	11.5	2.90	3.36	8.32	2.65	1.04	0.858	0.847	0.216
9/16	23.8	7.00	59.1	10.4	2.91	3.33	7.63	2.41	1.04	0.834	0.850	0.218
1/2	21.3	6.25	53.2	9.34	2.92	3.31	6.92	2.17	1.05	0.810	0.854	0.220
L8 × 8 × 1 1/8	56.9	16.7	98.0	17.5	2.42	2.41	98.0	17.5	2.42	2.41	1.56	1.000
1	51.0	15.0	89.0	15.8	2.44	2.37	89.0	15.8	2.44	2.37	1.56	1.000
7/8	45.0	13.2	79.6	14.0	2.45	2.32	79.6	14.0	2.45	2.32	1.57	1.000
3/4	38.9	11.4	69.7	12.2	2.47	2.28	69.7	12.2	2.47	2.28	1.58	1.000
5/8	32.7	9.61	59.4	10.3	2.49	2.23	59.4	10.3	2.49	2.23	1.58	1.000
9/16	29.6	8.68	54.1	9.34	2.50	2.21	54.1	9.34	2.50	2.21	1.59	1.000
1/2	26.4	7.75	48.6	8.36	2.50	2.19	48.6	8.36	2.50	2.19	1.59	1.000
L8 × 6 × 1	44.2	13.0	80.8	15.1	2.49	2.65	38.8	8.92	1.73	1.65	1.28	0.543
7/8	39.1	11.5	72.3	13.4	2.51	2.61	34.9	7.94	1.74	1.61	1.28	0.547
3/4	33.8	9.94	63.4	11.7	2.53	2.56	30.7	6.92	1.76	1.56	1.29	0.551
5/8	28.5	8.36	54.1	9.87	2.54	2.52	26.3	5.88	1.77	1.52	1.29	0.554
9/16	25.7	7.56	49.3	8.95	2.55	2.50	24.0	5.34	1.78	1.50	1.30	0.556
1/2	23.0	6.75	44.3	8.02	2.56	2.47	21.7	4.79	1.79	1.47	1.30	0.558
7/16	20.2	5.93	39.2	7.07	2.57	2.45	19.3	4.23	1.80	1.45	1.31	0.560
L8 × 4 × 1	37.4	11.0	69.6	14.1	2.52	3.05	11.6	3.94	1.03	1.05	0.846	0.247
3/4	28.7	8.44	54.9	10.9	2.55	2.95	9.36	3.07	1.05	0.953	0.852	0.258
9/16	21.9	6.43	42.8	8.35	2.58	2.88	7.43	2.38	1.07	0.882	0.861	0.265
1/2	19.6	5.75	38.5	7.49	2.59	2.86	6.74	2.15	1.08	0.859	0.865	0.267
L7 × 4 × 3/4	26.2	7.69	37.8	8.42	2.22	2.51	9.05	3.03	1.09	1.01	0.860	0.324
5/8	22.1	6.48	32.4	7.14	2.24	2.46	7.84	2.58	1.10	0.963	0.865	0.329
1/2	17.9	5.25	26.7	5.81	2.25	2.42	6.53	2.12	1.11	0.917	0.872	0.335
3/8	13.6	3.98	20.6	4.44	2.27	2.37	5.10	1.63	1.13	0.870	0.880	0.340

(continues)

TABLE B-10 Properties of Equal and Unequal Angle Sections: U.S. Customary Units (continued)

Size and thickness (in.)	Weight per foot (lb)	Area (in.²)	Axis X-X				Axis Y-Y					Axis Z-Z	
			I (in.⁴)	$S = I/c$ (in.³)	$r = \sqrt{I/A}$ (in.)	y (in.)	I (in.⁴)	$S = I/c$ (in.³)	$r = \sqrt{I/A}$ (in.)	x (in.)	$r = \sqrt{I/A}$ (in.)	tan α
L6 × 6 × 1	37.4	11.0	35.5	8.57	1.80	1.86	35.5	8.57	1.80	1.86	1.17	1.000
7/8	33.1	9.73	31.9	7.63	1.81	1.82	31.9	7.63	1.81	1.82	1.17	1.000
3/4	28.7	8.44	28.2	6.66	1.83	1.78	28.2	6.66	1.83	1.78	1.17	1.000
5/8	24.2	7.11	24.2	5.66	1.84	1.73	24.2	5.66	1.84	1.73	1.18	1.000
9/16	21.9	6.43	22.1	5.14	1.85	1.71	22.1	5.14	1.85	1.71	1.18	1.000
1/2	19.6	5.75	19.9	4.61	1.86	1.68	19.9	4.61	1.86	1.68	1.18	1.000
7/16	17.2	5.06	17.7	4.08	1.87	1.66	17.7	4.08	1.87	1.66	1.19	1.000
3/8	14.9	4.36	15.4	3.53	1.88	1.64	15.4	3.53	1.88	1.64	1.19	1.000
5/16	12.4	3.65	13.0	2.97	1.89	1.62	13.0	2.97	1.89	1.62	1.20	1.000
L6 × 4 × 7/8	27.2	7.98	27.7	7.15	1.86	2.12	9.75	3.39	1.11	1.12	0.857	0.421
3/4	23.6	6.94	24.5	6.25	1.88	2.08	8.68	2.97	1.12	1.08	0.860	0.428
5/8	20.0	5.86	21.1	5.31	1.90	2.03	7.52	2.54	1.13	1.03	0.864	0.435
9/16	18.1	5.31	19.3	4.83	1.90	2.01	6.91	2.31	1.14	1.01	0.866	0.438
1/2	16.2	4.75	17.4	4.33	1.91	1.99	6.27	2.08	1.15	0.987	0.870	0.440
7/16	14.3	4.18	15.5	3.83	1.92	1.96	5.60	1.85	1.16	0.964	0.873	0.443
3/8	12.3	3.61	13.5	3.32	1.93	1.94	4.90	1.60	1.17	0.941	0.877	0.446
5/16	10.3	3.03	11.4	2.79	1.94	1.92	4.18	1.35	1.17	0.918	0.882	0.448
L6 × 3 1/2 × 1/2	15.3	4.50	16.6	4.24	1.92	2.08	4.25	1.59	0.972	0.833	0.759	0.344
3/8	11.7	3.42	12.9	3.24	1.94	2.04	3.34	1.23	0.988	0.787	0.767	0.350
5/16	9.8	2.87	10.9	2.73	1.95	2.01	2.85	1.04	0.996	0.763	0.772	0.352
L5 × 5 × 7/8	27.2	7.98	17.8	5.17	1.49	1.57	17.8	5.17	1.49	1.57	0.973	1.000
3/4	23.6	6.94	15.7	4.53	1.51	1.52	15.7	4.53	1.51	1.52	0.975	1.000
5/8	20.0	5.86	13.6	3.86	1.52	1.48	13.6	3.86	1.52	1.48	0.978	1.000
1/2	16.2	4.75	11.3	3.16	1.54	1.43	11.3	3.16	1.54	1.43	0.983	1.000
7/16	14.3	4.18	10.0	2.79	1.55	1.41	10.0	2.79	1.55	1.41	0.986	1.000
3/8	12.3	3.61	8.74	2.42	1.56	1.39	8.74	2.42	1.56	1.39	0.990	1.000
5/16	10.3	3.03	7.42	2.04	1.57	1.37	7.42	2.04	1.57	1.37	0.994	1.000
L5 × 3 1/2 × 3/4	19.8	5.81	13.9	4.28	1.55	1.75	5.55	2.22	0.977	0.996	0.748	0.464
5/8	16.8	4.92	12.0	3.65	1.56	1.70	4.83	1.90	0.991	0.951	0.751	0.472
1/2	13.6	4.00	9.99	2.99	1.58	1.66	4.05	1.56	1.01	0.906	0.755	0.479
7/16	12.0	3.53	8.90	2.64	1.59	1.63	3.63	1.39	1.01	0.883	0.758	0.482
3/8	10.4	3.05	7.78	2.29	1.60	1.61	3.18	1.21	1.02	0.861	0.762	0.486
5/16	8.7	2.56	6.60	1.94	1.61	1.59	2.72	1.02	1.03	0.838	0.766	0.489
1/4	7.0	2.06	5.39	1.57	1.62	1.56	2.23	0.830	1.04	0.814	0.770	0.492

Designation													
L5 × 3 × 5/8	15.7	4.61	11.4	3.55	1.57	1.80	3.06	1.39	0.815	0.796	0.644		0.349
1/2	12.8	3.75	9.45	2.91	1.59	1.75	2.58	1.15	0.829	0.750	0.648		0.357
7/16	11.3	3.31	8.43	2.58	1.60	1.73	2.32	1.02	0.837	0.727	0.651		0.361
3/8	9.8	2.86	7.37	2.24	1.61	1.70	2.04	0.888	0.845	0.704	0.654		0.364
5/16	8.2	2.40	6.26	1.89	1.61	1.68	1.75	0.753	0.853	0.681	0.658		0.368
1/4	6.6	1.94	5.11	1.53	1.62	1.66	1.44	0.614	0.861	0.657	0.663		0.371
L4 × 4 × 3/4	18.5	5.44	7.67	2.81	1.19	1.27	7.67	2.81	1.19	1.27	0.778		1.000
5/8	15.7	4.61	6.66	2.40	1.20	1.23	6.66	2.40	1.20	1.23	0.779		1.000
1/2	12.8	3.75	5.56	1.97	1.22	1.18	5.56	1.97	1.22	1.18	0.782		1.000
7/16	11.3	3.31	4.97	1.75	1.23	1.16	4.97	1.75	1.23	1.16	0.785		1.000
3/8	9.8	2.86	4.36	1.52	1.23	1.14	4.36	1.52	1.23	1.14	0.788		1.000
5/16	8.2	2.40	3.71	1.29	1.24	1.12	3.71	1.29	1.24	1.12	0.791		1.000
1/4	6.6	1.94	3.04	1.05	1.25	1.09	3.04	1.05	1.25	1.09	0.795		1.000
L4 × 3 1/2 × 5/8	14.7	4.30	6.37	2.35	1.22	1.29	4.52	1.84	1.03	1.04	0.719		0.745
1/2	11.9	3.50	5.32	1.94	1.23	1.25	3.79	1.52	1.04	1.00	0.722		0.750
7/16	10.6	3.09	4.76	1.72	1.24	1.23	3.40	1.35	1.05	0.978	0.724		0.753
3/8	9.1	2.67	4.18	1.49	1.25	1.21	2.95	1.17	1.06	0.955	0.727		0.755
5/16	7.7	2.25	3.56	1.25	1.26	1.18	2.55	0.994	1.07	0.932	0.730		0.757
1/4	6.2	1.81	2.91	1.03	1.27	1.16	2.09	0.808	1.07	0.909	0.734		0.759
L4 × 3 × 5/8	13.6	3.98	6.03	2.30	1.23	1.37	2.87	1.35	0.849	0.871	0.637		0.534
1/2	11.1	3.25	5.05	1.89	1.25	1.33	2.42	1.12	0.864	0.827	0.639		0.543
7/16	9.8	2.87	4.52	1.68	1.25	1.30	2.18	0.992	0.871	0.804	0.641		0.547
3/8	8.5	2.48	3.96	1.46	1.26	1.28	1.92	0.866	0.879	0.782	0.644		0.551
5/16	7.2	2.09	3.38	1.23	1.27	1.26	1.65	0.734	0.887	0.759	0.647		0.554
1/4	5.8	1.69	2.77	1.00	1.28	1.24	1.36	0.599	0.896	0.736	0.651		0.558
L3 1/2 × 3 1/2 × 1/2	11.1	3.25	3.64	1.49	1.06	1.06	3.64	1.49	1.06	1.06	0.683		1.000
7/16	9.8	2.87	3.26	1.32	1.07	1.04	3.26	1.32	1.07	1.04	0.684		1.000
3/8	8.5	2.48	2.87	1.15	1.07	1.01	2.87	1.15	1.07	1.01	0.687		1.000
5/16	7.2	2.09	2.45	0.976	1.08	0.990	2.45	0.976	1.08	0.990	0.690		1.000
1/4	5.8	1.69	2.01	0.794	1.09	0.968	2.01	0.794	1.09	0.968	0.694		1.000
L3 1/2 × 3 × 1/2	10.2	3.00	3.45	1.45	1.07	1.13	2.33	1.10	0.881	1.06	0.621		0.714
7/16	9.1	2.65	3.10	1.29	1.08	1.10	2.09	0.975	0.889	1.04	0.622		0.718
3/8	7.9	2.30	2.72	1.13	1.09	1.08	1.85	0.851	0.897	1.01	0.625		0.721
5/16	6.6	1.93	2.33	0.954	1.10	1.06	1.58	0.722	0.905	0.990	0.627		0.724
1/4	5.4	1.56	1.91	0.776	1.11	1.04	1.30	0.589	0.914	0.968	0.631		0.727

(continues)

TABLE B-10 Properties of Equal and Unequal Angle Sections: U.S. Customary Units (continued)

Size and thickness (in.)	Weight per foot (lb)	Area (in.²)	Axis X-X					Axis Y-Y					Axis Z-Z	
			I (in.⁴)	$S=I/c$ (in.³)	$r=\sqrt{I/A}$ (in.)	y (in.)		I (in.⁴)	$S=I/c$ (in.³)	$r=\sqrt{I/A}$ (in.)	x (in.)		$r=\sqrt{I/A}$ (in.)	tan α
L3½ × 2½ × ½	9.4	2.75	3.24	1.41	1.09	1.20		1.36	0.760	0.704	0.705		0.534	0.486
⁷⁄₁₆	8.3	2.43	2.91	1.26	1.09	1.18		1.23	0.677	0.711	0.682		0.535	0.491
⅜	7.2	2.11	2.56	1.09	1.10	1.16		1.09	0.592	0.719	0.660		0.537	0.496
⁵⁄₁₆	6.1	1.78	2.19	0.927	1.11	1.14		0.939	0.504	0.727	0.637		0.540	0.501
¼	4.9	1.44	1.80	0.755	1.12	1.11		0.777	0.412	0.735	0.614		0.544	0.506
L3 × 3 × ½	9.4	2.75	2.22	1.07	0.898	0.932		2.22	1.07	0.898	0.932		0.584	1.000
⁷⁄₁₆	8.3	2.43	1.99	0.954	0.905	0.910		1.99	0.954	0.905	0.910		0.585	1.000
⅜	7.2	2.11	1.76	0.833	0.913	0.888		1.76	0.833	0.913	0.888		0.587	1.000
⁵⁄₁₆	6.1	1.78	1.51	0.707	0.922	0.865		1.51	0.707	0.922	0.865		0.589	1.000
¼	4.9	1.44	1.24	0.577	0.930	0.842		1.24	0.577	0.930	0.842		0.592	1.000
³⁄₁₆	3.71	1.09	0.962	0.441	0.939	0.820		0.962	0.441	0.939	0.820		0.596	1.000
L3 × 2½ × ½	8.5	2.50	2.08	1.04	0.913	1.00		1.30	0.744	0.722	0.750		0.520	0.667
⁷⁄₁₆	7.6	2.21	1.88	0.928	0.920	0.978		1.18	0.664	0.729	0.728		0.521	0.672
⅜	6.6	1.92	1.66	0.810	0.928	0.956		1.04	0.581	0.736	0.706		0.522	0.676
⁵⁄₁₆	5.6	1.62	1.42	0.688	0.937	0.933		0.898	0.494	0.744	0.683		0.525	0.680
¼	4.5	1.31	1.17	0.561	0.945	0.911		0.743	0.404	0.753	0.661		0.528	0.684
³⁄₁₆	3.39	0.996	0.907	0.430	0.954	0.888		0.577	0.310	0.761	0.638		0.533	0.688
L3 × 2 × ½	7.7	2.25	1.92	1.00	0.924	1.08		0.672	0.474	0.546	0.583		0.428	0.414
⁷⁄₁₆	6.8	2.00	1.73	0.894	0.932	1.06		0.609	0.424	0.553	0.561		0.429	0.421
⅜	5.9	1.73	1.53	0.781	0.940	1.04		0.543	0.371	0.559	0.539		0.430	0.428
⁵⁄₁₆	5.0	1.46	1.32	0.664	0.948	1.02		0.470	0.317	0.567	0.516		0.432	0.435
¼	4.1	1.19	1.09	0.542	0.957	0.993		0.392	0.260	0.574	0.493		0.435	0.440
³⁄₁₆	3.07	0.902	0.842	0.415	0.966	0.970		0.307	0.200	0.583	0.470		0.439	0.446
L2½ × 2½ × ½	7.7	2.25	1.23	0.724	0.739	0.806		1.23	0.724	0.739	0.806		0.487	1.000
⅜	5.9	1.73	0.984	0.566	0.753	0.762		0.984	0.566	0.753	0.762		0.487	1.000
⁵⁄₁₆	5.0	1.46	0.849	0.482	0.761	0.740		0.849	0.482	0.761	0.740		0.489	1.000
¼	4.1	1.19	0.703	0.394	0.769	0.717		0.703	0.394	0.769	0.717		0.491	1.000
³⁄₁₆	3.07	0.902	0.547	0.303	0.778	0.694		0.547	0.303	0.778	0.694		0.495	1.000
L2½ × 2 × ⅜	5.3	1.55	0.912	0.547	0.768	0.831		0.514	0.363	0.577	0.581		0.420	0.614
⁵⁄₁₆	4.5	1.31	0.788	0.466	0.776	0.809		0.446	0.310	0.584	0.559		0.422	0.620
¼	3.62	1.06	0.654	0.381	0.784	0.787		0.372	0.254	0.592	0.537		0.424	0.626
³⁄₁₆	2.75	0.809	0.509	0.293	0.793	0.764		0.291	0.196	0.600	0.514		0.427	0.631
L2 × 2 × ⅜	4.7	1.36	0.479	0.351	0.594	0.636		0.479	0.351	0.594	0.636		0.389	1.000
⁵⁄₁₆	3.92	1.15	0.416	0.300	0.601	0.614		0.416	0.300	0.601	0.614		0.390	1.000
¼	3.19	0.938	0.348	0.247	0.609	0.592		0.348	0.247	0.609	0.592		0.391	1.000
³⁄₁₆	2.44	0.715	0.272	0.190	0.617	0.569		0.272	0.190	0.617	0.569		0.394	1.000
⅛	1.65	0.484	0.190	0.131	0.626	0.546		0.190	0.131	0.626	0.546		0.398	1.000

Answers to Even-Numbered Problems

CHAPTER 1

1.2 58.3 MPa
1.4 $\sigma_{br} = 12\,000$ psi (C), $\sigma_{al} = 6000$ psi (C), $\sigma_{st} = 12\,000$ psi (T)
1.6 5.70 in.
1.8 24.0 kN
1.10 32.7 mm^2
1.12 9220 lb
1.14 87.1 MPa
1.16 $P = 50.2$ kN, $x = 602$ mm
1.18 $A_{AG} = 1.167$ in.^{2}, $A_{BC} = 5.15$ in.^{2}
1.20 $A_{BE} = 625$ mm^2, $A_{BF} = 427$ mm^2, $A_{CF} = 656$ mm^2
1.22 4060 lb
1.24 $\sigma = 11.91$ psi, $\tau = 44.4$ psi
1.26 550 kN
1.28 29.1 mm
1.30 (a) 53.1 MPa; (b) 33.3 MPa; (c) 18.18 MPa
1.32 17.46 mm
1.34 3190 lb
1.36 (a) 19.92 mm; (b) 84.3 MPa
1.38 19 770 lb
1.40 $b = 12.15$ in., $t = 0.510$ in.
1.42 51 500 lb·in.
1.44 70.8 mm
1.46 4 rivets
1.48 9.77 mm
1.50 324 kN
1.52 14.72 km
1.54 (a) 166.7 MPa; (b) 101.9 MPa; (c) 166.7 MPa
1.56 2250 lb

CHAPTER 2

2.2 (a) 58 ksi; (b) 10.5×10^6 psi; (c) 69 ksi; (d) 74 ksi; (e) 68 ksi
2.4 54.3 mm
2.6 1139 lb
2.8 53.3 kN
2.10 545×10^{-6} in.
2.12 4000 lb
2.14 1.607 mm (contraction)
2.16 18.0 kN
2.18 4830 lb
2.20 2.81 mm
2.22 115.0×10^{-3} in.
2.24 2.37 mm
2.26 53.0×10^{-3} in.
2.28 (No answer)
2.30 $\sigma_x = 73.6$ MPa (T), $\sigma_y = 53.3$ MPa (T)
2.32 $PL(\nu^2 - 1)/(EA)$
2.34 (No answer)
2.36 0.326
2.38 $\gamma = 9.98 \times 10^{-3}$ rad, $\epsilon_{AC} = 7.50 \times 10^{-3}$; $\epsilon_{BD} = -2.50 \times 10^{-3}$
2.40 $\epsilon_{AC} = \gamma/2$, $\epsilon_{BD} = -\gamma/2$
2.42 (No answer)
2.44 $\sigma_{st} = 83.5$ MPa (C), $\sigma_{ci} = 41.8$ MPa (C)
2.46 0.365 in.
2.48 288 mm
2.50 22.5 MPa (C)
2.52 15.58 kips
2.54 3.63 mm
2.56 22.2 MPa (T)
2.58 42.2 kips
2.60 $P_{AB} = 148.9$ kN (C), $P_{BC} = 1.11$ kN (T), $P_{CD} = 91.1$ kN (T)
2.62 $\sigma_{st} = 18.26$ ksi (T), $\sigma_{br} = 9.13$ ksi (C)
2.64 136.8×10^{-3} in.
2.66 $P_A = 239$ kN, $P_B = 184.2$ kN, $P_C = 177.2$ kN
2.68 $P_{AB} = P_{AD} = 2.48$ kN, $P_{AC} = 3.01$ kN
2.70 $P_{st} = 6370$ lb, $P_{al} = 5130$ lb
2.72 755 lb
2.74 (a) 18.0 ksi (T); (b) 95.5°F
2.76 (a) 40.6°C; (b) 60.0 MPa (C)
2.78 50.6°C
2.80 (No answer)
2.82 $\sigma_{cu} = 6.71$ MPa (T), $\sigma_{al} = 16.77$ MPa (C)
2.84 28.3°C (decrease)
2.86 $\sigma_{st} = 20.4$ ksi (C), $\sigma_{al} = 5.59$ ksi (C)
2.88 $\sigma_{AD} = \sigma_{CD} = 112.3$ MPa (C), $\sigma_{BD} = 112.3$ MPa (T)
2.90 $\sigma_{st} = 118.0$ MPa (T), $\sigma_{al} = 59.0$ MPa (T), $\sigma_{br} = 29.5$ MPa (T)
2.92 4680 psi (T)

Answers to Even-Numbered Problems

2.94 0.1988 in.
2.96 $A_{al}/A_{st} = 3.87$
2.98 (a) 7.12 in.^{2}; (b) 75.7×10^{-3} in.
2.100 15.16×10^{-3} mm
2.102 (a) 6130 psi (T); (b) 6930 psi (T)
2.104 563°F
2.106 $\sigma_A = 21.8$ ksi (T), $\sigma_C = 14.55$ ksi (T)
2.108 56.0°C (decrease)

CHAPTER 3

3.2 (a) 21.1 ksi; (b) 4.84°
3.4 352 mm
3.6 (a) 10.03 ksi; (b) 3.03°
3.8 (a) 114.0 mm; (b) 41.3 MPa
3.10 (No solution)
3.12 4.00 kN·m
3.14 594 N·m
3.16 51.9 mm
3.18 (a) $\tau_{st} = \tau_{br} = 18.11$ MPa, $\tau_{al} = 28.0$ MPa; (b) 5.29°
3.20 6.34°, clockwise when viewed from D toward A
3.22 $T_0 L/(2GJ)$
3.24 $T = 6.93$ kN·m, $b/a = 1.186$
3.26 $0.671L$
3.28 $T_C = -754$ lb·ft and $T_C = -135.4$ lb·ft
3.30 4.23 kip·ft
3.32 $\tau_{AB} = 201$ MPa, $\tau_{BC} = 12.84$ MPa, $\tau_{CD} = 125.4$ MPa
3.34 Left shaft: $\tau = 6.24$ ksi, right shaft: 14.80 ksi
3.36 (No solution)
3.38 (a) $(1.509 \times 10^{-3})T/t^3$; (b) $(1.444 \times 10^{-3})T/t^3$, 4.31%
3.40 1885 lb·ft
3.42 (a) 0.785; (b) 0.617
3.44 (a) 7020 lb·in.; (b) 2.12°
3.46 0.903°
3.48 (No solution)
3.50 4.20 kN·m
3.52 679 N·m
3.54 $\tau_{al} = 10.79$ MPa, $\tau_{cu} = 15.16$ MPa
3.56 (a) 62.8 in.; (b) 30.0°
3.58 2.12 in.
3.60 Circular tube: 38.8%, square tube: 61.2%

CHAPTER 4

4.2 $V = C_0/L$, $M = -C_0 + C_0 x/L$
4.4 $V = -w_0 x + w_0 x^2/(2L)$, $M = -w_0 x^2/2 + w_0 x^3/(6L)$
4.6 AB: $V = Pb/(a+b)$, $M = Pbx/(a+b)$;
BC: $V = -Pa/(a+b)$, $M = Pa[1 - x/(a+b)]$
4.8 AB: $V = -120x$ lb, $M = -60x^2$ lb·ft;
BC: $V = -960$ lb, $M = -960x + 3840$ lb·ft
4.10 AB: $V = -120x$ lb, $M = -60x^2$ lb·ft;
BC: $V = -120x + 810$ lb,
$M = -60x^2 + 810x - 2430$ lb·ft
4.12 AB: $V = -8x + 29$ kN, $M = -4x^2 + 29x$ kN·m;
BC: $V = -11$ kN, $M = -11x + 88$ kN·m
4.14 AB: $V = P/3$, $M = Px/3$;
BC: $V = -2P/3$, $M = P(L - 2x)/3$;
CD: $V = P/3$, $M = -P(L - x)/3$
4.16 AB: $V = 12$ kips, $M = 12x$ kip·ft;
BC: $V = 0$, $M = 48$ kip·ft;
CD: $V = 0$, $M = 0$
4.18 AB: $V = -60x + 670$ lb, $M = -30x^2 + 670x$ lb·ft;
BC: $V = -60x - 230$ lb,
$M = -30x^2 - 230x + 3600$ lb·ft;
CD: $V = -60x + 1480$ lb,
$M = -30x^2 + 1480x - 16920$ lb·ft
4.20 AB: $V = (P/2) \sin \theta$, $M = (PR/2)(1 - \cos \theta)$;
BC: $V = -(P/2) \sin \theta$, $M = (PR/2)(1 + \cos \theta)$
4.22 $V_{max} = -30$ kN, $M_{max} = 37.5$ kN·m
4.24 $V_{max} = \pm 20$ kN, $M_{max} = -20$ kN·m
4.26 $V_{max} = -3$ kips, $M_{max} = \pm 10$ kip·ft
4.28 $V_{max} = 72$ kN, $M_{max} = -72$ kN·m
4.30 $V_{max} = 60$ kN, $M_{max} = 80$ kN·m
4.32 $V_{max} = \pm 1920$ lb, $M_{max} = 4190$ lb·ft
4.34 $V_{max} = -700$ lb, $M_{max} = -1000$ lb·ft
4.36 $V_{max} = -w_0 L/3$, $M_{max} = w_0 L^2/(9\sqrt{3})$
4.38 $V_{max} = \pm w_0 L/4$, $M_{max} = w_0 L^2/24$
4.40 $V_{max} = \pm 50$ kN, $M_{max} = 42.5$ kN·m
4.42 $V_{max} = -4000$ lb, $M_{max} = -16 \times 10^3$ lb·ft
4.44 $V_{max} = -1380$ lb, $M_{max} = 3600$ lb·ft
4.46 $V_{max} = 80$ kN, $M_{max} = 160$ kN·m
4.48 (No solution)
4.50 $V_{max} = 12.27$ kips, $M_{max} = 127.6$ kip·ft
4.52 $V_{max} = -39$ kN, $M_{max} = -42$ kN·m
4.54 $V_{max} = \pm 960$ lb, $M_{max} = \pm 1920$ lb·ft
4.56 $V_{max} = 6$ kN, $M_{max} = 9$ kN·m
4.58 $V_{max} = -8$ kN, $M_{max} = -10$ kN·m
4.60 $V_{max} = -3840$ lb, $M_{max} = -20.5 \times 10^3$ lb·ft
4.62 $V_{max} = \pm 800$ lb, $M_{max} = 1600$ lb·ft
4.64 $M_{max} = 64$ kN·m
4.66 $V_{max} = 2360$ lb, $M_{max} = 6300$ lb·ft

CHAPTER 5

5.2 (a) 10.48 ksi (T); (b) 7.86 ksi (C); (c) 9.07 ksi (T)
5.4 $(\sigma_T)_{max} = 50.0$ MPa, $(\sigma_C)_{max} = 100.0$ MPa
5.6 1.758 in.
5.8 (a) 10.12 ksi; (b) 5.06 ksi (C)
5.10 1.676 kN/m
5.12 75.0 MPa (just to the right of B)
5.14 32.5 kN
5.16 6680 lb
5.18 75.0 mm

5.20	22.0 kN/m
5.22	166.7 lb/ft
5.24	11.55 ft
5.26	(a) 4850 N/m; (b) 185.3 N/m
5.28	21.5 kN/m
5.30	2.95 kN/m
5.32	$w_0 = 800$ lb/ft, $P = 7200$ lb
5.34	$(\sigma_T)_{max} = 32.0$ MPa, $(\sigma_C)_{max} = 80.0$ MPa
5.36	$(\sigma_T)_{max} = 16.0$ ksi, $(\sigma_C)_{max} = 8.0$ ksi
5.38	$(\sigma_T)_{max} = 4800$ psi, $(\sigma_C)_{max} = 9600$ psi
5.40	26.8 kN
5.42	320 lb/ft
5.44	0.707
5.46	W460 × 52, 106.1 MPa
5.48	W24 × 68, 19.79 ksi
5.50	W24 × 68, 15.64 ksi
5.52	W8 × 48, 16.20 ksi
5.54	DE: W310 × 28, 109.3 MPa; ABD: W690 × 125, 119.7 MPa
5.56	(No solution)
5.58	12.64 ft
5.60	2550 psi
5.62	20×10^{-3} in.
5.64	(a) 25.4 MPa; (b) 92.4%
5.66	505 lb/ft
5.68	$\tau_{max} = 351$ psi
5.70	4.60 kN/m
5.72	3.18 MPa
5.74	163.4 MPa
5.76	1250 lb/ft
5.78	2220 lb
5.80	$L = 4.56$ m, $\sigma_{max} = 5.83$ MPa, $\tau_{max} = 729$ kPa
5.82	741 lb
5.84	(b) 4010 lb
5.86	11.73 kN
5.88	144.2 mm square
5.90	(b) 3.19 kN
5.92	(a) 1369 lb; (b) 533 psi
5.94	237 lb
5.96	166.1 mm
5.98	(a) 4.91 in.; (b) 5890 psi
5.100	$\sigma_T = 3.38$ MPa, $\sigma_C = 1.688$ MPa
5.102	11.52 ksi
5.104	$(\sigma_T)_{max} = 12.00$ ksi, $(\sigma_C)_{max} = 6.00$ ksi
5.106	48.1 kN
5.108	889 N/m
5.110	W610 × 125, 113.5 MPa
5.112	(a) 3.28 MPa; (b) 31.8 MPa
5.114	131.0 mm
5.116	3670 lb
5.118	$\tau_{max} = 68.1$ psi, $e = 2.29$ in.
5.120	(a) 8950 N/m; (b) 94.6 mm

CHAPTER 6

6.2	(a) $EIv = \dfrac{w_0}{24}(2Lx^3 - x^4 - L^3 x)$; (b) $\dfrac{5w_0 L^4}{384EI} \downarrow$
6.4	$\dfrac{M_0 L^2}{8EI} \downarrow$ at $x = \dfrac{L}{2}$
6.6	For left half: $\sqrt{3} M_0 L^2/216 \uparrow$ at $x = \sqrt{3} L/6$
6.8	$EIv = -\dfrac{w_0}{120L}(x^5 - 5L^4 x + 4L^5)$
6.10	$\dfrac{3w_0 L^4}{640 EI} \downarrow$
6.12	$\dfrac{Pa^2(3L-a)}{6EI} \downarrow$
6.14	(a) 500 N·m³ $\downarrow$; (b) 4.50×10^{-6} m⁴
6.16	$v' = -0.00875$, $v = 7.5$ mm
6.18	3.35 kN·m³ $\downarrow$
6.20	$\dfrac{61 w_0 a^4}{36 EI} \downarrow$
6.22	619 mm
6.24	1981 lb·ft³ $\downarrow$
6.26	3.56 kN·m³ $\uparrow$
6.28	(a) $EIv = -50x^3 + 900\langle x-2\rangle^2 - \dfrac{25}{3}\langle x-4\rangle^4 + \dfrac{550}{3}\langle x-6\rangle^3 - 578x$ N·m³; (b) 2180 N·m³ $\downarrow$
6.30	(a) $EIv = 54x^2 - \dfrac{2}{3}x^3 - \dfrac{5}{3}\langle x-6\rangle^3 - 6264$ kip·ft³; (b) 6246 kip·ft³ $\downarrow$
6.32	$\dfrac{Pa^2}{12EI}(4a - 3b) \downarrow$
6.34	(a) 54.0 kN·m³ $\downarrow$; (b) 54.8 kN·m³ $\downarrow$
6.36	176.5 lb
6.38	114.3×10^3 kip·ft³ $\downarrow$
6.40	2750 N·m³ $\downarrow$
6.42	$0.0630 \dfrac{w_0 L^4}{EI} \downarrow$
6.44	$EIv = \dfrac{275}{3}x^3 - \dfrac{25}{3}x^4 - \dfrac{10}{9}\langle x-1\rangle^5 - \dfrac{5195}{6}x$ N·m³
6.46	423 kip·ft³ $\uparrow$
6.48	0.0998 in. $\downarrow$
6.50	(a) $w_0 L/10$; (b) $w_0 L^3/120$ $\circlearrowright$
6.52	853 lb·ft³ $\downarrow$
6.54	$\dfrac{\sqrt{3} M_0 L^2}{216 EI} \uparrow$
6.56	(a) $\dfrac{M_0 a}{3L}(L^2 - 3La + 2a^2) \uparrow$; (b) $\dfrac{M_0}{3L}(L^2 - 3La + 3a^2)$ $\circlearrowright$
6.58	64 000 lb·ft³ $\downarrow$
6.60	62.0 mm $\downarrow$
6.62	1440 lb·ft³ $\downarrow$
6.64	428 N·m³ $\downarrow$
6.66	$\dfrac{M_0 L^2}{128 EI_0} \uparrow$
6.68	$EI\delta_B = 1330$ lb·ft³ $\downarrow$, $EI\delta_C = 1716$ lb·ft³ $\downarrow$

Answers to Even-Numbered Problems

6.70	55.5 mm ↓
6.72	1733 N·m³ ↓
6.74	5.45 kips
6.76	135.4 mm
6.78	203 lb
6.80	266 kN·m³ ↓
6.82	$\dfrac{4Wa^3}{3EI}$ ↓
6.84	114.3×10^3 kip·ft³ ↓
6.86	$0.0630\dfrac{w_0 L^4}{EI}$
6.88	$\dfrac{11 w_0 b^4}{12 EI_0}$ ↓
6.90	$\dfrac{5PL}{24}$
6.92	$\dfrac{M_0 a}{2EI}(2b + a)$ ↓
6.94	$\dfrac{5 w_0 a^4}{8 EI_0}$ ↓
6.96	$EIv = \dfrac{w_0 x}{24}(-x^3 + 4Lx^2 - 8L^3)$
6.98	$EIv = \dfrac{w_0 x}{360L}(-3x^4 - 15Lx^3 + 40L^2 x^2 - 22L^4)$
6.100	$EIv_{AB} = \dfrac{w_0 x}{24}(-x^3 + 4ax^2 - 20a^3)$, $EIv_{BC} = \dfrac{w_0 a^2}{24}(6x^2 - 24ax + a^2)$
6.102	3240 lb·ft³ ↑
6.104	$41 w_0 L^4/384$ ↓
6.106	$\dfrac{w_0 b^3}{24}(3b + 4a)$ ↓
6.108	1271 lb·ft³ ↓
6.110	$\dfrac{Pa^3}{6E}\left(\dfrac{1}{I_1} + \dfrac{7}{I_2}\right)$ ↓

CHAPTER 7

7.2	$R_A = w_0 L/24$, $R_B = 7 w_0 L/24$, $M_B = w_0 L^2/24$
7.4	(a) $R_A = R_B = P/2$, $M_A = M_B = PL/8$
7.6	(a) $R_A = R_B = w_0 L/2$, $M_A = M_B = w_0 L^2/12$
7.8	$R_A = 3 w_0 L/20$, $R_B = 7 w_0 L/20$, $M_A = w_0 L^2/30$, $M_B = w_0 L^2/20$
7.10	(a) $R_A = R_B = P$, $M_A = M_B = \dfrac{2Pa}{3}$; (b) $\dfrac{5 Pa^3}{24 EI}$ ↓
7.12	$R_A = R_B = 34.6$ kN, $M_A = 14.40$ kN·m, $M_B = 38.4$ kN·m
7.14	$R_A = 40/3$ kN, $R_B = 14/3$ kN, $M_A = 8$ kN·m, $M_B = 4$ kN·m
7.16	$R_A = 334$ lb, $R_B = 823$ lb, $R_C = 283$ lb, $M_A = 309$ lb·ft
7.18	$R_A = 84.8$ kN, $R_B = 115.2$ kN, $M_B = 60.9$ kN·m
7.20	$10P/27$
7.22	$R_A = P$, $M_A = 3PL/8$, $M_B = PL/8$
7.24	$R_A = 42$ lb, $R_B = 1250$ lb, $R_C = 708$ lb
7.26	(a) $M_A = M_B = \dfrac{w_0 a^2}{6L}(3L - 2a)$; (b) $\dfrac{w_0 a^3}{24}(L - a)$ ↓
7.28	12.88 kN
7.30	308 lb
7.32	$R_A = w_0 L/16$ ↓, $R_B = 5 w_0 L/8$ ↑, $R_C = 7 w_0 L/16$ ↑
7.34	$R_A = 84.8$ kN, $R_B = 115.2$ kN, $M_B = 60.9$ kN·m
7.36	$R_A = P$, $M_A = 3PL/8$, $M_B = PL/8$
7.38	12.88 kN
7.40	$M_A = 3 w_0 L^2/16$, $M_C = 5 w_0 L^2/16$
7.42	0.947 N
7.44	$R_A = R_D = 2 w_0 L/5$, $R_B = R_C = 11 w_0 L/10$
7.46	$w_0 L^2/6$
7.48	$w_0 L/10$
7.50	$R_B = 1875$ lb ↑, $R_C = 1875$ lb ↓, $M_C = 3750$ lb·ft
7.52	$R_A = R_C = P\left(1 + \dfrac{3b}{2a}\right)$ ↑, $R_B = \dfrac{3Pb}{a}$ ↓
7.54	$\dfrac{7 w_0 L^4}{1152 EI}$
7.56	(No answer)

CHAPTER 8

8.2	208 psi
8.4	17.78 in.
8.6	4.44 m
8.8	52.5 ft
8.10	(a) 17 bolts; (b) 75.3 MPa
8.12	0.0498 mm
8.14	$28P$ psi (4 psi in straight rod)
8.16	218 lb
8.18	$\sigma_A = 22.7$ MPa (T), $\sigma_B = 17.33$ MPa (C)
8.20	13 220 psi (T)
8.22	246 kips, larger (265 kips) with no plate
8.24	$\sigma_{max} = 5000$ psi, $\sigma_{min} = 73$ psi
8.26	$\sigma_{max} = 2.64$ MPa, $\sigma_{min} = -17.72$ MPa
8.28	45.3 MPa (C)
8.30	2300 psi (T), 2200 psi (C)
8.32	$\sigma_A = 10.08$ MPa (C), $\sigma_B = 55.9$ MPa (T)
8.34	$\sigma_A = 29.2$ MPa (C), $\sigma_B = 4.17$ MPa (T)
8.36	25.4 MPa (T)
8.38	457 kN
8.40	$\sigma = -35$ MPa, $\tau = 43.3$ MPa
8.42	$\sigma = 7.71$ ksi, $\tau = 9.19$ ksi
8.44	$\sigma_{x'} = -6.93$ ksi, $\sigma_{y'} = 6.93$ ksi, $\tau_{x'y'} = 4.0$ ksi
8.46	$\sigma_{x'} = -16.66$ ksi, $\sigma_{y'} = 0.66$ ksi, $\tau_{x'y'} = 5.0$ ksi
8.48	$\sigma_1 = 72.4$ MPa, $\sigma_2 = -12.4$ MPa, $\theta_1 = 67.5°$
8.50	$\sigma_1 = -2.0$ ksi, $\sigma_2 = -22.0$ ksi, $\theta_1 = 45°$
8.52	$\tau_{max} = 11.18$ ksi, $\theta = -13.28°$
8.54	$\tau_{max} = 9.67$ ksi, $\theta = -3.56°$
8.56	$R = 14.14$ ksi, $\bar{\sigma} = 0$
8.58	$R = 0$, $\bar{\sigma} = -p$

8.60	$\sigma_{x'} = -6.93$ ksi, $\sigma_{y'} = 6.93$ ksi, $\tau_{x'y'} = 4.0$ ksi
8.62	$\sigma_{x'} = -16.66$ ksi, $\sigma_{y'} = 0.66$ ksi, $\tau_{x'y'} = 5.0$ ksi
8.64	$\sigma_{x'} = -3.27$ ksi, $\sigma_{y'} = -0.73$ ksi, $\tau_{x'y'} = 6.20$ ksi
8.66	$\sigma_{x'} = -81.8$ MPa, $\sigma_{y'} = 31.8$ MPa, $\tau_{x'y'} = 61.7$ MPa
8.68	(a) $\sigma_1 = 6.49$ ksi, $\sigma_2 = -10.49$ ksi, $\theta_1 = 22.5°$; (b) $\tau_{max} = 8.49$ ksi
8.70	(a) $\sigma_1 = -4.79$ ksi, $\sigma_2 = -19.21$ ksi, $\theta_2 = 73.2°$; (b) $\tau_{max} = 7.21$ ksi
8.72	(a) $\sigma_x = 3.46$ ksi, $\sigma_y = -3.46$ ksi, $\tau_{xy} = 5$ ksi; (b) $\sigma_1 = 6.08$ ksi, $\sigma_2 = -6.80$ ksi, $\theta_1 = 27.6°$
8.74	$\sigma_1 = 37.8$ MPa, $\sigma_2 = -27.8$ MPa, $\theta_1 = 26.2°$
8.76	12 ksi
8.78	$\sigma = 18.18$ ksi (T), $\tau = 21.7$ ksi
8.80	(a) 20 MPa; (b) 40 MPa
8.82	(a) 54.1 MPa; (b) 54.1 MPa
8.84	(a) 29.2 MPa; (b) 42.1 MPa
8.86	10 MPa
8.88	17.5 ksi
8.90	$\sigma_{max} = 78.6$ MPa (C), $\tau_{max} = 40.4$ MPa
8.92	$\sigma_{max} = 13.72$ ksi (C), $\tau_{max} = 8.18$ ksi
8.94	8450 N·m
8.96	5070 N·m
8.98	32.1 mm
8.100	$(\sigma_T)_{max} = 73.7$ MPa, $(\sigma_C)_{max} = 98.4$ MPa, $\tau_{max} = 59.9$ MPa
8.102	40 in. (inner diameter)
8.104	$\sigma_{max} = 2280$ psi, $\tau_{max} = 1223$ psi
8.106	(a) $\sigma_1 = 4870$ psi, $\sigma_2 = -1812$ psi, $\tau_{max} = 3340$ psi; (b) Stresses will be smaller.
8.108	46 mm
8.110	2.40 in.
8.112	$\sigma = 2.98$ MPa, $\tau = 11.51$ MPa
8.114	$\sigma_1 = 29.1$ MPa, $\sigma_2 = -1.24$ MPa, $\theta_1 = 11.7°$
8.116	3000 psi
8.118	(No answer)
8.120	(a) $\sigma_1 = 48.4$ MPa, $\sigma_2 = -105.5$ MPa, $\theta_2 = 26.6°$; (b) $\sigma = -104.9$ MPa, $\tau = 9.2$ MPa
8.122	$\sigma = -17200$ psi, $\tau = 6920$ psi
8.124	(No solution)
8.126	$\epsilon_b = (\epsilon_a + \epsilon_c)/2$
8.128	$\sigma_1 = 970$ psi, $\sigma_2 = -6970$ psi, $\tau_{max} = 3970$ psi
8.130	$\sigma_1 = 120.6$ MPa, $\sigma_2 = -6.3$ MPa, $\theta_1 = 38.0°$
8.132	$r/4$
8.134	8.66 ksi
8.136	$\sigma_x = -27.5$ MPa, $\sigma_y = 57.5$ MPa, $\tau_{xy} = 56.3$ MPa
8.138	(a) $\sigma_x = -72.0$ MPa, $\sigma_y = 32.0$ MPa; (b) $\theta_1 = 75°$
8.140	$\epsilon_2 = 60 \times 10^{-6}$, $|\gamma_{xy}| = 480 \times 10^{-6}$
8.142	(a) $\sigma_1 = 29.7$ ksi, $\sigma_2 = 0$; (b) $\sigma_1 = 3.25$ ksi, $\sigma_2 = -1.083$ ksi
8.144	(a) 9.04 MPa; (b) 8.08 MPa
8.146	1.144 in.

CHAPTER 9

9.2	15.38 kN·m
9.4	3640 lb
9.6	$\sigma_{wd} = 1100$ psi, $\sigma_{st} = 10530$ psi
9.8	62.8 mm
9.10	79.0 kN·m
9.12	$\sigma_T = 81.4$ MPa, $\sigma_C = 99.7$ MPa
9.14	5530 lb
9.16	74.8 MPa
9.18	Wood/steel: $0.0431V$ psi, wood/aluminum: $0.0414V$ psi
9.20	0.311 in.
9.22	$h = 5.57$ in., $A_{st} = 1.855$ in.^{2}
9.24	$\sigma_{st} = 129.6$ MPa, $\sigma_{co} = 6.92$ MPa
9.26	78.4 kN·m
9.28	$\sigma_{st} = 16230$ psi, $A_{st} = 3.80$ in.^{2}
9.30	$A_{st} = 1191$ mm^2, $M = 57.4$ kN·m
9.32	$b = 342$ mm, $d = 456$ mm, $A_{st} = 2190$ mm^2
9.34	269 kN·m
9.36	87.5 kN·m
9.38	$\sigma_{co} = 938$ psi, $(\sigma_{st})_T = 13070$ psi, $(\sigma_{st})_C = 4570$ psi

CHAPTER 10

10.2	3.85 in.
10.4	36.2 mm
10.6	(a) 9.86 m; (b) 738 kN
10.8	W250 × 67
10.10	1.751 in.
10.12	3.95 m
10.14	67.8°F
10.16	(a) 100.8; (b) 174.6
10.18	4.94 m
10.20	8.24 in.
10.22	(a) 106.1 kN; (b) 54.1 kN
10.24	67.1 kips
10.26	W12 × 79
10.28	(a) 335 kN; (b) 387 kN
10.30	(a) 22 in.; (b) 39 in.
10.32	(a) 15.35 ksi; (b) 2.46
10.34	$(\sigma_C)_{max} = 28.6$ MPa, $(\sigma_T)_{max} = 28.6$ MPa
10.36	8.26 in.
10.38	(a) 113.8 MPa; (b) 2.05
10.40	W18 × 76
10.42	5.84 kips (bending about z-axis)
10.44	45.3 kN (tension governs)

CHAPTER 11

11.2	$q_{max} = V/(\pi r)$
11.4	(a) $q_{max} = 9180$ N/m; (b) 35.3 mm
11.6	(a) $q_{max} = 771$ lb/in.; (b) 1.428 in.

Answers to Even-Numbered Problems

11.8 136.0 mm to the right of B
11.10 $0.720r$ to the left of B
11.12 $0.354b$ to the left of B
11.14 $0.7a$ to the left of B
11.16 21.8 mm to the left of B
11.18 (a) 74.4°; (b) 1.178 ksi
11.20 20.7 kips
11.22 (a) 73.95°; (b) 2.41 kips
11.24 1406 psi (C)
11.26 (a) ($y=0$, $z=83.3$ mm) and ($y=-93.8$ mm, $z=0$); (b) $(\sigma_t)_{max}=0.400$ MPa, $(\sigma_c)_{max}=2.40$ MPa
11.28 (No solution)
11.30 46.2 kN
11.32 1.537 in.
11.34 $(\sigma_t)_{max}=84.0$ MPa, $(\sigma_c)_{max}=98.2$ MPa
11.36 $(\sigma_t)_{max}=6.81$ ksi, $(\sigma_c)_{max}=6.11$ ksi

CHAPTER 12

12.2 $\delta_V = 0.0719$ in. $\downarrow$, $\delta_H = 0.0193$ in. $\rightarrow$
12.4 $5.83\dfrac{WL}{EA} \downarrow$
12.6 $\dfrac{7PL^2}{24EI} \circlearrowleft$
12.8 $\dfrac{Pb^2}{EI} \circlearrowleft$
12.10 $\dfrac{\pi}{2}T_0 R\left(\dfrac{1}{EI}+\dfrac{1}{GJ}\right)$
12.12 $\delta_H = \dfrac{Pba^2}{2EI} \rightarrow$, $\delta_V = \dfrac{Pb^2(3a+b)}{3EI} \downarrow$
12.14 $\delta_C = \dfrac{2PR^3}{EI} \leftarrow$, $\delta_B = \dfrac{PR^3}{EI}\left(\dfrac{\pi}{4}+1\right) \downarrow$
12.16 $\dfrac{Wb}{EA}\left(\sqrt{2}+\dfrac{3}{2}\right) \downarrow$
12.18 $7P/4$
12.20 $P_{AB} = P_{AD} = 1046$ lb (T), $P_{AC} = 1259$ lb (T)
12.22 188.0 lb
12.24 At A: 262 N $\leftarrow$, 750 N $\downarrow$; at D: 338 N $\leftarrow$, 750 N $\uparrow$
12.26 0.547 in.
12.28 21 500 psi
12.30 0.491 in.
12.32 $\sigma_{max} = 21\,100$ psi, $\delta_{max} = 0.359$ in.
12.34 $2v\sqrt{\dfrac{mE}{2AL}}$
12.36 Unsafe
12.38 1875 psi
12.40 (a) 114.3 MPa; (b) 131.5 MPa
12.42 (a) 1.313; (b) 1.327
12.44 3.93 kN
12.46 7.18 MN
12.48 1.397
12.50 4.74 kN·m
12.52 5280 lb·ft
12.54 2.69
12.56 (a) $3.02\dfrac{P}{Db}$; (b) $3.14\dfrac{P}{Db}$; (c) $3.73\dfrac{P}{Db}$
12.58 111.7 MPa
12.60 32.2 ksi
12.62 14.08 ksi
12.64 585 lb

CHAPTER 13

13.2 1.169
13.4 (a) 1.698; (b) 1.273
13.6 1.108
13.8 1.398
13.10 57.3 kN·m
13.12 540 kip·in.
13.14 $P_{st} = 54.7$ kN (T), $P_{al} = 27.4$ kN (C)
13.16 $\sigma_{al} = 7.40$ ksi (T), $\sigma_{st} = 5.56$ ksi (T)
13.18 $\tau|_{\rho=r} = 7\tau_{yp}/24$, $\tau|_{\rho=r/2} = -17\tau_{yp}/48$
13.20 103.8°
13.22 15.1 MPa
13.24 $\sigma|_{y=2.5\,\text{in.}} = -10.47$ ksi, $\sigma|_{y=1.5\,\text{in.}} = 9.72$ ksi
13.26 96.4°
13.28 (a) 13.5 kN·m; (b) 1.286
13.30 9.16 kN·m
13.32 $\dfrac{2a+b}{ab}M_L$
13.34 7.50 kips
13.36 206 kN
13.38 $5K/4$
13.40 $11.66\,M_L/L^2$
13.42 $\dfrac{5M_L}{2a}$

APPENDIX A

A.2 $A = 150 \times 10^3$ mm^2, $\bar{k}_x = 57.7$ mm
A.4 $\bar{y} = 3.50$ in., $\bar{I}_x = 291$ in.^{4}
A.6 12.96×10^6 mm^4
A.8 42.1 in.^{4}
A.10 17.31×10^6 mm^4
A.12 (a) 616×10^3 mm^4; (b) 576×10^3 mm^4
A.14 $0.433R$
A.16 $\bar{k}_x = 194.3$ mm
A.18 $\bar{y} = 9.03$ in., $\bar{I}_x = 477$ in.^{4}
A.20 $3.07R^4$
A.22 -323×10^3 mm^4
A.24 792 in.^{4}

A.26 $I_u = 2.00 \times 10^6$ mm^4, $I_v = 28.0 \times 10^6$ mm^4, $I_{uv} = 0$
A.28 (a) 0.983×10^6 mm^4; (b) $\theta_1 = 28.5°$, $\theta_2 = 118.5°$
A.30 $I_u = 1057$ in.^{4}, $I_v = 3943$ in.^{4}, $I_{uv} = -899$ in.^{4}
A.32 $I_1 = 49.77 \times 10^6$ mm^4, $I_2 = 15.15 \times 10^6$ mm^4, $\theta_1 = -24.5°$, $\theta_2 = 65.5°$
A.34 $I_1 = 143.6 \times 10^6$ mm^4, $I_2 = 19.79 \times 10^6$ mm^4, $\theta_1 = 41.4°$, $\theta_2 = 131.4°$
A.36 $I_u = I_v = 16.94$ in.^{4}, $I_{uv} = 0$
A.38 $I_1 = 913$ in.^{4}, $I_2 = 132.1$ in.^{4}, $\theta_1 = -27.5°$, $\theta_2 = 62.5°$

Photo Credits

Chapter 1, page 1:   Copyright © Streano/Havens
Chapter 2, page 29:   Copyright © TestResources.com
Chapter 3, page 73:   Copyright © Harald Sund/TheImageBank/GettyImages
Chapter 4, page 99:   Copyright © Streano/Havens
Chapter 5, page 135:   Copyright © Tony Freeman/PhotoEdit
Chapter 6, page 191:   Copyright © Michael Rosenfield/Stone Collection/GettyImages
Chapter 7, page 243:   Copyright © Matt Brown Photographer
Chapter 8, page 271:   Copyright © Streano/Havens
Chapter 9, page 343:   Copyright © J Silver/SuperStock
Chapter 10, page 363:   Supplied by Jaan Kiusalaas
Chapter 11, page 389:   Copyright © Michael Rosenfeld/Stone Collection/GettyImages
Chapter 12, page 417:   Copyright © Sacramento Bee/John Trotter
Chapter 13, page 455:   Courtesy of ASM International

Index

A

Absolute maximum shear stress, 310–312
Allowable stress. *see* Working stress
Angle of twist
    in circular bars, 76
    in thin-walled tubes, 90
    per unit length, 75
Area method, 111–116
    concentrated forces and couples, 114–116
    distributed loading, 112–114
Axial loading combined with lateral loading, 277–279
Axially loaded bars
    centroidal loading and, 4–5
    Saint Venant's principle and, 5–6
    strain, 34–35
    strain energy in, 418–419
    stresses on inclined planes for, 6–7

B

Balanced-stress reinforcement, 355
Beams
    defined, 99
    deflection of. *see* Deflection of beams
    design
        fasteners in built-up beams, 180–181
        for flexure and shear, 173
    shear and bending moment. *see* Shear and bending moment, beams
    stresses. *see* Stresses, beams
    types of, 100–101
Beams, composite. *see* Composite beams
Beams, curved. *see* Curved beams
Beams, statically indeterminate. *see* Statically indeterminate beams
Beams, thin-walled. *see* Thin-walled beams
Bearing stress, defined and examples, 18
Bending moment. *see* Shear and bending moment, beams
Bending moment diagrams, area method, 111
Bending moment diagrams by parts, 218–220
    cantilever beams, 220–221
    simply supported beams, 221–222
Bending stress. *see also* Unsymmetrical bending
    assumptions in analysis of, 136
    economic sections
        selecting, 156
        standard structural shapes and, 155

    flexure formula, 139
        section modulus, 139–140
        units in, 141
    limit analysis and, 471
    maximum stress
        symmetric cross sections, 140–141
        unsymmetric cross sections, 141
    residual stresses and, 464–465
    strain energy and, 419
Biaxial loading, 44–45
Biaxial state of stress. *see also* state of stress
    Hooke's Law and, 44
Bracket functions. *see* Macaulay bracket functions
Brittle materials,
    maximum normal stress theory of failure, 437
    Mohr's theory of failure, 437–438
Buckling load. *see* Critical load
Built-up (fabricated) beams, fasteners, 180–181

C

Cantilever beams, 100
Cantilever columns, critical load, 366
Castigliano's theorem, deflections by, 420
Centroids of areas, 480
Centroidal (axial) loading. *see* Axially loaded bars
Channel sections
    properties, in SI units, 512–513
    properties, in U.S. Customary units, 525–526
Circular shafts, 74–78
    assumptions in analysis of, 74
    power transmission, 77–78
    statically indeterminate problems, 78
    torsion formulas for, 76–77
Circumferential stress (hoop stress), 272–273
Coefficient of thermal expansion, 60, 503–504
Collapse mechanism, 471
Columns, 363–387
    critical load
        defined, 364–365
        discussion of, 367–368
        Euler's formula, 365–366
    defined, 363
    intermediate columns
        AISC specifications, 373–374
        tangent modulus theory, 372–373
    secant formula for eccentric loading, 378–382

Columns (continued)
    application of, 380–382
    derivation, 379–380
  types of, 363–364
Compatibility equations
  static indeterminacy and, 51
  thermal stresses and, 61
Complementary planes and stresses, 7
Composite area, method, 483–485
Composite beams, 343–362
  deflection, 350
  flexure formula, 344–345
  reinforced concrete
    analysis of, 353–355
    balanced-stress reinforcement, 355
  shear stress, 349–350
Concrete beams, reinforced
  analysis, 353–355
  balanced-stress reinforcement, 355
Constraints, support reactions and, 244
Continuous beams, 101
Coordinate transformation, 402
Critical load
  defined, 364–365
  discussion, 367–368
  Euler's formula, 365–366
Critical stress
  for columns, 367
  vs. slenderness ratio, 373
Cross sections
  least radius of gyration of, 367
  neutral axis of, 136–137
  principal axes of inertia, 493–494
  shear center of, 392–393
  standard structural shapes, 155
Curved beams, 407–415
  formula for, 410–411
Cylindrical thin-walled vessels, 272–273

D

Dead loads, 34
Deflection, dynamic, 430
Deflection formulas
  cantilever beams, 232
  simply supported beams, 233
Deflection of beams, 191–242
  Castigliano's theorem for, 420–421
  composite beams, 350
  double-integration method for
    differential equation of elastic curve, 192–194
    procedure, 194–195
    double-integration using bracket functions, 205–208
  method of superposition for, 231–233
  moment-area method, 215–222
    bending moment diagrams by parts, 218–220

    cantilever beams, 220–221
    simply supported beams, 221–222
    theorems, 216–218
Degree of indeterminacy, 243
Design
  beams
    fasteners, 180–181
    for flexure and shear, 173
    intermediate columns
      AISC specifications, 373–374
      formulas for, 372–374
      tangent modulus theory, 372–373
Differential equation of elastic curve, 192–194
Differential equation of equilibrium, 112
Direct shear, 17
Discontinuity functions, 192, 205–208. *see also* Macaulay bracket functions
Displacement, as magnitude of deflection, 195
Distortion strain energy, 438
Distributed loading, 100, 112–114
Double-integration method, 192–195, 244–245
  differential equation of elastic curve by, 192–195
  procedure for, 194–195
Double-integration method using bracket functions, 205–208, 250
Double shear, 17
Ductile materials, 438–439
  impact resistance of, 432
  limit loading for, 456
  yield criteria:
    maximum distortion energy theory, 438–439
    maximum shear stress theory, 438
Ductility, as mechanical property, 30
Dynamic deflection, 430
Dynamic loading, 429–432
  assumptions in analysis of, 429
  elastic bodies, 431
  impact factor, 430
  mass-spring model, 430–431
  modulus of resilience, modulus of toughness, 431–432

E

Eccentric load. *see* Secant formula for eccentric loading
Eccentricity ratio, 380
Elastic curve, of beams
  defined, 192
  differential equation of, 192–194
Elastic limit, 33
Elastic-perfectly plastic material, 456, 463
Elastic spring-back, 465–466
Elastic unloading, 463, 465
Elasticity, modulus of. *see* Modulus of elasticity
Electrical-resistance strain gages, 333–334
Elongation
  axially loaded bars, 30–35

per unit length (normal strain), 30
Endurance limit, S-N diagrams, 451
Energy methods
    Castigliano's theorem, 420–421
    work and strain energy, 418–419
Engineering mechanics, 1
Equal angle sections
    properties, in SI units, 514–515
    properties, in U.S. Customary units, 527–530
Equation of the elastic curve, 194–195
Euler, Leonhard, 365
Euler angle, 380
Euler's formula, 365–366

F
Fabricated (built-up) beams, 180–181
Factor of safety. see Safety, factor of
Failure criteria. see Theories of failure
Fasteners, designing, 180–181
Fatigue limit, S-N diagrams, 451
Fatigue tests, 450
Fatigue under repeated loading, 450–452
First Moment-Area Theorem, 216–217
First moments of area
    defined, 479
    locating centroid of area, 480
Fixed beams, 101
Flanges, defined, 154
Flexural rigidity of beams, 193
Flexure formula
    assumptions, 136–137
    composite beams, 341–345
    curved beams, 407–411
    derived, 136–140
    section modulus, 139–140
    see also Bending stress
Force, axial. see Axial force
Force, shear. see Shear force
Forces
    bearing force, 18
    calculating work done by, 418
    concentrated, 114–116
    external, 1–2
    internal, 1–3
    normal, 3–4
Fracture mechanics, 451
Free-body diagrams
    determining internal forces with, 2
Fully plastic state, 464–465

G
Gage length, 31, 334
General (three-dimensional) state of stress, 311–312
Generalized Hooke's Law, 44–45

Global bending moment equations, 207–208
Gyration, radius of, 367, 483

H
Hooke's Law
    generalized, 44–45
    shear loading, 45
    triaxial loading, 45
    uniaxial loading, 44
Hoop stress (circumferential stress), 272–273
Horizontal shear stress, 161–163
Huber-Hencky-von Mises yield criterion, 438

I
I-beams. see also S-shapes, 155
    properties of, in SI units, 510–511
    properties of, in U.S. Customary units, 523–524
Impact factor, 430
Inclination of the neutral axis, 401–402
Indeterminancy, degree of, 243
Induced shear, 17
Inelastic action, 455–478
    limit analysis
        axial loading, 469–470
        bending, 471
        torsion, 470–471
    limit moment, 458–460
    limit torque, 456–457
    residual stresses
        bending, 464–465
        elastic spring-back, 465–466
        loading-unloading cycle, 463
        torsion, 463–464
Intermediate columns, 364
    AISC specifications, 373–374
    design formulas for, 372–374
    tangent modulus theory and, 372–373
Internal couples, 3
Internal forces
    components of, 3
    using equilibrium analysis in computation of, 2

L
Lateral bucking, 155
Lateral and axial loading, 277–289
Least radius of gyration, 367
Limit analysis
    axial loading, 469–470
    bending, 471
    torsion, 470–471
Limit load, 456
Limit moment, 458–460
Limit torque, 456–457
    limit analysis and, 470
    residual stresses and, 464

Line loads, 100
Load, axial. *see* Axial (centroidal) loading
Load, critical. *see* Critical load
Load, eccentric. *see* Secant formula for eccentric loading
Load, limit. *see* Limit load
Loading-unloading cycle, 463
Loads
 combined. *see* Stresses, combined loads
 concentrated vs. distributed, 100
 moving across beams, 125–126
 relationship to shear force and bending moment, 113–114
Long columns, 363–364
 compared with intermediate columns, 373
 critical stress for, 367
Longitudinal stress in pressure vessels, 272–273

M

Macaulay bracket functions, 205–208
 definition of, 207
 integration of, 207
Mass-spring model, dynamic loading, 430–431
Maximum distortion energy theory, 438–439
Maximum in-plane shear stress, 294–295
 compared with absolute maximum shear stress, 310
 computing, 296
Maximum normal stress theory of yielding, 437
Maximum shear stress theory of yielding, 438
Mechanical properties in tension, 30–33
Median line, thin-walled tubes, 89
Metals, properties of, 503–504
Method of composite area, 483–485
Method of double integration. *see* Double-integration method
Method of superposition, 231–233
 principle of superposition and, 231
 statically indeterminate beams, 260
Middle surface, thin-walled tubes, 88
Modulus of elasticity
 Hooke's Law and, 32
 for metals, 503–504
 relationship to shear modulus, 337–338
Modulus of resilience, 431–432
Modulus of rigidity. *see* Shear modulus
Modulus of toughness, 431–432
Mohr's circle, for plane stress, 301–304
 construction of, 302
 properties of, 303–304
 verification of, 304
Mohr's circle, for plane strain, 328–329
Mohr's circle, for second moments of area, 494
Mohr's theory of failure, 437–438
Moment-area method, 215–222
 bending moment diagrams by parts for, 218–220
 cantilever beams, 220–221
 simply supported beams, 221–222
 statically indeterminate beams, 254

 theorems, 216–218
Moment-curvature relationship, 193
Moments of inertia. *see* Second moments of area
Multiaxial loading
 biaxial loading, 44–45
 triaxial loading, 45

N

Necking, 33
Neutral axis, 136–137
 inclination of, 401–403, 405
Neutral surface, 136–137
Nominal strain, 30–31
Nominal stress, 32
Normal force, 3
Normal strain, 30–31
 transformation equations for, 327–328
Normal stress
 concentrated loading and, 4–5
 definition of, 3
 Saint Venant's principle, 5–6
 stress concentrations for, 444–445
Notch sensitivity, 452

O

Octahedral shear theory, 438
Offset method for yield point, 33
Outward normal of a plane, defined, 291
Overhanging beams, 100

P

Parallel-axis theorems
 second moments of area, 481–482
 polar moment of inertia, 482–483
 products of inertia, 482
Perfectly plastic material. *see* Elastic-perfectly plastic material
Permanent set
 elastic limit and, 33
 residual stresses and, 463
Plane areas
 first moments, 479–480
 second moments, 480–485
  method of composite areas, 483–485
  moments and product of inertia, 480–481
  parallel-axis theorems, 481–483
  radius of gyration, 483
 transformation of second moments
  compared with stress transformation, 493
  Mohr's circle, 494
  moments and products of inertia, 492–493
  principal moments of inertia and principal axes, 493–494
Plane stress. *see* State of Stress (plane stress)
Plastic hinges, 459, 471
Poisson's ratio, 44

Polar moment of inertia
    parallel-axis theorems for, 481–483
    torsion of circular shafts and, 75
Power transmission, shafts, 77–78
Principal axes of inertia, 493–494
Principal directions for stress, 293
Principal moments of inertia, 493–494
Principal planes for stress, 293–294
Principal stresses, 293–294
Principle of superposition, 231
Products of inertia for area
    defined, 480
    parallel-axis theorem for, 482
    transformation equation for, 492–493
Properties
    channel sections
        in SI units, 512–513
        in U.S. Customary units, 525–526
    equal angle sections
        in SI units, 514–515
        in U.S. Customary units, 527–530
    I-beams (S shapes)
        in SI units, 510–511
        in U.S. Customary units, 523–524
    metals, 503–504
    unequal angle sections
        in SI units, 516–517
        in U.S. Customary units, 527–530
    W-shapes (wide flange sections)
        in SI units, 505–509
        in U.S. Customary units, 518–522
Proportional limit
    Hooke's Law and, 32–33
    metals, 503–504
Propped cantilever beam, 100–101
Propped cantilever columns, 366

R

Radius of gyration, 367, 483
Radius, of thin-walled vessels, 274
Rankine, W., 437
Rectangular bars, stress concentration factors, 446–447
Rectangular cross sections, shear stress, 164–165
Redundant reactions, 243, 421
Reference planes for stress, 289–290
Reinforced concrete beams. see also Composite beams
    analysis of, 353–355
    balanced-stress reinforcement, 355
Residual stresses
    bending, 464–465
    defined, 463
    elastic spring-back, 465–466
    loading-unloading cycle and, 463
    torsion, 463–464
Resilience, modulus of, 431–432
Rigidity, modulus of. see Shear modulus
Rivets, shear stress in, 17
Roller supports, 100
Rupture stress, 33

S

S-N diagrams, 450–451
S-shape beams. see also I-beams, 155
Safety, factor of
    for columns, 374
    secant formula and, 382
    working stress and, 34
Saint Venant's principle, 5–6
Secant formula for eccentric loading, 378–382
Second Moment-Area Theorem, 217–218
Second moments of area
    definitions, 480–481
    method of composite areas, 483–485
    moments and product of inertia, definitions, 480–481
    parallel-axis theorems for, 481–483
    radius of gyration of an area, 483
Section modulus, 139–140
Shafts, circular. see Circular shafts
Shear and bending moment in beams, 99–133
    area method
        concentrated forces and couples, 114–116
        distributed loading, 112–114
        procedure, 116
    equations and diagrams
        procedures for determining, 102
        sign conventions, 101–102
    moving loads, 125–126
    supports and loads, 100–101
Shear center, 392–393
Shear, deformations, 3
Shear flow
    in thin-walled beams, 390–393
    in thin-walled tubes, 88–89
Shear flow diagrams, 391–392
Shear force
    defined, 3
    diagrams, 102
    relationship to load and bending moment, 113–114
Shear modulus, 45
    relationship to modulus of elasticity, 337–338
Shear strain, 45, 326
    relationship to shear stress, 45
    torsion of circular shafts and, 75
    transformation equations for, 328
Shear stress
    absolute maximum shear stress, 310–314
    composite beams and, 349–350
    direct shear, 17
    horizontal shear stress, 161–163
    inclined planes, 6–7
    maximum in-plane shear stress, 294–295
    procedure for analysis in beams, 165

Shear stress *(continued)*
    rectangular and wide-flange sections, 164–165
    relationship to shear strain, 45
    torsion formulas, 75–77
    vertical shear stress in beams, 163
Short columns, 363, 372
Sign conventions
    axial forces, 4
    moment-area theorems and, 218
    shear force and bending moment, 101–102
    stress at a point (plane stress), 290–291
    torque and angle of twist, 76
Simply supported beams
    defined, 100
    deflection formulas, 233
    loads, moving, 125–126
Simply supported columns, 364–366
Single shear, 17
Singularity functions, 207
Slenderness ratio, 367
    vs. critical stress in columns, 373
    vs. working stress in columns, 374
Spherical thin-walled vessels, 273–274
Spring-back, 465–466
Standard structural shapes, 155
State of strain
    strain gages, 333–334
    strain rosette, 334–335
    transformation of strain, 325–329
        equations, 327–328
        Mohr's circle, 328–329
State of stress (plane stress)
    absolute maximum shear stress, 310–312
    defined, 290
    Mohr's circle
        construction of, 302
        properties of, 303–304
        verification of, 304
    reference planes, 289–290
    sign conventions and subscript notation, 290–291
    transformation
        equations, 291–293
        maximum in-plane shear stress, 294–295
        principal stresses and principal planes, 293–294
        procedures for computing, 295–296
Static deformation of a spring, 430
Statically indeterminate beams, 243–269
    double-integration method, 244–245
    double integration method using bracket functions, 250
    method of superposition, 260
    moment-area method, 254
Statically indeterminate problems
    axial load problems, 51
    beam problems, 243–269
    solving by Castigliano's theorem, 421
    torsion problems, 78

Stiffness, as mechanical property, 29
Strain, 29–63
    axially loaded bars, 34–35
    generalized Hooke's Law, 44–45
    normal strain, 30–31
    tension tests and, 31–33
    thermal stresses, 60–61
    transformation. *see* Transformation of strain
Strain at a point. *see* State of strain
Strain energy
    bars and beams, 418–419
    Castigliano's theorem and, 420–421
    defined, 418
    density, 438
Strain gages, 333–334
Strain-hardening, 456
Strain rosette, 334–335
Stress. *see also* State of stress (plane stress)
    axial, direct shear and bearing stresses, 1–27
    beam stresses. *see* Stresses, beams
    combined loads. *see* Stresses, combined loads
    as force intensity at a point, 3
    reference planes for, 289–290
    transformation. *see* Transformation of stress
Stress amplitude, 450
Stress at a point (plane stress). *see* State of stress (plane stress)
Stress concentration, 444–447
Stress concentration factor
    defined, 445
    fatigue strength and, 452
    for rectangular and circular bars, 446–447
Stress-strain diagrams
    comparing grades of steel, 432
    ductile materials, 32, 463
    elastic limit, 33
    elastic-perfectly plastic material, 456
    proportional limit, 32–33
    rupture stress, 33
    slope (tangent modulus) of, 372–373
    ultimate stress, 33
    yield point, 33
Stress transformation. *see* Transformation of stress
Stress vector, 3
Stresses, beams, 135–190
    bending stress
        economic sections, 154–156
        flexure formulas, 136–141
        at a given point, 140
        maximums, 140–141
    shear stress
        horizontal shear stress in beams, 161–163
        procedure for analyzing in beams, 165
        rectangular and wide-flange sections, 164–165
        vertical shear stress in beams, 163
Stresses, combined loads, 271–342